Engineering Vibration Analysis with Application to Control Systems

Engineering Vibration Analysis with Application to Control Systems

C. F. Beards BSc, PhD, C Eng, MRAeS, MIOA

Consultant in Dynamics, Noise and Vibration
Formerly of the Department of Mechanical Engineering
Imperial College of Science, Technology and Medicine
University of London

Edward Arnold
A member of the Hodder Headline Group
LONDON SYDNEY AUCKLAND

First published in Great Britain 1995 by
Edward Arnold, a division of Hodder Headline PLC,
338 Euston Road, London NW1 3BH

British Library Cataloguing in Publication Data
A catalogue record for this book is available from the British Library

ISBN 0 340 63183 X

1 2 3 4 5 95 96 97 98 99

Typeset in 10 on 12pt Times by
PPS Limited, London Road, Amesbury, Wilts.
Printed and bound in Great Britain by
J W Arrowsmith Ltd, Bristol

'Learning without thought is labour lost;
thought without learning is perilous.'

Confucius, 551–479 BC

Contents

Preface

The high cost and questionable supply of many materials, land and other resources, together with the sophisticated analysis and manufacturing methods now available, have resulted in the construction of many highly stressed and lightweight machines and structures, frequently with high energy sources, which have severe vibration problems. Often, these dynamic systems also operate under hostile environmental conditions and with minimum maintenance. It is to be expected that even higher performance levels will be demanded of all dynamic systems in the future, together with increasingly stringent performance requirement parameters such as low noise and vibration levels, ideal control system responses and low costs. In addition it is widely accepted that low vibration levels are necessary for the smooth and quiet running of machines, structures and all dynamic systems. This is a highly desirable and sought after feature which enhances any system and increases its perceived quality and value, so it is essential that the causes, effects and control of the vibration of engineering systems are clearly understood in order that effective analysis, design and modification may be carried out. That is, the demands made on many present day systems are so severe, that the analysis and assessment of the dynamic performance is now an essential and very important part of the design. Dynamic analysis is performed so that the system response to the expected excitation can be predicted and modifications made as required. This is necessary to control the dynamic response parameters such as vibration levels, stresses, fatigue, noise and resonance. It is also necessary to be able to analyse existing systems when considering the effects of modifications and searching for performance improvement.

There is therefore a great need for all practising designers, engineers and scientists, as well as students, to have a good understanding of the analysis methods used for predicting the vibration response of a system, and methods for determining control system performance. It is also essential to be able to understand, and contribute to, published and quoted data in this field including the use of, and understanding of, computer programs.

There is great benefit to be gained by studying the analysis of vibrating systems and control system dynamics together, and in having this information in a single text, since the analyses of the vibration of elastic systems and the dynamics of control systems are closely linked. This is because in many cases the same equations of motion occur in the analysis of vibrating systems as in control systems, and thus the techniques and results developed in the analysis of one system may be applied to the other. It is therefore a very efficient way of studying vibration and control.

This has been successfully demonstrated in my previous books *Vibration Analysis and Control System Dynamics* (1981) and *Vibrations and Control Systems* (1988). Favourable reaction to these books and friendly encouragement from fellow academics, co-workers, students and my publisher has led me to write *Engineering Vibration Analysis with Application to Control Systems.*

Whilst I have adopted a similar approach in this book to that which I used previously, I have taken the opportunity to revise, modify, update and expand the material and the title reflects this. This new book discusses very comprehensively the analysis of the vibration of dynamic systems and then shows how the techniques and results obtained in vibration analysis may be applied to the study of control system dynamics. There are now 75 worked examples included, which amplify and demonstrate the analytical principles and techniques so that the text is at the same time more comprehensive and even easier to follow and understand than the earlier books. Furthermore, worked solutions and answers to most of the 130 or so problems set are included. (I trust that readers will try the problems *before* looking up the worked solutions in order to gain the greatest benefit from this.)

Excellent advanced specialised texts on engineering vibration analysis and control systems are available, and some are referred to in the text and in the bibliography, but they require advanced mathematical knowledge and understanding of dynamics, and often refer to idealised systems rather than to mathematical models of real systems. This book links basic dynamic analysis with these advanced texts, paying particular attention to the mathematical modelling and analysis of real systems and the interpretation of the results. It therefore gives an introduction to advanced and specialised analysis methods, and also describes how system parameters can be changed to achieve a desired dynamic performance.

The book is intended to give practising engineers, and scientists as well as students of engineering and science to first degree level, a thorough understanding of the principles and techniques involved in the analysis of vibrations and how they can also be applied to the analysis of control system dynamics. In addition it provides a sound theoretical basis for further study.

Chris Beards
January 1995

Acknowledgements

Some of the problems first appeared in University of London B.Sc. (Eng) Degree Examinations, set for students of Imperial College, London. The section on random vibration has been reproduced with permission from the *Mechanical Engineers Reference Book*, 12th edn, Butterworth – Heinemann, 1993.

General notation

a	damping factor, dimension, displacement.
b	circular frequency (rad/s), dimension, port coefficient.
c	coefficient of viscous damping, velocity of propagation of stress wave.
c_c	coefficient of critical viscous damping $= 2\sqrt{(mk)}$.
c_d	equivalent viscous damping coefficient for dry friction damping $= 4F_d/\pi\omega X$.
c_H	equivalent viscous damping coefficient for hysteretic damping $= \eta k/\omega$.
d	diameter.
f	frequency (Hz), exciting force.
f_s	Strouhal frequency (Hz).
g	acceleration constant.
h	height, thickness.
j	$\sqrt{-1}$.
k	linear spring stiffness, beam shear constant, gain factor.
k_T	torsional spring stiffness.
k^*	complex stiffness $= k(1 + \mathrm{j}\eta)$.

l	length.
m	mass.
q	generalized coordinate.
r	radius.
s	Laplace operator $= a + jb$.
t	time.
u	displacement.
v	velocity, deflection.
x	displacement.
y	displacement.
z	displacement.
A	amplitude, constant, cross-sectional area.
B	constant.
$C_{1,2,3,4}$	constants,
D	flexural rigidity $= Eh^3/12(1 - v^2)$, hydraulic mean diameter,
D	derivative w.r.t. time.
E	modulus of elasticity.
E'	in-phase, or storage modulus.
E''	quadrature, or loss modulus.
E^*	complex modulus $= E' + \mathrm{j}E''$.
F	exciting force amplitude.
F_{d}	Coulomb (dry) friction force (μN).
F_{T}	transmitted force.
G	centre of mass, modulus of rigidity, gain factor.
I	mass moment of inertia.
J	second moment of area, moment of inertia.
K	stiffness, gain factor.
L	length.
$\mathscr{L}$	Laplace transform.
M	mass, moment, mobility.
N	applied normal force, gear ratio.
P	force.
Q	factor of damping, flow rate.

Q_i	generalized external force.
R	radius of curvature.
$[S]$	system matrix.
T	kinetic energy,
	tension,
	time constant.
T_R	transmissibility $= F_T/F$.
V	potential energy,
	speed.
X	amplitude of motion.
$\{X\}$	column matrix.
X_s	static deflection $= F/k$.
X/X_s	dynamic magnification factor.
Z	impedance.
α	coefficient,
	influence coefficient,
	phase angle,
	receptance.
β	coefficient,
	receptance.
γ	coefficient,
	receptance.
δ	deflection.
ε	short time,
	strain.
ε_0	strain amplitude.
ζ	damping ratio $= c/c_c$.
η	loss factor $= E''/E'$.
θ	angular displacement,
	slope.
λ	matrix eigenvalue,
	$[\rho A\omega^2/EI]^{1/4}$.
μ	coefficient of friction,
	mass ratio $= m/M$.
ν	Poisson's ratio,
	circular exciting frequency (rad/s).
ρ	material density.
σ	stress.
σ_0	stress amplitude.
τ	period of vibration $= 1/f$.
τ_d	period of dry friction damped vibration.
τ_v	period of viscous damped vibration.
ϕ	phase angle,
	function of time,
	angular displacement.

ψ	phase angle.
ω	undamped circular frequency (rad/s).
ω_{d}	dry friction damped circular frequency.
ω_{v}	viscous damped circular frequency $= \omega\sqrt{(1 - \zeta^2)}$.
Λ	logarithmic decrement $= \ln X_1/X_{11}$.
Φ	transfer function.
Ω	natural circular frequency (rad/s).

1

Introduction

The vibration which occurs in most machines, vehicles, structures, buildings and dynamic systems is undesirable, not only because of the resulting unpleasant motions and the dynamic stresses which may lead to fatigue and failure of the structure or machine, and the energy losses and reduction in performance which accompany vibrations, but also because of the noise produced. Noise is generally considered to be unwanted sound, and since sound is produced by some source of motion or vibration causing pressure changes which propagate through the air or other transmitting medium, vibration control is of fundamental importance to sound attenuation. Vibration analysis of machines and structures is therefore often a necessary prerequisite for controlling not only vibration but also noise.

Until early this century, machines and structures usually had very high mass and damping, because heavy beams, timbers, castings and stonework were used in their construction. Since the vibration excitation sources were often small in magnitude, the dynamic response of these highly damped machines was low. However, with the development of strong lightweight materials, increased knowledge of material properties and structural loading, and improved analysis and design techniques, the mass of machines and structures built to fulfil a particular function has decreased. Furthermore, the efficiency and speed of machinery have increased so that the vibration exciting forces are higher, and dynamic systems often contain high energy sources which can create intense noise and vibration problems. This process of increasing excitation with reducing machine mass and damping has continued at an increasing rate to the present day when few, if any, machines can be designed without carrying out the necessary vibration analysis, if their dynamic performance is to be acceptable. The demands made on machinery, structures, and dynamic systems are also increasing, so that the dynamic performance requirements are always rising.

There have been very many cases of systems failing or not meeting performance targets because of resonance, fatigue, excessive vibration of one component or another, or high noise levels. Because of the very serious effects which unwanted vibrations can have on dynamic systems, it is essential that vibration analysis be carried out as an inherent part of their design, when necessary modifications can most easily be made to eliminate vibration, or at least to reduce it as much as possible. However, it must also be recognized that it may sometimes be necessary to reduce the vibration of an existing machine, either because of inadequate initial design, or by a change in function of the machine, or by a change in environmental conditions or performance requirements, or by a revision of acceptable noise levels. Therefore techniques for the analysis of vibration in dynamic systems should be applicable to existing systems as well as those in the design stage; it is the solution to the vibration or noise problem which may be different, depending on whether or not the system already exists.

There are two factors which control the amplitude and frequency of vibration of a dynamic system: these are the excitation applied and the dynamic characteristics of the system. Changing either the excitation or the dynamic characteristics will change the vibration response stimulated in the system. The excitation arises from external sources, and these forces or motions may be periodic, harmonic or random in nature, or arise from shock or impulsive loadings.

To summarize, present-day machines and structures often contain high-energy sources which create intense vibration excitation problems, and modern construction methods result in systems with low mass and low inherent damping. Therefore careful design and analysis is necessary to avoid resonance or an undesirable dynamic performance.

The demands made on automatic control systems are also increasing. Systems are becoming larger and more complex, whilst improved performance criteria, such as reduced response time and error, are demanded. Whatever the duty of the system, from the control of factory heating levels to satellite tracking, or from engine fuel control to controlling sheet thickness in a steel rolling mill, there is continual effort to improve performance whilst making the system cheaper, more efficient, and more compact. These developments have been greatly aided in recent years by the wide availability of microprocessors. Accurate and relevant analysis of control system dynamics is necessary in order to determine the response of new system designs, as well as to predict the effects of proposed modifications on the response of an existing system, or to determine the modifications necessary to enable a system to give the required response.

There are two reasons why it is desirable to study vibration analysis and the dynamics of control systems together as dynamic analysis. Firstly, because control systems can then be considered in relation to mechanical engineering using mechanical analogies, rather than as a specialized and isolated aspect of electrical engineering, and secondly, because the basic equations governing the behaviour of vibration and control systems are the same: different emphasis is placed on the different forms of the solution available, but they are all dynamic systems. Each analysis system benefits from the techniques developed in the other.

Dynamic analysis can be carried out most conveniently by adopting the following three-stage approach:

Stage I. Devise a mathematical or physical model of the system to be analysed.
Stage II. From the model, write the equations of motion.
Stage III. Evaluate the system response to relevant specific excitation.

These stages will now be discussed in greater detail.

Stage I. The mathematical model

Although it may be possible to analyse the complete dynamic system being considered, this often leads to a very complicated analysis, and the production of much unwanted information. A simplified mathematical model of the system is therefore usually sought which will, when analysed, produce the desired information as economically as possible and with acceptable accuracy. The derivation of a simple mathematical model to represent the dynamics of a real system is not easy, if the model is to give useful and realistic information.

However, to model any real system a number of simplifying assumptions can often be made. For example, a distributed mass may be considered as a lumped mass, or the effect of damping in the system may be ignored particularly if only resonance

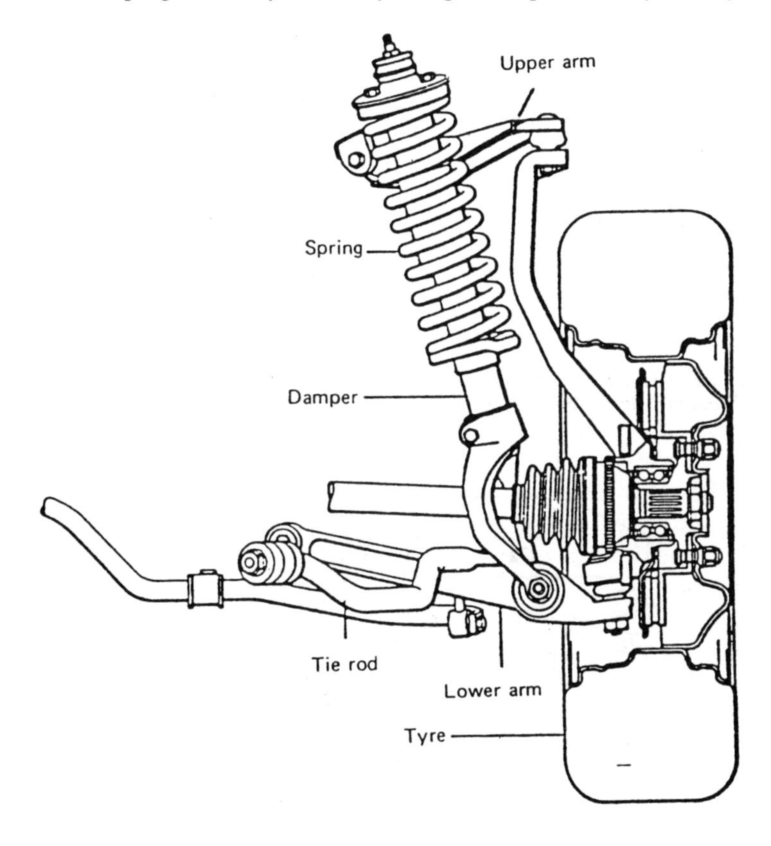

Fig. 1.1. Rover 800 front suspension. (By courtesy of Rover Group.)

frequencies are needed or the dynamic response required at frequencies well away from a resonance, or a non-linear spring may be considered linear over a limited range of extension, or certain elements and forces may be ignored completely if their effect is likely to be small. Furthermore, the directions of motion of the mass elements are usually restrained to those of immediate interest to the analyst.

Thus the model is usually a compromise between a simple representation which is easy to analyse but may not be very accurate, and a complicated but more realistic model which is difficult to analyse but gives more useful results. Consider for example, the analysis of the vibration of the front wheel of a motor car. Fig. 1.1 shows a typical suspension system. As the car travels over a rough road surface, the wheel moves up and down, following the contours of the road. This movement is transmitted to the upper and lower arms, which pivot about their inner mountings, causing the coil

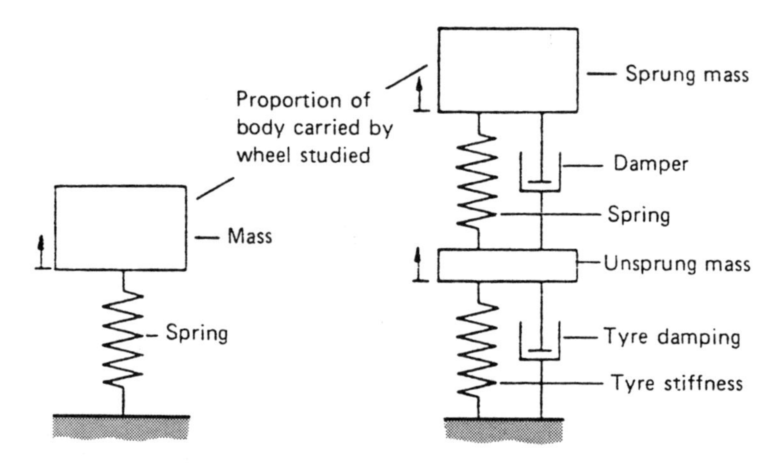

Fig. 1.2(a). Simplest model – motion in a vertical direction only can be analysed.

Fig. 1.2(b). Motion in a vertical direction only can be analysed.

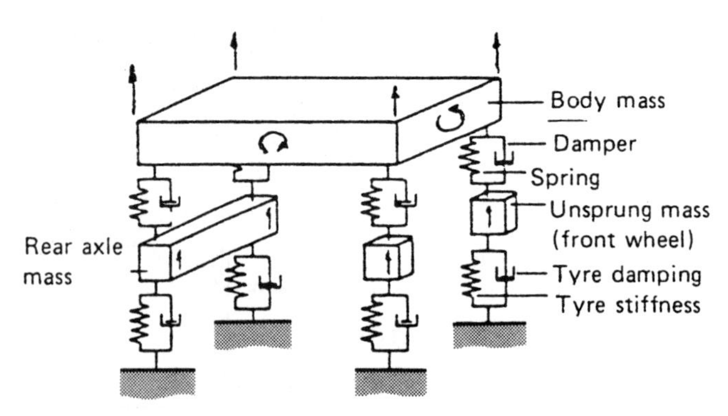

Fig. 1.2(c). Motion in a vertical direction, roll, and pitch can be analysed.

spring to compress and extend. The action of the spring isolates the body from the movement of the wheel, with the shock absorber or damper absorbing vibration and sudden shocks. The tie rod controls longitudinal movement of the suspension unit.

Fig. 1.2(a) is a very simple model of this same system, which considers translational motion in a vertical direction only: this model is not going to give much useful information, although it is easy to analyse. The more complicated model shown in Fig. 1.2(b) is capable of producing some meaningful results at the cost of increased labour in the analysis, but the analysis is still confined to motion in a vertical direction only. A more refined model, shown in Fig. 1.2(c), shows the whole car considered, translational and rotational motion of the car body being allowed.

If the modelling of the car body by a rigid mass is too crude to be acceptable, a finite element analysis may prove useful. This technique would allow the body to be represented by a number of mass elements.

The vibration of a machine tool such as a lathe can be analysed by modelling the machine structure by the two degree of freedom system shown in Fig. 1.3. In the simplest analysis the bed can be considered to be a rigid body with mass and inertia, and the headstock and tailstock are each modelled by lumped masses. The bed is supported by springs at each end as shown. Such a model would be useful for determining the lowest or fundamental natural frequency of vibration. A refinement to this model, which may be essential in some designs of machine where the bed cannot be considered rigid, is to consider the bed to be a flexible beam with lumped masses attached as before.

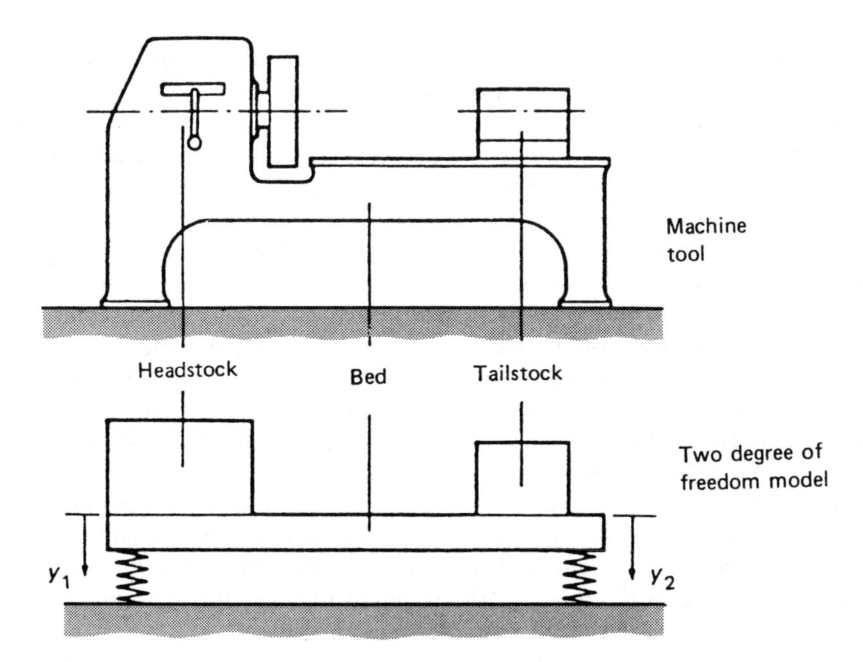

Fig. 1.3. Machine tool vibration analysis model.

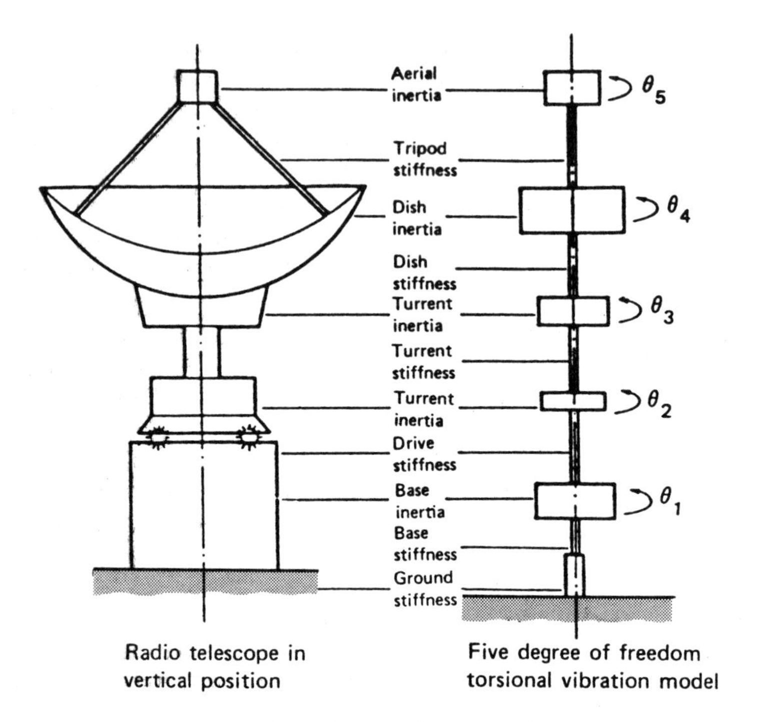

Fig. 1.4. Radio telescope vibration analysis model.

To analyse the torsional vibration of a radio telescope when in the vertical position a five degree of freedom model, as shown in Fig. 1.4, can be used. The mass and inertia of the various components may usually be estimated fairly accurately, but the calculation of the stiffness parameters at the design stage may be difficult; fortunately the natural frequencies are proportional to the square root of the stiffness. If the structure, or a similar one, is already built, the stiffness parameters can be measured. A further simplification of the model would be to put the turret inertia equal to zero, so that a three degree of freedom model is obtained. Such a model would be easy to analyse and would predict the lowest natural frequency of torsional vibration with fair accuracy, providing the correct inertia and stiffness parameters were used. It could not be used for predicting any other modes of vibration because of the coarseness of the model. However, in many structures only the lowest natural frequency is required, since if the structure can survive the amplitudes and stresses at this frequency it will be able to survive other natural frequencies too.

None of these models include the effect of damping in the structure. Damping in most structures is very low so that the difference between the undamped and the damped natural frequencies is negligible. It is usually only necessary to include the effects of damping in the model if the response to a specific excitation is sought, particularly at frequencies in the region of a resonance.

A block diagram model is usually used in the analysis of control systems. For example, a system used for controlling the rotation and position of a turntable about

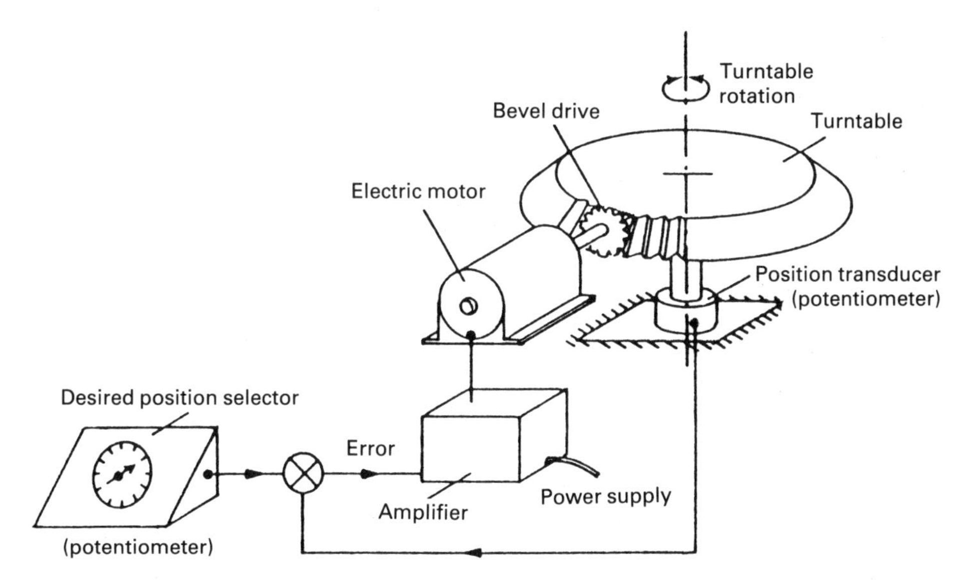

Fig. 1.5. Turntable position control system.

a vertical axis is shown in Fig. 1.5. The turntable can be used for mounting a telescope or gun, or if it forms part of a machine tool it can be used for mounting a workpiece for machining. Fig. 1.6 shows the block diagram used in the analysis.

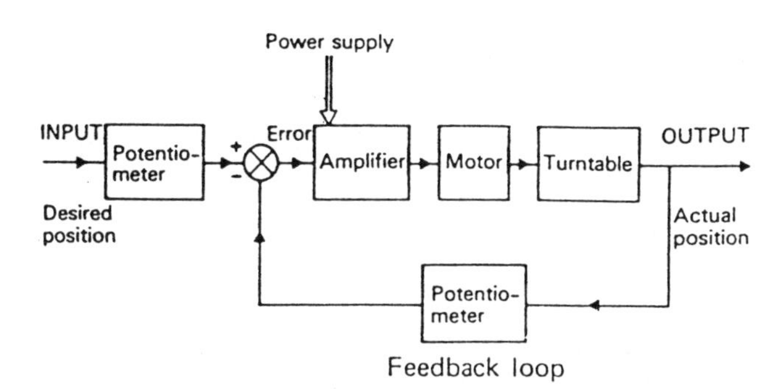

Fig. 1.6. Turntable position control system: block diagram model.

It can be seen that the feedback loop enables the input and output positions to be compared, and the error signal, if any, is used to activate the motor and hence rotate the turntable until the error signal is zero; that is, the actual position and the desired position are the same.

The model parameters

Because of the approximate nature of most models, whereby small effects are neglected and the environment is made independent of the system motions, it is usually reasonable to assume constant parameters and linear relationships. This means that the coefficients in the equations of motion are constant and the equations themselves are linear: these are real aids to simplifying the analysis. Distributed masses can often be replaced by lumped mass elements to give ordinary rather that partial differential equations of motion. Usually the numerical value of the parameters can, substantially, be obtained directly from the system being analysed. However, model system parameters are sometimes difficult to assess, and then an intuitive estimate is required, engineering judgement being of the essence.

It is not easy to create a relevant mathematical model of the system to be analysed, but such a model does have to be produced before Stage II of the analysis can be started. Most of the material in subsequent chapters is presented to make the reader competent to carry out the analyses described in Stages II and III. A full understanding of these methods will be found to be of great help in formulating the mathematical model referred to above in Stage I.

Stage II. The equations of motion

Several methods are available for obtaining the equations of motion from the mathematical model, the choice of method often depending on the particular model and personal preference. For example, analysis of the free-body diagrams drawn for each body of the model usually produces the equations of motion quickly: but it can be advantageous in some cases to use an energy method such as the Lagrange equation.

From the equations of motion the characteristic or frequency equation is obtained, yielding data on the natural frequencies, modes of vibration, general response, and stability.

Stage III. Response to specific excitation

Although Stage II of the analysis gives much useful information on natural frequencies, response, and stability, it does not give the actual system response to specific excitations. It is necessary to know the actual response in order to determine such quantities as dynamic stress, noise, output position, or steady-state error for a range of system inputs, either force or motion, including harmonic, step and ramp. This is achieved by solving the equations of motion with the excitation function present.

Remember:

Stage I.	Model
Stage II.	Equations
Stage III.	Excitation

A few examples have been given above to show how real systems can be modelled, and the principles of their analysis. To be competent to analyse system models it is first necessary to study the analysis of damped and undamped, free and forced vibration of single degree of freedom systems such as those discussed in Chapter 2. This not only allows the analysis of a wide range of problems to be carried out, but it is also essential background to the analysis of systems with more than one degree of freedom, which is considered in Chapter 3. Systems with distributed mass, such as beams, are analysed in Chapter 4. Some aspects of automatic control system analysis which require special consideration, particularly their stability and system frequency response, are discussed in Chapter 5. Each of these chapters includes worked examples to aid understanding of the theory and techniques described, whilst Chapter 6 contains a number of problems for the reader to try. Chapter 7 contains answers and worked solutions to most of the problems in Chapter 6. A comprehensive bibliography and an index are included.

2

The vibrations of systems having one degree of freedom

All real systems consist of an infinite number of elastically connected mass elements and therefore have an infinite number of degrees of freedom; and hence an infinite number of coordinates are needed to describe their motion. This leads to elaborate equations of motion and lengthy analyses. However, the motion of a system is often such that only a few coordinates are necessary to describe its motion. This is because the displacements of the other coordinates are restrained or not excited, so that they are so small that they can be neglected. Now, the analysis of a system with a few degrees of freedom is generally easier to carry out than the analysis of a system with many degrees of freedom, and therefore only a simple mathematical model of a system is desirable from an analysis viewpoint. Although the amount of information that a simple model can yield is limited, if it is sufficient then the simple model is adequate for the analysis. Often a compromise has to be reached, between a comprehensive and elaborate multi-degree of freedom model of a system, which is difficult and costly to analyse but yields much detailed and accurate information, and a simple few degrees of freedom model that is easy and cheap to analyse but yields less information. However, adequate information about the vibration of a system can often be gained by analysing a simple model, at least in the first instance.

The vibration of some dynamic systems can be analysed by considering them as a one degree or single degree of freedom system; that is a system where only one coordinate is necessary to describe the motion. Other motions may occur, but they are assumed to be negligible compared to the coordinate considered.

A system with one degree of freedom is the simplest case to analyse because only one coordinate is necessary to completely describe the motion of the system. Some real systems can be modelled in this way, either because the excitation of the system is such that the vibration can be described by one coordinate although the system could vibrate in other directions if so excited, or the system really is simple, as for example a clock pendulum. It should also be noted that a one degree of freedom model of a complicated system can often be constructed where the analysis of a particular mode of vibration is to be carried out. To be able to analyse one degree of freedom systems is therefore an essential ability in vibration analysis. Furthermore, many of the techniques developed in single degree of freedom analysis are applicable to more complicated systems.

2.1. FREE UNDAMPED VIBRATION

2.1.1 Translation vibration

In the system shown in Fig. 2.1 a body of mass m is free to move along a fixed horizontal surface. A spring of constant stiffness k which is fixed at one end is attached at the other end to the body. Displacing the body to the right (say) from the equilibrium position causes a spring force to the left (a restoring force). Upon release this force gives the body an acceleration to the left. When the body reaches its equilibrium position the spring force is zero, but the body has a velocity which carries it further to the left although it is retarded by the spring force which now acts to the right. When the body is arrested by the spring the spring force is to the right so that the body moves to the right, past its equilibrium position, and hence reaches its initial displaced position. In practice this position will not quite be reached because damping in the system will have dissipated some of the vibrational energy. However, if the damping is small its effect can be neglected.

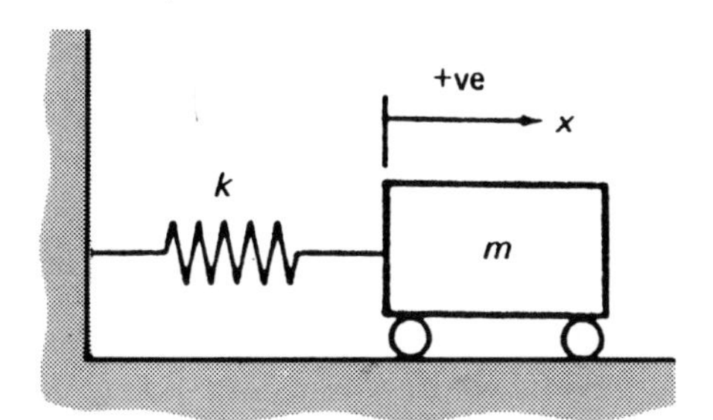

Fig. 2.1. Single degree of freedom model – translation vibration.

If the body is displaced a distance x_0 to the right and released, the free-body diagrams (FBD's) for a general displacement x are as shown in Figs. 2.2(a) and (b).

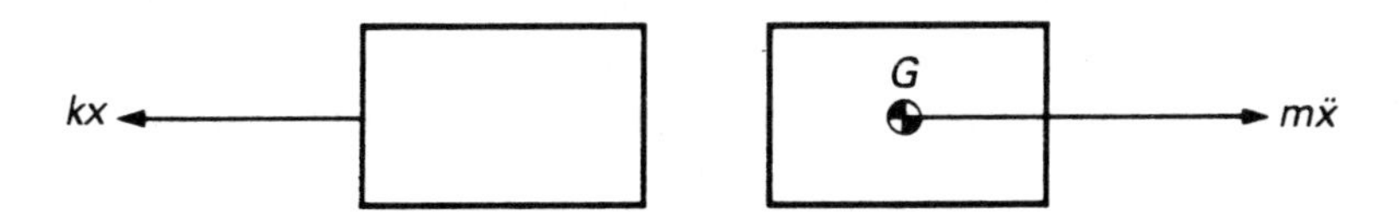

Fig. 2.2. (a) Applied force; (b) effective force.

The effective force is always in the direction of positive x. If the body is being retarded $\ddot{x}$ will be calculated to be negative. The mass of the body is assumed constant: this is usually so, but not always, as for example in the case of a rocket burning fuel. The spring stiffness k is assumed constant: this is usually so within limits; see section 2.1.3. It is assumed that the mass of the spring is negligible compared to the mass of the body; cases where this is not so are considered in section 4.3.1.

From the free-body diagrams the equation of motion for the system is

$$m\ddot{x} = -kx \qquad \text{or} \qquad \ddot{x} + (k/m)x = 0. \tag{2.1}$$

This will be recognized as the equation for simple harmonic motion. The solution is

$$x = A \cos \omega t + B \sin \omega t, \tag{2.2}$$

where A and B are constants which can be found by considering the initial conditions, and ω is the circular frequency of the motion. Substituting (2.2) into (2.1) we get

$$-\omega^2 (A \cos \omega t + B \sin \omega t) + (k/m)(A \cos \omega t + B \sin \omega t) = 0.$$

Since $(A \cos \omega t + B \sin \omega t) \neq 0$ (otherwise no motion),

$$\omega = \sqrt{(k/m)} \text{ rad/s},$$

and

$$x = A \cos \sqrt{(k/m)}t + B \sin \sqrt{(k/m)}t.$$

Now

$$x = x_0 \text{ at } t = 0,$$

thus

$$x_0 = A \cos 0 + B \sin 0, \quad \text{and therefore } x_0 = A,$$

and

$$\dot{x} = 0 \text{ at } t = 0,$$

thus

$$0 = -A\sqrt{(k/m)} \sin 0 + B\sqrt{(k/m)} \cos 0, \quad \text{and therefore } B = 0;$$

that is,

$$x = x_0 \cos \sqrt{(k/m)}t. \tag{2.3}$$

The system parameters control ω and the type of motion but not the amplitude x_0, which is found from the initial conditions. The mass of the body is important, its weight is not, so that for a given system, ω is independent of the local gravitational field.

The frequency of vibration, f, is given by

$$f = \frac{\omega}{2\pi}, \quad \text{or} \quad f = \frac{1}{2\pi}\sqrt{\left(\frac{k}{m}\right)} \text{Hz}. \tag{2.4}$$

The motion is as shown in Fig. 2.3.

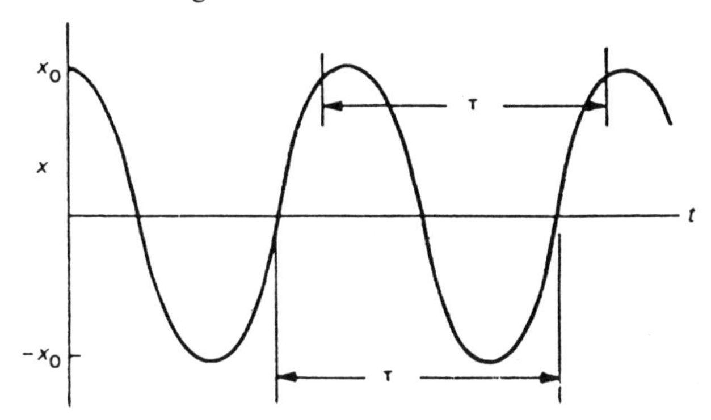

Fig. 2.3. Simple harmonic motion.

The period of the oscillation, τ, is the time taken for one complete cycle so that

$$\tau = \frac{1}{f} = 2\pi\sqrt{(m/k)} \text{ seconds.} \tag{2.5}$$

The analysis of the vibration of a body supported to vibrate only in the vertical or y direction can be carried out in a similar way to that above. Fig. 2.4 shows the system.

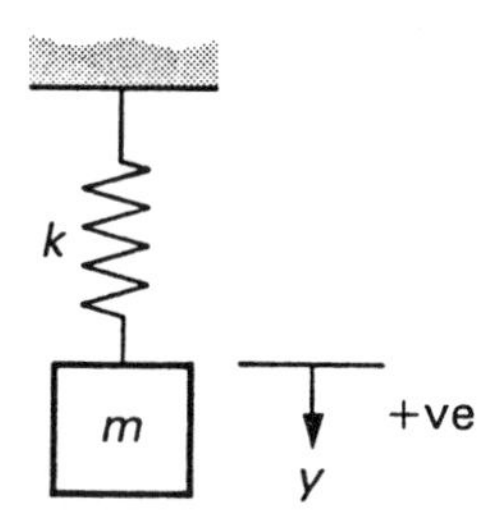

Fig. 2.4. Vertical motion.

The spring extension δ when the body is fastened to the spring is given by $k\delta = mg$. When the body is given an additional displacement y_0 and released the FBDs for a general displacement y, are as in Fig. 2.5.

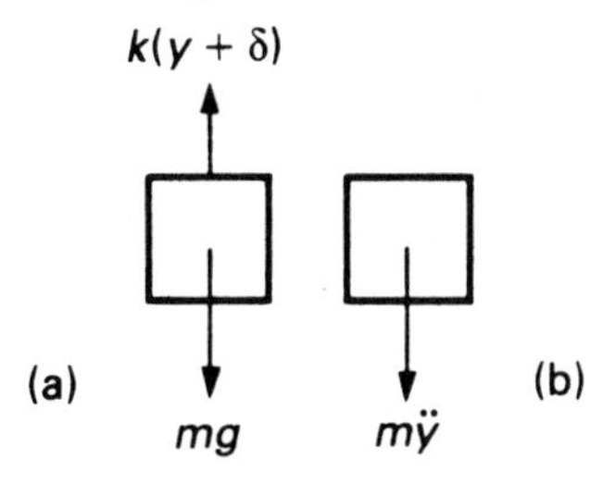

Fig. 2.5. (a) Applied forces; (b) effective force.

The equation of motion is

$$m\ddot{y} = mg - k(y + \delta)$$
$$= mg - ky - k\delta = -ky, \quad \text{since } mg = k\delta.$$

that is,

$$\ddot{y} + (k/m)y = 0. \tag{2.6}$$

This is similar to equation (2.1), so that the general solution can be written as

$$y = A \cos \sqrt{(k/m)}t + B \sin \sqrt{(k/m)}t. \tag{2.7}$$

Note that $\sqrt{(k/m)} = \sqrt{(g/\delta)}$, because $mg = k\delta$. That is, if δ is known, then the frequency of vibration can be found.

For the initial conditions $y = y_0$ at $t = 0$ and $\dot{y} = 0$ at $t = 0$,

$$y = y_0 \cos \sqrt{(k/m)}t. \tag{2.8}$$

Comparing (2.8) with (2.3) shows that for a given system the frequency of vibration is the same whether the body vibrates in a horizontal or vertical direction.

Sometimes more than one spring acts in a vibrating system. The spring, which is considered to be an elastic element of constant stiffness, can take many forms in practice; for example, it may be a wire coil, rubber block, beam or air bag. Combined spring units can be replaced in the analysis by a single spring of equivalent stiffness as follows.

(1) Springs connected in series

The three-spring system of Fig. 2.6.(a) can be replaced by the equivalent spring of Fig. 2.6(b).

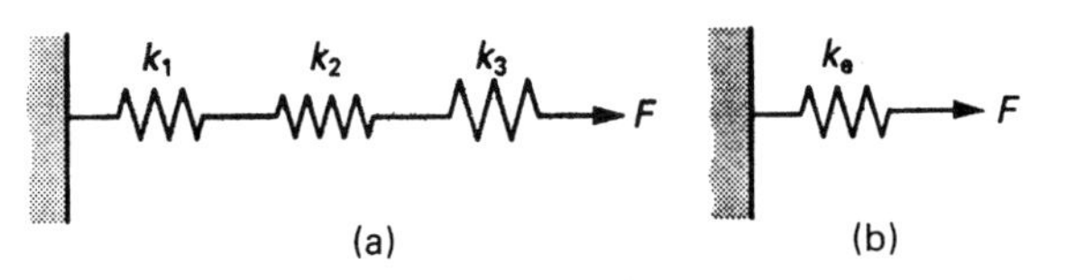

Fig. 2.6. Spring systems.

If the deflection at the free end, δ, experienced by applying the force F is to be the same in both cases,

$$\delta = F/k_e = F/k_1 + F/k_2 + F/k_3,$$

that is,

$$1/k_e = \sum_1^3 1/k_i.$$

In general, the reciprocal of the equivalent stiffness of springs connected in series is obtained by summing the reciprocal of the stiffness of each spring.

(2) Springs connected in parallel
The three-spring system of Fig. 2.7(a) can be replaced by the equivalent spring of Fig. 2.7(b).

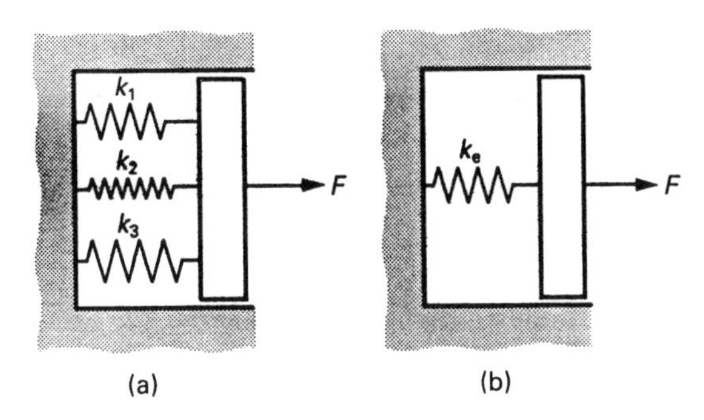

Fig. 2.7. Spring systems.

Since the defection δ must be the same in both cases, the sum of the forces exerted by the springs in parallel must equal the force exerted by the equivalent spring. Thus

$$F = k_1\delta + k_2\delta + k_3\delta = k_e\delta,$$

that is,

$$k_e = \sum_{i=1}^{3} k_i.$$

In general, the equivalent stiffness of springs connected in parallel is obtained by summing the stiffness of each spring.

2.1.2 Torsional vibration

Fig. 2.8 shows the model used to study torsional vibration.

A body with mass moment of inertia I about the axis of rotation is fastened to a bar of torsional stiffness k_T. If the body is rotated through an angle θ_o and released, torsional vibration of the body results. The mass moment of inertia of the shaft about the axis of rotation is usually negligible compared with I.

For a general displacement θ the FBDs are as given in Figs. 2.9(a) and (b). Hence the equation of motion is

$$I\ddot{\theta} = -k_T\theta,$$

or

$$\ddot{\theta} + \left(\frac{k_T}{I}\right)\theta = 0.$$

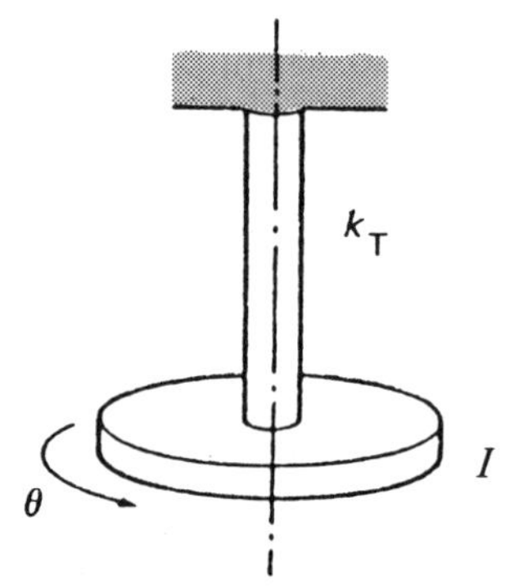

Fig. 2.8. Single degree of freedom model – torsional vibration.

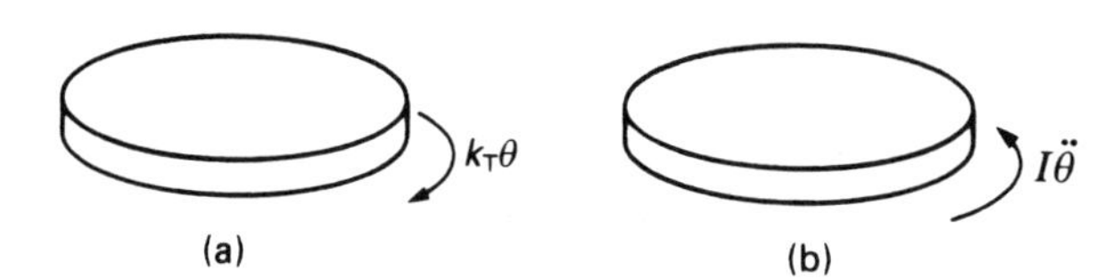

Fig. 2.9. (a) Applied torque; (b) effective torque.

This is of a similar form to equation (2.1). That is, the motion is simple harmonic with frequency $1/2\pi\,\sqrt{(k_T/I)}$ Hz.

The torsional stiffness of the shaft, k_T, is equal to the applied torque divided by the angle of twist.
Hence

$$k_T = \frac{GJ}{l},\ \text{for a circular section shaft,}$$

where G = modulus of rigidity for shaft material,
J = second moment of area about the axis of rotation, and
l = length of shaft.

Hence

$$f = \frac{\omega}{2\pi} = \frac{1}{2\pi}\sqrt{(GJ/Il)}\ \text{Hz},$$

and

$$\theta = \theta_0 \cos\sqrt{(GJ/Il)}t,$$

when $\theta = \theta_0$ and $\dot{\theta} = 0$ at $t = 0$.

If the shaft does not have a constant diameter, it can be replaced analytically by an equivalent shaft of different length but with the same stiffness and a constant diameter.

For example, a circular section shaft comprising a length l_1 of diameter d_1 and a length l_2 of diameter d_2 can be replaced by a length l_1 of diameter d_1 and a length l of diameter d_1 where, for the same stiffness,

$$(GJ/l)_{\text{length } l_2 \text{ diameter } d_2} = (GJ/l)_{\text{length } l/\text{diameter } d_1}$$

that is, for the same shaft material, $d_2^4/l_2 = d_1{}^4/l$.
Therefore the equivalent length l_e of the shaft of constant diameter d_1 is given by

$$l_e = l_1 + (d_1/d_2)^4 l_2.$$

It should be noted that the analysis techniques for translational and torsional vibration are very similar, as are the equations of motion.

The torsional vibration of a geared system

Consider the system shown in Fig. 2.10. The mass moments of inertia of the shafts

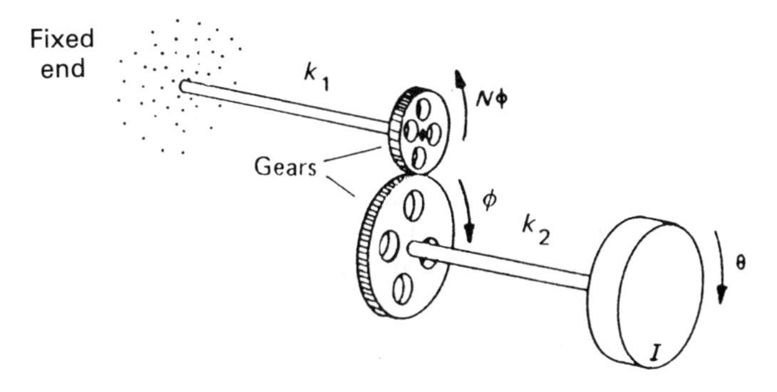

Fig. 2.10. Geared system.

and gears about their axes of rotation are considered negligible. The shafts are supported in bearings which are not shown, and the gear ratio is $N:1$.

From the FBDs, T_2, the torque in shaft 2 is $T_2 = k_2\,(\theta - \phi) = -\,I\ddot{\theta}$ and T_1, the torque in shaft 1, is $T_1 = k_1 N\phi$; since $NT_1 = T_2$, $T_2 = k_1 N^2\phi$ and $\phi = k_2\theta/(k_2 + k_1 N^2)$. Thus the equation of motion becomes

$$I\ddot{\theta} + \left(\frac{k_2 k_1 N^2}{k_2 + k_1 N^2}\right)\theta = 0,$$

and

$$f = \frac{1}{2\pi}\sqrt{\frac{[(k_2 k_1 N^2)/(k_2 + k_1 N^2)]}{I}}\,\text{Hz}.$$

that is, k_{eq}, the equivalent stiffness referred to shaft 2, is $(k_1 k_2 N^2)/(k_1 N^2 + k_2)$, or $1/k_{eq} = 1/k_2 + 1/(k_1 N^2)$.

2.1.3 Non-linear spring elements

Any spring elements have a force–deflection relationship which is linear only over a limited range of deflection. Fig. 2.11 shows a typical characteristic.

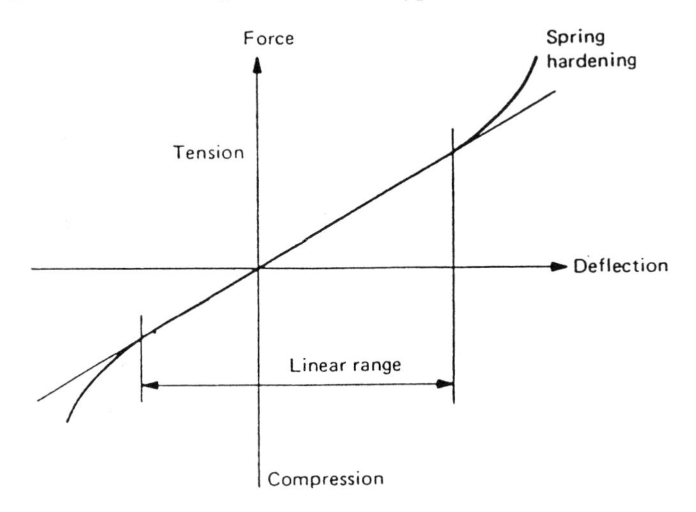

Fig. 2.11. Non-linear spring characteristic.

The non-linearities in this characteristic may be caused by physical effects such as the contacting of coils in a compressed coil spring, or by excessively straining the spring material so that yielding occurs. In some systems the spring elements do not act at the same time, as shown in Fig. 2.12(a), or the spring is designed to be non-linear as shown in Figs. 2.12(b) and (c).

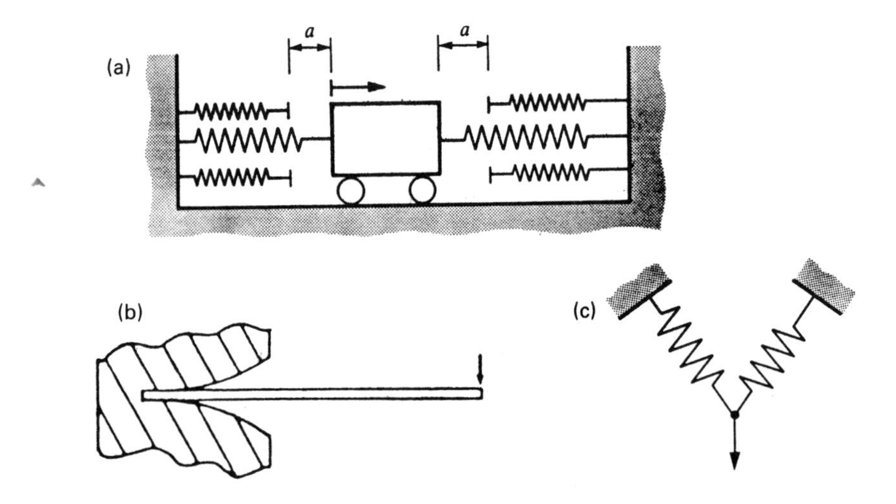

Fig. 2.12. Non-linear spring systems.

Analysis of the motion of the system shown in Fig. 2.12(a) requires analysing the motion until the half-clearance a is taken up, and then using the displacement and velocity at this point as initial conditions for the ensuing motion when the extra springs are operating. Similar analysis is necessary when the body leaves the influence of the extra springs. See Example 4.

2.1.4 Energy methods for analysis

For undamped free vibration the total energy in the vibrating system is constant throughout the cycle. Therefore the maximum potential energy V_{max} is equal to the maximum kinetic energy T_{max} although these maxima occur at different times during the cycle of vibration. Furthermore, since the total energy is constant,

$$T + V = \text{constant},$$

and thus

$$\frac{\mathrm{d}}{\mathrm{d}t}(T + V) = 0.$$

Applying this method to the case, already considered, of a body of mass m fastened to a spring of stiffness k, when the body is displaced a distance x from its equilibrium position,

$$\text{strain energy (SE) in spring} = \tfrac{1}{2}\,kx^2.$$

$$\text{kinetic energy (KE) of body} = \tfrac{1}{2}\,m\dot{x}^2.$$

Hence

$$V = \tfrac{1}{2}kx^2,$$

and

$$T = \tfrac{1}{2}m\dot{x}^2.$$

Thus

$$\frac{\mathrm{d}}{\mathrm{d}t}\left(\tfrac{1}{2}m\dot{x}^2 + \tfrac{1}{2}kx^2\right) = 0,$$

that is

$$m\ddot{x}\dot{x} + k\dot{x}x = 0,$$

or

$$\ddot{x} + \left(\frac{k}{m}\right)x = 0, \text{ as before, equation (2.1).}$$

This is a very useful method for certain types of problem in which it is difficult to apply Newton's laws of motion.

Alternatively, assuming SHM, if $x = x_0 \cos \omega t$,

The maximum SE, $V_{max} = \frac{1}{2}kx_0^2$,

and the maximum KE, $T_{max} = \frac{1}{2}m(x_0\omega)^2$.

Thus, since $T_{max} = V_{max}$,

$\frac{1}{2}kx_0^2 = \frac{1}{2}mx_0^2\omega^2$,

or $\omega = \sqrt{(k/m)}$ rad/s.

Example 1

A link AB in a mechanism is a rigid bar of uniform section 0.3 m long. It has a mass of 10 kg, and a concentrated mass of 7 kg is attached at B. The link is hinged at A and is supported in a horizontal position by a spring attached at the mid point of the bar. The stiffness of the spring is 2 kN/m. Find the frequency of small free oscillations of the system. The system is as shown below.

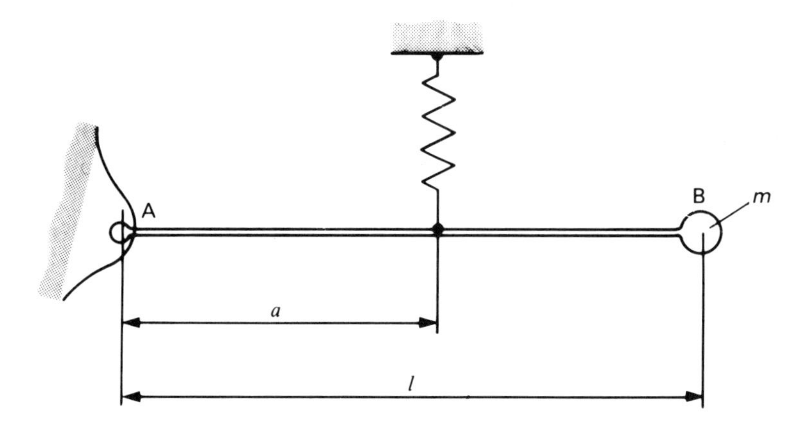

For rotation about A the equation of motion is

$$I_A\ddot{\theta} = -ka^2\theta,$$

that is,

$$\ddot{\theta} + (ka^2/I_A)\theta = 0.$$

This is SHM with frequency

$$\frac{1}{2\pi}\sqrt{(ka^2/I_A)}\text{Hz}.$$

In this case

$$a = 0.15\text{m}, \quad l = 0.3\text{m}, \quad k = 2000\text{N/m},$$

and

$$I_A = 7(0.3)^2 + \tfrac{1}{3} \times 10(0.3)^2 = 0.93 \text{ kg m}^2.$$

Hence

$$f = \frac{1}{2\pi}\sqrt{\left(\frac{2000\,(0.15)^2}{0.93}\right)} = 1.1 \text{ Hz}.$$

Example 2

A small turbo-generator has a turbine disc of mass 20 kg and radius of gyration 0.15 m driving an armature of mass 30 kg and radius of gyration 0.1 m through a steel shaft 0.05 m diameter and 0.4 m long. The modulus of rigidity for the shaft steel is $86 \times 10^9 \text{N/m}^2$. Determine the natural frequency of torsional oscillation of the system, and the position of the node. The shaft is supported by bearings which are not shown.

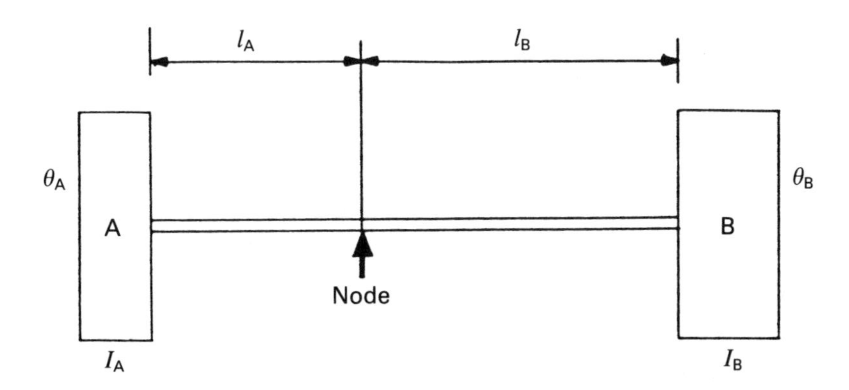

The rotors must twist in opposite directions to each other; that is, along the shaft there is a section of zero twist: this is called a node. The frequency of oscillation of each rotor is the same, and since there is no twist at the node,

$$f = \frac{1}{2\pi}\sqrt{\left(\frac{GJ}{I_A l_A}\right)} = \frac{1}{2\pi}\sqrt{\left(\frac{GJ}{I_B l_B}\right)},$$

that is,

$$I_A l_A = I_B l_B.$$

Now

$$I_A = 20(0.15)^2 = 0.45 \text{ kg m}^2, \text{ and } I_B = 30(0.1)^2 = 0.3\text{kg m}^2.$$

Thus

$$l_A = (0.3/0.45) l_B.$$

Since $l_A + l_B = 0.4$ m, $\quad l_A = 0.16$m,

that is, the node is 0.16 m from the turbine disc.

Hence

$$f = \frac{1}{2\pi}\sqrt{\left(\frac{86 \times 10^9 \;\; \pi(0.05)^4}{0.45 \times 0.16 \times 32}\right)} = 136 \text{ Hz } (= 136 \times 60 = 8160 \text{ rev/min}).$$

If the generator is run near to 8160 rev/min this resonance will be excited causing high dynamic stresses and probable fatigue failure of the shaft at the node.

Example 3

A uniform cylinder of mass m is rotated through a small angle θ_0 from the equilibrium position and released. Determine the equation of motion and hence obtain the frequency of free vibration. The cylinder rolls without slipping.

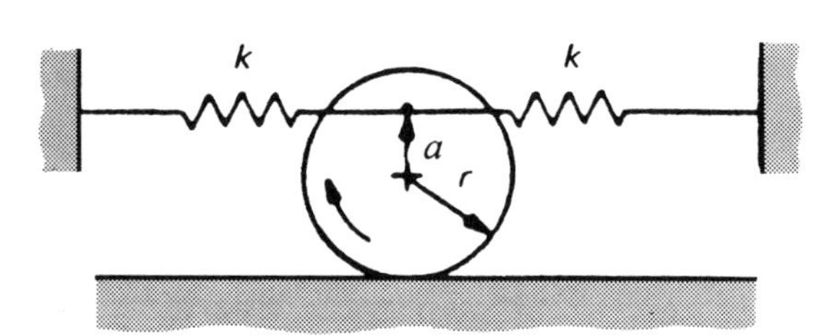

If the axis of the cylinder moves a distance x and turns through an angle θ so that $x = r\theta$,

$$\text{KE} = \tfrac{1}{2}\, m\dot{x}^2 + \tfrac{1}{2}I\dot{\theta}^2, \quad \text{where} \quad I = \tfrac{1}{2}\, mr^2.$$

Hence

$$\text{KE} = \tfrac{3}{4}\, mr^2\dot{\theta}^2.$$
$$\text{SE} = 2 \times \tfrac{1}{2} \times k[(r + a)\theta]^2 = k(r + a)^2\theta^2.$$

Now, energy is conserved, so $(\tfrac{3}{4}\, mr^2\dot{\theta}^2 + k(r + a)^2\theta^2)$ is constant, that is,

$$\frac{\mathrm{d}}{\mathrm{d}t}(\tfrac{3}{4}\, mr^2\dot{\theta}^2 + k(r + a)^2\theta^2) = 0$$

or

$$\tfrac{3}{4}mr^2 2\dot{\theta}\ddot{\theta} + k(r + a)^2 2\theta\dot{\theta} = 0.$$

Thus the equation of the motion is

$$\ddot{\theta} + \frac{k(r + a)^2\theta}{(3/4)mr^2} = 0.$$

Hence the frequency of free vibration is

$$\frac{1}{2\pi}\sqrt{\left[\frac{4k(r + a)^2}{3mr^2}\right]}\text{Hz.}$$

Example 4

In the system shown in Fig. 2.12(a) find the period of free vibration of the body if it is displaced a distance x_0 from the equilibrium position and released, when x_0 is greater than the half clearance a. Each spring has a stiffness k, and damping is negligible.

Initially the body moves under the action of four springs so that the equation of motion is

$$m\ddot{x} = -2kx - 2k(x - a),$$

that is,

$$m\ddot{x} + 4kx = 2ka.$$

The general solution comprises a complementary function and a particular integral such that

$$x = A \cos \omega t + B \sin \omega t + \frac{a}{2}.$$

With initial conditions $x = x_0$ at $t = 0$, and $\dot{x} = 0$ at $t = 0$, the solution is

$$x = \left(x_0 - \frac{a}{2}\right) \cos \omega t + \frac{a}{2}, \quad \text{where} \quad \omega = \sqrt{\left(\frac{4k}{m}\right)} \text{ rad/s}.$$

When $x = a$, $t = t_1$ and

$$a = \left(x_0 - \frac{a}{2}\right) \cos \omega t_1 + \frac{a}{2}.$$

so that

$$t_1 = \sqrt{\left(\frac{m}{4k}\right)} \cos^{-1} \frac{a}{2(x_0 - a/2)}.$$

This is the time taken for the body to move from $x = x_0$ to $x = a$.
If a particular value of x_0 is chosen, $x_0 = 2a$ say, then

$$x = \frac{3a}{2} \cos \omega t + \frac{a}{2} \quad \text{and} \quad t_1 = \sqrt{\left(\frac{m}{4k}\right)} \cos^{-1}\left(\frac{1}{3}\right),$$

and when $x = a$,

$$\dot{x} = -\frac{3a}{2}\omega \sin \omega t_1 = -a\sqrt{(8k/m)}.$$

For motion from $x = a$ to $x = 0$ the body moves under the action of two springs only, with the initial conditions

$$x_0 = a \quad \text{and} \quad \dot{x} = -a\sqrt{\left(\frac{8k}{m}\right)}$$

The equation of motion is $m\ddot{x} + 2kx = 0$ for this interval, the general solution to which is $x = C \cos \Omega t + D \sin \Omega t$, where $\Omega = \sqrt{(2k/m)}$ rad/s.

Since $x_0 = a$ at $t = 0$, $C = a$
and because

$$\dot{x} = -a\sqrt{\left(\frac{8k}{m}\right)} \text{ at } t = 0$$

and

$$\dot{x} = -C\Omega \sin \Omega t + D\Omega \cos \Omega t$$

then $D = -2a$. That is,

$$x = a \cos \Omega t - 2a \sin \Omega t.$$

When $x = 0$, $t = t_2$ and $\cos \Omega t_2 = 2 \sin \Omega t_2$ or $\tan \Omega t_2 = \frac{1}{2}$, that is

$$t_2 = \sqrt{\left(\frac{m}{2k}\right)} \tan^{-1}\left(\frac{1}{2}\right).$$

Thus the time for $\frac{1}{4}$ cycle is $(t_1 + t_2)$, and the time for one cycle, that is the period of free vibration when $x_0 = 2a$ is $4(t_1 + t_2)$ or

$$4\left[\sqrt{\left(\frac{m}{4k}\right)} \cos^{-1}\left(\frac{1}{3}\right) + \sqrt{\left(\frac{m}{2k}\right)} \tan^{-1}\left(\frac{1}{2}\right)\right] \text{s.}$$

Example 5

A uniform wheel of radius R can roll without slipping on an inclined plane. Concentric with the wheel, and fixed to it, is a drum of radius r around which is wrapped one end of a string. The other end of the string is fastened to an anchored spring, of stiffness k, as shown. Both spring and string are parallel to the plane. The total mass of the wheel/drum assembly is m and its moment of inertia about the axis through the centre of the wheel O is I. If the wheel is displaced a small distance from its equilibrium position and released, derive the equation describing the ensuing motion and hence calculate the frequency of the oscillations. Damping is negligible.

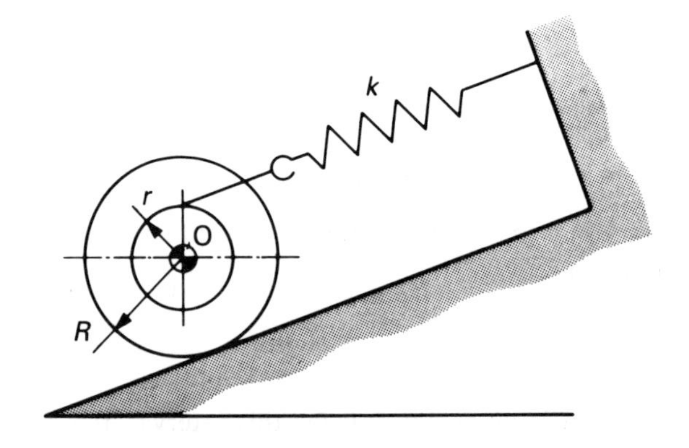

If the wheel is given an anti-clockwise rotation θ from the equilibrium position, the spring extension is $(R + r)\,\theta$ so that the restoring spring force is $k(R + r)\theta$.

The FBDs are

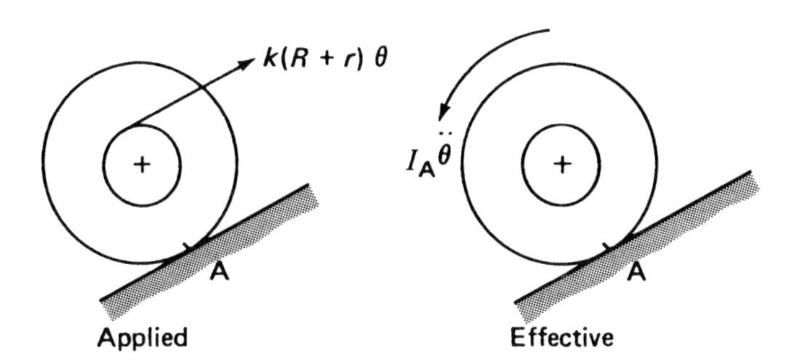

The rotation is instantaneously about the contact point A so that taking moments about A gives the equation of motion as

$$I_A\ddot{\theta} = -\,k(R + r)^2\theta.$$

(The moment due to the weight cancels with the moment due to the initial spring tension.)

Now $I_A = I + mR^2$, so

$$\ddot{\theta} + \left(\frac{k(R + r)^2}{I + mR^2}\right)\theta = 0,$$

and the frequency of oscillation is

$$\frac{1}{2\pi}\sqrt{\left(\frac{k(R + r)^2}{I + mR^2}\right)}\text{Hz}.$$

An alternative method for obtaining the frequency of oscillation is to consider the energy in the system.

Now

$$\text{SE, } V = \tfrac{1}{2}k(R + r)^2\theta^2,$$

and

$$\text{KE, } T = \tfrac{1}{2}\text{I}_A\dot{\theta}^2.$$

(Weight and initial spring tension effects cancel.)

So $T + V = \frac{1}{2}I_A\dot{\theta}^2 + \frac{1}{2}k(R + r)^2\theta^2$,

and

$$\frac{\text{d}}{\text{d}t}(T + V) = \tfrac{1}{2}I_A 2\dot{\theta}\ddot{\theta} + \tfrac{1}{2}k(R + r)^2 2\dot{\theta}\theta = 0.$$

Hence

$$I_A\ddot{\theta} + k(R + r)^2\theta = 0,$$

which is the equation of motion.

Or, we can put $V_{max} = T_{max}$, and if $\theta = \theta_0 \sin \omega t$ is assumed,

$$\tfrac{1}{2}k(R + r)^2\theta_0^2 = \tfrac{1}{2}I_A\omega^2\theta_0^2,$$

so that

$$\omega = \sqrt{\left(\frac{k(R + r)^2}{I_A}\right)} \text{ rad/s},$$

where

$$I_A = I + mr^2 \quad \text{and} \quad f = (\omega/2\pi) \text{ Hz}.$$

Example 6

A uniform building of height $2h$ and mass m has a rectangular base $a \times b$ which rests on an elasic soil. The stiffness of the soil, k, is expressed as the force per unit area required to produce unit deflection.

Find the lowest frequency of free low-amplitude swaying oscillation of the building.

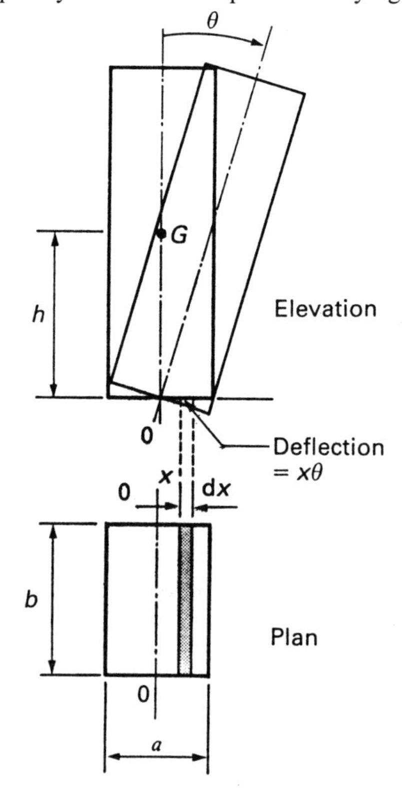

The lowest frequency of oscillation about the axis O–O through the base of the building is when the oscillation occurs about the shortest side, length a.

I_0 is the mass moment of inertia of the building about axis O–O.

The FBDs are

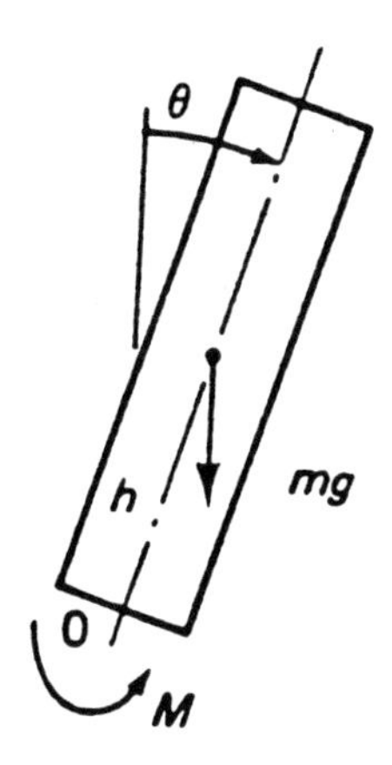

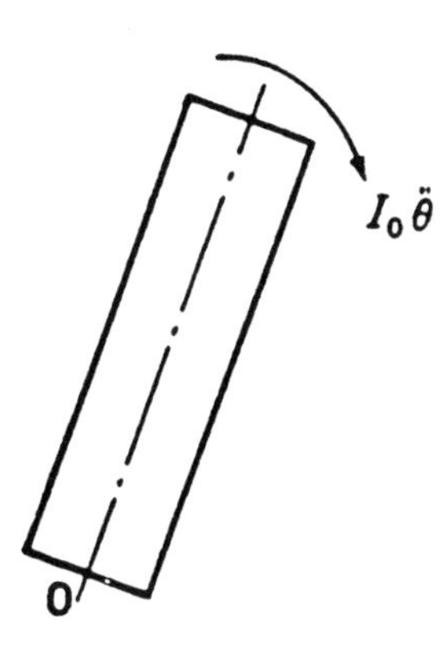

and the equation of motion for small θ is given by

$$I_0\ddot{\theta} = mgh\theta - M,$$

where M is the restoring moment from the elastic soil.

For the soil, k = force/(area × deflection), so considering an element of the base as shown, the force on element $= kb\ \mathrm{d}x \times x\theta$, and the moment of this force about axis O–O $= kb\mathrm{d}x \times x\theta \times x$. Thus the total restoring moment M, assuming the soil acts similarly in tension and compression is

$$M = 2\int_0^{a/2} kbx^2\theta \mathrm{d}x$$

$$= 2kb\theta \frac{(a/2)^3}{3} = \frac{ka^3b}{12}\theta.$$

Thus the equation of motion becomes

$$I_0\ddot{\theta} + \left(\frac{ka^3b}{12} - mgh\right)\theta = 0.$$

Motion is therefore simple harmonic, with frequency

$$f = \frac{1}{2\pi}\sqrt{\left(\frac{ka^3b/12 - mgh}{I_0}\right)}\text{Hz.}$$

An alternative solution can be obtained by considering the energy in the system. In this case,

$$T = \tfrac{1}{2}I_0\dot{\theta}^2,$$

and

$$V = \tfrac{1}{2}2\int_0^{a/2} kb\mathrm{d}x \times x\theta \times x\theta - \frac{mgh\theta^2}{2},$$

where the loss in potential energy of building weight is given by $mgh\ (1 - \cos\theta) \simeq mgh\theta^2/2$, since $\cos\theta \simeq 1 - \theta^2/2$ for small values of θ. Thus

$$V = \left(\frac{ka^3b}{24} - \frac{mgh}{2}\right)\theta^2.$$

Assuming simple harmonic motion, and putting $T_{max} = V_{max}$, gives

$$\omega^2 = \left(\frac{ka^3b/12 - mgh}{I_0}\right),$$

as before.
Note that for stable oscillation, $\omega > 0$, so that

$$\left(\frac{ka^3b}{12} - mgh\right) > 0.$$

That is $ka^3b > 12mgh$.

This expression gives the minimum value of k, the soil stiffness, for stable oscillation of a particular building to occur. If k is less that 12 mgh/a^3b the building will fall over when disturbed.

2.2 FREE DAMPED VIBRATION

All real systems dissipate energy when they vibrate. The energy dissipated is often very small, so that an undamped analysis is sometimes realistic; but when the damping is significant its effect must be included in the analysis, particularly when the amplitude of vibration is required. Energy is dissipated by frictional effects, for example that occurring at the connection between elements, internal friction in deformed members, and windage. It is often difficult to model damping exactly because many mechanisms may be operating in a system. However, each type of damping can be analysed, and since in many dynamic systems one form of damping predominates, a reasonably accurate analysis is usually possible.

The most common types of damping are viscous, dry friction and hysteretic. Hysteretic damping arises in structural elements due to hysteresis losses in the material.

The type and amount of damping in a system has a large effect on the dynamic response levels.

2.2.1 Vibration with viscous damping

Viscous damping is a common form of damping which is found in many engineering systems such as instruments and shock absorbers. The viscous damping force is proportional to the first power of the velocity across the damper, and it always opposes the motion, so that the damping force is a linear continuous function of the velocity. Because the analysis of viscous damping leads to the simplest mathematical treatment, analysts sometimes approximate more complex types of damping to the viscous type.

Consider the single degree of freedom model with viscous damping shown in Fig. 2.13.

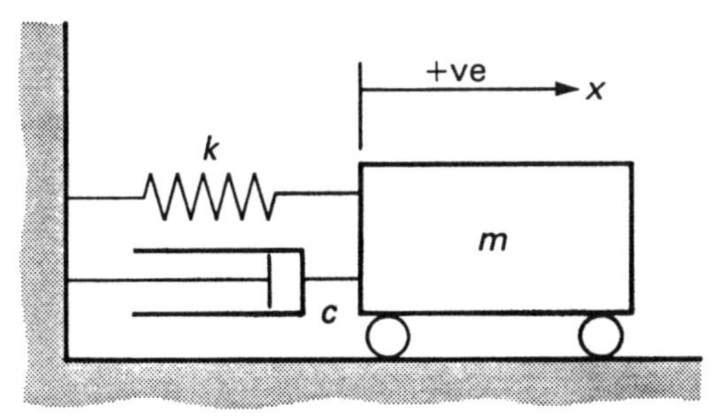

Fig. 2.13. Single degree of freedom model with viscous damping.

The only unfamiliar element in the system is the viscous damper with coefficient c. This coefficient is such that the damping force required to move the body with a velocity $\dot{x}$ is $c\dot{x}$.

For motion of the body in the direction shown, the free body diagrams are as in

Fig. 2.14. (a) Applied force; (b) effective force.

Fig. 2.14(a) and (b). The equation of motion is therefore

$$m\ddot{x} + c\dot{x} + kx = 0. \tag{2.9}$$

This equation of motion pertains to the whole of the cycle: the reader should verify that this is so. (Note: displacements to the left of the equilibrium position are negative, and velocities and accelerations from right to left are also negative).

Equation (2.9) is a second-order differential equation which can be solved by assuming a solution of the form $x = Xe^{st}$. Substituting this solution into equation (2.9) gives

$$(ms^2 + cs + k)Xe^{st} = 0.$$

Since $Xe^{st} \neq 0$ (otherwise no motion), $\quad ms^2 + cs + k = 0,$

so, $$s_{1,2} = -\frac{c}{2m} \pm \frac{\sqrt{(c^2 - 4mk)}}{2m}.$$

Hence

$$x = X_1 e^{s_1 t} + X_2 e^{s_2 t},$$

where X_1 and X_2 are arbitrary constants found from the initial conditions. The system response evidently depends on whether c is positive or negative, and on whether c^2 is greater than, equal to, or less than $4mk$.

The dynamic behaviour of the system depends on the numerical value of the radical, so we define critical damping as that value of $c(c_c)$ which makes the radical zero: that is,

$$c_c = 2\sqrt{(km)}.$$

Hence

$$c_c/2m = \sqrt{(k/m)} = \omega, \qquad \text{the undamped natural frequency, and}$$

$$c_c = 2\sqrt{(km)} = 2m\omega.$$

The actual damping in a system can be specified in terms of c_c by introducing the damping ratio ζ.

Thus $$\zeta = c/c_c,$$

and $$s_{1,2} = [-\zeta \pm \sqrt{(\zeta^2 - 1)}]\omega. \qquad (2.10)$$

The response evidently depends on whether c is positive or negative, and on whether ζ is greater than, equal to, or less than unity. Usually c is positive, so we need consider only the other possibilities.

Case 1. $\zeta < 1$; that is, damping less than critical
From equation (2.10)

$$s_{1,2} = -\zeta\omega \pm j\sqrt{(1 - \zeta^2)}\omega, \quad \text{where} \quad j = \sqrt{(-1)},$$

so

$$x = e^{-\zeta\omega t}[X_1 e^{j\sqrt{(1-\zeta^2)}\omega t} + X_2 e^{-j\sqrt{(1-\zeta^2)}\omega t}]$$

and

$$x = Xe^{-\zeta\omega t} \sin(\sqrt{(1 - \zeta^2)}\,\omega t + \phi).$$

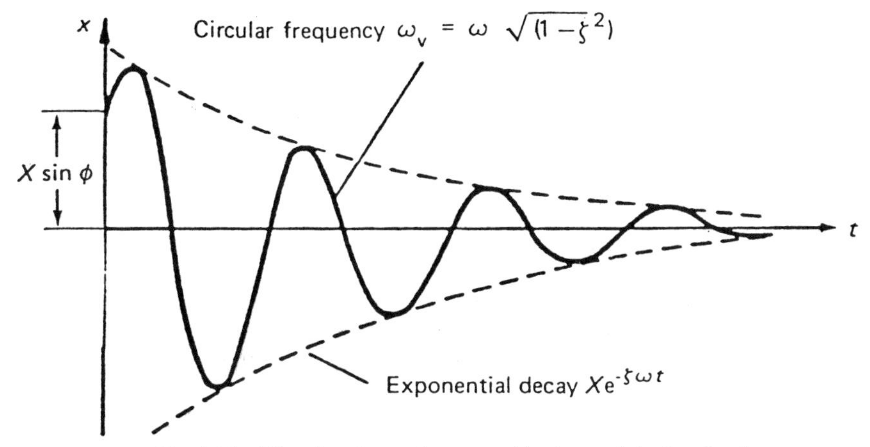

Fig. 2.15. Vibration decay of system with viscous damping, $\zeta < 1$.

The motion of the body is therefore an exponentially decaying harmonic oscillation with circular frequency $\omega_v = \omega\sqrt{(1-\zeta^2)}$, as shown in Fig. 2.15.

The frequency of the viscously damped oscillation ω_v, is given by $\omega_v = \omega\sqrt{(1-\zeta^2)}$, that is, the frequency of oscillation is reduced by the damping action. However, in many systems this reduction is likely to be small, because very small values of ζ are common; for example in most engineering structures ζ is rarely greater than 0.02. Even if $\zeta = 0.2$, $\omega_v = 0.98\ \omega$. This is not true in those cases where ζ is large, for example in motor vehicles where ζ is typically 0.7 for new shock absorbers.

Case 2. $\zeta = 1$; that is, critical damping
Both values of s are $-\ \omega$. However, two constants are required in the solution of equation (2.9), thus $x = (A + Bt)e^{-\omega t}$ may be assumed.

Critical damping represents the limit of periodic motion, hence the displaced body is restored to equilibrium in the shortest possible time, and without oscillation or overshoot. Many devices, particularly electrical instruments, are critically damped to take advantage of this property.

Case 3. $\zeta > 1$; that is, damping greater than critical
There are two real values of s, so $x = X_1e^{s_1t} + X_2e^{s_2t}$.

Since both values of s are negative the motion is the sum of two exponential decays, as shown in Fig. 2.16.

Logarithmic decrement Λ

A convenient way of determining the damping in a system is to measure the rate of decay of oscillation. It is usually not satisfactory to measure ω_v and ω, because unless $\zeta > 0.2$, $\omega \simeq \omega_v$.

The logarithmic decrement, Λ, is the natural logarithm of the ratio of any two successive amplitudes in the same direction, and so from Fig. 2.17.

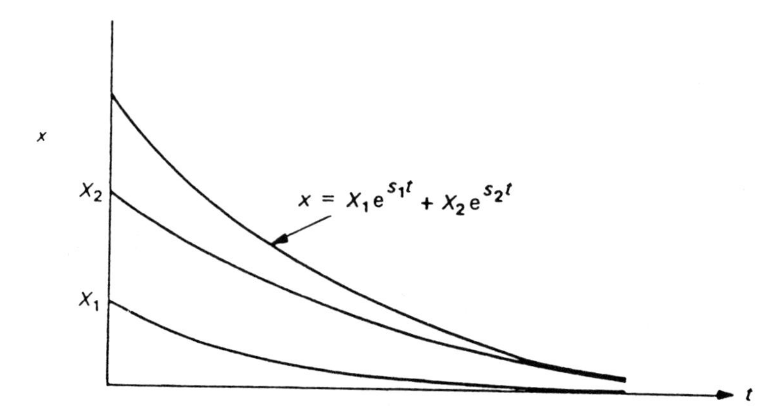

Fig. 2.16. Disturbance decay of system with viscous damping, $\zeta > 1$.

$$\Lambda = \ln \frac{X_1}{X_{11}}$$

Since

$$x = X e^{-\zeta \omega t} \sin (\omega_v t + \phi),$$

if

$$X_1 = X e^{-\zeta \omega t}, \qquad X_{11} = X e^{-\zeta \omega (t + \tau_v)},$$

where τ_v is the period of the damped oscillation.

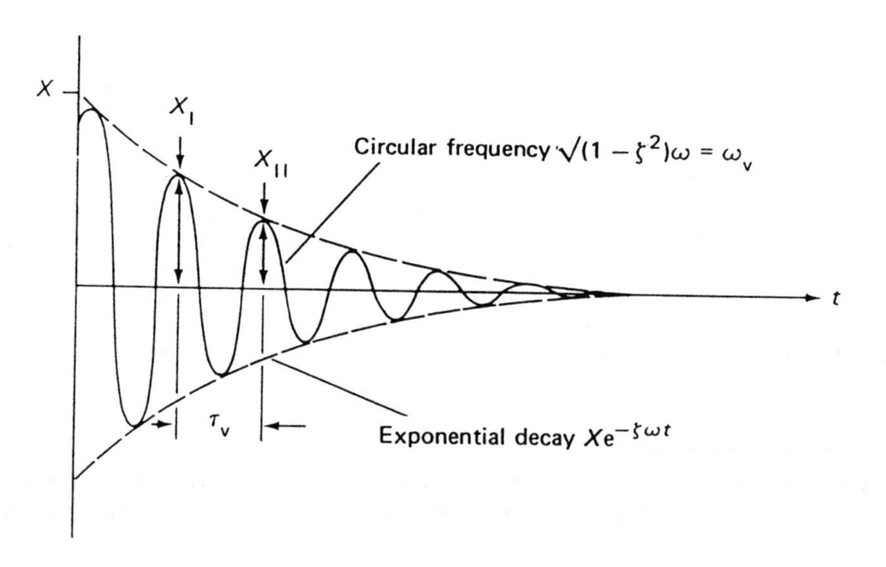

Fig. 2.17. Vibration decay.

Thus

$$\Lambda = \ln \frac{X e^{-\zeta \omega t}}{X e^{-\zeta \omega (t + \tau_v)}} = \zeta \omega \tau_v.$$

Since

$$\tau_v = \frac{2\pi}{\omega_v} = \frac{2\pi}{\omega\sqrt{(1 - \zeta^2)}},$$

$$\Lambda = \frac{2\pi\zeta}{\sqrt{(1 - \zeta^2)}} = \ln\left(\frac{X_1}{X_{11}}\right).$$

For small values of $\zeta (\ngtr 0.25)$, $\Lambda \simeq 2\pi\zeta$.

It should be noted that this analysis assumes that the point of maximum displacement in a cycle and the point where the envelope of the decay curve $Xe^{-\zeta\omega t}$ touches the decay curve itself, are coincident. This is usually very nearly so, and the error in making this assumption is usually negligible, except in those cases where the damping is high.

For low damping it is preferable to measure the amplitude of oscillations many cycles apart so that an easily measurable difference exists.

In this case $\Lambda = \ln\left(\frac{X_1}{X_{11}}\right) = \frac{1}{N} \times \ln\left(\frac{X_1}{X_{N+1}}\right)$ since

$$\frac{X_1}{X_{11}} = \frac{X_{11}}{X_{111}} \text{ etc.}$$

Example 7

Consider the transverse vibration of a bridge structure. For the fundamental frequency it can be considered as a single degree of freedom system. The bridge is deflected at mid-span (by winching the bridge down) and suddenly released. After the initial disturbance the vibration was found to decay exponentially from an amplitude of 10 mm to 5.8 mm in three cycles with a frequency of 1.62 Hz. The test was repeated with a vehicle of mass 40 000 kg at mid-span, and the frequency of free vibration was measured to be 1.54 Hz.

Find the effective mass, the effective stiffness, and the damping ratio of the structure.

Let m be the effective mass and k the effective stiffness. Then

$$f_1 = 1.62 = \frac{1}{2\pi}\sqrt{\left(\frac{k}{m}\right)} \text{ Hz},$$

and

$$f_2 = 1.54 = \frac{1}{2\pi}\sqrt{\left(\frac{k}{m + 40 \times 10^3}\right)} \text{ Hz},$$

if it is assumed that ζ is small enough for $f_v \simeq f$.

Thus $$\left(\frac{1.62}{1.54}\right)^2 = \frac{m + 40 \times 10^3}{m},$$

so

$$m = 375 \times 10^3 \text{ kg}.$$

Since

$$k = (2\pi f_1)^2 m,$$

$$k = 38\,850 \text{ kN/m}.$$

Now

$$\Lambda = \ln \frac{X_1}{X_{\text{II}}} = \tfrac{1}{3} \ln \frac{X_1}{X_{\text{IV}}} = \tfrac{1}{3} \ln \left(\frac{10}{5.8}\right)$$

$$= 0.182.$$

Hence

$$\Lambda = \frac{2\pi\zeta}{\sqrt{(1 - \zeta^2)}} = 0.182,$$

and so $\zeta = 0.029$. (This compares with a value of about 0.002 for cast iron material. The additional damping originates mainly in the joints of the structure.) This value of ζ confirms the assumption that $f_v \simeq f$.

Example 8

A light rigid rod of length L is pinned at one end O and has a body of mass m attached at the other end. A spring and viscous damper connected in parallel are fastened to the rod at a distance a from the support. The system is set up in a *horizontal* plane: a plan view is shown.

Assuming that the damper is adjusted to provide critical damping, obtain the motion of the rod as a function of time if it is rotated through a small angle θ_0 and then released. Given that $\theta_0 = 2°$ and the undamped natural frequency of the system is 3 rad/s, calculate the displacement 1 s after release.

Explain the term *logarithmic decrement* as applied to such a system and calculate its value assuming that the damping is reduced to 80% of its critical value.

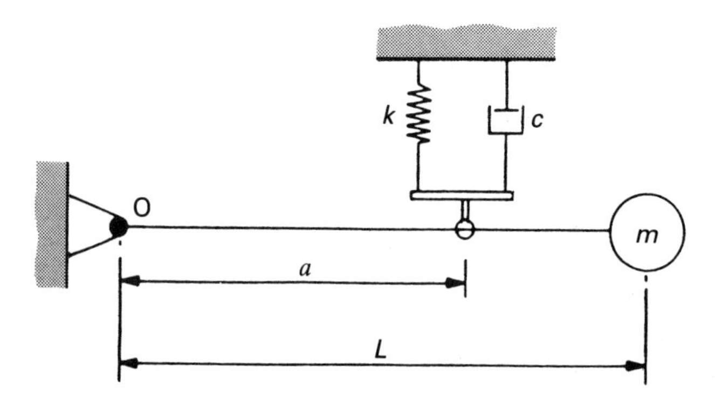

Let the rod turn through an angle θ from the equilibrium position.
Note that the system oscillates in the horizontal plane so that the FBDs are:

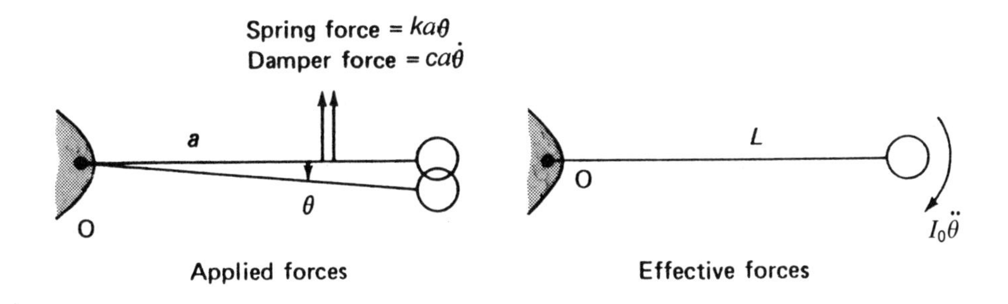

Taking moments about the pivot O gives

$$I_0\ddot{\theta} = -ca^2\dot{\theta} - ka^2\theta,$$

where $I_0 = ml^2$, so the equation of motion is

$$mL^2\ddot{\theta} + ca^2\dot{\theta} + ka^2\theta = 0.$$

Now the system is adjusted for critical damping, so that $\zeta = 1$.
The solution to the equation of motion is therefore of the form

$$\theta = (A + Bt)\mathrm{e}^{-\omega t}.$$

Now, $\theta = \theta_0$ when $t = 0$, and $\mathrm{d}\theta/\mathrm{d}t = 0$ when $t = 0$. Hence

$$\theta_0 = A,$$

and

$$0 = B\mathrm{e}^{-\omega t} + (A + Bt)(-\omega)\mathrm{e}^{-\omega t},$$

so that

$$B = \theta_0\omega.$$

Hence

$$\theta = \theta_0(1 + \omega)\mathrm{e}^{-\omega t}.$$

If $\omega = 3$ rad/s, $t = 1$ s and $\theta_0 = 2°$,

$$\theta = 2(1 + 3)\mathrm{e}^{-3} = 0.4°.$$

The logarithmic decrement

$$\Lambda = \ln\frac{X_1}{X_{11}} = \frac{2\pi\zeta}{\sqrt{(1 - \zeta^2)}},$$

so that if $\zeta = 0.8$,

$$\Lambda = \frac{5.027}{0.6} = 8.38$$

Root locus study of damping

It is often convenient to consider how the roots of equation (2.10) vary as ζ increases from zero. The roots of this equation are given by

$$s_{1,2}/\omega = -\zeta \pm j\sqrt{(1-\zeta^2)} \qquad \text{for } 1 > \zeta > 0$$

and

$$s_{1,2}/\omega = -\zeta \pm \sqrt{(\zeta^2 - 1)} \qquad \text{for } \zeta > 1.$$

These roots can be conveniently displayed in a plot of imaginary (s/ω) against real (s/ω); since for every value of ζ there are two values of (s/ω), the roots when plotted form two loci as shown in Fig. 2.18. The position of a root in the (s/ω) plane indicates the frequency of oscillation (Im (s/ω) axis) of the system, if any, and the rate of growth of decay of oscillation (Re (s/ω) axis).

Because the Re (s/ω) is negative on the left of the Im (s/ω) axis, all roots which lie to the left of the Im (s/ω) axis represent a decaying oscillation and therefore a stable system. Roots to the right of the Im (s/ω) axis represent a growing oscillation and an unstable system.

This root study of damping is a useful design technique, because the effects of changing the damping ratio on the response of a system can easily be seen. It also has important applications in the study of control system dynamics. Note that the damping ratio ζ is given by $\cos\theta$ in Fig. 2.18.

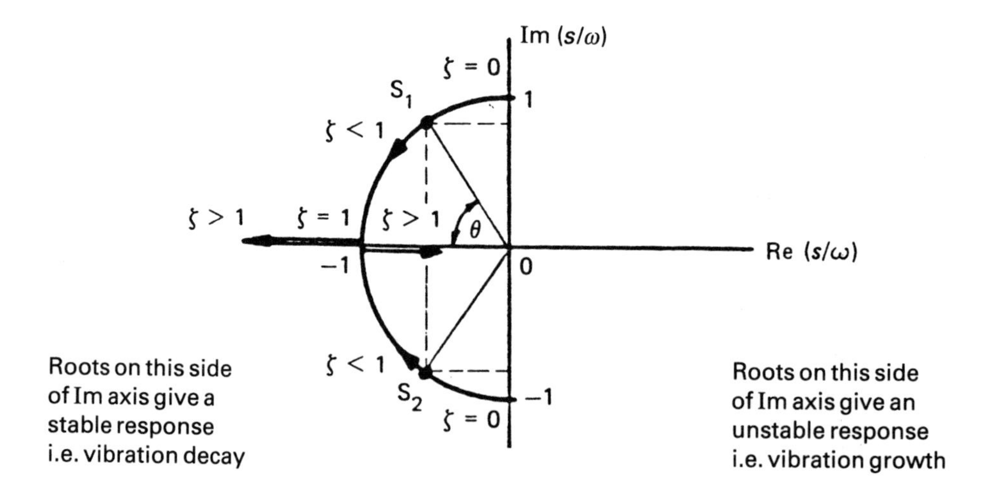

Fig. 2.18. Root locus plot.

2.2.2 Vibration with Coulomb (dry friction) damping

Steady friction forces occur in many systems when relative motion takes place between adjacent members. These forces are independent of amplitude and frequency; they always oppose the motion and their magnitude may, to a first approximation, be considered constant. Dry friction can, of course, just be one of the damping mechanisms present; however, in some systems it is the main source of damping. In these cases the damping can be modelled as in Fig. 2.19

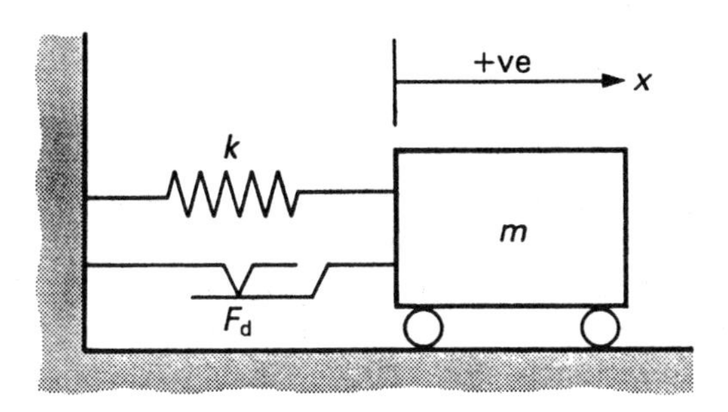

Fig. 2.19. System with Coulomb damping.

The constant friction force F_d always opposes the motion, so that if the body is displaced a distance x_0 to the right and released from rest we have, for motion from right to left only,

$$m\ddot{x} = F_d - kx$$

$$\text{or} \quad m\ddot{x} + kx = F_d. \tag{2.11}$$

The solution to the complementary function is $x = A \sin \omega t + B \cos \omega t$, and the complete solution is

$$x = A \sin \omega t + B \cos \omega t + \frac{F_d}{k} \tag{2.12}$$

where $\omega = \sqrt{(k/m)}$ rad/s.

Note. The particular integral may be found by using the D-operator. Thus equation (2.11) is

$$(D^2 + \omega^2)x = F_d/m$$

so

$$x = (1/\omega^2)[1 + (D^2/\omega^2)]^{-1}F_d/m$$
$$= [1 - (D^2/\omega^2) + \cdots]F_d/m\omega^2 = F_d/k.$$

The initial conditions were $x = x_0$ at $t = 0$, and $\dot{x} = 0$ at $t = 0$. Substitution into equation (2.12) gives

$$A = 0 \quad \text{and} \quad B = x_0 - \frac{F_d}{k}.$$

Hence

$$x = \left(x_0 - \frac{F_d}{k}\right) \cos \omega t + \frac{F_d}{k}. \tag{2.13}$$

At the end of the half cycle right to left, $\omega t = \pi$ and

$$x_{(t=\pi/\omega)} = -x_0 + \frac{2F_d}{k}.$$

That is, there is a reduction in amplitude of $2F_d/k$ per half cycle.

From symmetry, for motion from left to right when the friction force acts in the opposite direction to the above, the initial displacement is $(x_0 - 2F_d/k)$ and the final displacement is therefore $(x_0 - 4F_d/k)$, that is the reduction in amplitude is $4F_d/k$ per cycle. This oscillation continues until the amplitude of the motion is so small that the maximum spring force is unable to overcome the friction force F_d. This can happen whenever the amplitude is $\leqslant \pm (F_d/k)$. The manner of oscillation decay is shown in Fig. 2.20; the motion is sinusoidal for each half cycle, with successive half cycles centred on points distant $+ (F_d/k)$ and $- (F_d/k)$ from the origin. The oscillation ceases with $|x| \leqslant F_d/k$. The zone $x = \pm F_d/k$ is called the *dead zone.*

To determine the frequency of oscillation we rewrite the equation of motion (2.11) as

$$m\ddot{x} + k(x - (F_d/k)) = 0.$$

Now if $x' = x - (F_d/k)$, $\ddot{x}' = \ddot{x}$ so that $m\ddot{x}' + kx' = 0$, from which the frequency of oscillation is $(1/2\pi)\sqrt{(k/m)}$ Hz. That is, the frequency of oscillation is not affected by Coulomb friction.

Example 9

Part of a structure can be modelled as a torsional system comprising a bar of stiffness 10 kN m/rad and a beam of moment of inertia about the axis of rotation of 50 kg m^2. The bottom guide imposes a friction torque of 10 N m.

Given that the beam is displaced through 0.05 rad from its equilibrium position and released, find the frequency of the oscillation, the number of cycles executed before the beam motion ceases, and the position of the beam when this happens.

Now

$$\omega = \sqrt{\left(\frac{k_T}{I}\right)} = \sqrt{\frac{10 \times 10^3}{50}} = 14.14 \text{ rad/s}.$$

Hence

$$f = \frac{14.14}{2\pi} = 2.25 \text{ Hz}.$$

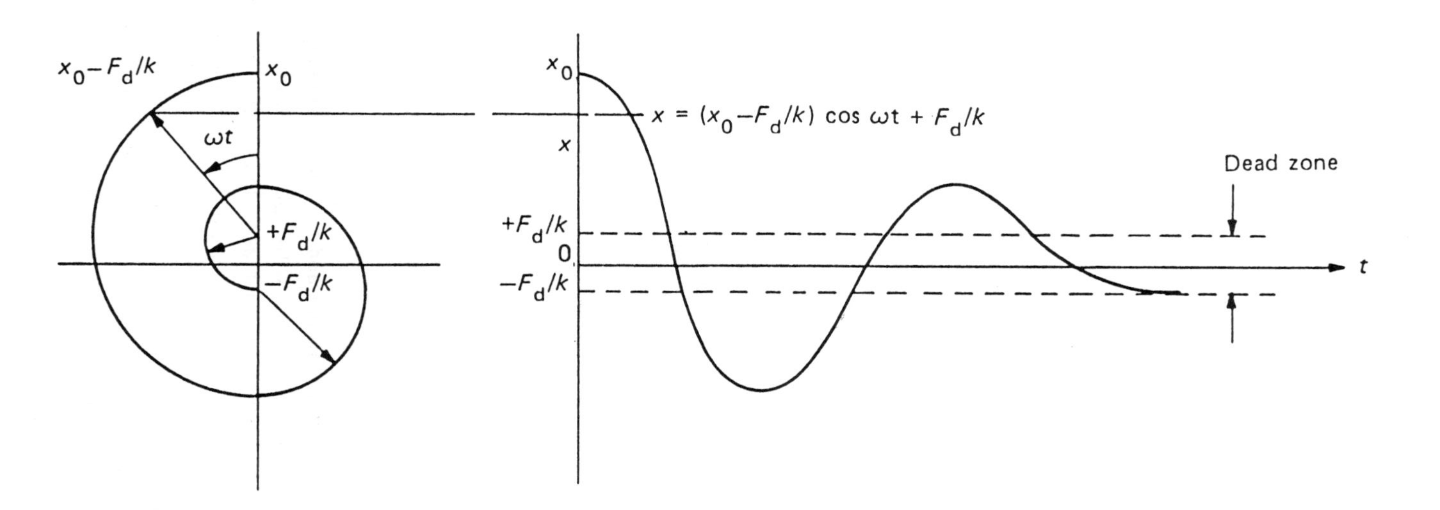

Fig. 2.20. Vibration decay of system with Coulomb damping

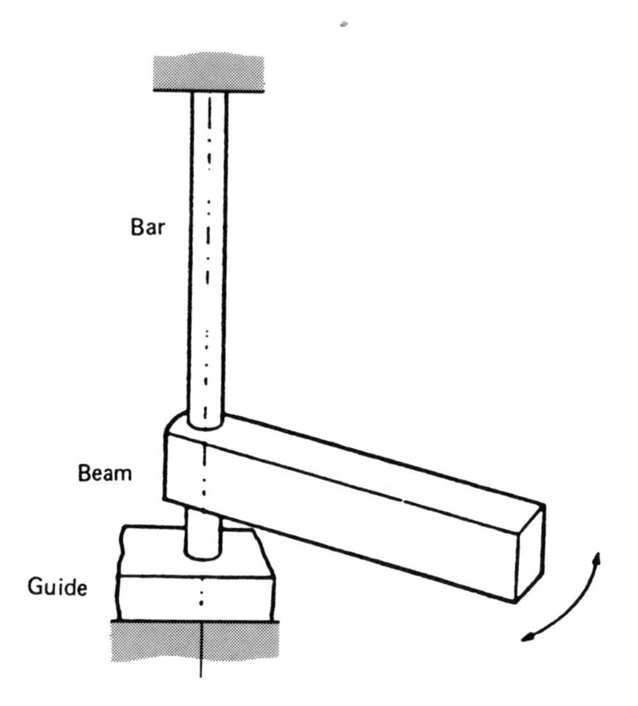

$$\text{Loss in amplitude/cycle} = \frac{4F_d}{k} = \frac{4 \times 10}{10^4} \text{ rad}$$

$$= 0.004 \text{ rad.}$$

Number of cycles for motion to cease

$$= \frac{0.05}{0.004} = 12\tfrac{1}{2}.$$

The beam is in the initial (equilibrium) position when motion ceases. The motion is shown opposite.

2.2.3 Vibration with combined viscous and Coulomb damping

The free vibration of dynamic systems with viscous damping is characterized by an exponential decay of the oscillation, whereas systems with Coulomb damping possess a linear decay of oscillation. Many real systems have both forms of damping, so that their vibration decay is a combination of exponential and linear functions.

The two damping actions are sometimes amplitude dependent, so that initially the decay is exponential, say, and only towards the end of the oscillation does the Coulomb effect show. In the analyses of these cases the Coulomb effect can easily be separated from the total damping to leave the viscous damping alone. The exponential decay with viscous damping can be checked by plotting the amplitudes on logarithmic-linear axes when the decay should be seen to be linear.

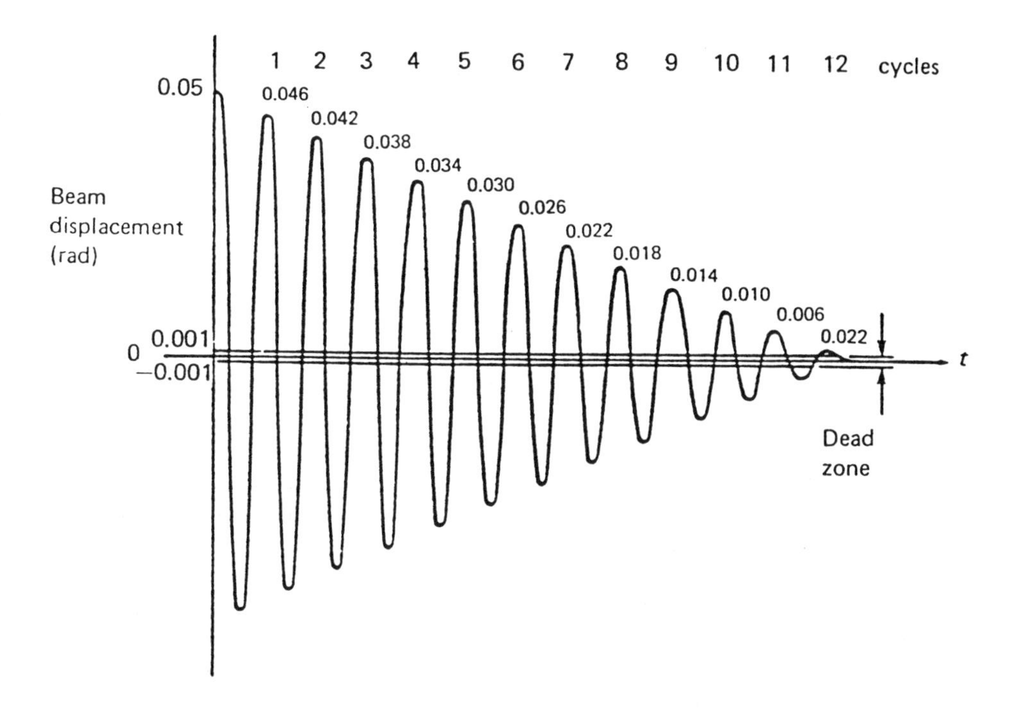

If the Coulomb and viscous effects cannot be separated in this way, a mixture of linear and exponential decay functions have to be found by trial and error in order to conform with the experimental data.

2.2.4 Vibration with hysteretic damping

Experiments on the damping that occurs in solid materials and structures which have been subjected to cyclic stressing have shown the damping force to be independent of frequency. This internal, or material, damping is referred to as hysteretic damping. Since the viscous damping force $c\dot{x}$ is dependent on the frequency of oscillation, it is not a suitable way of modelling the internal damping of solids and structures. The analysis of systems and structures with this form of damping therefore requires the damping force $c\dot{x}$ to be divided by the frequency of oscillation ω. Thus the equation of motion becomes $m\ddot{x} + (c/\omega)\dot{x} + kx = 0$.

However, it has been observed from experiments carried out on many materials and structures that under harmonic forcing the stress leads the strain by a constant angle, α.

Thus for an harmonic strain, $\varepsilon = \varepsilon_0 \sin \nu t$, where ν is the forcing frequency, the induced stress is $\sigma = \sigma_0 \sin (\nu t + \alpha)$. Hence

$$\sigma = \sigma_0 \cos \alpha \sin \nu t + \sigma_0 \sin \alpha \cos \nu t$$

$$= \sigma_0 \cos\alpha \sin \nu t + \sigma_o \sin\alpha \sin\left(\nu t + \frac{\pi}{2}\right).$$

The first component of stress is in phase with the strain ε, whilst the second component is in quadrature with ε and $\pi/2$ ahead. Putting $\mathrm{j} = \sqrt{(-1)}$,

$$\sigma = \sigma_0 \cos\alpha \sin\nu t + \mathrm{j}\sigma_0 \sin\alpha \sin\nu t.$$

Hence a complex modulus E^* can be formulated, where

$$E^* = \frac{\sigma}{\varepsilon} = \frac{\sigma_0}{e_0}\cos\alpha + \mathrm{j}\frac{\sigma_0}{\varepsilon_0}\sin\alpha$$

$$= E' + jE'',$$

where E' is the in-phase or storage modulus, and E'' is the quadrature or loss modulus.

The loss factor η, which is a measure of the hysteretic damping in a structure, is equal to E''/E', that is, $\tan\alpha$.

It is not usually possible to separate the stiffness of a structure from its hysteretic damping, so that in a mathematical model these quantities have to be considered together. The complex stiffness k^* is given by $k^* = k(1 + \mathrm{j}\eta)$, where k is the static stiffness and η the hysteretic damping loss factor.

The equation of free motion for a single degree of freedom system with hysteretic damping is therefore $m\ddot{x} + k^*x = 0$. Fig. 2.21 shows a single degree of freedom model with hysteretic damping of coefficient c_{H}.

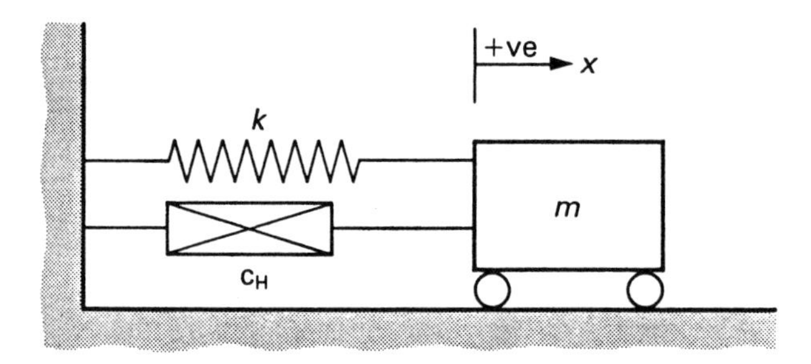

Fig. 2.21. Single degree of freedom model with hysteretic damping.

The equation of motion is

$$m\ddot{x} + (c_{\mathrm{H}}/\omega)\dot{x} + kx = 0.$$

Now if $x = X\mathrm{e}^{\mathrm{j}\omega t}$,

$$\dot{x} = \mathrm{j}\omega x \quad \text{and} \quad \left(\frac{c_{\mathrm{H}}}{\omega}\right)\dot{x} = \mathrm{j}c_{\mathrm{H}}x.$$

Thus the equation of motion becomes

$$m\ddot{x} + (k + \mathrm{j}c_{\mathrm{H}})x = 0.$$

Since

$$k + \mathrm{j}c_\mathrm{H} = k\left(1 + \frac{\mathrm{j}c_\mathrm{H}}{k}\right) = k(1 + \mathrm{j}\eta) = k^*,$$

we can write

$$m\ddot{x} + k^*x = 0.$$

That is, the combined effect of the elastic and hysteretic resistance to motion can be represented as a complex stiffness, k^*.

A range of values of η for some common engineering materials is given below. For more detailed information on material damping mechanisms and loss factors, see *Damping of Materials and Members in Structural Mechanics* by B. J. Lazan (Pergamon, 1968).

Material	Loss factor
Aluminium-pure	0.00002–0.002
Aluminium alloy-dural	0.0004–0.001
Steel	0.001–0.008
Lead	0.008–0.014
Cast iron	0.003–0.03
Manganese copper alloy	0.05–0.1
Rubber-natural	0.1–0.3
Rubber-hard	1.0
Glass	0.0006–0.002
Concrete	0.01–0.06

2.2.5 Energy dissipated by damping

The energy dissipated per cycle by the viscous damping force in a single degree of freedom vibrating system is approximately

$$4\int_0^X c\dot{x}\,\mathrm{d}x.$$

if $x = X\sin\omega t$ is assumed for the complete cycle. The energy dissipated is therefore

$$4\int_0^{\pi/2} cX^2\omega^2\cos\omega t\,\mathrm{d}t = \pi c\omega X^2.$$

The energy dissipated per cycle by Coulomb damping is $4F_\mathrm{d}X$ approximately. Thus an equivalent viscous damping coefficient for Coulomb damping c_d can be deduced, where

$$\pi c_\mathrm{d}\omega X^2 = 4F_\mathrm{d}X,$$

that is,

$$c_d = \frac{4F_d}{\pi\omega X}.$$

The energy dissipated per cycle by a force F acting on a system with hysteretic damping is $\int F\,dx$, where $F = k^*x = k(1 + j\eta)x$, and x is the displacement.

For harmonic motion $x = X \sin \omega t$,

so

$$\begin{aligned} F &= kX \sin \omega t + j\eta kX \sin \omega t \\ &= kX \sin \omega t + \eta kX \cos \omega t. \end{aligned}$$

Now

$$\sin \omega t = \frac{x}{X}, \qquad \text{therefore} \qquad \cos \omega t = \frac{\sqrt{(X^2 - x^2)}}{X}.$$

Thus

$$F = kx \pm \eta k\sqrt{(X^2 - x^2)}.$$

This is the equation of an ellipse as shown in Fig. 2.22. The energy dissipated is given by the area enclosed by the ellipse.

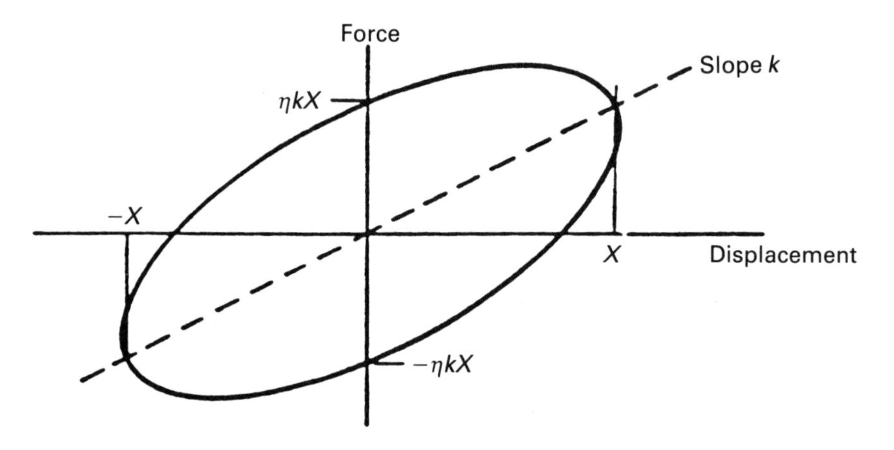

Fig. 2.22. Elliptical force–displacement relationship for a system with hysteretic damping.

Hence

$$\begin{aligned} \int F\mathrm{d}x &= \int_0^X (kx \pm \eta k\sqrt{(X^2 - x^2)})\mathrm{d}x \\ &= \pi X^2 \eta k. \end{aligned}$$

An equivalent viscous damping coefficient c_{H} is given by

$$\pi c_{\mathrm{H}} \omega X^2 = \pi \eta k X^2,$$

that is

$$c_{\mathrm{H}} = \frac{\eta k}{\omega},$$

Note also that $c = c_{\mathrm{H}}/\omega$.

Example 10

A single degree of freedom system has viscous damping, with $\zeta = 0.02$. Find the energy dissipated per cycle as a function of the energy in the system at the start of that cycle. Also find the amplitude of the 12th cycle if the amplitude of the 3rd cycle is 1.8 mm.

$$\zeta \ll 1, \text{ so } \ln (X_1/X_2) = 2\pi\zeta = 0.126.$$

Thus

$$X_1/X_2 = \mathrm{e}^{0.126} = 1.134.$$

Energy at start of cycle $= \frac{1}{2}kX_1{}^2$ (stored as strain energy in spring)

Energy at end of cycle $= \frac{1}{2}kX_2{}^2$

$$\frac{\text{Energy dissipated during cycle}}{\text{Energy at start of cycle}}$$

$$= \frac{\frac{1}{2}kX_1^2 - \frac{1}{2}kX_2^2}{\frac{1}{2}kX_1^2} = 1 - (X_2/X_1)^2 = 1 - 0.7773 = 0.223.$$

that is 22.3% of initial energy is dissipated in one cycle.
Now

$$X_1/X_2 = 1.134,\ X_2/X_3 = 1.134 \ldots (X_{n-1})/X_n = 1.134.$$

Therefore

$$X_3/X_{12} = (1.134)^9 = 3.107$$

that is

$$X_{12} = \frac{1.8}{3.107} = 0.579 \text{ mm}.$$

Example 11

A gun is designed so that when fired the barrel recoils against a spring. At the end of the recoil a viscous damper is engaged which allows the barrel to return to its equilibrium position in the minimum time without overshoot. Determine the spring constant and damping coefficient for a gun whose barrel has a mass of 500 kg if the initial recoil velocity is 30 m/s and the recoil distance is 1.6 m.

Find also the time required for the barrel to return to a position 0.1 m from its initial position.

Since the initial kinetic energy of the barrel must equal the maximum strain energy in the spring,

$$\tfrac{1}{2}k1.6^2 = \tfrac{1}{2}500(30)^2,$$

that is $k = 175$ kN/m.

Since critical damping is required, the damping coefficient is

$$2\sqrt{(500 \times 175 \times 10^3)} = 18\,700 \text{ Ns/m}.$$

For critical damping

$$x = (A + Bt)e^{-\omega t}$$

If $x = x_0$ and $\dot{x} = 0$ at $t = 0$,

$$x = x_0(1 + \omega t)e^{-\omega t}.$$

Now $\omega = \sqrt{(175 \times 10^3/500)} = 18.7$ rad/s and $x = 0.1$ m, so we require the solution to $0.1 = 1.6(1 + 18.7t)e^{-18.7t}$.

This equation can be solved in many ways, e.g. by trial and error or graphically, to give $t = 0.24$ s.

2.3 FORCED VIBRATION

Many real systems are subjected to periodic excitation. This may be due to unbalanced rotating parts, reciprocating components, or a shaking foundation. Sometimes large motions of the suspended body are desired as in vibratory feeders and compactors, but usually we require very low vibration amplitudes over a large range of exciting forces and frequencies. Some periodic forces are harmonic, but even if they are not, they can be represented as a series of harmonic functions using Fourier analysis techniques. Because of this the response of elastically supported bodies to harmonic exciting forces and motions must be studied.

2.3.1 Response of a viscous damped system to a simple harmonic exciting force with constant amplitude

In the system shown in Fig. 2.23, the body of mass m is connected by a spring and viscous damper to a fixed support, whilst an harmonic force of circular frequency ν and amplitude F acts upon it, in the line of motion.

The equation of motion is

$$m\ddot{x} + c\dot{x} + kx = F \sin \nu t. \tag{2.14}$$

The solution to $m\ddot{x} + c\dot{x} + kx = 0$, which has already been studied, is the complementary function; it represents the initial vibration which quickly dies away. The sustained motion is given by the particular solution. A solution $x = X(\sin \nu t - \phi)$ can be assumed, because this represents simple harmonic motion at the frequency of

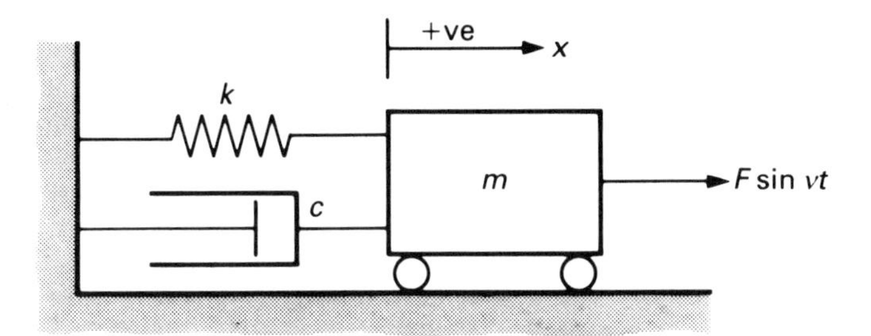

Fig. 2.23. Single degree of freedom model of a forced system with viscous damping.

the exciting force with a displacement vector which lags the force vector by ϕ, that is the motion occurs after the application of the force.

Assuming $x = X \sin(\nu t - \phi)$,

$$\dot{x} = X\nu \cos(\nu t - \phi) = X\nu \sin(\nu t - \phi + \tfrac{1}{2}\pi),$$

and

$$\ddot{x} = -X\nu^2 \sin(\nu t - \phi) = X\nu^2 \sin(\nu t - \phi + \pi).$$

The equation of motion (2.14) thus becomes

$$mX\nu^2 \sin(\nu t - \phi + \pi) + cX\nu \sin(\nu t - \phi + \pi/2) + kX \sin(\nu t - \phi)$$
$$= F \sin \nu t.$$

A vector diagram of these forces can now be drawn, Fig 2.24.

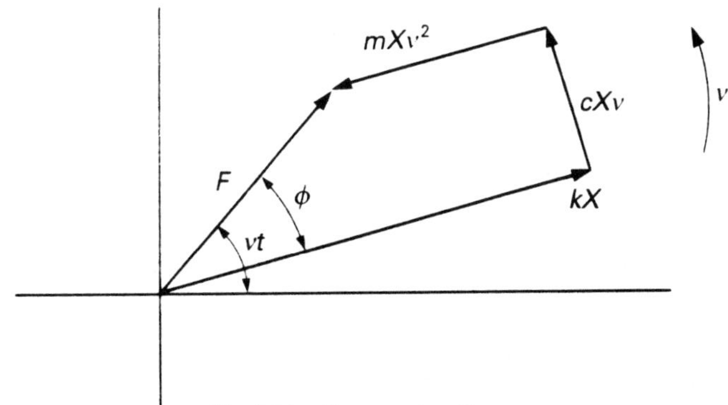

Fig. 2.24. Force vector diagram.

From the diagram,

$$F^2 = (kX - mX\nu^2)^2 + (cX\nu)^2,$$

or

$$X = F/\sqrt{((k - m\nu^2)^2 + (c\nu)^2)}, \tag{2.15}$$

and

$$\tan \phi = cXv/(kX - mXv^2).$$

Thus the steady-state solution to equation (2.14) is

$$x = \frac{F}{\sqrt{((k - mv^2)^2 + (cv)^2)}} \sin(vt - \phi),$$

where

$$\phi = \tan^{-1}\left(\frac{cv}{k - mv^2}\right).$$

The complete solution includes the transient motion given by the complementary function:

$$x = Ae^{-\zeta\omega t} \sin(\omega\sqrt{(1 - \zeta^2)}t + \alpha).$$

Fig. 2. 25 shows the combined motion.

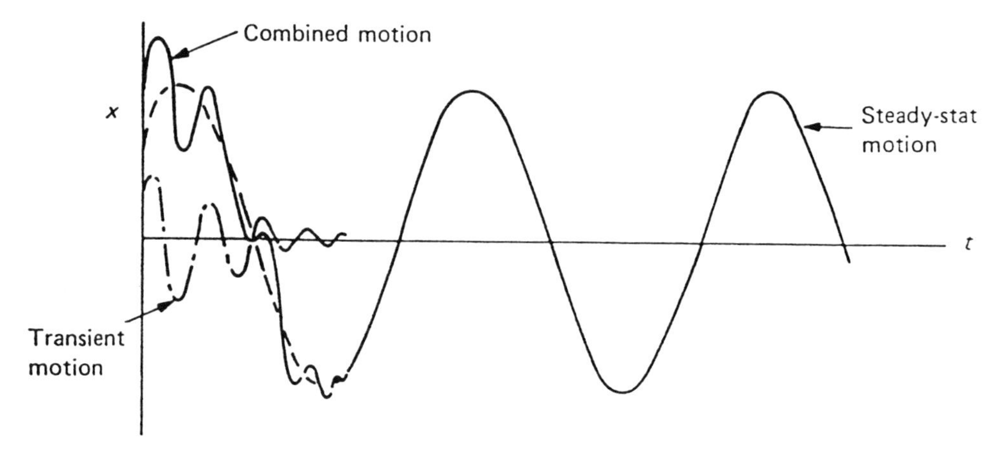

Fig. 2.25. Forced vibration, combined motion.

Equation (2.15) can be written in a more convenient form if we put

$$\omega = \sqrt{\left(\frac{k}{m}\right)} \text{ rad/s} \quad \text{and} \quad X_s = \frac{F}{k}.$$

Then

$$\frac{X}{X_s} = \frac{1}{\sqrt{\left\{\left[1 - \left(\frac{v}{\omega}\right)^2\right]^2 + \left[2\zeta\frac{v}{\omega}\right]^2\right\}}}, \tag{2.16}$$

$$\text{and } \phi = \tan^{-1}\left[\frac{2\zeta\,(v/\omega)}{1-(v/\omega)^2}\right]$$

X/X_s is known as the dynamic magnification factor, because X_s is the static deflection of the system under a steady force F, and X is the dynamic amplitude.

By considering different values of the frequency ratio v/ω, we can plot X/X_s and ϕ as functions of frequency for various values of ζ. Figs. 2.26 and 2.27 show the results.

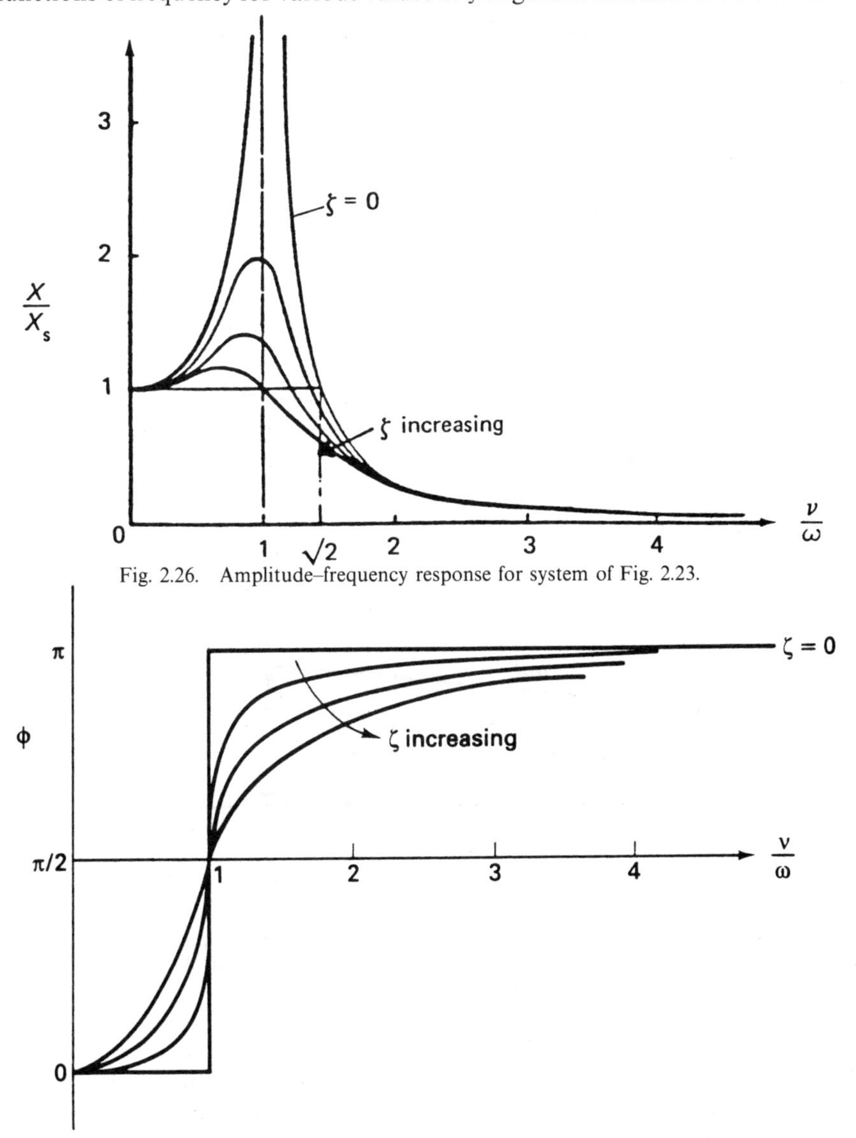

Fig. 2.26. Amplitude–frequency response for system of Fig. 2.23.

Fig. 2.27. Phase–frequency response for system of Fig. 2.23.

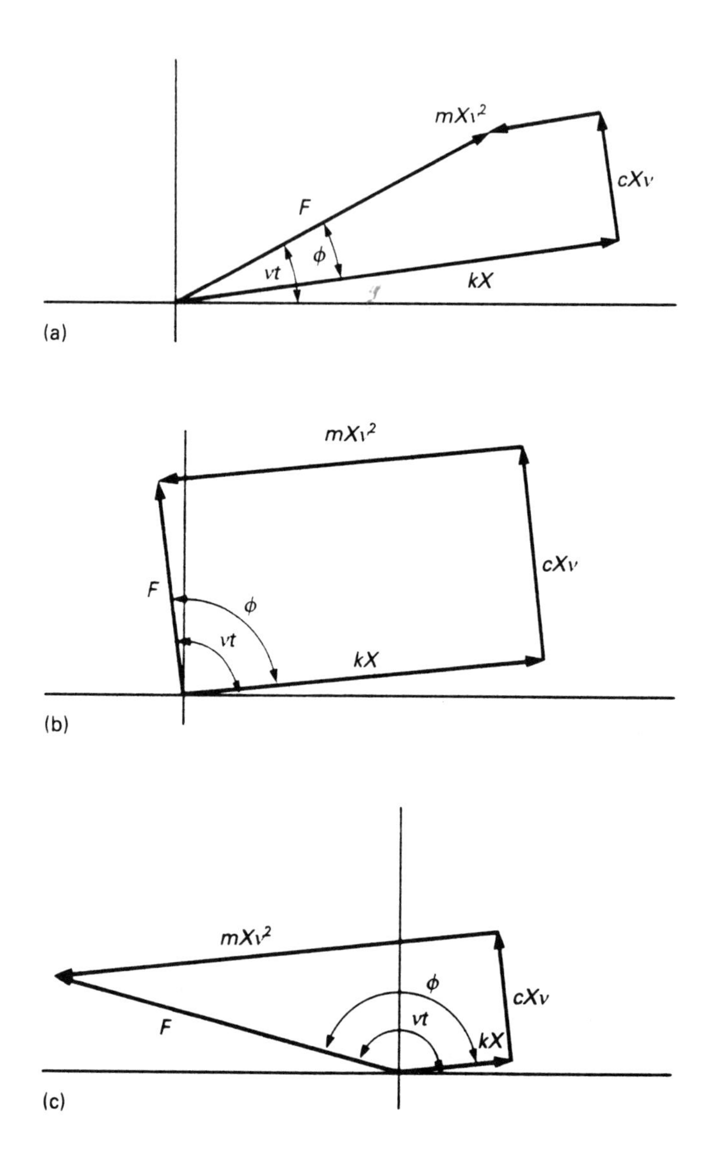

Fig. 2.28. Forced vibration vector diagrams: (a) $\nu/\omega \ll 1$ Exciting force approximately equal to spring force; (b) $\nu/\omega = 1$ Exciting force equal to damping force, and inertia force equal to spring force; (c) $\nu/\omega \gg 1$ Exciting force nearly equal to inertia force.

The effect of the frequency ratio on the vector diagram is shown in Fig. 2.28.

The importance of mechanical vibration arises mainly from the large values of X/X_s experienced in practice when ν/ω has a value near unity: this means that a small harmonic force can produce a large amplitude of vibration. The phenomenon known as **resonance** occurs when the forcing frequency is equal to the natural frequency, that is when $\nu/\omega = 1$. The maximum value of X/X_s actually occurs at values of ν/ω less than unity: the value can be found by differentiating equation (2.16) with respect to ν/ω. Hence

$$(\nu/\omega)_{(X/Xs)\max} = \sqrt{(1 - 2\zeta^2)} \simeq 1, \quad \text{for} \quad \zeta \text{ small,}$$

and

$$(X/X_s)_{\max} = 1/(2\zeta\sqrt{(1 - \zeta^2)}).$$

For small values of ζ, $(X/X_S)_{\max} \simeq 1/2\zeta$ which is the value pertaining to $\nu/\omega = 1$. $1/2\zeta$ is a measure of the damping in a system and is referred to as the Q factor.
Note. An alternative solution to the equation of motion can be obtained by putting $F \sin \nu t = \mathrm{Im}(F\mathrm{e}^{\mathrm{j}\nu t})$.
Then $m\ddot{x} + c\dot{x} + kx = F\mathrm{e}^{\mathrm{j}\nu t}$, and a solution $x = X\mathrm{e}^{\mathrm{j}\nu t}$ can be assumed.

Thus $(k - m\nu^2)X + \mathrm{j}c\nu X = F$, or $X = \dfrac{F}{(k - m\nu^2) + \mathrm{j}c\nu}$.

Hence $X = \dfrac{F}{\sqrt{((k - m\nu^2)^2 + (c\nu)^2)}}$, as before, equation (2.15).

Both *reciprocating and rotating unbalance* in a system produce an exciting force of the inertia type and result in the amplitude of the exciting force being proportional to the square of the frequency of excitation.

For an unbalanced body of mass m_r, at an effective radius r, rotating at an angular speed ν, the exciting force is therefore $m_r r\nu^2$. If this force is applied to a single degree of freedom system such as that in Fig. 2.23, the component of the force in the direction of motion is $m_r r\nu^2 \sin \nu t$, and the amplitude of vibration is

$$X = \frac{(m_r/m)r(\nu/\omega)^2}{\sqrt{((1 - (\nu/\omega)^2)^2 + (2\zeta\nu\omega)^2)}} \tag{2.17}$$

(see equation (2.16)).

The value of ν/ω for maximum X found by differentiating equation (2.17) is given by

$$(\nu/\omega)_{X\max} = 1/\sqrt{(1 - 2\zeta^2)}$$

that is, the peak of the response curve occurs when $\nu > \omega$. This is shown in Fig. 2.29

Also,

$$X_{\max} = (m_r/m)r/2\zeta\sqrt{(1 - \zeta^2)}.$$

It can be seen that away from the resonance condition ($\nu/\omega = 1$) the system response is not greatly affected by damping unless this happens to be large. Since in most mechanical systems the damping is small ($\zeta < 0.1$) it is often permissible to neglect the damping when evaluating the frequency for maximum amplitude and also the amplitude–frequency response away from the resonance condition.

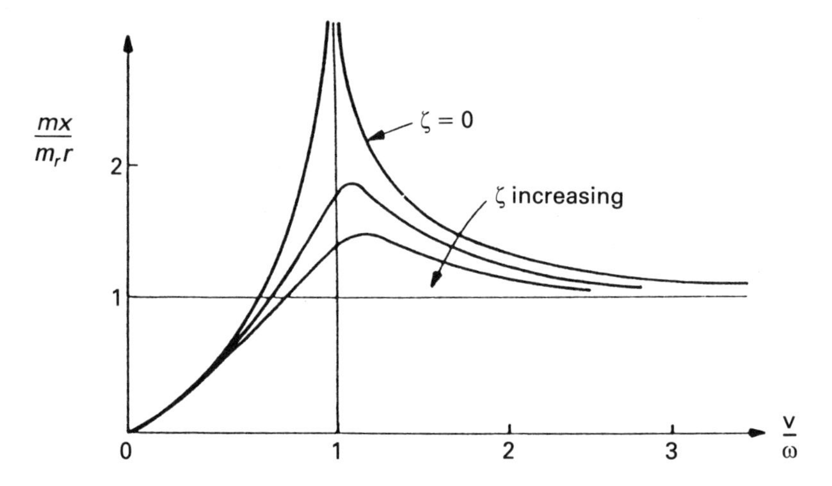

Fig. 2.29. Amplitude–frequency response, with rotating unbalance excitation.

Example 12

A turbo-charger rotor consists of a steel shaft 10 mm in diameter which is simply supported in two self-aligning bearings placed 100 mm apart. A turbine rotor of mass 0.2 kg is located at the mid-point of the shaft. There is negligible damping. The rotor is carefully balanced, the maximum out-of-balance being 10^{-5} kg m. In previous applications the turbo-charger has been run successfully at 35 000 rev/min and 55 000 rev/min. Calculate the amplitude of the deflection of the mid-point of the shaft, and its phase with respect to the out-of-balance excitation when it is running at these speeds.

In a new application it was proposed to run the turbo-charger at 45 000 rev/min. Comment on the implications of this proposal.

The stiffness, k of the shaft at its mid-point is $48EI/l^3$, where $E = 2 \times 10^{11}$ N/m^2, $I = \pi(0.01)^4 4/64$ m^4 and $l = 0.1$ m. Hence

$$k = \frac{48 \times 2 \times 10^{11}}{0.1^3}\left[\frac{\pi(0.01)^4}{64}\right] = 4.71 \times 10^6 \text{ N/m}.$$

The effective mass, m_{eff}, of the system can be taken to be the mass of the rotor plus one half of the mass of the shaft. Thus

$$m_{\text{eff}} = 0.2 + \tfrac{1}{2} \times 7850 \times 0.1\,\frac{\pi(0.01)^2}{4} = 0.231 \text{ kg}.$$

The natural frequency of the system is therefore

$$\sqrt{\left(\frac{4.71 \times 10^6}{0.231}\right)} = 4515 \text{ rad/s},$$

which corresponds to 43 100 rev/min.

The amplitude of shaft deflection is

$$\frac{F}{k\left(1 - \left(\dfrac{\nu}{\omega}\right)^2\right)},$$

where $F = mr\nu^2$ and $\nu = 2\pi \times 35\,000/60 = 3663$ rad/s. Thus

$$F = 10^{-5} \times 3663^2 = 134.2 \text{ N}, \quad \text{and} \quad \frac{\nu}{\omega} = \frac{3663}{4515} = 0.811.$$

Hence the amplitude at 35 000 rev/min

$$= \frac{134.2}{4.71 \times 10^6(1 - 0.811^2)} = 0.083 \text{ mm},$$

and is in phase with the excitation because the excitation is below the resonance frequency, and the amplitude is therefore positive.

At 55 000 rev/min, the amplitude is

$$\frac{331}{4.71 \times 10^6(1 - 1.27^2)} = -0.112 \text{ mm},$$

which is out of phase with the excitation.

Running the turbo-charger at 45 000 rev/min is potentially dangerous because of the proximity to resonance which occurs at 43 100 rev/min.

It can be seen from Fig. 2.26, 2.28 and 2.29 that the system response at low frequencies ($\ll \omega$) is stiffness-dependant, and that in the region of resonance the response is damping-dependant, whereas at high frequencies ($\gg \omega$) the response is governed by the system mass. It is most important to realize this when attempting to reduce the vibration of a system. For example the application of increased damping will have little effect if the excitation and response frequencies are in a region well away from resonance such as in that controlled by the system mass.

Before attempting to reduce the vibration levels in a machine or structure by increasing its damping, every effort should be made to reduce the vibration excitation at its source. It has to be accepted that many machines and processes generate a disturbing force of one sort or another, but the frequency of the disturbing force should not be at or near a natural frequency of the structure otherwise resonance will occur, with the resulting high amplitudes of vibration and dynamic stresses, and noise and fatigue problems. Resonance may also prevent the system fulfilling the desired function.

Some reduction in excitation can often be achieved by changing the machinery generating the vibration, but this can usually only be done at the design stage. Resiting equipment may also effect some improvement. However, structural vibration caused by external excitation sources such as ground vibration, cross winds or turbulence from adjacent buildings can only be controlled by damping.

In some machines vibrations are deliberately excited as part of the process, for example, in vibratory conveyors and compactors, and in ultrasonic welding. Naturally, nearby machines have to be protected from these vibrations.

Rotating machinery such as fans, turbines, motors and propellers can generate disturbing forces at several different frequencies such as the rotating speed and blade passing frequency. Reciprocating machinery such as compressors and engines can rarely be perfectly balanced, and an exciting force is produced at the rotating speed and at harmonics. Strong vibration excitation in structures can also be caused by pressure fluctuation in gases and liquids flowing in pipes; and intermittent loads such as those imposed by lifts in buildings.

There are two basic types of structural vibration; steady-state vibration caused by continually running machines such as engines, air-conditioning plants and generators either within the structure or situated in a neighbouring structure, and transient vibration caused by a short-duration disturbance such as a lorry or train passing over an expansion joint in a road or over a bridge.

Some relief from steady-state vibration excitation can often be gained by moving the source of the excitation, since the mass of the vibration generator has some effect on the natural frequencies of the supporting structure. For example, in a building it may be an advantage to move mechanical equipment to a lower floor, and in a ship re-siting propulsion or service machinery may prove effective. The effect of local stiffening of the structure may prove to be disappointing, however, because by increasing the stiffness the mass is also increased, so that the change in the $\sqrt{(k/m)}$ may prove to be very small.

Occasionally a change in the vibration generating equipment can reduce vibration levels. For example a change in gear ratios is a mechanical drive system, or a change from a four-bladed to a three-bladed propeller in a ship propulsion system will alter the excitation frequency provided the speed of rotation is not changed. However, in many cases the running speeds of motors and engines are closely controlled as in electric generator sets, so there is no opportunity for changing the excitation frequency.

If vibration excitation cannot be reduced to acceptable levels, so that the system response is still too large, some measure of vibration isolation may be necessary; see section 2.3.2.1 below.

2.3.2 Response of a viscous damped system supported on a foundation subjected to harmonic vibration

The system considered is shown in Fig. 2.30. The foundation is subjected to harmonic vibration $A \sin \nu t$ and it is required to determine the response, x, of the body.

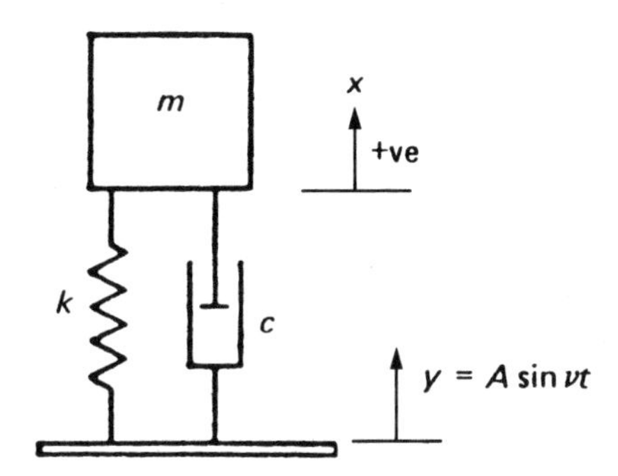

Fig. 2.30. Single degree of freedom model of a vibrated system with viscous damping.

The equation of motion is

$$m\ddot{x} = c(\dot{y} - \dot{x}) + k(y - x). \tag{2.18}$$

If the displacement of the body relative to the foundation, u, is required, we may write $u = x - y$, and equation (2.18) becomes

$$m\ddot{u} + c\dot{u} + ku = -m\ddot{y} = m\nu^2 A \sin \nu t.$$

This equation is similar to (2.14) so that the solution may be written directly as

$$u = \frac{A(\nu/\omega)^2}{\sqrt{((1-(\nu/\omega)^2)^2 + (2\zeta\nu/\omega)^2)}} \sin\left\{\nu t - \tan^{-1}\frac{2\zeta(\nu/\omega)}{1-(\nu/\omega)^2}\right\}.$$

If the absolute motion of the body is required we rewrite equation (2.18) as

$$\begin{aligned} m\ddot{x} + c\dot{x} + kx &= c\dot{y} + ky \\ &= cA\nu \cos \nu t + kA \sin \nu t \\ &= A\sqrt{(k^2 + (c\nu)^2)} \sin(\nu t + \alpha) \end{aligned}$$

where

$$\alpha = \tan^{-1}\frac{c\nu}{k}.$$

Hence, from the previous result,

$$x = \frac{A\sqrt{(k^2 + (c\nu)^2)}}{\sqrt{((k - m\nu^2)^2 + (c\nu)^2)}} \sin(\nu t - \phi + \alpha).$$

The motion transmissibility is defined as the ratio of the amplitude of the absolute body vibration to the amplitude of the foundation vibration. Thus,

$$\text{motion transmissibility} = \frac{X}{A}$$

$$= \frac{\sqrt{\left(1 + \left(2\zeta\frac{\nu}{\omega}\right)^2\right)}}{\sqrt{\left(\left[1 - \left(\frac{\nu}{\omega}\right)^2\right]^2 + \left[2\zeta\frac{\nu}{\omega}\right]^2\right)}}.$$

2.3.2.1 Vibration isolation

The dynamic forces produced by machinery are often very large. However, the force transmitted to the foundation or supporting structure can be reduced by using flexible mountings with the correct properties; alternatively a machine can be isolated from foundation vibration by using the correct flexible mountings. Fig. 2.31 shows a model of such a system.

The force transmitted to the foundation is the sum of the spring force and the

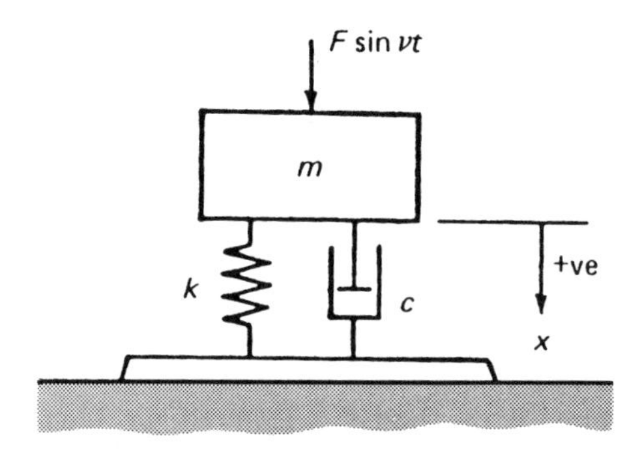

Fig. 2.31. Single degree of freedom system with foundation.

damper force. Thus the transmitted force $= kx + c\dot{x}$ and F_T, the amplitude of the transmitted force is given by

$$F_T = \sqrt{[(kX)^2 + (c\nu X)^2]}.$$

The force transmission ratio or transmissibility, T_R, is given by

$$T_R = \frac{F_T}{F} = \frac{X\sqrt{[k^2 + (c\nu)^2]}}{F}$$

since

$$X = \frac{F/k}{\sqrt{\left(\left[1 - \left(\frac{\nu}{\omega}\right)^2\right]^2 + \left[2\zeta\frac{\nu}{\omega}\right]^2\right)}},$$

$$T_R = \frac{\sqrt{\left(1 + \left(2\zeta\frac{\nu}{\omega}\right)^2\right)}}{\sqrt{\left(\left[1 - \left(\frac{\nu}{\omega}\right)^2\right]^2 + \left[2\zeta\frac{\nu}{\omega}\right]^2\right)}}$$

Therefore the force and motion transmissibilities are the same.

The effect of ν/ω on T_R is shown in Fig. 2.32. It can be seen that for good isolation $\nu/\omega > \sqrt{2}$, hence a low value of ω is required which implies a low stiffness; that is, a flexible mounting: this may not always be acceptable in practice where a certain minimum stiffness is usually necessary to satisfy operating criteria.

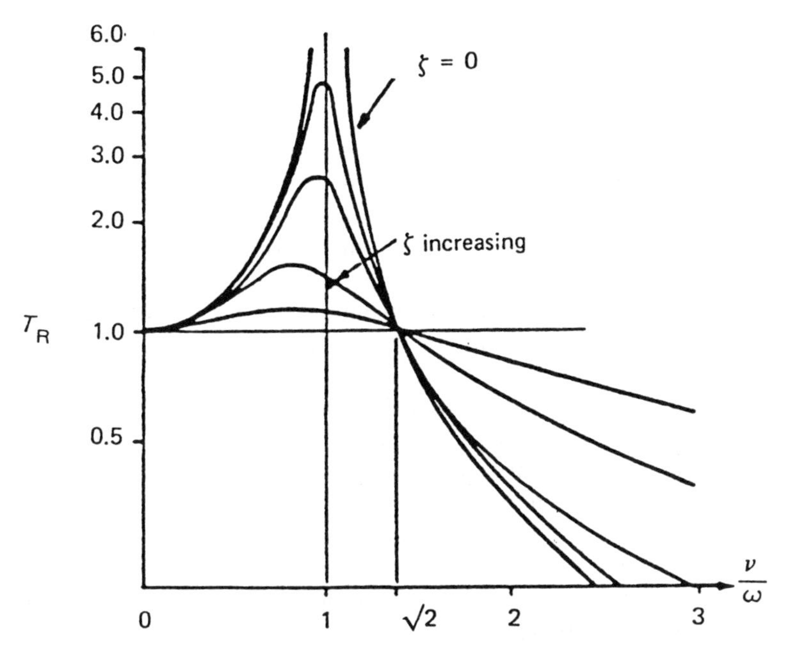

Fig. 2.32. Transmissibility-frequency ratio response.

It is particularly important to be able to isolate vibration sources because structure borne vibration can otherwise be easily transmitted to parts which radiate well, and serious noise problems can occur. Theoretically, low stiffness isolators are desirable to give a low natural frequency. However, this often results in isolators which are too soft and stability problems may arise. The system can be attached rigidly to a large block which effectively increases its mass so that stiffer isolators can be used. The centre of mass of the combined system is also lowered, giving improved stability.

For the best response a mounting system may be designed with snubbers which control the large amplitudes while providing little or no damping when the amplitudes are small.

There are four types of spring material commonly used for resilient mountings and vibration isolation: air, metal, rubber and cork. Air springs can be used for very low-frequency suspensions: resonance frequencies as low as 1 Hz can be achieved whereas metal springs can only be used for resonance frequencies greater than about 1.3 Hz. Metal springs can transmit high frequencies, however, so rubber or felt pads are often used to prohibit metal-to-metal contact between the spring and the structure. Different forms of spring element can be used such as coil, torsion, cantilever and beam. Rubber can be used in shear or compression but rarely in tension. It is important to determine the dynamic stiffness of a rubber isolator because this is generally much greater than the static stiffness. Also rubber possesses some inherent damping although this may be sensitive to amplitude, frequency and temperature. Natural frequencies from 5 Hz upwards can be achieved. Cork is one of the oldest materials used for vibration isolation. It is usually used in compression and natural frequencies of 25 Hz upwards are typical.

It is shown in section 2.3.2 that good vibration isolation, that is low force and motion transmissibility, can be achieved by supporting the vibration generator on a flexible low-frequency mounting. Thus although disturbing forces are generated, only a small proportion of them are transmitted to the supporting structure. However, this theory assumes that a mode of vibration is excited by a harmonic force passing through the centre of mass of the installation; although this is often a reasonable approximation it rarely actually occurs in practice because, due to a lack of symmetry of the supported machine, several different mountings may be needed to achieve a level installation, and the mass centre is seldom in the same plane as the tops of the mountings.

Thus the mounting which provides good isolation against a vertical exciting force may allow excessive horizontal motion, because of a frequency component close to the natural frequency of the horizontal mode of vibration. Also a secondary exciting force acting eccentrically from the centre of mass can excite large rotation amplitudes when the frequency is near to that of a rocking mode of an installation.

To limit the motion of a machine installation that generates harmonic forces and moments the mass and inertia of the installation supported by the mountings may have to be increased; that is an inertia block may have to be added to the installation. If non-metallic mountings are used the dynamic stiffness at the frequencies of interest will have to be found, probably by carrying out further dynamic tests in which the mounting is correctly loaded; they may also possess curious damping characteristics which may be included in the analysis by using the concept of complex stiffness, as discussed in section 2.2.4.

Air bags or bellows are sometimes used for very low-frequency mountings where some swaying of the supported system is acceptable. This is an important consideration because if the motion of the inertia block and the machinery is large, pipework and other services may be overstressed, which can lead to fatigue failure of these components. Approximate analysis shows that the natural frequency of a body supported on bellows filled with air under pressure is inversely proportional to the square root of the volume of the bellows, so that a change in natural frequency can be effected simply by a change in bellows volume. This can easily be achieved by opening or closing valves connecting the bellows to additional receivers, or by adding a liquid to the bellows. Natural frequencies of 0.5 Hz, or even less, are obtainable . An additional advantage of air suspension is that the system can be made self-levelling, when fitted with suitable valves and an air supply. Air pressures of about 5 to 10 times atmospheric pressure are usual.

Greater attenuation of the exciting force at high frequencies can be achieved by using a two-stage mounting. In this arrangement the machine is set on flexible mountings on an inertia block, which is itself supported by flexible mountings. This may not be too expensive to install since in many cases an existing sub-frame or structure can be used as the inertia block. If a floating floor in a building is used as the inertia block, some allowance must be made for the additional stiffness arising from the air space below it. This can be found by measuring the dynamic stiffness of the floor by means of resonance tests.

Naturally, techniques used for isolating structures from exciting forces arising in machinery and plant can also be used for isolating delicate equipment from vibrations in the structure. For example, sensitive electrical equipment in ships can be isolated from hull vibration, and operating tables and metrology equipment can be isolated from building vibration.

The above isolation systems are all *passive*; an *active* isolation system is one in which the exciting force or moment is applied by an externally powered force or couple. The opposing force or moment can be produced by means such as hydraulic rams, out-of-balance rotating bodies or electro-magnetism. Naturally it is essential to have accurate phase and amplitude control, to ensure that the opposing force is always equal, and opposite, to the exciting force. Although active isolation systems can be expensive to install, excellent results are obtainable so that the supporting structure is kept almost completely still. However it must be noted that force actuators such as hydraulic rams must react on another part of the system

If, after careful selection and design of machinery and equipment, careful installation and commissioning, and carrying out isolation as necessary the vibration levels in the system are still too large, then some increase in the damping is necessary. This is also the case when excitation occurs from sources beyond the designers' control such as cross winds, earthquakes and currents.

Example 13

A spring-mounted body moves with a velocity v along an undulating surface, as shown.

The body has a mass m and is connected to the wheel by a spring of stiffness k, and a viscous damper whose damping coefficient is c. The undulating surface has a wavelength L and an amplitude h.

Derive an expression for the ratio of amplitudes of the absolute vertical displacement of the body to the surface undulations.

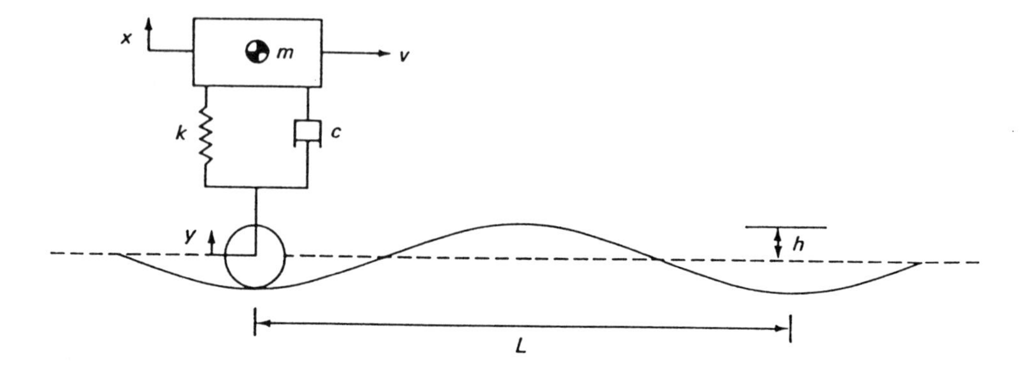

The system can be considered to be

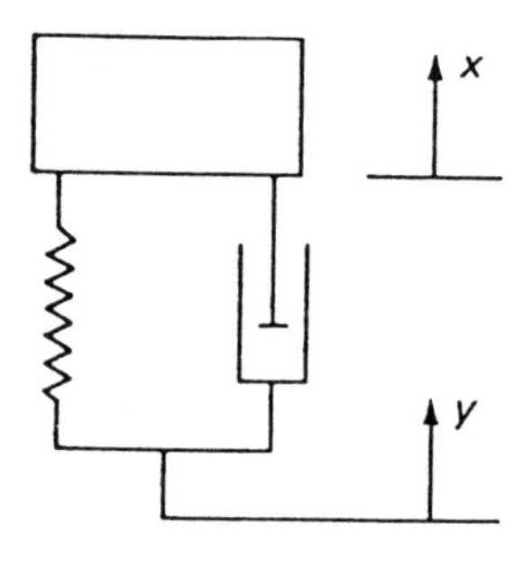

where $y = h \cos \dfrac{2\pi z}{L}$ and $z = vt$.

Hence $y = h \cos \dfrac{2\pi v}{L} t = h \cos \nu t$, where $\nu = \dfrac{2\pi v}{L}$.

The FBDs are

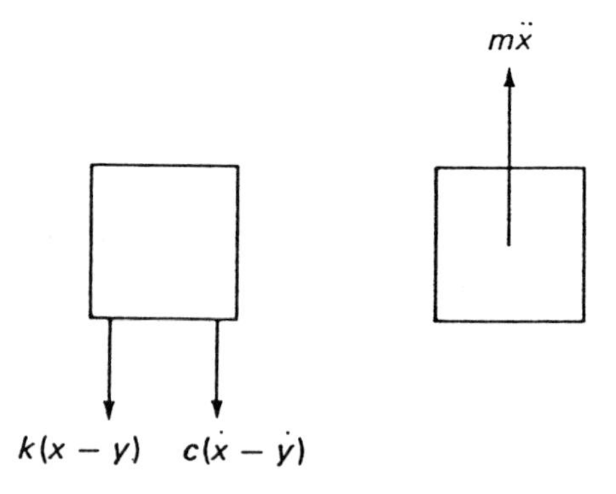

Hence the equation of motion is

$$m\ddot{x} = -k(x - y) - c(\dot{x} - \dot{y}),$$

or

$$m\ddot{x} + c\dot{x} + kx = c\dot{y} + ky.$$

Now

$$y = h \cos \nu t, \text{ so}$$

$$m\ddot{x} + c\dot{x} + kx = \sqrt{[k^2 + (c\nu)^2]}h \sin (\nu t + \phi).$$

Hence, if

$$x = X_0 \sin (\nu t + \alpha),$$

$$X_0 = \frac{h\sqrt{[k^2 + (c\nu)^2]}}{\sqrt{[(k - m\nu^2)^2 + (c\nu)^2]}}$$

So,

$$\frac{X_0}{h} = \frac{\sqrt{\left[k^2 + \left(\frac{2\pi v}{L}c\right)^2\right]}}{\sqrt{\left[\left(k - \left(\frac{2\pi v}{L}\right)^2 m\right)^2 + \left(\frac{2\pi v}{L}c\right)^2\right]}}$$

Example 14

The vibration on the floor in a building is SHM at a frequency in the range 15–60 Hz. It is desired to install sensitive equipment in the building which must be insulated from floor vibration. The equipment is fastened to a small platform which is supported by three similar springs resting on the floor, each carrying an equal load. Only vertical motion occurs. The combined mass of the equipment and platform is 40 kg, and the equivalent viscous damping ratio of the suspension is 0.2.

Find the maximum value for the spring stiffness, if the amplitude of transmitted vibration is to be less than 10% of the floor vibration over the given frequency range.

$$T_R = \frac{\sqrt{\left[1 + \left(2\zeta\frac{\nu}{\omega}\right)^2\right]}}{\sqrt{\left(\left[1 - \left(\frac{\nu}{\omega}\right)^2\right]^2 + \left[2\zeta\frac{\nu}{\omega}\right]^2\right)}}$$

$$T_R = 0.1 \text{ with } \zeta = 0.2 \text{ is required.}$$

Thus

$$\left[1 - \left(\frac{\nu}{\omega}\right)^2\right]^2 + \left[0.4\left(\frac{\nu}{\omega}\right)\right]^2 = 100\left[1 + \left(0.4\frac{\nu}{\omega}\right)^2\right]$$

that is

$$\left(\frac{\nu}{\omega}\right)^4 - 17.84\left(\frac{\nu}{\omega}\right)^2 - 99 = 0.$$

Hence

$$\frac{\nu}{\omega} = 4.72.$$

When

$$\nu = 15.2\pi \text{ rad/s}, \ \omega = 19.97 \text{ rad/s}.$$

Since

$$\omega = \sqrt{\left(\frac{k}{m}\right)} \quad \text{and} \quad m = 40 \text{ kg},$$

total $k = 15\,935$ N/m,

that is, stiffness of each spring = 15 935/3 N/m = 5.3 kN/m.

The amplitude of the transmitted vibration will be less than 10% at frequencies above 15 Hz.

Example 15

A machine of mass m generates a disturbing force $F \sin \nu t$; to reduce the force transmitted to the supporting structure, the machine is mounted on a spring of stiffness k with a damper in parallel. Compare the effectiveness of this isolation system for viscous and hysteretic damping.

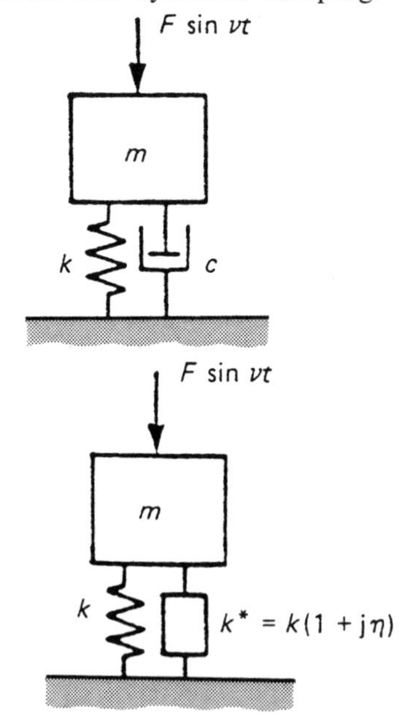

Viscous damping. From section 2.3.2.1,

$$T_R = \frac{F_T}{F} = \frac{\sqrt{\left[1 + \left(2\zeta\frac{\nu}{\omega}\right)^2\right]}}{\sqrt{\left\{\left[1 - \left(\frac{\nu}{\omega}\right)^2\right]^2 + \left[2\zeta\frac{\nu}{\omega}\right]^2\right\}}}$$

Hysteretic damping. From section 2.2.5,

Putting $\eta = \frac{c\nu}{k} = 2\zeta\frac{\nu}{\omega}$,

$$T_R = \frac{F_T}{F} = \frac{\sqrt{(1 + \eta^2)}}{\sqrt{\left\{\left[1 - \left(\frac{\nu}{\omega}\right)^2\right]^2 + \eta^2\right\}}}.$$

The effectiveness of these isolators can be compared using these expressions for T_R. The results are given in the table below.

It can be seen that the isolation effects are similar for the viscous and hysteretically damped isolators, except at high frequency ratios when the hysteretic damping gives much better attentuation of T_R. At these frequencies it is better to decouple the viscous damped isolator by attaching small springs or rubber bushes at each end.

	Viscously damped isolator	Hysteretically damped isolator
Value of T_R when $v = 0^+$	1	1
Frequency ratio v/ω for resonance	1	1
Value of T_R at resonance	$\frac{\sqrt{[1 + (2\zeta)^2]}}{2\zeta} \simeq \frac{1}{2\zeta}$	$\frac{\sqrt{(1 + \eta^2)}}{\eta} \simeq \frac{1}{\eta}$
Value of T_R when $v/\omega = \sqrt{2}$	1	1
Frequency ratio v/ω for isolation.	$>\sqrt{2}$	$>\sqrt{2}$
High frequency, $v/\omega \gg 1$, attenuation of T_R	$\frac{2\zeta}{v/\omega}$	$\frac{1}{(v/\omega)^2}$

Example 16

A single degree of freedom vibrational system of very small viscous damping ($\zeta < 0.1$) is excited by an harmonic force of frequency v and amplitude F. Show that the Q factor of the system is equal to the reciprocal of twice the damping ratio ζ. The Q factor is equal to $(X/X_S)_{max}$.

It is sometimes difficult to measure ζ in this way because the static deflection X_S of the body under a force F is very small. Another way is to obtain the two frequencies p_1 and p_2 (one either side of the resonance frequency ω) at the half-power points.

Show that $Q = 1/2\zeta = \omega/(p_2 - p_1) = \omega/\Delta\omega$.

(The half-power points are those points on the response curve with an amplitude $1/\sqrt{2}$ times the amplitude at resonance).

From equation (2.16) above

$$X = (F/k)/\sqrt{((1 - (v/\omega)^2)^2 + (2\zeta v/\omega)^2)}$$

If $v = 0$, $\quad X_S = F/k$, and at resonance $v = \omega$,

so

$$X_{max} = (F/k)/2\zeta = X_S/2\zeta,$$

that is

$$Q = (X/X_{\text{static}})_{\text{max}} = 1/2\zeta.$$

If X_s cannot be determined, the Q factor can be found by using the half-power point method. This method requires very accurate measurement of the vibration amplitude for excitation frequencies in the region of resonance. Once X_{max} and ω have been located, the so called half-power points are found when the amplitude is $X_p = X_{\text{max}}/\sqrt{2}$ and the corresponding frequencies either side of ω, p_1 and p_2 determined. Since the energy dissipated per cycle is proportional to X^2, the energy dissipated is reduced by 50% when the amplitude is reduced by a factor of $1/\sqrt{2}$.

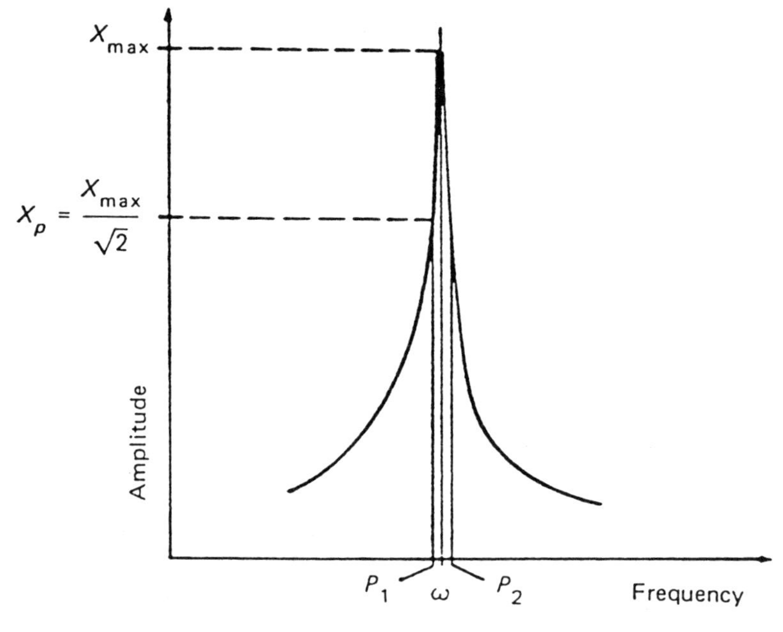

Amplitude–frequency response

Now

$$X = \frac{F/k}{\sqrt{\left\{\left[1 - \left(\frac{\nu}{\omega}\right)^2\right]^2 + \left[2\zeta\frac{\nu}{\omega}\right]^2\right\}}}$$

Thus

$$X_{\text{max}} = \frac{F/k}{2\zeta} \quad \left(\zeta \text{ small, so } X_{\text{max}} \text{ occurs at } \frac{\nu}{\omega} \simeq 1\right),$$

and

$$X_p = \frac{X_{max}}{\sqrt{2}} = \frac{F/k}{\sqrt{2}.2\zeta} = \frac{F/k}{\sqrt{\left\{\left[1 - \left(\frac{p}{\omega}\right)^2\right]^2 + \left[2\zeta\frac{p}{\omega}\right]^2\right\}}}$$

Hence

$$\left[1 - \left(\frac{p}{\omega}\right)^2\right]^2 + \left[2\zeta\frac{p}{\omega}\right]^2 = 8\zeta^2,$$

and

$$\left(\frac{p}{\omega}\right)^2 = (1 - 2\zeta^2) \pm 2\zeta\sqrt{(1 - \zeta^2)}.$$

That is

$$\frac{p_2^2 - p_1^2}{\omega^2} = 4\zeta\sqrt{(1 - \zeta^2)} \simeq 4\zeta, \qquad \text{if } \zeta \text{ is small.}$$

Now

$$\frac{p_2^2 - p_1^2}{\omega^2} = \left(\frac{p_2 - p_1}{\omega}\right)\left(\frac{p_2 + p_1}{\omega}\right) = 2\left(\frac{p_2 - p_1}{\omega}\right),$$

because $(p_1 + p_2)/\omega = 2$. That is, a symmetrical response curve is assumed for small ζ. Thus

$$\frac{p_2 - p_1}{\omega} = 2\zeta = \frac{\Delta\omega}{\omega} = \frac{1}{Q},$$

where $\Delta\omega$ is the frequency bandwidth at the half-power points.

Therefore, for light damping, the damping ratio ζ, and hence the Q factor, associated with any mode of vibration can be found from the amplitude-frequency measurements at resonance and the half-power points. Care is needed to ensure that the exciting device does not load the system and alter the frequency response and the damping. It should be noted that some difficulty is often encountered in measuring X_{max} accurately.

In real systems and structures a very high Q at a low frequency, or a very low Q at a high frequency, seldom occur, but it can be appreciated from the above that very real measuring difficulties can be encountered when trying to measure bandwidths of only a few Hz accurately, even if the amplitude of vibration can be determined. The table below shows the relationship between Q and Δf for different values of frequency.

Q factor	500	50	5
Resonance frequency (Hz)		Frequency bandwidth (Hz)	
10	0.02	0.2	2
100	0.2	2	20
1000	2	20	200

An improvement in accuracy in determining Q can often be obtained by measuring both amplitude and phase of the response for a range of exciting frequencies. Consider a single degree of freedom system under forced excitation $Fe^{j\nu t}$. The equation of motion is

$$m\ddot{x} + c\dot{x} + kx = Fe^{j\nu t}.$$

A solution $x = Xe^{j\nu t}$ can be assumed, so that

$$-m\nu^2 X + jc\nu X + kX = F.$$

Hence

$$\frac{X}{F} = \frac{1}{(k - m\nu^2) + jc\nu}$$

$$= \frac{k - m\nu^2}{(k - m\nu^2)^2 + (c\nu)^2} - j\frac{c\nu}{(k - m\nu^2)^2 + (c\nu)^2}$$

That is, X/F consists of two vectors, $\mathrm{Re}(X/F)$ in phase with the force, and $\mathrm{Im}(X/F)$ in quadrature with the force. The locus of the end point of vector X/F as ν varies is shown opposite for a given value of c. This is obtained by calculating real and imaginary components of X/F for a range of frequencies.
Experimentally this curve can be obtained by plotting the measured amplitude and phase of (X/F) for each exciting frequency.
Since

$$\tan\phi = \frac{k - m\nu^2}{c\nu},$$

when $\phi = 45°$ and $135°$,

$$1 = \frac{k - mp_1^2}{cp_1} \quad \text{and} \quad -1 = \frac{k - mp_2^2}{cp_2}.$$

Hence

$$mp_1^2 + cp_1 - k = 0 \quad \text{and} \quad mp_2^2 - cp_2 - k = 0.$$

Subtracting one equation from the other gives $p_2 - p_1 = c/m$, or

$$\frac{p_2 - p_1}{\omega} = \frac{\Delta\omega}{\omega} = 2\zeta = \frac{1}{Q}.$$

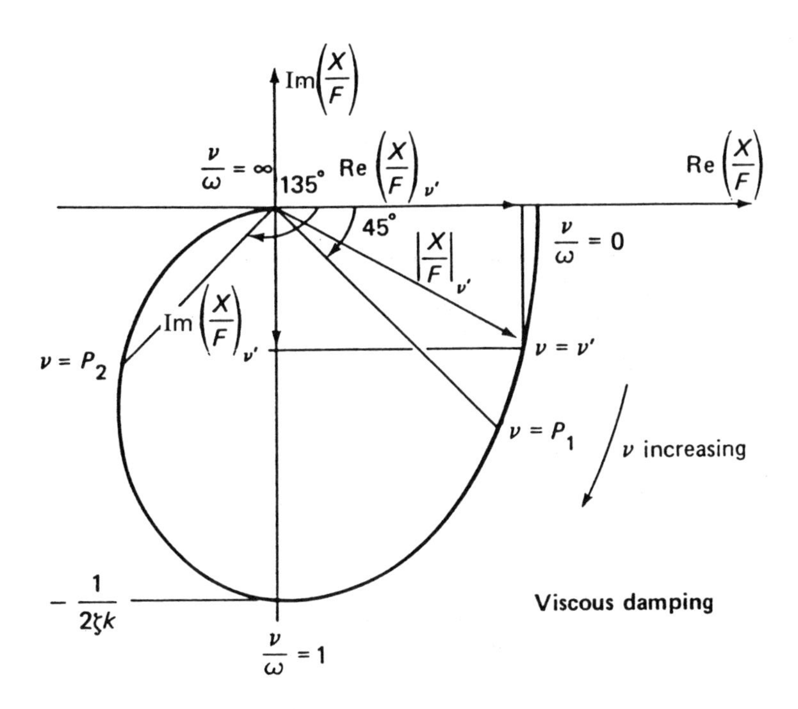

That is, X/F at resonance lies along the imaginary axis, and the half-power points occur when $\phi = 45°$ and $135°$. If experimental results are plotted on these axes a smooth curve can be drawn through them so that the half-power points can be accurately located.

The method is also effective when the damping is hysteretic, because in this case

$$\frac{X}{F} = \frac{1}{(k - m\nu^2) + j\eta k},$$

so that

$$\mathrm{Re}\left(\frac{X}{F}\right) = \frac{k - m\nu^2}{(k - m\nu^2)^2 + (\eta k)^2} \quad \text{and} \quad \mathrm{Im}\left(\frac{X}{F}\right) = \frac{-\eta k}{(k - m\nu^2)^2 + (\eta k)^2}.$$

Thus

$$\left[\mathrm{Re}\left(\frac{X}{F}\right)\right]^2 + \left[\mathrm{Im}\left(\frac{X}{F}\right)\right]^2 = \frac{1}{(k - m\nu^2)^2 + (\eta k)^2},$$

or

$$\left[\mathrm{Re}\left(\frac{X}{F}\right)\right]^2 + \left[\mathrm{Im}\left(\frac{X}{F}\right) - \frac{1}{2\eta k}\right]^2 = \left(\frac{1}{2\eta k}\right)^2.$$

That is, the locus of (X/F) as ν increases from zero is part of a circle, centre $(0, -1/2\eta k)$ and radius $1/2\eta k$, as shown.

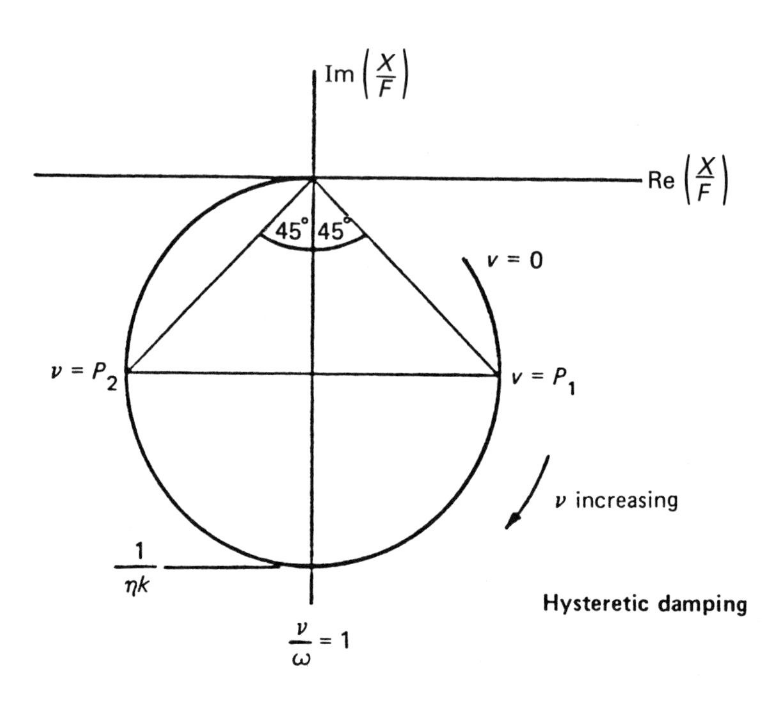

Hysteretic damping

In this case therefore it is particularly easy to draw an accurate locus from a few experimental results, and p_1 and p_2 are located on the horizontal diameter of the circle.

This technique is known variously as a frequency locus plot, Kennedy–Pancu diagram, or Nyquist diagram.

It must be realized that the assessment of damping can only be approximate. It is difficult to obtain accurate, reliable, experimental data, particularly in the region of resonance; the analysis will depend on whether viscous or hysteretic damping is assumed, and some non-linearity may occur in a real system. These effects may cause the frequency-locus plot to rotate and translate in the $\text{Re}(X/F)$, $\text{Im}(X/F)$ plane. In these cases the resonance frequency can be found from that part of the plot where the greatest rate of change of phase with frequency occurs.

Example 17

A machine of mass 550 kg is flexibly supported on rubber mountings which provide a force proportional to displacement of 210 kN/m, together with a viscous damping force. The machine gives an exciting force of the form $R\nu^2 \cos \nu t$, where R is a constant. At very high speeds of rotation, the measured amplitude of vibration is 0.25 mm, and the maximum amplitude recorded as the speed is slowly increased from zero is 2 mm. Find the value of R and the damping ratio.

Now,

$$X = Rv^2/(\sqrt{(k - mv^2)^2 + c^2v^2})$$

If v is large,

$$X \to Rv^2/(\sqrt{(m^2v^4)}) = R/m.$$

Hence

$$R = mX = (550 \times 0.25)/1000 = 0.1375 \text{ kg m}.$$

For maximum X, $\mathrm{d}X/\mathrm{d}v = 0$, hence $v^2 = 2k^2/(2mk - c^2)$, and

$$X_{\max} = \frac{R/2m}{\zeta\sqrt{(1 - \zeta^2)}}.$$

So

$$\zeta\sqrt{(1 - \zeta^2)} = R/(2mX_{\max}) = 0.1375/(2 \times 550 \times 2 \times 10^{-3})$$
$$= 0.062,$$

that is,

$$\zeta = 0.062.$$

2.3.3 Response of a Coulomb damped system to a simple harmonic exciting force with constant amplitude

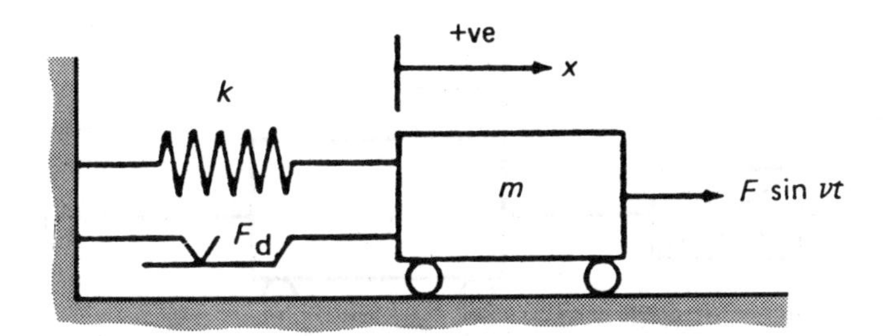

Fig. 2.33. Single degree of freedom model of a forced system with Coulomb damping.

In the system shown in Fig. 2.33 the damper relies upon dry friction.

The equation of motion is non-linear because the constant friction force F_d always opposes the motion:

$$m\ddot{x} + kx \pm F_d = F \sin vt.$$

If F_d is large compared to F, discontinuous motion will occur, but in most systems F_d is usually small so that an approximate continuous solution is valid. The approximate solution is obtained by linearizing the equation of motion; this can be done by expressing F_d in terms of an equivalent viscous damping coefficient, c_d. From section 2.2.5,

$$c_d = \frac{4F_d}{\pi v X}.$$

The solution to the linearized equation of motion gives the amplitude X of the motion as

$$X = \frac{F}{\sqrt{[(k - mv^2)^2 + (c_d v)^2]}}.$$

Hence

$$X = \frac{F}{\sqrt{(k - mv^2)^2 + (4F_d/\pi X)^2]}}.$$

That is,

$$\frac{X}{X_s} = \frac{\sqrt{(1 - (4F_d/\pi F)^2)}}{1 - (v/\omega)^2}.$$

This expression is satisfactory for small damping forces, but breaks down if $4F_d/\pi F < 1$, that is $F_d > (\pi/4)F$.

At resonance the amplitude is not limited by Coulomb friction.

2.3.4 Response of a hysteretically damped system to a simple harmonic exciting force with constant amplitude

In the single degree of freedom model shown in Fig. 2.34 the damping is hysteretic.

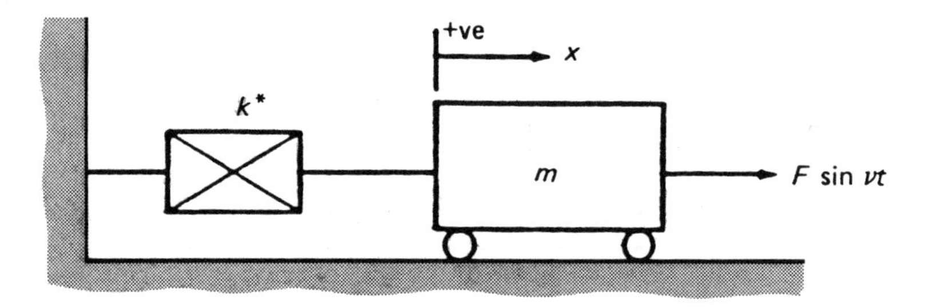

Fig. 2.34. Single degree of freedom model of a forced system with hysteric damping.

The equation of motion is

$$m\ddot{x} + k^*x = F \sin vt.$$

Since

$$k^* = k(1 + j\eta),$$

$$x = \frac{F \sin vt}{(k - mv^2) + j\eta k}.$$

and

$$\frac{X}{X_s} = \frac{1}{\sqrt{([1 - (v/\omega)^2]^2 + \eta^2)}}.$$

This result can also be obtained from the analysis of a viscous damped system by substituting $c = \eta k/v$.

It should be noted that if $c = \eta k/v$, at resonance $c = \eta\sqrt{(km)}$, that is $\eta = 2\zeta = 1/Q$.

2.3.5 Response of a system to a suddenly applied force

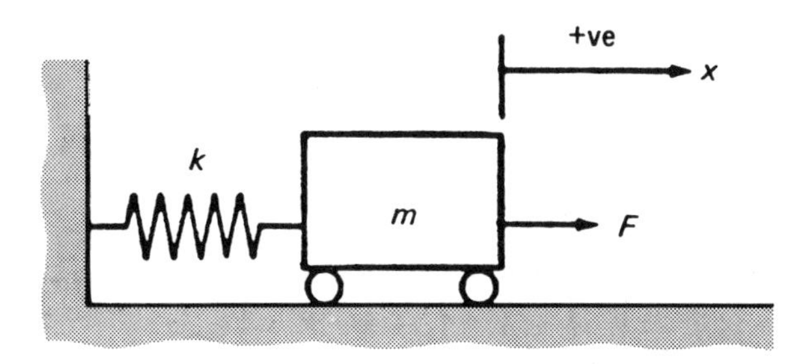

Fig. 2.35. Single degree of freedom model with constant exciting force.

Consider a single degree of freedom undamped system, such as the system shown in Fig. 2.35, which has been subjected to a suddenly applied force F. The equation of motion is $m\ddot{x} + kx = F$. The solution to this equation comprises a complementary function $A \sin \omega t + B \cos \omega t$, where $\omega = \sqrt{(\mathrm{k/m})}$ rad/s together with a particular solution. The particular solution may be found by using the D-operator. Thus the equation of motion can be written

$$\left(1 + \frac{\mathrm{D}^2}{\omega^2}\right)x = \frac{F}{k},$$

and

$$x = \left(1 + \frac{\mathrm{D}^2}{\omega^2}\right)^{-1}\frac{F}{k} = \left(1 - \frac{\mathrm{D}^2}{\omega^2} - \cdots\right)\frac{F}{k} = \frac{F}{k}.$$

That is, the complete solution to the equation of motion is

$$x = A \sin \omega t + B \cos \omega t + \frac{F}{k}.$$

If the initial conditions are such that $x = \dot{x} = 0$ at $t = 0$, then $B = -F/k$ and $A = 0$. Hence

$$x = \frac{F}{k}(1 - \cos \omega t).$$

The motion is shown in Fig. 2.36. It will be seen that the maximum dynamic displacement is twice the static displacement occurring under the same load. This is an important consideration in systems subjected to suddenly applied loads.

If the system possesses viscous damping of coefficient c, the solution to the equation of motion is $x = Xe^{-\zeta\omega t} \sin(\omega_v t + \phi) + F/k$.

With the same initial conditions as above,

$$x = \frac{F}{k}\left[1 - \frac{e^{-\zeta\omega t}}{\sqrt{(1-\zeta^2)}} \sin\left(\omega\sqrt{(1-\zeta^2)}t + \tan^{-1}\frac{\sqrt{(1-\zeta^2)}}{\zeta}\right)\right]$$

This reduces to the undamped case if $\zeta = 0$. The response of the damped system is shown in Fig. 2.37.

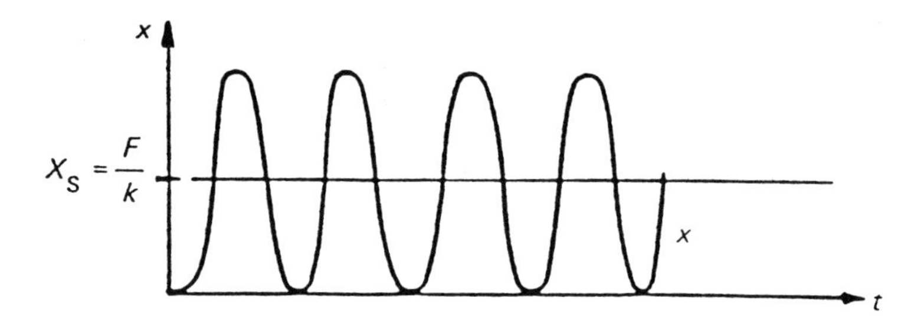

Fig. 2.36. Displacement–time response for system shown in Fig. 2.35.

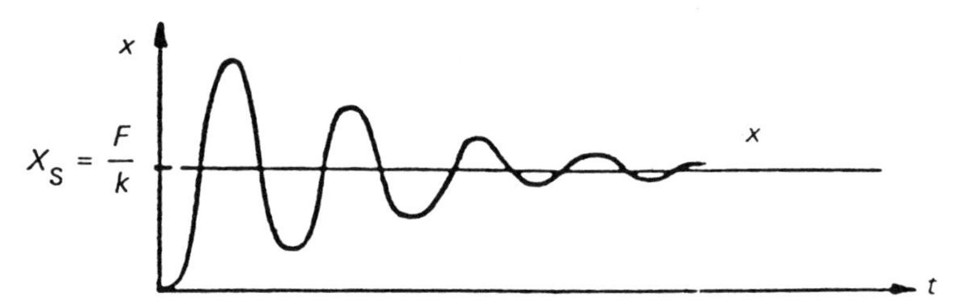

Fig. 2.37. Displacement–time response for single degree of freedom system with viscous damping.

2.3.6 Shock excitation

Some systems are subjected to shock or impulse loads arising from suddenly applied, non-periodic, short duration exciting forces.

The impulsive force shown in Fig. 2.38 consists of a force of magnitude $F_{\max}/\varepsilon$ which has a time duration of ε.

The impulse is equal to

$$\int_t^{t+\varepsilon} \left(\frac{F_{\max}}{\varepsilon}\right) dt.$$

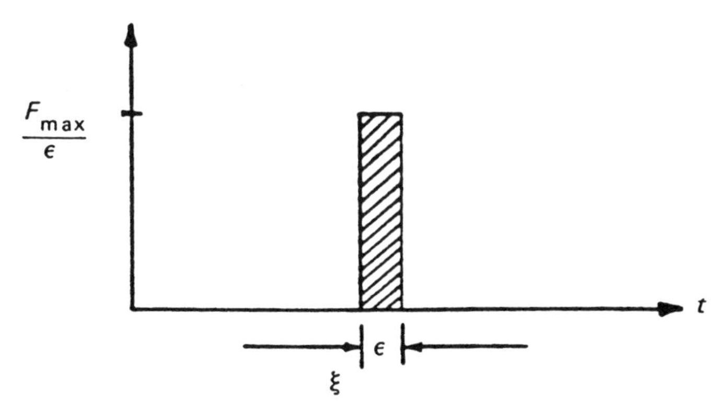

Fig. 2.38. Impulse.

When F_{max} is equal to unity, the force in the limiting case $\varepsilon \rightarrow 0$ is called either the unit impulse or the delta function, and is identified by the symbol $\delta(t - \xi)$, where

$$\int_0^\infty \delta(t - \xi)\mathrm{d}\xi = 1.$$

Since $F\,\mathrm{d}t = m\,\mathrm{d}v$, the impulse F_{max} acting on a body will result in a sudden change in its velocity without an appreciable change in its displacement. Thus the motion of a single degree of freedom system excited by an impulse F_{max} corresponds to free vibration with initial conditions $x = 0$ and $\dot{x} = v_0 = F_{max}/m$ at $t = 0$.

Once the response $g(t)$ say, to a unit impulse excitation is known, it is possible to establish the equation for the response of a system to an arbitrary exciting force $F(t)$. For this the arbitrary pulse in considered to comprise a series of impulses as shown in Fig. 2.39.

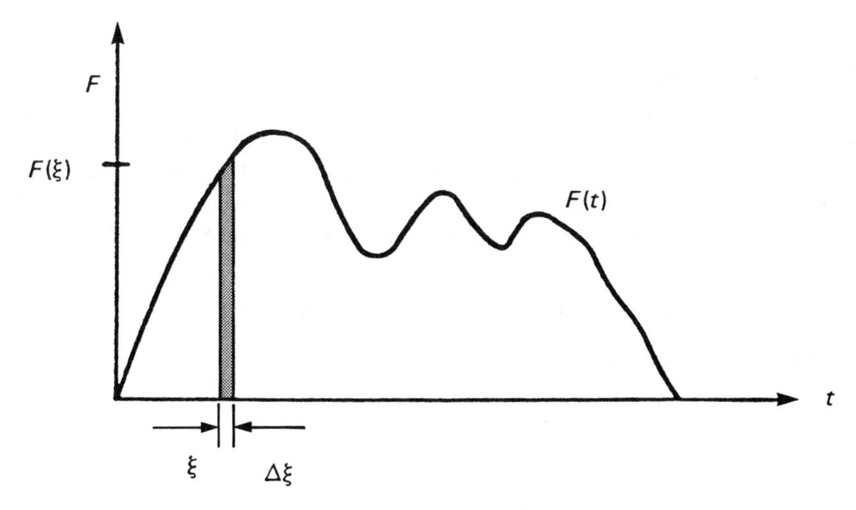

Fig. 2.39. Force–time pulse.

If one of the impulses is examined which starts at time ξ, its magnitude is $F(\xi)\Delta\xi$, and its contribution to the system response at time t is found by replacing the time with the elapsed time $(t - \xi)$ as shown in Fig. 2.40.

If the system can be assumed to be linear, the principle of superposition can be applied, so that

$$x(t) = \int_0^t F(\xi)g(t - \xi)\mathrm{d}\xi.$$

This is known as the Duhamel integral.

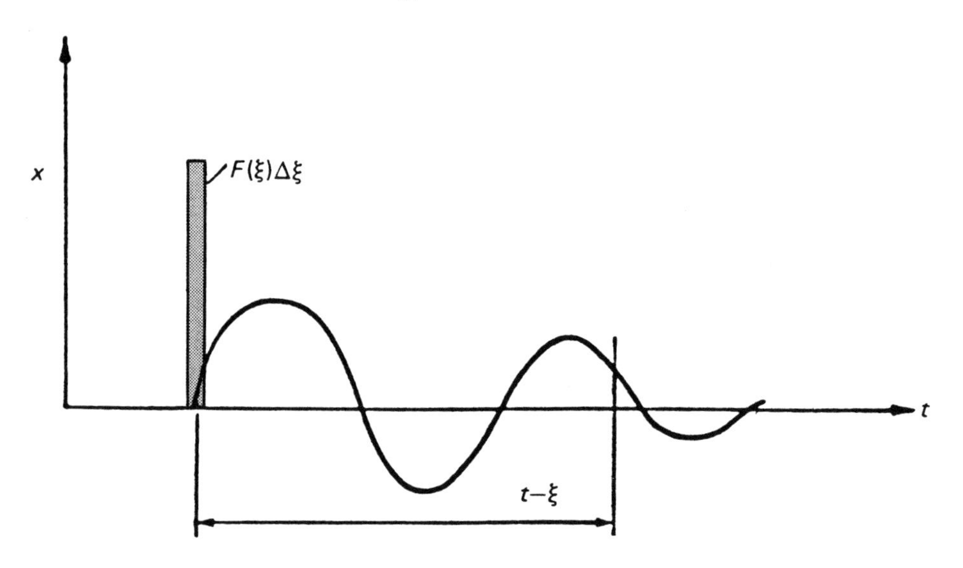

Fig. 2.40. Displacement–time response to impulse.

2.3.7 Harmonic analysis

A function which is periodic but not harmonic can be represented by the sum of a number of terms, each term representing some multiple of the fundamental frequency. In a *linear* system each of these harmonic terms acts as if it alone were exciting the system, and the system response is the sum of the excitation of all the harmonics.

For example, if the periodic forcing function of a single degree of freedom undamped system is

$$F_1 \sin(vt + \alpha_1) + F_2 \sin(2vt + \alpha_2) + F_3 \sin(3vt + \alpha_3) + \cdots + F_n \sin(nvt + \alpha_n),$$

the steady-state response to $F_1 \sin(vt + \alpha_1)$ is

$$x_1 = \frac{F_1}{k\left(1 - \left(\dfrac{v}{\omega}\right)^2\right)} \sin(vt + \alpha_1),$$

and the response to $F_2 \sin(2vt + \alpha_2)$ is

$$x_2 = \frac{F_2}{k\left(1 - \left(\frac{2\nu}{\omega}\right)^2\right)} \sin(2\nu t + \alpha_2),$$

and so on, so that

$$x = \sum_{n=1}^{n} \frac{F_n}{k\left(1 - \left(\frac{n\nu}{\omega}\right)^2\right)} \sin(n\nu t + \alpha_n).$$

Clearly that harmonic which is closest to the system natural frequency will most influence the response.

A periodic function can be written as the sum of a number of harmonic terms by writing a Fourier series for the function. A Fourier series can be written

$$F(t) = \frac{a_0}{2} + \sum_{n=1}^{\infty} (a_n \cos n\nu t + b_n \sin n\nu t),$$

where

$$a_0 = \frac{2}{\tau} \int_0^\tau F(t)\mathrm{d}t,$$

$$a_n = \frac{2}{\tau} \int_0^\tau F(t) \cos \nu t \mathrm{d}t, \text{ and}$$

$$b_n = \frac{2}{\tau} \int_0^\tau F(t) \sin \nu t \mathrm{d}t.$$

For example, consider the first four terms of the Fourier series representation of the square wave shown in Fig. 2.41 to be required; $\tau = 2\pi$ so $\nu = 1$ rad/s.

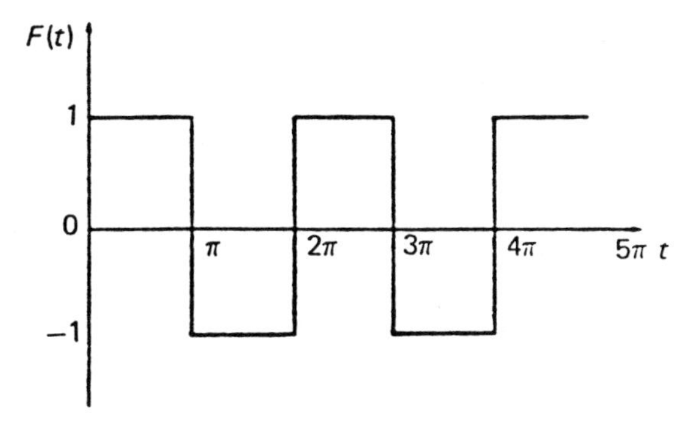

Fig. 2.41. Square wave.

$$F(t) = \frac{a_0}{2} + a_1 \cos vt + a_2 \cos 2vt + \cdots$$

$$+ b_1 \sin vt + b_2 \sin 2vt + \cdots$$

$$a_0 = \frac{2}{\tau}\int_0^\tau F(t)dt = \frac{2}{2\pi}\int_0^\pi 1\ dt + \frac{2}{2\pi}\int_\pi^{2\pi} -1\ dt = 0.$$

$$a_1 = \frac{2}{\tau}\int_0^\tau F(t)\cos vtdt$$

$$= \frac{2}{2\pi}\int_0^\pi \cos vtdt + \frac{2}{2\pi}\int_\pi^{2\pi} -\cos vtdt = 0.$$

Similarly

$$a_2 = a_3 = \cdots = 0.$$

$$b_1 = \frac{2}{\tau}\int_0^\tau F(t)\sin vtdt$$

$$= \frac{2}{2\pi}\int_0^\pi \sin vtdt + \frac{2}{2\pi}\int_\pi^{2\pi} -\sin vtdt$$

$$= \frac{1}{\pi v}[-\cos vt]_0^\pi + \frac{1}{\pi v}[\cos vt]_\pi^{2\pi} = \frac{4}{\pi v}.$$

Since $v = 1$ rad/s,

$$b_1 = \frac{4}{\pi}.$$

It is found that $b_2 = 0$, $b_3 = 4/3\pi$ and so on.

Thus $F(t) = \frac{4}{\pi}\left[\sin t + \frac{1}{3}\sin 3t + \frac{1}{5}\sin 5t + \frac{1}{7}\sin 7t + \cdots\right]$ is the series representation of the square wave shown.

If this stimulus is applied to a simple undamped system with $\omega = 4$ rad/s, say, the steady-state response is given by

$$x = \frac{\frac{4}{\pi}\sin t}{1-(\frac{1}{4})^2} + \frac{\frac{4}{3\pi}\sin 3t}{1-(\frac{3}{4})^2} + \frac{\frac{4}{5\pi}\sin 5t}{1-(\frac{5}{4})^2} + \frac{\frac{4}{7\pi}\sin 7t}{1-(\frac{7}{4})^2}\cdots$$

that is, $x = 1.36 \sin t + 0.97 \sin 3t - 0.45 \sin 5t - 0.09 \sin 7t - \cdots$

Usually three or four terms of the series dominate the predicted response.

It is worth sketching the components of $F(t)$ above to show that they produce a reasonable square wave, whereas the components of x do not. This is an important result.

2.3.8 Random vibration

If the vibration response parameters of a dynamic system are accurately known as functions of time, the vibration is said to be *deterministic*. However, in many systems and processes responses cannot be accurately predicted; these are called *random processes*. Examples of a random process are turbulence, fatigue, the meshing of imperfect gears, surface irregularities, the motion of a car running along a rough road, and building vibration excited by an earthquake. Fig. 2.42 shows a random process.

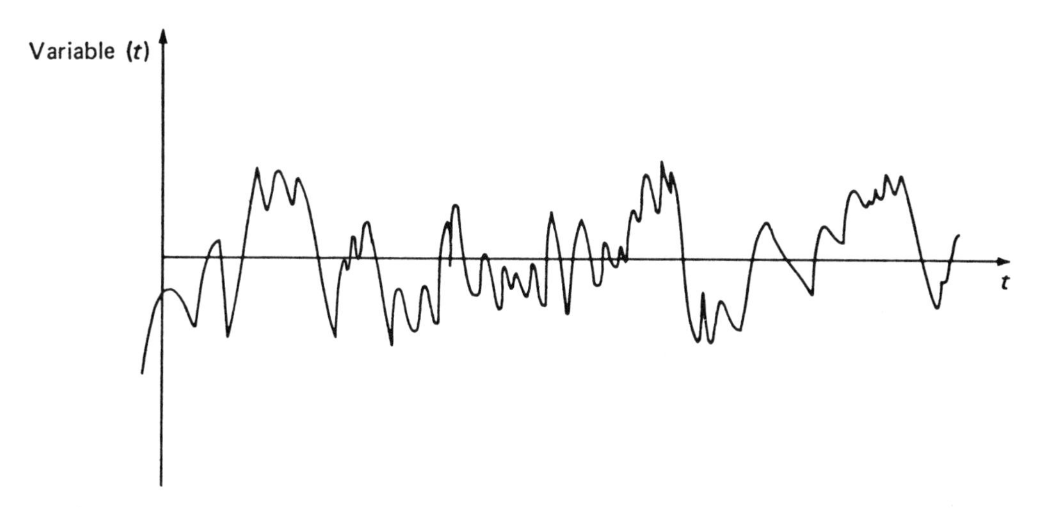

Fig. 2.42. Example random process variable as $f(t)$.

A collection of sample functions $x_1(t)$, $x_2(t)$, $x_3(t)$, ..., $x_n(t)$ which make up the random process $x(t)$ is called an *ensemble*, as shown in Fig. 2.43. These functions may comprise, for example, records of noise, pressure fluctuations or vibration levels, taken under the same conditions but at different times.

Any quantity which cannot be precisely predicted is non-deterministic and is known as a *random variable* or a *probabilistic quantity*. That is, if a series of tests is conducted to find the value of a particular parameter x and that value is found to vary in an unpredictable way that is not a function of any other parameter, than x is a random variable.

2.3.8.1 Probability distribution

If n experimental values of a variable x are $x_1, x_2, x_3, \ldots, x_n$, the probability that the value of x will be less than x' is n'/n, where n' is the number of x values which are less than or equal to x'. That is,

$$\text{Prob}\ (x \leqslant x') = n'/n.$$

When n approaches ∞ this expression is the probability distribution function of x, denoted by $P(x)$, so that

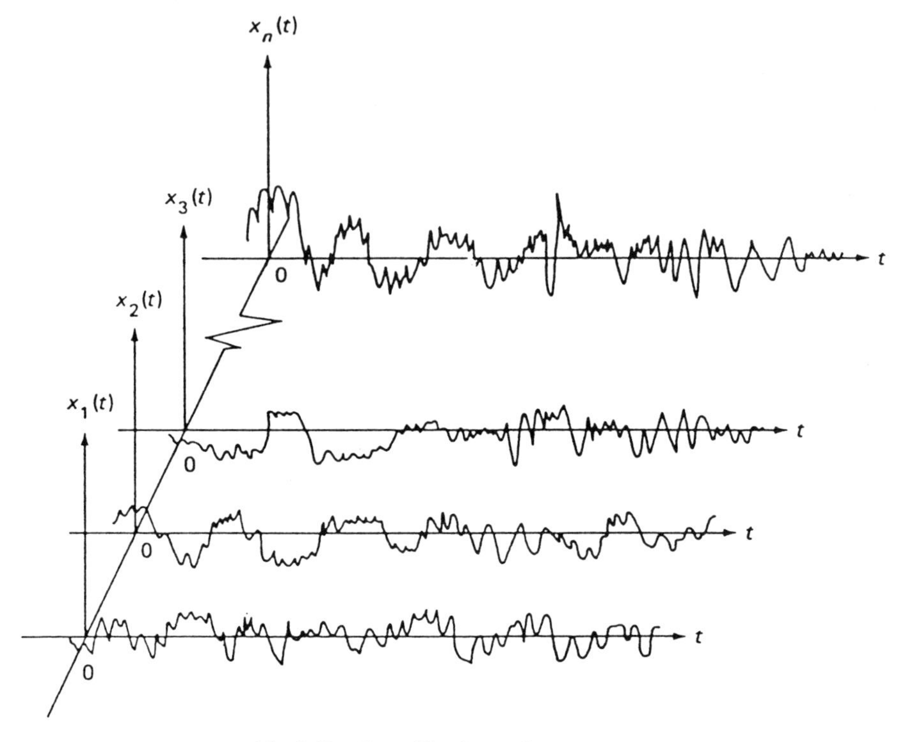

Fig. 2.43. Ensemble of a random process.

$$P(x) = \operatorname*{Lt}_{n \to \infty} (n'/n)$$

The typical variation of $P(x)$ with x is shown in Fig. 2.44. Since $x(t)$ denotes a physical quantity,

$$\text{Prob } (x < -\infty) = 0, \text{ and Prob } (x < +\infty) = 1.$$

The *probability density function* is the derivative of $P(x)$ with respect to x and this is denoted by $p(x)$. That is,

$$p(x) = \frac{\mathrm{d}P(x)}{\mathrm{d}(x)}$$

$$= \operatorname*{Lt}_{\Delta x \to 0} \left[\frac{P(x + \Delta x) - P(x)}{\Delta x} \right],$$

where $P(x + \Delta x) - P(x)$ is the probability that the value of $x(t)$ will lie between x and $x + \Delta x$ (Fig. 2.44). Now,

$$p(x) = \frac{\mathrm{d}P(x)}{\mathrm{d}x},$$

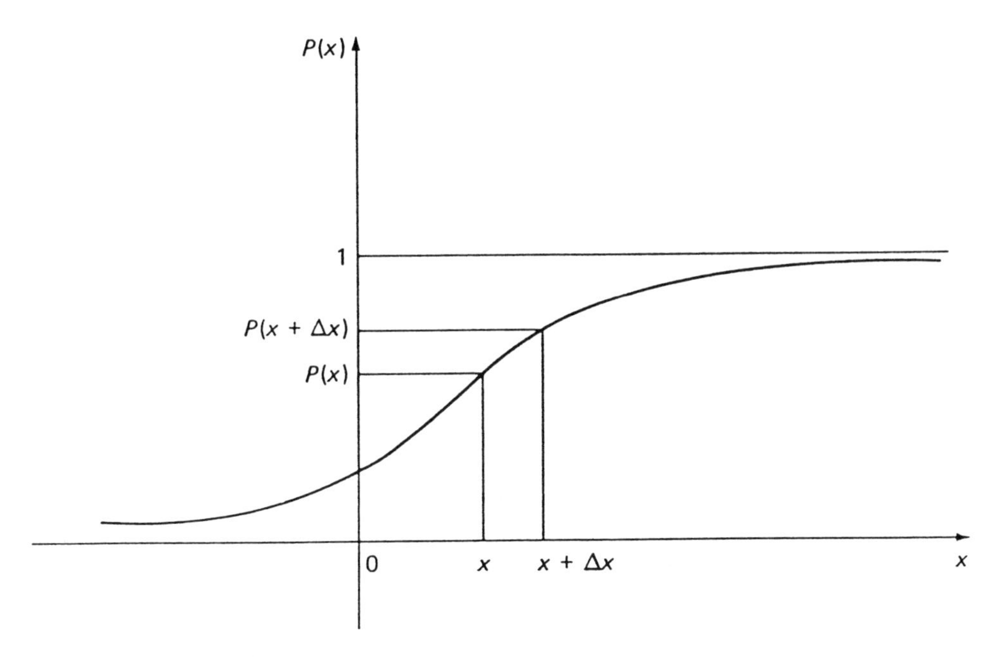

Fig. 2.44. Probability distribution function as $f(x)$.

so that

$$P(x) = \int_{-\infty}^{x} p(x)\mathrm{d}x.$$

Hence

$$P(\infty) = \int_{-\infty}^{\infty} p(x)\mathrm{d}x = 1,$$

so that the area under the probability density function curve is unity.

A random process is *stationary* if the joint probability density

$$p(x(t_1), x(t_2), x(t_3), \ldots)$$

depends only on the time differences $t_2 - t_1$, $t_3 - t_2$ and so on, and not on the actual time instants. That is, the ensemble will look just the same if the time origin is changed. A random process is *ergodic* if every sample function is typical of the entire group.

The expected value of $f(x)$, which is written

$E[f(x)]$ or $\overline{f(x)}$, is

$$E[f(x)] = \overline{f(x)} = \int_{-\infty}^{\infty} f(x)p(x)\mathrm{d}x,$$

so that the expected value of a stationary random process $x(t)$ is

$$E[x(t_1)] = E[x(t_1 + t)]$$

for any value of t.

If $f(x) = x$, the expected value or *mean value* of x,

$E[x]$ or $\bar{x}$, is

$$E[x] = \bar{x} = \int_{-\infty}^{\infty} xp(x)\mathrm{d}x.$$

In addition, if $f(x) = x^2$, the *mean square value* $\bar{x}^2$ of x is

$$E[x^2] = \bar{x}^2 = \int_{-\infty}^{\infty} x^2 p(x)\mathrm{d}x.$$

The *variance* of x, σ^2 is the mean square value of x about the mean, that is,

$$\sigma^2 = E[(x - \bar{x})^2] = \int_{-\infty}^{\infty} (x - \bar{x})^2 p(x)\mathrm{d}x = (\bar{x}^2) - (\bar{x})^2.$$

σ is the *standard deviation* of x, hence

variance = (standard deviation)^2 = {mean square - (mean)^2}

If two or more random variables x_1 and x_2, represent a random process at two different instants of time, then

$$E[f(x_1,x_2)] = \int_{-\infty}^{\infty}\int_{-\infty}^{\infty} f(x_1,x_2)p(x_1,x_2)\mathrm{d}x_1\mathrm{d}x_2,$$

and if t_1 and t_2 are the two instants of time,

$$E[x(t_1),x(t_2)] = R(t_1,t_2),$$

which is the *auto-correlation function* for the random process (Fig 2.45).

For random processes which are stationary,

$$E[x(t_1),\ x(t_2)] = R(t_1,\ t_2) = R(t_2 - t_1) = R(\tau),$$

say, since the average depends only on time differences. If the process is also ergodic, then

$$R(\tau) = \operatorname*{Lt}_{T\to\infty} \frac{1}{2\pi} \int_{-T}^{T} x(t)x(t + \tau)\mathrm{d}t.$$

It is worth noting that

$$R(0) = E[x(t)^2] = \operatorname*{Lt}_{T\to\infty} \frac{1}{2T} \int_{-T}^{T} x^2(t)\mathrm{d}t,$$

which is the average power in a sample function.

2.3.8.2 Random processes

The most important random process is the *Gaussian*, or normal random process. This is because a wide range of physically observed random waveforms can be represented

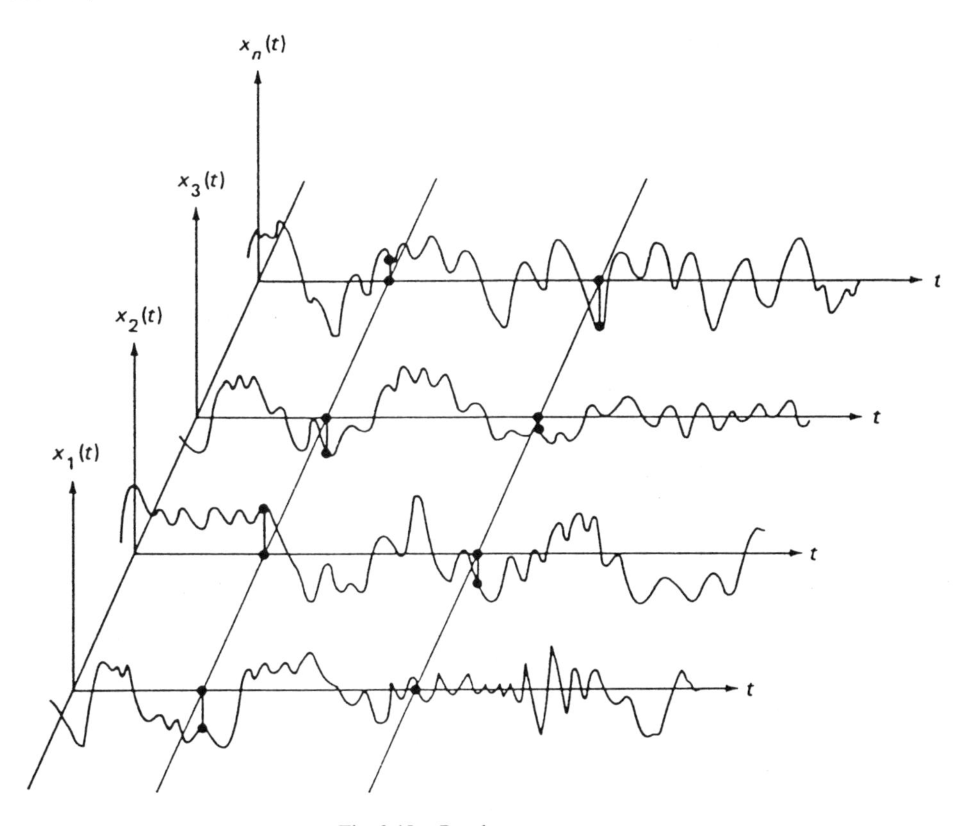

Fig. 2.45. Random processes.

as Gaussian processes, and the process has mathematical features which make analysis relatively straightforward.

The probability density function of a Gaussian process $x(t)$ is

$$p(x) = \frac{1}{\sqrt{(2\pi)}\sigma} \exp\left[-\tfrac{1}{2}\left(\frac{x - \bar{x}}{\sigma}\right)^2\right],$$

where σ is the standard deviation of x, and $\bar{x}$ is the mean value of x.

The values of σ and $\bar{x}$ may vary with time for a non-stationary process but are independant of time if the process is stationary.

One of the most important features of the Gaussian process is that the response of a linear system to this form of excitation is usually another, but still Guassian, random process. The only changes are that the magnitude and standard deviation of the response may differ from those of the excitation.

A Gaussian probability density function is shown in Fig. 2.46. It can be seen to be symmetric about the mean value $\bar{x}$, and the standard deviation σ controls the spread.

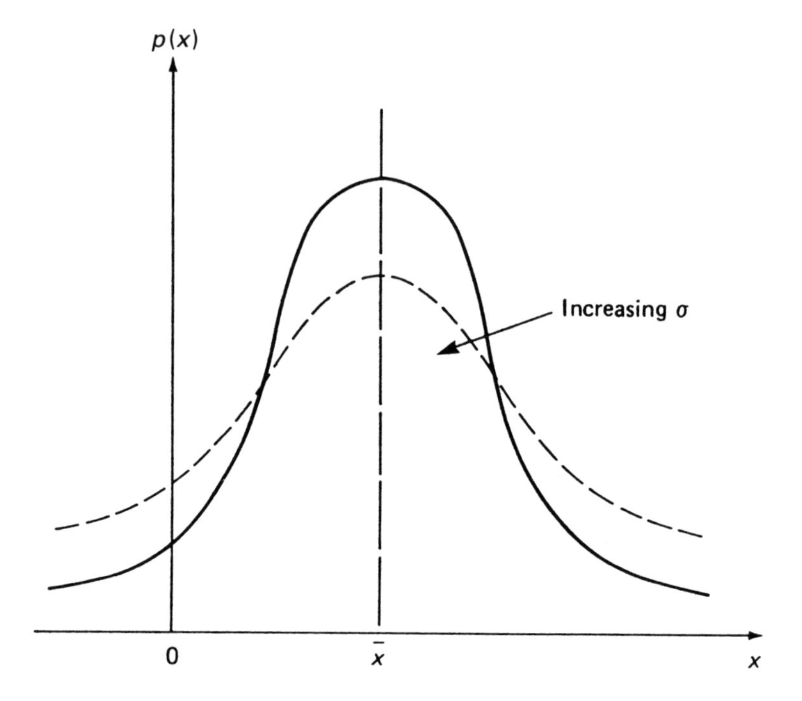

Fig. 2.46. Gaussian probability density function.

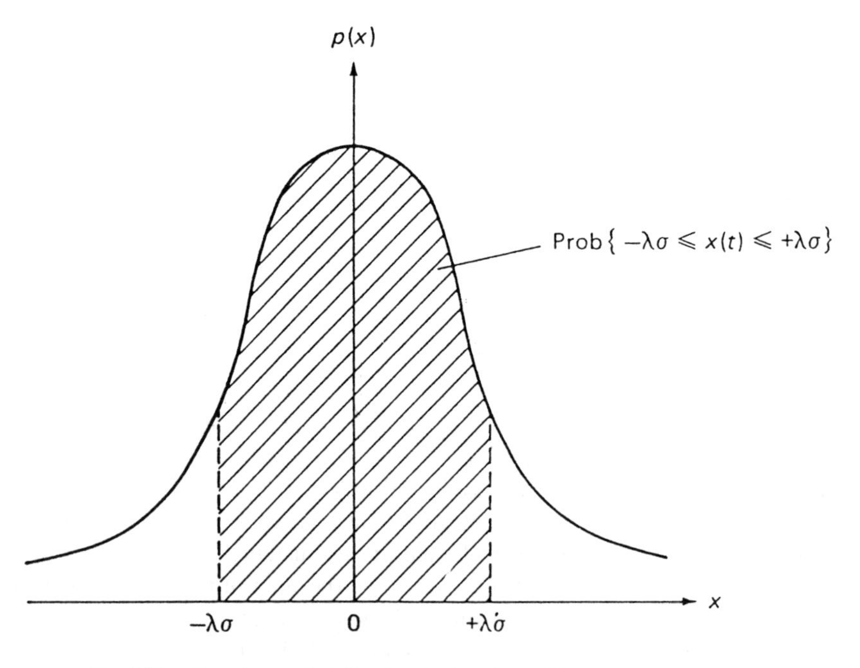

Fig. 2.47. Gaussian probability density function with zero mean.

The probability that $x(t)$ lies between $-\lambda\sigma$ and $+\lambda\sigma$, where λ is a positive number, can be found since, if $\bar{x} = 0$,

$$\text{Prob}\{-\lambda\sigma \leqslant x(t) \leqslant +\lambda\sigma\} = \int_{-\lambda\sigma}^{+\lambda\sigma} \frac{1}{\sqrt{(2\pi)}\sigma} \exp\left(-\frac{1}{2}\frac{x^2}{\sigma^2}\right) dx.$$

Fig. 2.47 shows the Gaussian probability density function with zero mean.

This integral has been calculated for a range of values of λ; the results are given below. The probability that $x(t)$ lies outside the range $-\lambda\sigma$ to $+\lambda\sigma$ is 1 minus the value of the above integral; this probability is also given.

Value of λ	Prob $[-\lambda\sigma \leqslant x(t) \leqslant \lambda\sigma]$	Prob $[\lvert x(t)\rvert > \lambda\sigma]$
0	0	1.0000
0.2	0.1585	0.8415
0.4	0.3108	0.6892
0.6	0.4515	0.5485
0.8	0.5763	0.4237
1.0	0.6827	0.3173
1.2	0.7699	0.2301
1.4	0.8586	0.1414
1.6	0.8904	0.1096
1.8	0.9281	0.0719
2.0	0.9545	0.0455
2.2	0.9722	0.0278
2.4	0.9835	0.0164
2.6	0.9907	0.0093
2.8	0.9949	0.0051
3.0	0.9973	0.0027
3.2	0.9986	0.00137
3.4	0.9993	0.00067
3.6	0.9997	0.00032
3.8	0.9998	0.00014
4.0	0.9999	0.00006

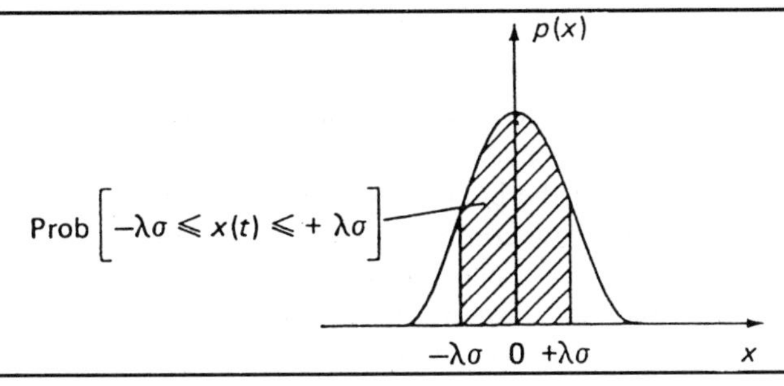

2.3.8.3 *Spectral density*

The spectral density $S(\omega)$ of a stationary random process is the Fourier transform of the autocorrelation function $R(\tau)$. It is given by

$$S(\omega) = \frac{1}{2\pi}\int_{-\infty}^{\infty} R(\tau)e^{-j\omega\tau}d\tau.$$

The inverse, which also holds true, is

$$R(\tau) = \int_{-\infty}^{\infty} S(\omega)e^{-j\omega\tau}d\omega$$

If $\tau = 0$

$$R(0) = \int_{-\infty}^{\infty} S(\omega)d\omega = E[x^2]$$

That is, the mean square value of a stationary random process x is the area under the $S(\omega)$ against frequency curve. A typical spectral density function is shown in Fig. 2.48.

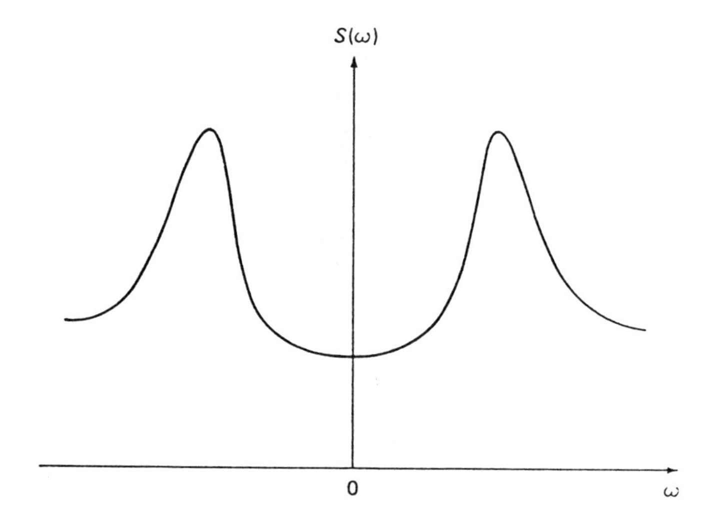

Fig. 2.48. Typical spectral density function.

A random process whose spectral density is constant over a very wide frequency range is called *white noise*. If the spectral density of a process has a significant value over a narrower range of frequencies, but one which is nevertheless still wide compared with the centre frequency of the band, it is termed a *wide-band process* (Fig. 2.49). If the frequency range is narrow compared with the centre frequency it is termed a *narrow-band process* (Fig. 2.50). Narrow-band processes frequently occur in engineering practice because real systems often respond strongly to specific exciting frequencies and thereby effectively act as a filter.

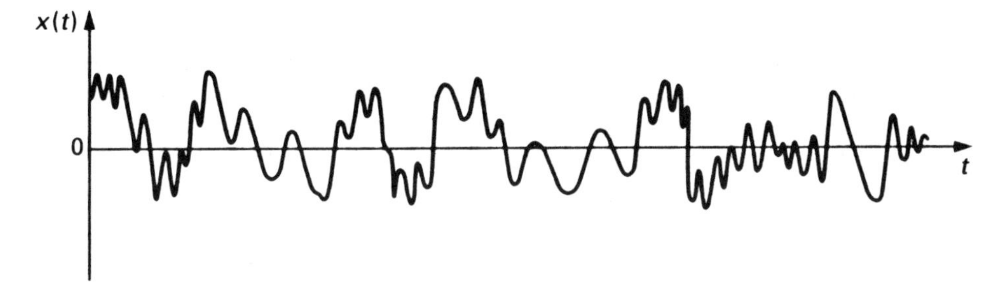

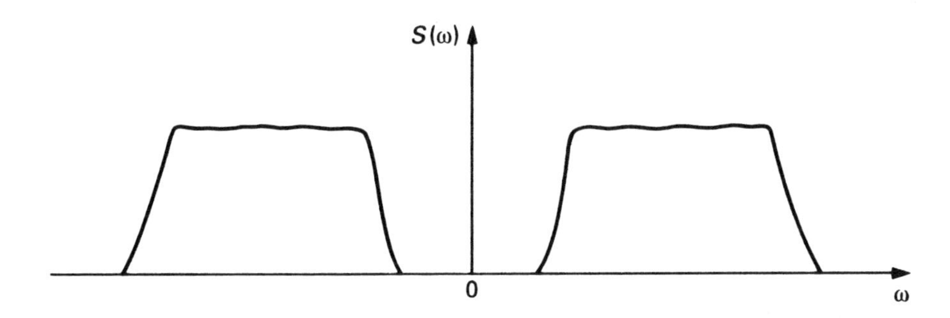

Fig. 2.49. Wide-band process.

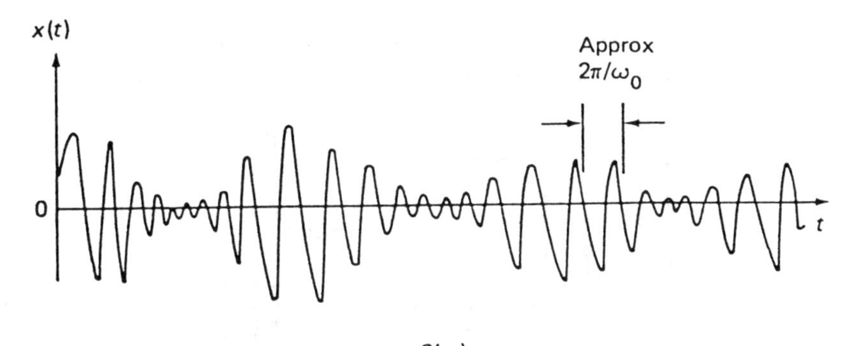

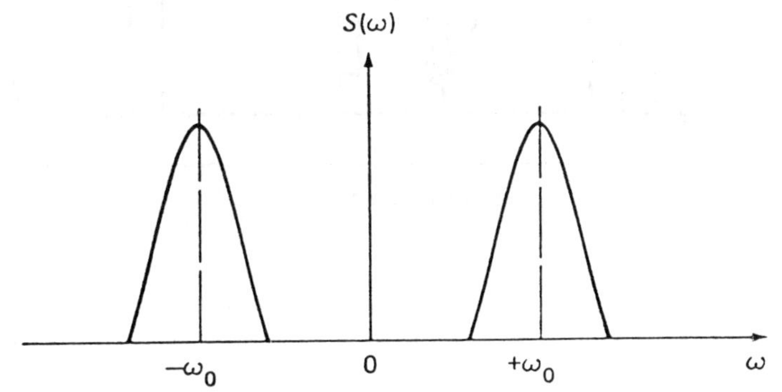

Fig. 2.50. Narrow band process.

2.3.9 The measurement of vibration

The most commonly used device for vibration measurement is the piezoelectric accelerometer, which gives an electric signal proportional to the vibration acceleration. This signal can readily be amplified, analysed, displayed, recorded, and so on. The principles of this device can be studied by referring to Fig. 2.51 which shows a body of mass m supported by an elastic system of stiffness k and effective viscous damping of coefficient c.

This dynamic system is usually enclosed in a case which is fastened to the surface whose vibration is to be measured. The body has a pointer fixed to it, which moves over a scale fastened to the case: that is it measures u, the motion of the suspended body relative to that of the vibrating surface so that $u = x - y$.

Now from section 2.3.2, the amplitude of u is

$$U = \frac{A\left(\frac{\nu}{\omega}\right)^2}{\sqrt{\left(\left[1 - \left(\frac{\nu}{\omega}\right)^2\right]^2 + \left[2\zeta\left(\frac{\nu}{\omega}\right)\right]^2\right)}}.$$

so that if ω is low and $\nu \gg \omega$,

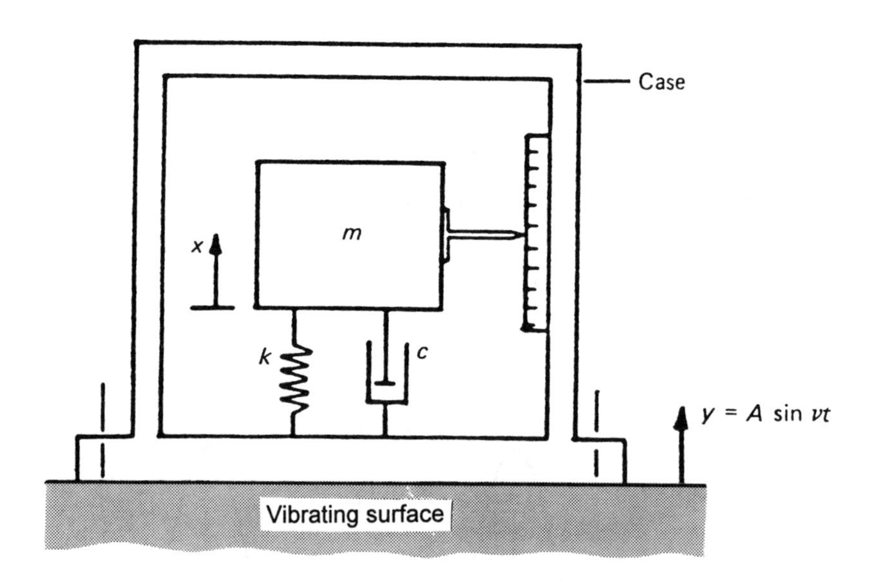

Fig. 2.51. Vibration measuring device.

$$U \simeq \frac{A(\nu/\omega)^2}{(\nu/\omega)^2} = A.$$

that is, the device measures the input vibration amplitude; when operating in this mode it is called a *vibrometer*, and if ω is high so that $\omega \gg \nu$, then

$$U \simeq \frac{A(\nu/\omega)^2}{1} = \frac{1}{\omega^2} A\nu^2,$$

that is, the device measures the input vibration acceleration amplitude; when operating in this mode it is called an *accelerometer*.

By adjusting the system parameters correctly it is possible to make

$$\sqrt{\left\{\left[1 - \left(\frac{\nu}{\omega}\right)^2\right]^2 + \left[2\zeta\left(\frac{\nu}{\omega}\right)\right]^2\right\}}$$

have a value close to unity for exciting frequencies ν up to about 0.3ω. Commercial accelerometers usually have piezoelectric elements instead of a spring and damper, so that the electric signal produced is proportional to the relative motion u, above.

3

The vibrations of systems having more than one degree of freedom

Many real systems can be represented by a single degree of freedom model. However, most actual systems have several bodies and several restraints and therefore several degrees of freedom. The number of degrees of freedom that a system possesses is equal to the number of independent coordinates necessary to describe the motion of the system. Since no body is completely rigid, and no spring is without mass, every real system has more than one degree of freedom, and sometimes it is not sufficiently realistic to approximate a system by a single degree of freedom model. Thus, it is necessary to study the vibration of systems with more than one degree of freedom.

Each flexibly connected body in a multi-degree of freedom system can move independently of the other bodies, and only under certain conditions will all bodies undergo an harmonic motion at the same frequency. Since all bodies move with the same frequency, they all attain their amplitudes at the same time even if they do not all move in the same direction. When such motion occurs the frequency is called a *natural frequency* of the system, and the motion is a principal mode of vibration: the number of natural frequencies and principal modes that a system possesses is equal to the number of degrees of freedom of that system. The deployment of the system at its lowest or first natural frequency is called its first mode, at the next highest or second natural frequency it is called the second mode and so on.

A two degree of freedom system will be considered initially. This is because the addition of more degrees of freedom increases the labour of the solution procedure but does not introduce any new analytical principles.

Initially, we will obtain the equations of motion for a two degree of freedom model, and from these find the natural frequencies and corresponding mode shapes.

Some examples of two degree of freedom models of vibrating systems are shown in Figs 3.1(a)–(h).

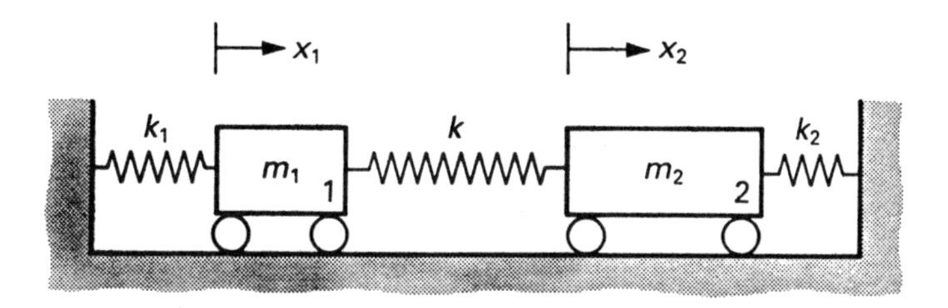

Fig. 3.1(a) Linear undamped system, horizontal motion. Coordinates x_1 and x_2.

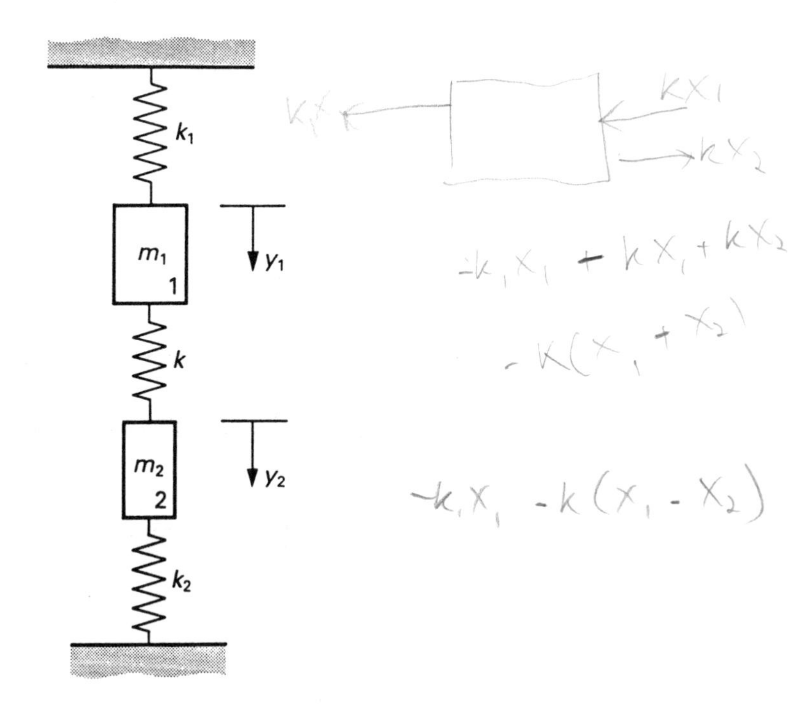

Fig. 3.1(b) Linear undamped system, vertical motion. Coordinates y_1 and y_2.

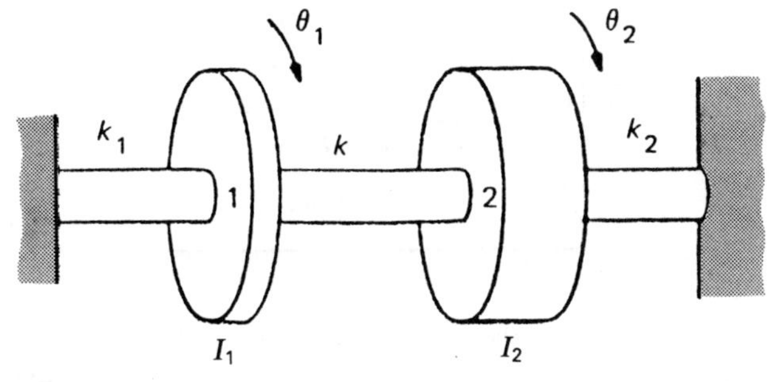

Fig. 3.1(c) Torsional undamped system. Coordinates θ_1 and θ_2.

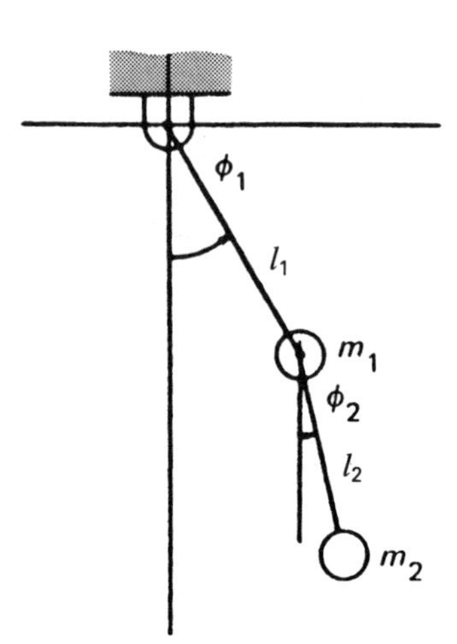

Fig. 3.1(d) Coupled pendula. Coordinates ϕ_1 and ϕ_2.

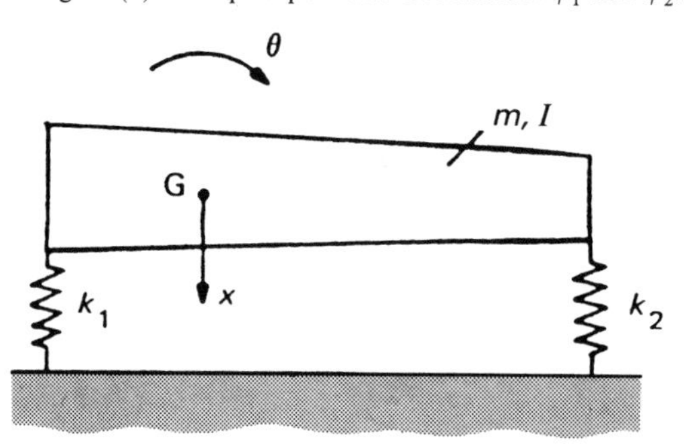

Figt. 3.1(e) System with combined translation and rotation. Coordinates x and θ

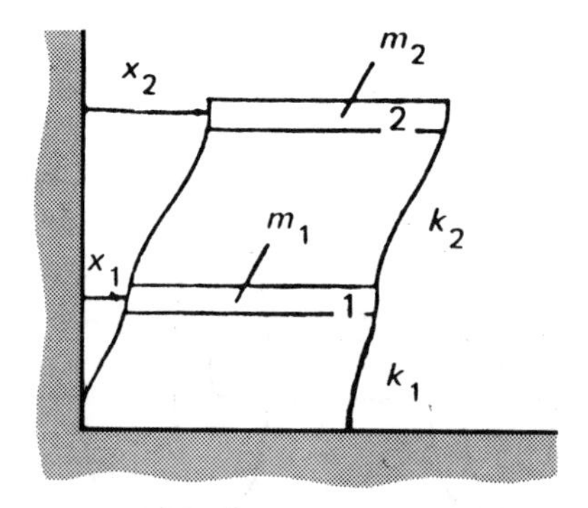

Fig. 3.1(f) Shear frame. Coordinates x_1 and x_2.

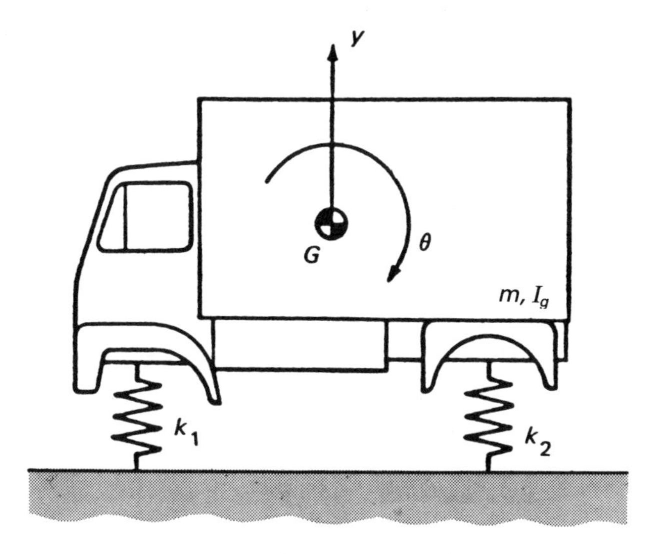

Fig. 3.2(g) Two degree of freedom model, rotation plus translation. Coordinates y and θ.

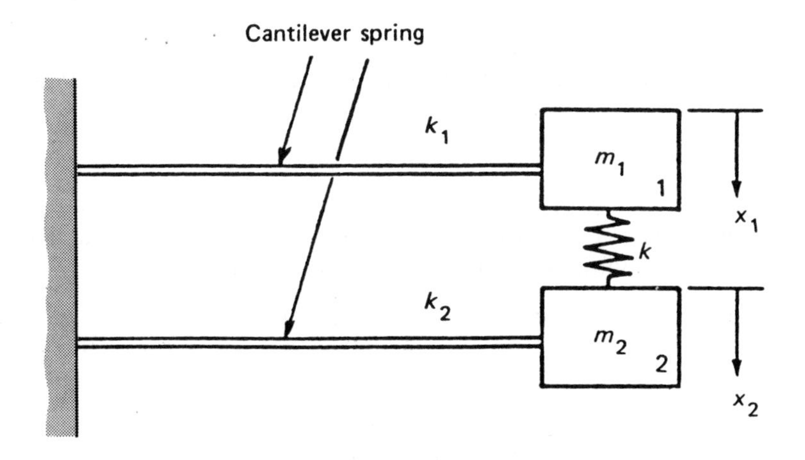

Fig. 3.1(h) Two degree of freedom model, translation vibration. Coordinates x_1 and x_2.

3.1 THE VIBRATION OF SYSTEMS WITH TWO DEGREES OF FREEDOM

3.1.1 Free vibration of an undamped system

Of the examples of two degree of freedom models shown in Figs. 3.1(a)–(h), consider the system shown in Fig. 3.1(a). If $x_1 > x_2$ the FBDs are as shown in Figure 3.2.

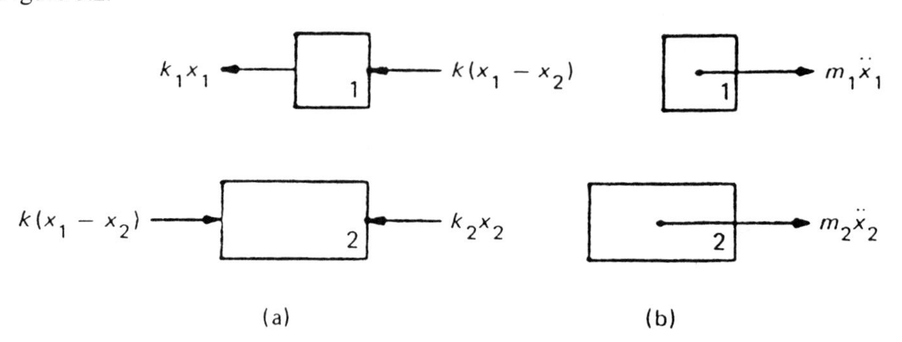

Fig. 3.2. (a) Applied forces; (b) effective forces.

The equations of motion are therefore,

$$m_1\ddot{x}_1 = -k_1x_1 - k(x_1 - x_2) \qquad \text{for body 1,} \tag{3.1}$$

and

$$m_2\ddot{x}_2 = k(x_1 - x_2) - k_2x_2 \qquad \text{for body 2.} \tag{3.2}$$

The same equations are obtained if $x_1 < x_2$ is assumed because the direction of the central spring force is then reversed.

Equations (3.1) and (3.2) can be solved for the natural frequencies and corresponding mode shapes by assuming a solution of the form

$$x_1 = A_1\sin(\omega t + \psi) \text{ and } x_2 = A_2\sin(\omega t + \psi).$$

This assumes that x_1 and x_2 oscillate with the same frequency ω and are either in phase or π out of phase. This is a sufficient condition to make ω a natural frequency.

Substituting these solutions into the equations of motion gives

$$-m_1A_1\omega^2\sin(\omega t + \psi) = -k_1A_1\sin(\omega t + \psi) - k(A_1 - A_2)\sin(\omega t + \psi)$$

and

$$-m_2A_2\omega^2\sin(\omega t + \psi) = k(A_1 - A_2)\sin(\omega t + \psi) - k_2A_2\sin(\omega t + \psi).$$

Since these solutions are true for all values of t,

$$A_1(k + k_1 - m_1\omega^2) + A_2(-k) = 0 \tag{3.3}$$

and

$$A_1(-k) + A_2(k_2 + k - m_2\omega^2) = 0. \tag{3.4}$$

A_1 and A_2 can be eliminated by writing

$$\begin{vmatrix} k + k_1 - m_1\omega^2 & -k \\ -k & k + k_2 - m_2\omega^2 \end{vmatrix} = 0 \tag{3.5}$$

This is the *characteristic* or *frequency equation.* Alternatively, we may write

$$\left.\begin{aligned} A_1/A_2 &= k/(k + k_1 - m_1\omega^2) \quad \text{from (3.3)} \\ &\text{and} \\ A_1/A_2 &= (k_2 + k - m_2\omega^2)/k \quad \text{from (3.4)} \end{aligned}\right\} \tag{3.6}$$

Thus

$$k/(k + k_1 - m_1\omega^2) = (k_2 + k - m_2\omega^2)/k$$

and

$$(k + k_1 - m_1\omega^2)(k_2 + k - m_2\omega^2) - k^2 = 0. \tag{3.7}$$

This result is the frequency equation and could also be obtained by multiplying out the above determinant, equation (3.5).

The solutions to equation (3.7) give the natural frequencies of free vibration for the system in Fig. 3.1(a). The corresponding mode shapes are found by substituting these frequencies, in turn, into either of equations (3.6).

Consider the case when $k_1 = k_2 = k$, and $m_1 = m_2 = m$. The frequency equation is $(2k - m\omega^2)^2 - k^2 = 0$; that is,

$$m^2\omega^4 - 4mk\omega^2 + 3k^2 = 0, \quad \text{or} \quad (m\omega^2 - k)(m\omega^2 - 3k) = 0.$$

Therefore, either $m\omega^2 - k = 0$, or $m\omega^2 - 3k = 0$.
Thus

$$\omega_1 = \sqrt{(k/m)}\text{rad/s} \quad \text{and} \quad \omega_2 = \sqrt{(3\text{k/m})}\ \text{rad/s}.$$

If

$$\omega = \sqrt{(k/m)}\text{rad/s}, \quad (A_1/A_2)_{\omega=\sqrt{(k/m)}} = +1,$$

and if

$$\omega = \sqrt{(3k/m)}\text{rad/s}, \quad (A_1/A_2)_{\omega=\sqrt{(3k/m)}} = -1. \quad \text{(from 3.6)}$$

This gives the mode shapes corresponding to the frequencies ω_1 and ω_2. Thus, the first mode of free vibration occurs at a frequency $(1/2\pi)\sqrt{(k/m)}$Hz and $(A_1/A_2)^{\text{I}} = 1$, that is, the bodies move in phase with each other and with the same amplitude as if connected by a rigid link, Fig. 3.3. The second mode of free vibration occurs at a frequency $(1/2\pi)\sqrt{(3k/m)}$Hz and $(A_1/A_2)^{\text{II}} = -1$, that is, the bodies move exactly out of phase with each other, but with the same amplitude, see Fig. 3.3.

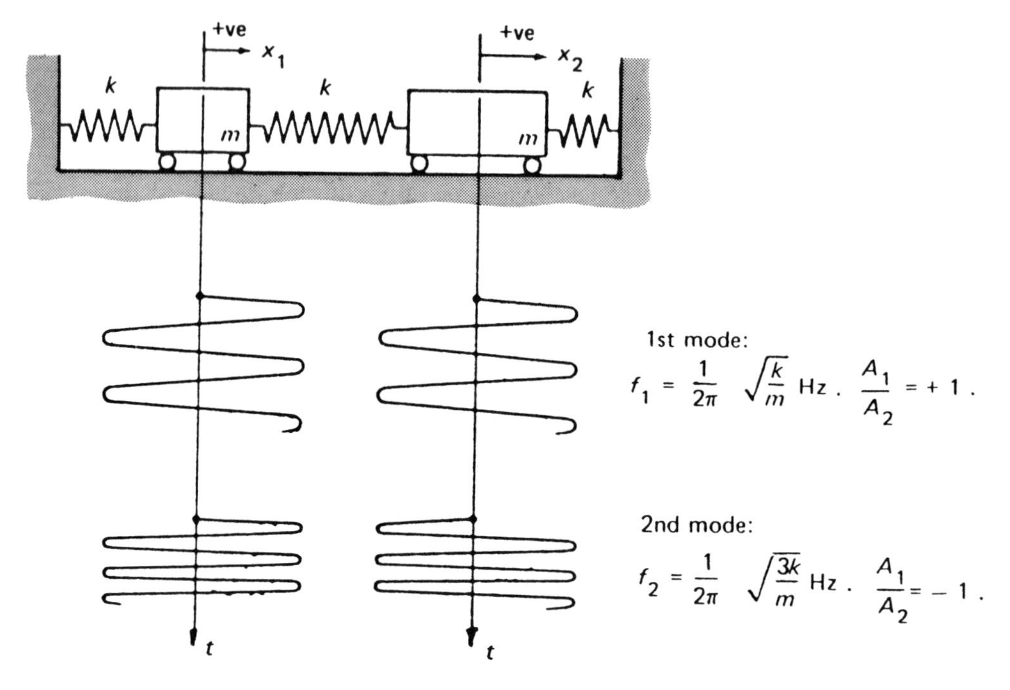

Fig. 3.3. Natural frequencies and mode shapes for two degree of freedom translation vibration system. Bodies of equal mass and springs of equal stiffness.

3.1.2 Free motion

The two modes of vibration can be written

$$\begin{Bmatrix} x_1 \\ x_2 \end{Bmatrix}^{\mathrm{I}} = \begin{Bmatrix} A_1 \\ A_2 \end{Bmatrix}^{\mathrm{I}} \sin(\omega_1 t + \psi_1),$$

and

$$\begin{Bmatrix} x_1 \\ x_2 \end{Bmatrix}^{\mathrm{II}} = \begin{Bmatrix} A_1 \\ A_2 \end{Bmatrix}^{\mathrm{II}} \sin(\omega_2 t + \psi_2),$$

where the ratio A_1/A_2 is specified for each mode.

Since each solution satisfies the equation of motion, the general solution is

$$\begin{Bmatrix} x_1 \\ x_2 \end{Bmatrix} = \begin{Bmatrix} A_1 \\ A_2 \end{Bmatrix}^{\mathrm{I}} \sin(\omega_1 t + \psi_1) + \begin{Bmatrix} A_1 \\ A_2 \end{Bmatrix}^{\mathrm{II}} \sin(\omega_2 t + \psi_2),$$

where A_1, A_2, ψ_1, ψ_2 are found from the initial conditions.

For example, for the system considered above, if one body is displaced a distance X and released,

$$x_1(0) = X \quad \text{and} \quad x_2(0) = \dot{x}_1(0) = \dot{x}_2(0) = 0$$

where $x_1(0)$ means the value of x_1 when $t = 0$, and similarly for $x_2(0)$, $\dot{x}_1(0)$ and $\dot{x}_2(0)$. Remembering that in this system $\omega_1 = \sqrt{(k/m)}$, $\omega_2 = \sqrt{(3k/m)}$, and

$$\left(\frac{A_1}{A_2}\right)_{\omega_1} = +1 \quad \text{and} \quad \left(\frac{A_1}{A_2}\right)_{\omega_2} = -1,$$

we can write

$$x_1 = \sin(\sqrt{(k/m)}t + \psi_1) + \sin(\sqrt{(3k/m)}t + \psi_2),$$

and

$$x_2 = \sin(\sqrt{(k/m)}t + \psi_1) - \sin(\sqrt{(3k/m)}t + \psi_2).$$

Substituting the initial conditions $x_1(0) = X$ and $x_2(0) = 0$ gives

$$X = \sin \psi_1 + \sin \psi_2,$$

and

$$0 = \sin \psi_1 - \sin \psi_2,$$

that is,

$$\sin \psi_1 = \sin \psi_2 = X/2.$$

The remaining conditions give $\cos \psi_1 = \cos \psi_2 = 0$.
Hence

$$x_1 = (X/2) \cos \sqrt{(k/m)}t + (X/2) \cos \sqrt{(3k/m)}t,$$

and

$$x_2 = (X/2) \cos \sqrt{(k/m)}t - (X/2) \cos \sqrt{(3k/m)}t.$$

That is, *both* natural frequencies are excited and the motion of each body has two harmonic components.

3.1.3 Coordinate coupling

In some systems the motion is such that the coordinates are coupled in the equations of motion. Consider the system shown in Fig. 3.1(e); only motion in the plane of the figure is considered, horizontal motion being neglected because the lateral stiffness of the springs is assumed to be negligible. The coordinates of rotation, θ, and translation, x, are coupled as shown in Fig. 3.4. G is the centre of mass of the rigid beam of mass m and moment of inertia I about G.

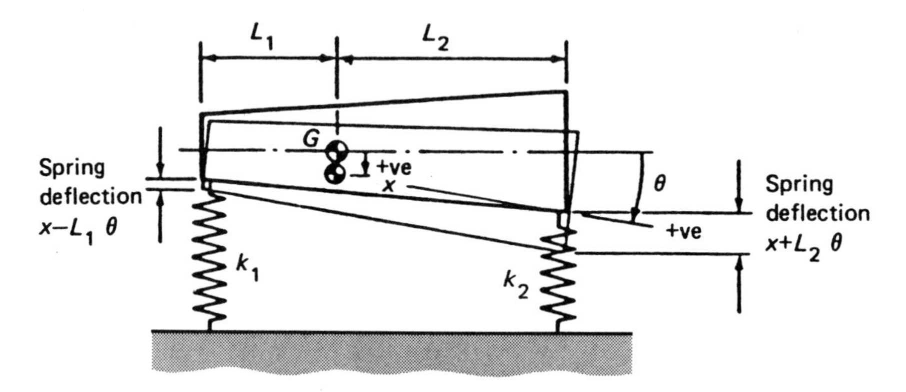

Fig. 3.4. Two degree of freedom model, rotation plus translation.

The FBDs are shown in Fig. 3.5; since the weight of the beam is supported by the springs, both the initial spring forces and the beam weight may be omitted.

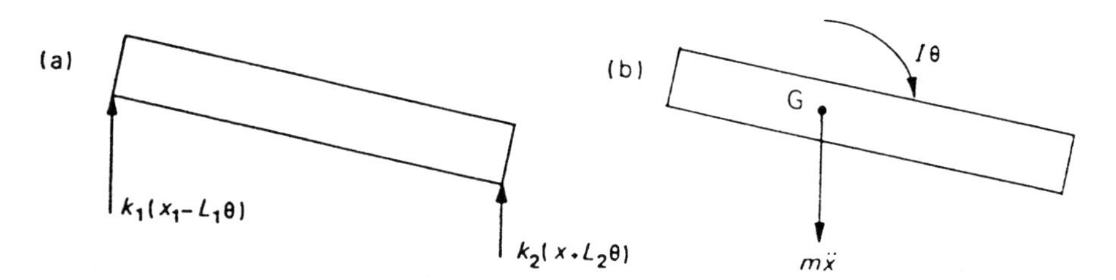

Fig. 3.5. (a) Applied forces. (b) Effective force and moment.

For small amplitudes of oscillation (so that $\sin\theta \simeq \theta$) the equations of motion are

$$m\ddot{x} = -k_1(x - L_1\theta) - k_2(x + L_2\theta)$$

and

$$I\ddot{\theta} = k_1(x - L_1\theta)L_1 - k_2(x + L_2\theta)L_2,$$

that is,

$$m\ddot{x} + (k_1 + k_2)x - (k_1L_1 - k_2L_2)\theta = 0$$

and

$$I\ddot{\theta} - (k_1L_1 - k_2L_2)x + (k_1L_1^2 + k_2L_2^2)\theta = 0.$$

It will be noticed that these equations can be uncoupled by making $k_1L_1 = k_2L_2$; if this is arranged, translation (x motion) and rotation (θ motion) can take place independently. Otherwise translation and rotation occur simultaneously.

Assuming $x_1 = A_1 \sin(\omega t + \psi)$ and $\theta = A_2 \sin(\omega t + \psi)$, substituting into the equations of motion gives

$$-m\omega^2 A_1 + (k_1 + k_2)A_1 - (k_1L_1 - k_2L_2)A_2 = 0,$$

and

$$-I\omega^2 A_2 - (k_1L_1 - k_2L_2)A_1 + (k_1L_1^2 + k_2L_2^2)A_2 = 0.$$

That is,

$$A_1(k_1 + k_2 - m\omega^2) + A_2(-(k_1L_1 - k_2L_2)) = 0,$$

and

$$A_1(-(k_1L_1 - k_2L_2)) + A_2(k_1L_1^2 + k_2L_2^2 - I\omega^2) = 0.$$

Hence the frequency equation is

$$\begin{vmatrix} k_1 + k_2 - m\omega^2 & -(k_1L_1 - k_2L_2) \\ -(k_1L_1 - k_2L_2) & k_1L_1^2 + k_2L_2^2 - I\omega^2 \end{vmatrix} = 0.$$

For each natural frequency, there is a corresponding mode shape, given by A_1/A_2.

Example 18

A system is modelled as a straight link AB 1 m long of mass 10 kg, supported horizontally by springs A and B of stiffness k_1 and k_2 respectively. The moment of inertia of AB about its centre of mass G is 1 kg m^2, and G is located at a distance a from A and b from B.

Find the relationship between k_1, k_2, a, and b so that one mode of free vibration shall be translational motion only and the other mode rotation only.

If $a = 0.3$ m and $k_1 = 13$ kN/m, find k_2 to give these two modes of vibration and calculate the two natural frequencies.

The frequency equation is (section 3.1.3)

$$\begin{vmatrix} k_1 + k_2 - m\omega^2 & -(k_1a - k_2b) \\ -(k_1a - k_2b) & k_1a^2 + k_2b^2 - I\omega^2 \end{vmatrix} = 0.$$

For the modes to be uncoupled $k_1a = k_2b$ and then

$$\omega_1 = \sqrt{((k_1 + k_2)/m)}, \qquad \text{(bouncing mode)}$$

and

$$\omega_2 = \sqrt{((k_1a^2 + k_2b^2)/I)} \qquad \text{(rocking mode)}.$$

Since $a = 0.3$ m, $b = 0.7$ m and $k_1 = 13$ kN/m,

$k_2 = (13 \times 0.3)/0.7 = 5.57\,\mathrm{kN/m}$ and then

$\omega_1 = \sqrt{(18570/10)} = 43.1\,\mathrm{rad/s}$; so $f_1 = 6.86\,\mathrm{Hz}$.

and $\omega_2 = \sqrt{((13 \times 0.3^2 + 5.57 \times 0.7^2)10^3)} = 62.45\,\mathrm{rad/s}$; so $f_2 = 9.94\,\mathrm{Hz}$.

Example 19

When transported, a space vehicle is supported in a horizontal position by two springs, as shown. The vehicle can be considered to be a rigid body of mass m and radius of gyration h about an axis normal to the plane of the figure through the mass centre G. The rear support has a stiffness k_1 and is at a distance a from G while the front support has a stiffness k_2 and is at a distance b from G. The only motions possible for the vehicle are vertical translation and rotation in the vertical plane.

Write the equations of small amplitude motion of the vehicle and obtain the frequency equation in terms of the given parameters.

Given that $k_1a = k_2b$, determine the natural frequencies of the free vibrations of the vehicle and sketch the corresponding modes of vibration. Also state or sketch the modes of vibration if $k_1a \neq k_2b$.

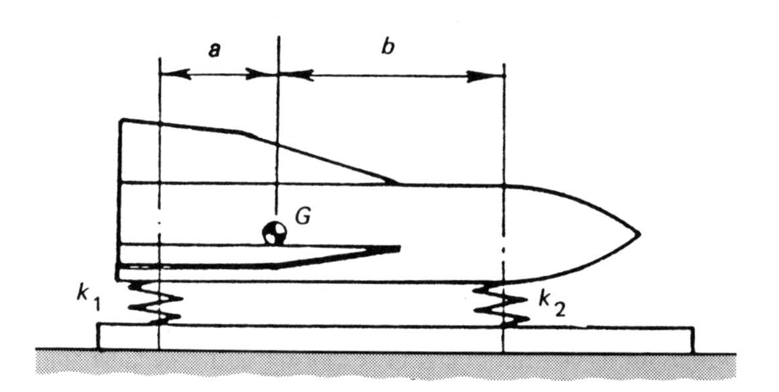

The FBDs are as below:

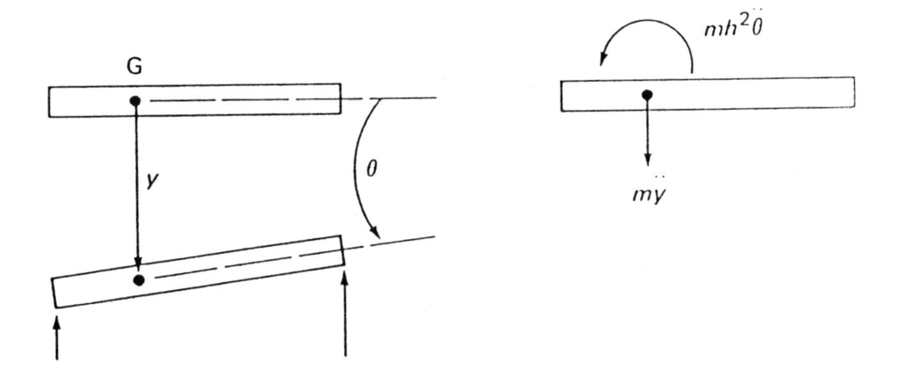

The equations of motion are

$$k_1(y + a\theta) + k_2(y - b\theta) = -m\ddot{y},$$

and

$$k_1(y + a\theta)a - k_2(y - b\theta)b = -mh^2\ddot{\theta}.$$

Assuming

$$y = Y \sin \nu t \qquad \text{and} \qquad \theta = \Theta \sin \nu t,$$

these give

$$Y(k_1 + k_2 - m\nu^2) + \Theta(k_1 a - k_2 b) = 0,$$

and

$$Y(k_1 a - k_2 b) + \Theta(k_1 a^2 + k_2 b^2 - mh^2\nu^2) = 0.$$

The frequency equation is, therefore,

$$(k_1 + k_2 - m\nu^2)(k_1 a^2 + k_2 b^2 - mh^2\nu^2) - (k_1 a - k_2 b)^2 = 0.$$

If $k_1 a = k_2 b$, motion is uncoupled so

$$\nu_1 = \sqrt{\left(\frac{k_1 + k_2}{m}\right)} \text{ rad/s} \qquad \text{and} \qquad \nu_2 = \sqrt{\left(\frac{k_1 a^2 + k_2 b^2}{mh^2}\right)} \text{ rad/s}.$$

ν_1 is the frequency of a bouncing or translation mode (no rotation):

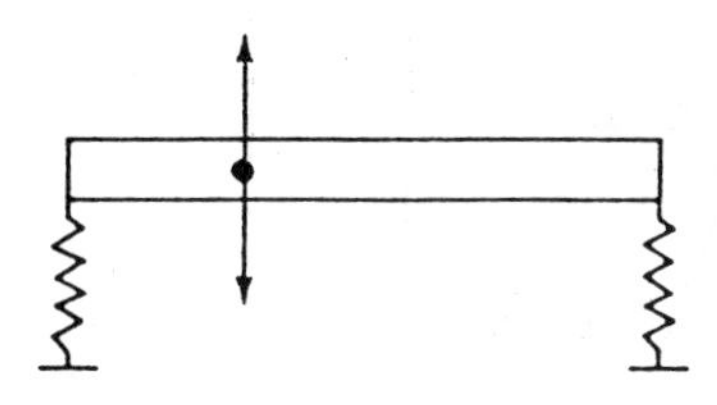

ν_2 is the frequency of a rotation mode (no bounce)

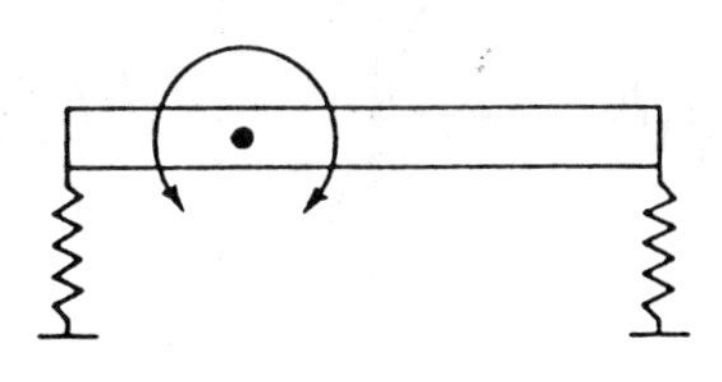

If $k_1a \neq k_2b$, the modes are coupled:

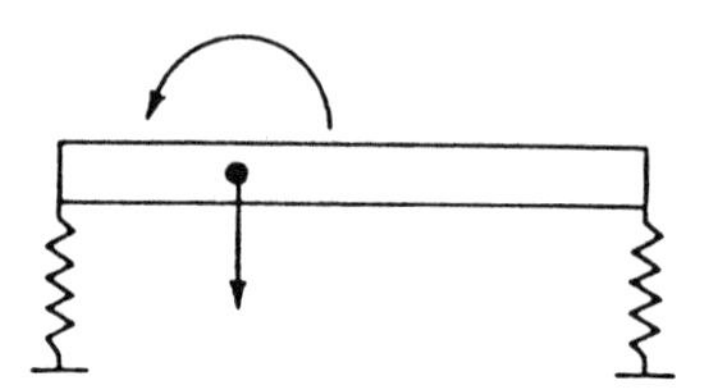

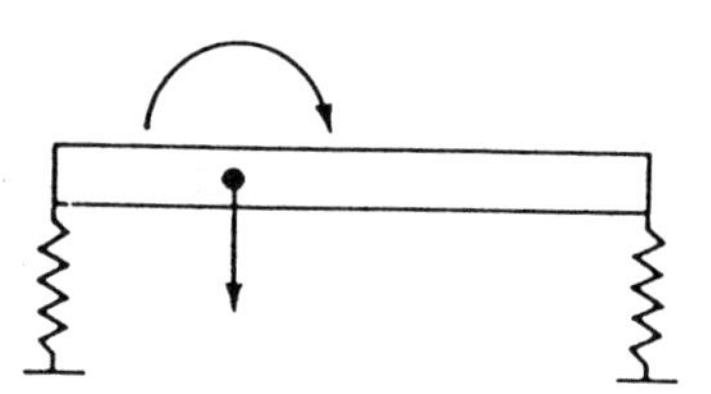

Example 20

In a study of earthquakes, a building is idealized as a rigid body of mass M supported on two springs, one giving translational stiffness k and the other rotational stiffness k_T, as shown.

Given that I_G is the mass moment of inertia of the building about its mass centre G, write down the equations of motion using coordinates x for the translation from the equilibrium position, and θ for the rotation of the building.

Hence determine the frequency equation of the motion.

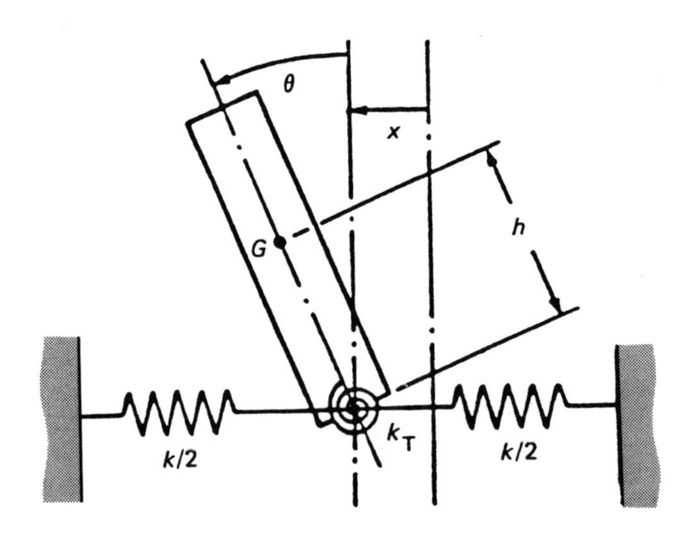

The FBDs are as follows.

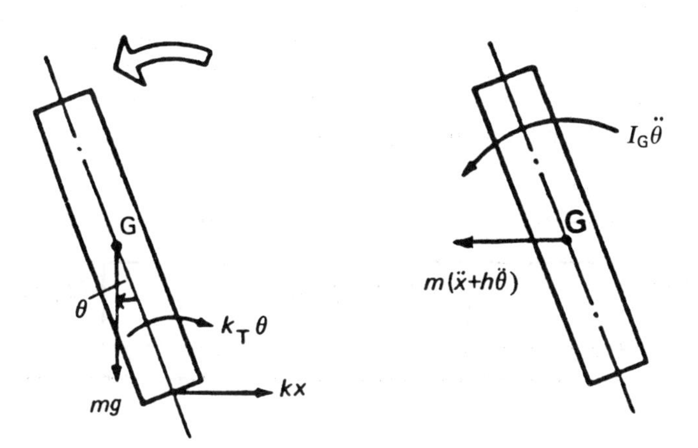

Assume small θ (earthquakes), hence

$$m(\ddot{x} + h\ddot{\theta}) = -kx,$$

and

$$I_G\ddot{\theta} + m(\ddot{x} + h\ddot{\theta})h = -k_T\theta + mgh\theta.$$

The equations of motion are therefore

$$mh\ddot{\theta} + m\ddot{x} + kx = 0,$$

and

$$mh\ddot{x} + (mh^2 + I_G)\ddot{\theta} - (mgh - k_T)\theta = 0.$$

If

$$\theta = A_1 \sin \omega t \quad \text{and} \quad x = A_2 \sin \omega t,$$

$$-mh\omega^2 A_1 - m\omega^2 A_2 + kA_2 = 0,$$

and

$$+mh\omega^2 A_2 + (mh^2 + I_G)\omega^2 A_1 + (mgh - k_T)A_1 = 0.$$

The frequency equation is

$$\begin{vmatrix} -mh\omega^2 & k - m\omega^2 \\ (mh^2 + I_G)\omega^2 + (mgh - k_T) & mh\omega^2 \end{vmatrix} = 0,$$

That is, $(mh\omega^2)^2 + (k - m\omega^2)[(mh^2 + I_G)\omega^2 + (mgh - k_T)] = 0$, or,

$$mI_G\omega^4 - \omega^2[mkh^2 + I_Gk - m^2gh + mk_T] - mghk + kk_T = 0.$$

3.1.4 Forced vibration

Harmonic excitation of vibration in a system may be generated in a number of ways, for example by unbalanced rotating or reciprocating machinery, or it may arise from periodic excitation containing a troublesome harmonic component.

A two degree of freedom model of a dynamic system excited by an harmonic force $F \sin \nu t$ is shown in Fig. 3.6. Damping is assumed to be negligible. The force has a constant amplitude F and a frequency $\nu/2\pi$ Hz.

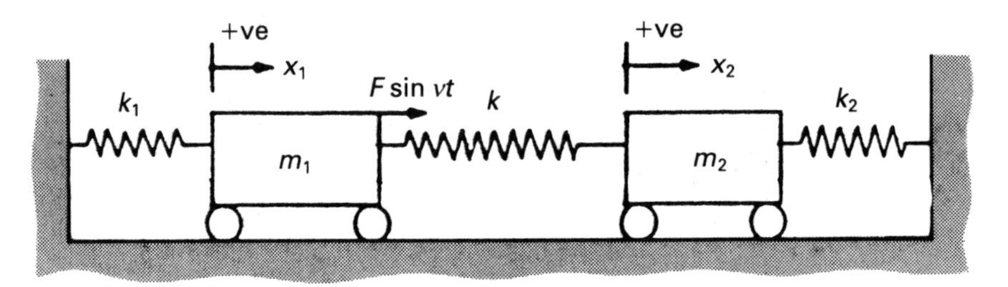

Fig. 3.6. Two degree of freedom model with forced excitation.

The equations of motion are

$$m_1\ddot{x}_1 = -k_1x_1 - k(x_1 - x_2) + F \sin \nu t,$$

and

$$m_2\ddot{x}_2 = k(x_1 - x_2) - k_2x_2.$$

Since there is zero damping, the motions are either in phase or π out of phase with the driving force, so that the following solutions may be assumed:

$$x_1 = A_1 \sin \nu t \qquad \text{and} \qquad x_2 = A_2 \sin \nu t.$$

Substituting these solutions into the equations of motion gives

$$A_1(k_1 + k - m_1\nu^2) + A_2(-k) = F,$$

and

$$A_1(-k) + A_2(k_2 + k - m_2\nu^2) = 0.$$

Thus

$$A_1 = \frac{F(k_2 + k - m_2\nu^2)}{\Delta},$$

and

$$A_2 = \frac{Fk}{\Delta},$$

where

$$\Delta = (k_2 + k - m_2 v^2)(k_1 + k - m_1 v^2) - k^2,$$

and

$$\Delta = 0 \text{ is the frequency equation.}$$

Hence the response of the system to the exciting force is determined.

Example 21

A two-wheel trailer is drawn over an undulating surface in such a way that the vertical motion of the tyre may be regarded as sinusoidal, the pitch of the undulations being 5 m. The combined stiffness of the tyres is 170 kN/m and that of the main springs is 60 kN/m; the axle and attached parts have a mass of 400 kg, and the mass of the body is 500 kg. Find (a) the critical speeds of the trailer in km/h, and (b) the amplitude of the trailer body vibration if the trailer is drawn at 50 km/h, and the amplitude of the undulations is 0.1 m.

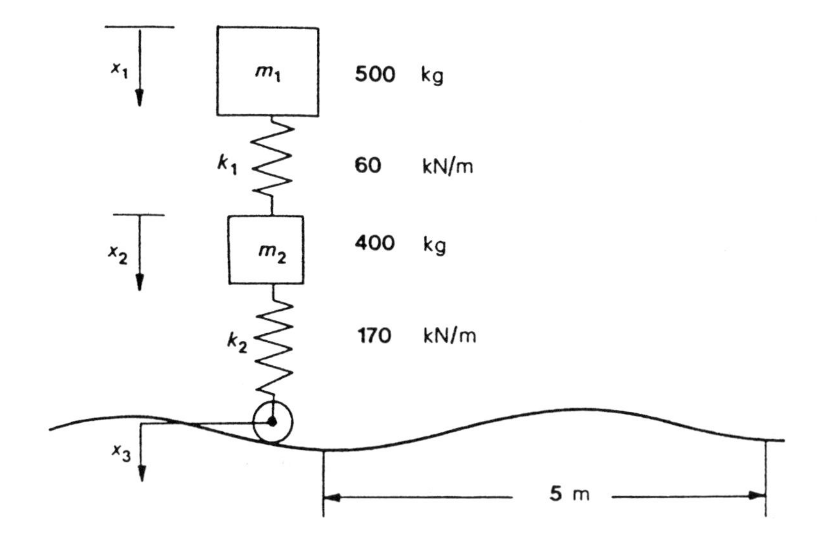

The equations of motion are

$$m_1 \ddot{x}_1 = -k_1(x_1 - x_2),$$

and

$$m_2 \ddot{x}_2 = k_1(x_1 - x_2) - k_2(x_2 - x_3).$$

Assuming $x_1 = A_1 \sin vt$, $x_2 = A_2 \sin vt$, and $x_3 = A_3 \sin vt$,

$$A_1(k_1 - m_1 v^2) + A_2(-k_1) = 0$$

and

$$A_1(-k_1) + A_2(k_1 - k_2 m_2 v^2) = k_2 A_3.$$

The frequency equation is

$$(k_1 + k_2 - m_2 v^2)(k_1 - m_1 v^2) - k_1^2 = 0.$$

The critical speeds are those which correspond to the natural frequencies and hence excite resonances. The frequency equation simplifies to

$$m_1 m_2 v^4 - (m_1 k_1 + m_1 k_2 + m_2 k_1) v^2 + k_1 k_2 = 0.$$

Hence substituting the given data,

$$500 \times 400 \times v^4 - (500 \times 60 + 500 \times 170 + 400 \times 60)\, 10^3 v^2 + 60 \times 170 \times 10^6 = 0.$$

That is $0.2v^4 - 139v^2 + 10\,200 = 0$, which can be solved by the formula. Thus $v = 16.3$ rad/s or 20.78 rad/s, and $f = 2.59$ Hz or 3.3 Hz.

Now if the trailer is drawn at v km/h, or $v/3.6$ m/s, the frequency is $v/(3.6 \times 5)$Hz. Therefore the critical speeds are

$$v_1 = 18 \times 2.59 = 46.6 \text{ km/h},$$

and

$$v_2 = 18 \times 3.3 = 59.4 \text{ km/h}.$$

Towing the trailer at either of these speeds will excite a resonance in the system. From the equations of motion,

$$A_1 = \left\{ \frac{k_1 k_2}{(k_1 + k_2 - m_2 v^2)(k_1 - m_1 v^2) - k_1^2} \right\} A_3,$$

$$= \left\{ \frac{10\,200}{0.2v^4 - 139v^2 + 10\,200} \right\} A_3.$$

At 50 km/h, $v = 17.49$ rad/s.

Thus $A_1 = -0.749\, A_3$. Since $A_3 = 0.1$ m, the amplitude of the trailer vibration is 0.075 m. This motion is π out of phase with the road undulations.

3.1.5 The undamped dynamic vibration absorber

If a single degree of freedom system or mode of a multi-degree of freedom system is excited into resonance, large amplitudes of vibration result with accompanying high dynamic stresses and noise and fatigue problems. In most mechanical systems this is not acceptable.

If neither the excitation frequency nor the natural frequency can conveniently be altered, this resonance condition can often be successfully controlled by adding a further single degree of freedom system. Consider the model of the system shown in Fig. 3.7, where K and M are the effective stiffness and mass of the primary system when vibrating in the troublesome mode.

The absorber is represented by the system with parameters k and m. From section 3.1.4 it can be seen that the equations of motion are

$$M\ddot{X} = -KX - k(X - x) + F \sin vt, \qquad \text{for the primary system}$$

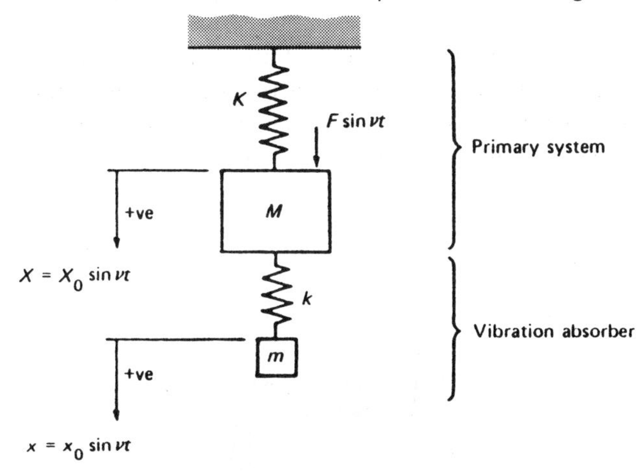

Fig. 3.7. System with undamped vibration absorber.

and

$$m\ddot{x} = k(X - x),$$

for the vibration absorber. Substituting

$$X = X_0 \sin \nu t \quad \text{and} \quad x = x_0 \sin \nu t$$

gives

$$X_0(K + k - M\nu^2) + x_0(-k) = F,$$

and

$$X_0(-k) + x_0(k - m\nu^2) = 0.$$

Thus

$$X_0 = \frac{F(k - m\nu^2)}{\Delta},$$

and

$$x_0 = \frac{Fk}{\Delta},$$

where

$$\Delta = (k - m\nu^2)(K + k - M\nu^2) - \mathrm{k}^2,$$

and $\Delta = 0$ is the frequency equation.

It can be seen that not only does the system now possess two natural frequencies, Ω_1 and Ω_2 instead of one, but by arranging for $k - m\nu^2 = 0$, X_0 can be made zero.

Thus if $\sqrt{(k/m)} = \sqrt{(K/M)}$, the response of the primary system at its original resonance frequency can be made zero. This is the usual tuning arrangement for an undamped absorber because the resonance problem in the primary system is only severe when $\nu \simeq \sqrt{(K/M)}$ rad/s. This is shown in Fig. 3.8.

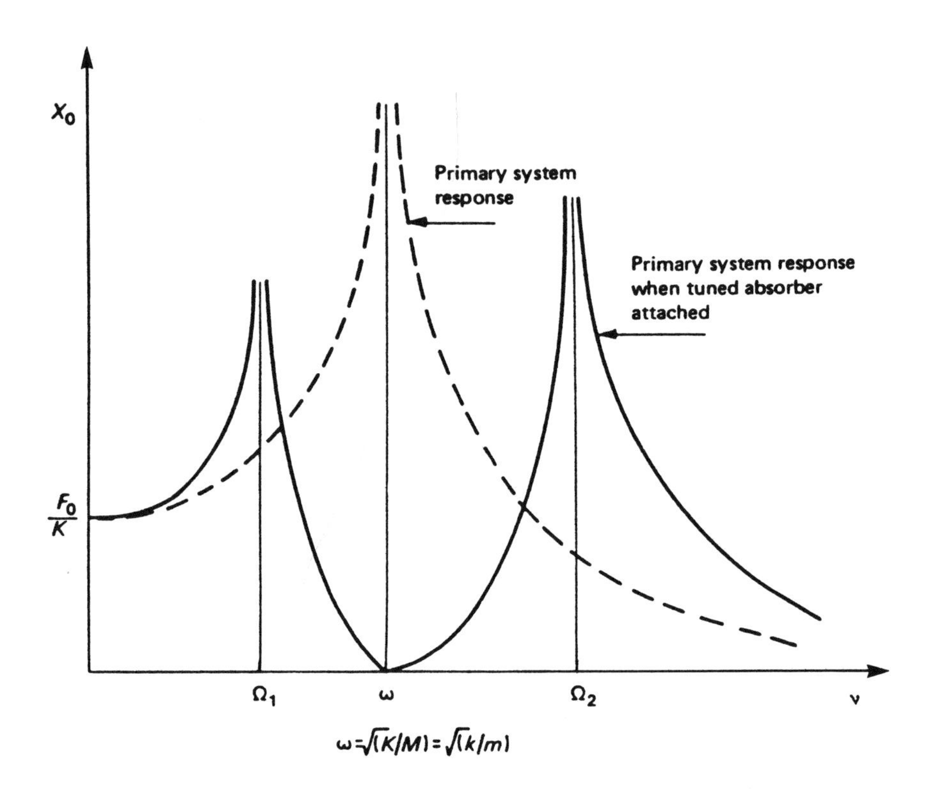

Fig. 3.8. Amplitude-frequency response for system with and without tuned absorber.

When $X_0 = 0$, $x_0 = -F/k$, so that the force in the absorber spring, kx_0 is $-F$; thus the absorber applies a force to the primary system which is equal and opposite to the exciting force. Hence the body in the primary system has a net zero exciting force acting on it and therefore zero vibration amplitude.

If an absorber is correctly tuned $\omega^2 = K/M = k/m$, and if the mass ratio $\mu = m/M$, the frequency equation $\Delta = 0$ is

$$\left(\frac{\nu}{\omega}\right)^4 - (2 + \mu)\left(\frac{\nu}{\omega}\right)^2 + 1 = 0.$$

This is a quadratic equation in $(\nu/\omega)^2$. Hence

$$\left(\frac{\nu}{\omega}\right)^2 = \left(1+\frac{\mu}{2}\right) \pm \sqrt{\left(\mu+\frac{\mu^2}{4}\right)},$$

and the natural frequencies Ω_1 and Ω_2 are found to be

$$\frac{\Omega_{1,2}}{\omega} = \left[\left(1+\frac{\mu}{2}\right) \pm \sqrt{\left(\mu+\frac{\mu^2}{4}\right)}\right]^{1/2}$$

For a small μ, Ω_1 and Ω_2 are very close to each other, and near to ω; increasing μ gives better separation between Ω_1 and Ω_2 as shown in Fig. 3.9.

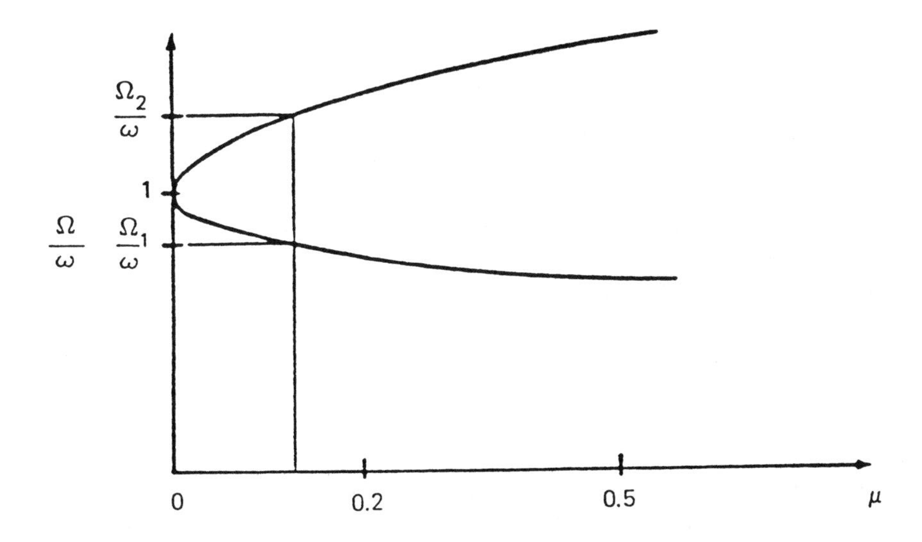

Fig. 3.9. Effect of absorber mass ratio on natural frequencies.

This effect is of great importance in those systems where the excitation frequency may vary; if μ is small, resonances at Ω_1 or Ω_2 may be excited. It should be noted that since

$$\left(\frac{\Omega_1}{\omega}\right)^2 = \left(1+\frac{\mu}{2}\right) - \sqrt{\left(\mu+\frac{\mu^2}{4}\right)}$$

and

$$\left(\frac{\Omega_2}{\omega}\right)^2 = \left(1+\frac{\mu}{2}\right) + \sqrt{\left(\mu+\frac{\mu^2}{4}\right)}, \qquad \text{Then}$$

$$\frac{\Omega_1^2\Omega_2^2}{\omega^4} = \left(1+\frac{\mu}{2}\right)^2 - \left(\mu+\frac{\mu^2}{4}\right) = 1.$$

That is,

$$\Omega_1 \cdot \Omega_2 = \omega^2.$$

Also

$$\left(\frac{\Omega_1}{\omega}\right)^2 + \left(\frac{\Omega_2}{\omega}\right)^2 = 2 + \mu.$$

These relationships are very useful when designing absorbers. If the proximity of Ω_1 and Ω_2 to ω is likely to be a hazard, damping can be added in parallel with the absorber spring, to limit the response at these frequencies. Unfortunately, if damping is added, the response at frequency ω will no longer be zero. A design criterion that has to be carefully considered is the possible fatigue and failure of the absorber spring: this could have severe consequences. In view of this, some damped absorber systems dispense with the absorber spring and sacrifice some of the absorber effectiveness. This is particularly important in torsional systems, where the device is known as a *Lanchester damper*.

Example 22

The figure represents a pump of mass m_1 which rests on springs having a stiffness k_1, so that only vertical motion can occur. Given that the damping is negligible and the mass m_2 is ignored, derive an expression for the frequency of the harmonic disturbing force at which the pump will execute vertical oscillations of very large—theoretically infinite—amplitude.

Given that an undamped dynamic absorber of mass m_2 is then connected to the pump by a spring of stiffness k_2, as shown, prove that the amplitude of the oscillations of the pump is reduced to zero when

$$\frac{k_2}{m_2} = \nu^2,$$

where ν is the natural frequency of the free vibrations of the pump in the absence of the dynamic vibration absorber.

The pump has a mass of 130 kg and rotates at a constant speed of 2400 revolutions per minute but due to a rotating unbalance very large amplitudes of pump vibration on the spring supports result. An undamped vibration absorber is to be fitted so that the nearest natural frequency of the system is at least 20 per cent removed from the running speed of 2400 revolutions per minute. Find the smallest absorber mass necessary and the corresponding spring stiffness.

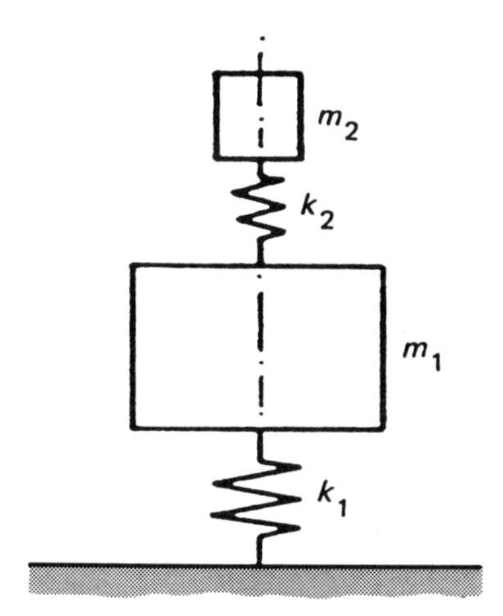

The pump can be modelled as below:

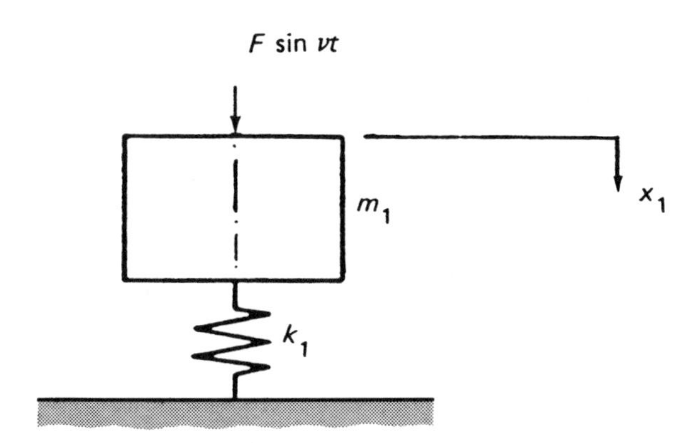

The equation of motion is

$$m_1\ddot{x}_1 + k_1x_1 = F\sin\nu t,$$

so that if

$$x_1 = X_1\sin\nu t, \quad x_1 = \frac{F}{k_1 - m_1\nu^2}.$$

When

$$\nu = \sqrt{\frac{k_1}{m_1}}, \qquad X_1 = \infty.$$

That is, resonance occurs when $\nu = \omega = \sqrt{(k_1/m_1)}$. With a vibration absorber added, the system is

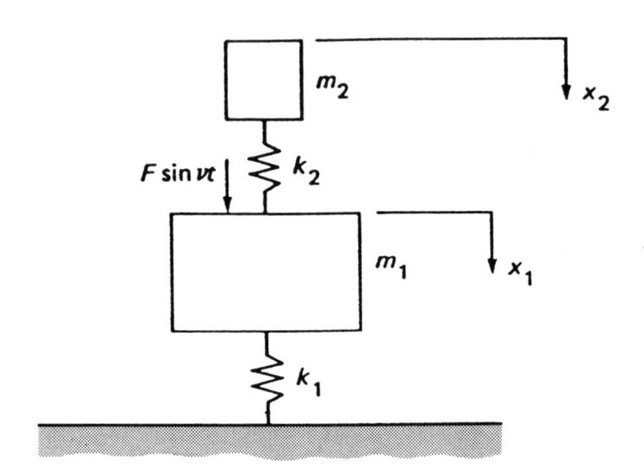

The FBDs are therefore, if $x_2 > x_1$ is assumed,

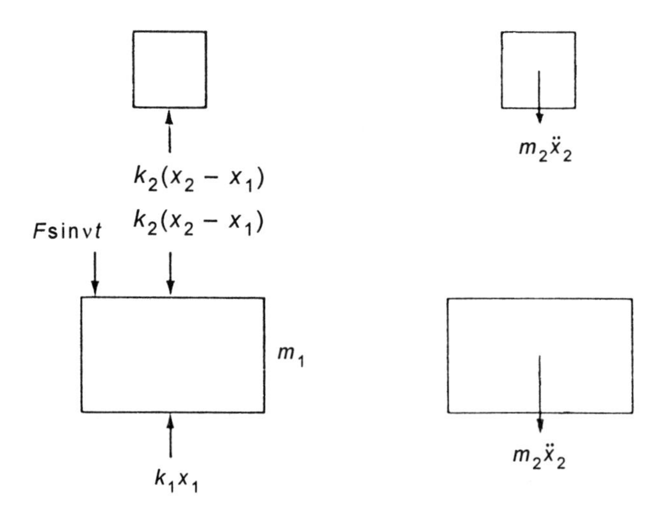

The equations of motion are thus

$$m_2\ddot{x}_2 = -k_2(x_2 - x_1),$$

or

$$m_2\ddot{x}_2 + k_2x_2 - k_2x_1 = 0,$$

and

$$m_1\ddot{x}_1 = k_2(x_2 - x_1) - k_1x_1 + F \sin \nu t,$$

or

$$m_1\ddot{x}_1 + (k_1 + k_2)x_1 - k_2x_2 = F \sin \nu t.$$

Assuming $x_1 = X_1 \sin \nu t$ and $x_2 = X_2 \sin \nu t$, these equations give

$$X_1[k_1 + k_2 - m_1\nu^2] + X_2[-k_2] = F,$$

and

$$X_1[-k_2] + X_2[k_2 - m_2\nu^2] = 0.$$

That is

$$X_1 = \frac{F(k_2 - m_2\nu^2)}{[(k_1 + k_2) - m_1\nu^2][k_2 - m_2\nu^2] - k_2^2}.$$

Thus if

$$\nu^2 = \frac{k_2}{m_2}, \qquad X_1 = 0.$$

Now the frequency equation is

$$[(k_1 + k_2) - m_1\nu^2][k_2 - m_2\nu^2] - k_2^2 = 0.$$

If we put

$$\mu = \frac{m_2}{m_1} = \frac{k_2}{k_1} \quad \text{and} \quad \Omega = \sqrt{\frac{k_2}{m_2}} = \sqrt{\frac{k_1}{m_1}},$$

this becomes

$$\nu^4 - \nu^2(\mu\Omega^2 + 2\Omega^2) + \Omega^4 = 0,$$

or

$$\left(\frac{\nu}{\Omega}\right)^4 - \left(\frac{\nu}{\Omega}\right)^2(2 + \mu) + 1 = 0$$

so that

$$\left(\frac{\nu}{\Omega}\right)^2 = \frac{2 + \mu}{2} \pm \sqrt{\left(\frac{\mu^2 + 4\mu}{4}\right)}.$$

The limiting condition for the smallest absorber mass is $(\nu_1/\Omega) = 0.8$ because then $(\nu_2/\Omega) = 1.25$, which is acceptable. Thus

$$\left(0.64 - \frac{\mu + 2}{2}\right)^2 = \frac{\mu^2 + 4\mu}{4}$$

and

$$\mu = 0.2.$$

Hence

$$m_2 = 0.2 \times 130 = 26 \text{ kg},$$

and

$$k_2 = (80\pi)^2\, m = 1642 \text{ kN/m}.$$

Example 23

A system has a violent resonance at 79 Hz. As a trial remedy a vibration absorber is attached which results in a resonance frequency of 65 Hz. How many such absorbers are required if no resonance is to occur between 60 and 120 Hz?

Since

$$\left(\frac{\Omega_1}{\omega}\right)^2 + \left(\frac{\Omega_2}{\omega}\right)^2 = 2 + \mu,$$

and

$$\Omega_1\Omega_2 = \omega^2$$

in the case of one absorber, with $\omega = 79$ Hz and $\Omega_1 = 65$ Hz,

$$\Omega_2 = \frac{79^2}{65} = 96 \text{ Hz}.$$

Also

$$\left(\frac{65}{79}\right)^2 + \left(\frac{96}{79}\right)^2 = 2 + \mu,$$

so $\mu = 0.154$.

In the case of n absorbers, if

$$\Omega_1 = 60 \text{ Hz}, \qquad \Omega_2 = \frac{79^2}{60} = 104 \text{ Hz (too low)}.$$

So require $\Omega_2 = 120$ Hz and then $\Omega_1 = (79^2/120) = 52$ Hz. Hence

$$\left(\frac{52}{79}\right)^2 + \left(\frac{120}{79}\right)^2 = 2 + \mu'.$$

Thus

$$\mu' = 0.74 = n.\mu \qquad \text{and} \qquad n = \frac{0.74}{0.154} = 4.82.$$

Thus five absorbers are required.

Example 24

A machine tool of mass 3000 kg has a large resonance vibration in the vertical direction at 120 Hz. To control this resonance, an undamped vibration absorber of mass 600 kg is fitted tuned to 120 Hz. Find the frequency range in which the amplitude of the machine vibration is less with the absorber fitted than without.

If (X_0) with absorber $= (X_0)$ without absorber,

$$\frac{F(k - m\nu^2)}{(K + k - M\nu^2)(k - m\nu^2) - k^2} = -\frac{F}{K - M\nu^2} \quad \text{(phase requires } - \text{ve sign).}$$

Multiplying out and putting $\mu = m/M$ gives

$$2\left(\frac{\nu}{\omega}\right)^4 - (4 + \mu)\left(\frac{\nu}{\omega}\right)^2 + 2 = 0.$$

Since

$$\mu = \frac{600}{3000} = 0.2,$$

$$\left(\frac{\nu}{\omega}\right)^2 = \frac{4 + \mu}{4} \pm \tfrac{1}{4}\sqrt{(\mu^2 + 8\mu)} = 1.05 \pm 0.32.$$

Thus

$$\frac{\nu}{\omega} = 1.17 \text{ or } 0.855,$$

and

$$f = 102 \text{ Hz} \quad \text{or} \quad 140 \text{ Hz}, \quad \text{where} \quad \nu = 2\pi f.$$

Thus the required frequency range is 102–140 Hz.

3.1.6 System with viscous damping

If a system possesses damping of a viscous nature, the damping can be modelled similarly to that in the system shown in Fig. 3.10.

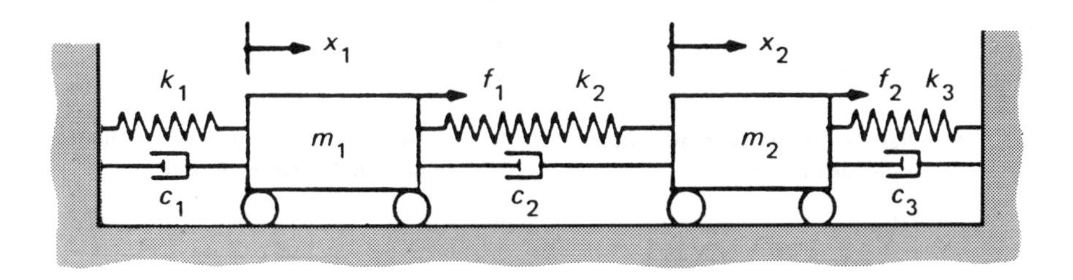

Fig. 3.10. Two degree of freedom viscous damped model with forced excitation.

For this system the equations of motion are

$$m_1\ddot{x}_1 + k_1x_1 + k_2(x_1 - x_2) + c_1\dot{x}_1 + c_2(\dot{x}_1 - \dot{x}_2) = f_1,$$

and

$$m_2\ddot{x}_2 + k_2(x_2 - x_1) + k_3x_2 + c_2(\dot{x}_2 - \dot{x}_1) + c_3\dot{x}_2 = f_2.$$

Solutions of the form $x_1 = A_1\mathrm{e}^{st}$ and $x_2 = A_2\mathrm{e}^{st}$ can be assumed, where the Laplace operator s is equal to $a + \mathrm{j}b$, $\mathrm{j} = \sqrt{(-1)}$, and a and b are real—that is each solution contains an harmonic component of frequency b, and a vibration decay component of damping factor a. By substituting these solutions into the equations of motion a frequency equation of the form

$$s^4 + \alpha s^3 + \beta s^2 + \gamma s + \delta = 0$$

can be deduced, where α, β, γ, and δ are real coefficients. From this equation four roots and thus four values of s can be obtained. In general the roots form two complex conjugate pairs such as $a_1 \pm \mathrm{j}b_1$, and $a_2 \pm \mathrm{j}b_2$. These represent solutions of the form $x = \mathrm{Re}(Xe^{at}\,.\,\mathrm{e}^{\mathrm{j}bt}) = Xe^{at}\cos bt$. That is, the motion of the bodies is harmonic, and decays exponentially with time. The parameters of the system determine the magnitude of the frequency and the decay rate.

It is often convenient to plot these roots on a complex plane as shown in Fig. 3.11. This is known as the s-plane.

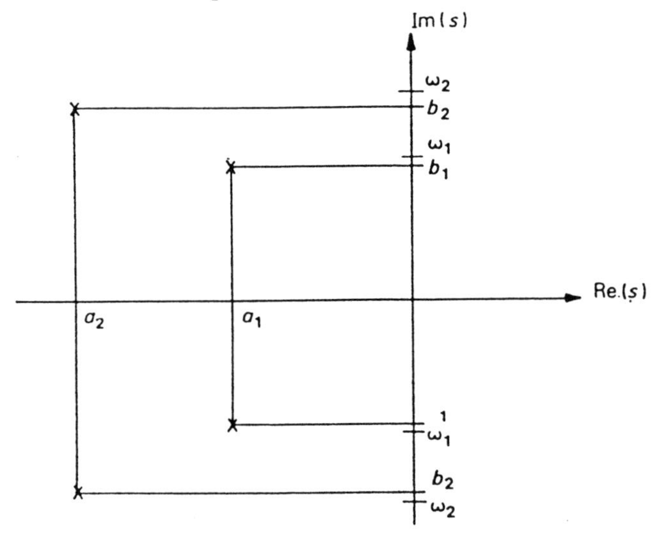

LHS has −ve exponent so oscillation decay. Stable motion

RHS has +ve exponent so oscillation growth. Unstable motion

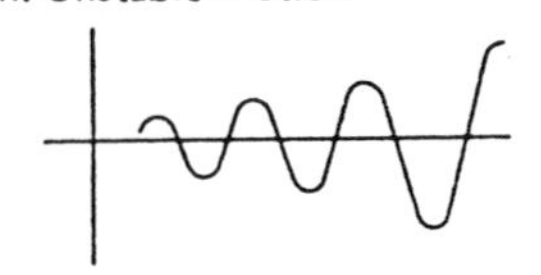

Fig. 3.11. s-plane.

For light damping the damped frequency for each mode is approximately equal to the undamped frequency, that is, $b_1 \simeq \omega_1$ and $b_2 \simeq \omega_2$.

The right-hand side of the s-plane (Re(s) + ve) represents a root with a positive exponent, that is, a term which grows with time, so unstable motion may exist.

The left-hand side contains roots with a negative exponent so stable motion exists. See also Fig. 2.20.

All passive systems have negative real parts and are therefore stable but some systems such as rolling wheels and rockets can be become unstable, and thus it is important that the stability of a system is considered. This can be conveniently done by plotting the roots of the frequency equation on the s-plane, as above. This technique has particular application in the study of control system stability.

3.2 THE VIBRATION OF SYSTEMS WITH MORE THAN TWO DEGREES OF FREEDOM

The vibration analysis of a dynamic system with three or more degrees of freedom can be carried out in the same way as the analysis given above for two degrees of freedom. However, the method becomes tedious for many degrees of freedom, and numerical methods may have to be used to solve the frequency equation. A computer can, of course, be used to solve the frequency equation and determine the corresponding mode shapes. Although computational and computer techniques are extensively used in the analysis of multi-degree of freedom systems, it is essential for the analytical and numerical bases of any program used to be understood, to ensure its relevance to the problem considered, and that the program does not introduce unacceptable approximations and calculation errors. For this reason it is necessary to derive the basic theory and equations for multi-degree of freedom systems. Computational techniques are essential, and widely used, for the analysis of the sophisticated structural models often devised and considered necessary, and computer packages are available for routine analyses. However, considerable economies in writing the analysis and performing the computations can be achieved, by adopting a matrix method for the analysis. Alternatively an energy solution can be obtained by using the Lagrange equation, or some simplification in the analysis achieved by using the receptance technique. The matrix method will be considered first.

3.2.1 The matrix method

The matrix method for analysis is a convenient way of handling several equations of motion. Furthermore, specific information about a system, such as its lowest natural frequency, can be obtained without carrying out a complete and detailed analysis. The matrix method of analysis is particularly important because it forms the basis of many computer solutions to vibration problems. The method can best be demonstrated by means of an example. For a full description of the matrix method see *Mechanical Vibrations: Introduction to Matrix Methods*, by J. M. Prentis & F. A. Leckie (Longmans, 1963).

Example 25

A structure is modelled by the three degree of freedom system shown. Determine the highest natural frequency of free vibration and the associated mode shape.

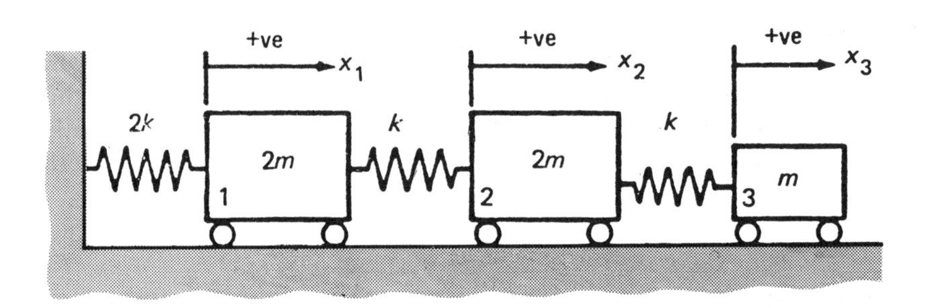

The equations of motion are

$$2m\ddot{x}_1 + 2kx_1 + k(x_1 - x_2) = 0,$$

$$2m\ddot{x}_2 + k(x_2 - x_1) + k(x_2 - x_3) = 0, \qquad \text{and}$$

$$m\ddot{x}_3 + k(x_3 - x_2) = 0.$$

If x_1, x_2 and x_3 take the form $X \sin \omega t$ and $\lambda = m\omega^2/k$, these equations can be written

$$\tfrac{3}{2}X_1 - \tfrac{1}{2}X_2 = \lambda X_1,$$

$$-\tfrac{1}{2}X_1 + X_2 - \tfrac{1}{2}X_3 = \lambda X_2, \qquad \text{and}$$

$$-X_2 + X_3 = \lambda X_3.$$

that is,

$$\begin{bmatrix} 1.5 & -0.5 & 0 \\ -0.5 & 1 & -0.5 \\ 0 & -1 & 1 \end{bmatrix} \begin{Bmatrix} X_1 \\ X_2 \\ X_3 \end{Bmatrix} = \lambda \begin{Bmatrix} X_1 \\ X_2 \\ X_3 \end{Bmatrix}$$

or

$$[\mathbf{S}]\{\mathbf{X}\} = \lambda\{\mathbf{X}\}$$

where $[\mathbf{S}]$ is the system matrix, $\{\mathbf{X}\}$ is a column matrix, and the factor λ is a scalar quantity.

This matrix equation can be solved by an iteration procedure. This procedure is started by assuming a set of deflections for the column matrix $\{\mathbf{X}\}$ and multiplying by $[\mathbf{S}]$; this results in a new column matrix. This matrix is normalized by making one of the amplitudes unity and dividing each term in the column by the particular amplitude which was put equal to unity. The procedure is repeated until the amplitudes stabilize to a definite pattern. Convergence is always to the highest value of λ and its associated column matrix. Since $\lambda = m\omega^2 k$, this means that the highest natural frequency is found. Thus to start the iteration a reasonable assumed mode would be

$$\begin{Bmatrix} X_1 \\ X_2 \\ X_3 \end{Bmatrix} = \begin{Bmatrix} 1 \\ -1 \\ 2 \end{Bmatrix}.$$

Now

$$\begin{bmatrix} 1.5 & -0.5 & 0 \\ -0.5 & 1 & -0.5 \\ 0 & -1 & 1 \end{bmatrix} \begin{Bmatrix} 1 \\ -1 \\ 2 \end{Bmatrix} = \begin{Bmatrix} 2 \\ -2.5 \\ 3 \end{Bmatrix} = 3 \begin{Bmatrix} 0.67 \\ -0.83 \\ 1.00 \end{Bmatrix}$$

Using this new column matrix gives

$$\begin{bmatrix} 1.5 & -0.5 & 0 \\ -0.5 & 1 & -0.5 \\ 0 & -1 & 1 \end{bmatrix} \begin{Bmatrix} 0.67 \\ -0.83 \\ 1.00 \end{Bmatrix} = \begin{Bmatrix} 1.415 \\ -1.665 \\ 1.83 \end{Bmatrix} = 1.83 \begin{Bmatrix} 0.77 \\ -0.91 \\ 1.00 \end{Bmatrix}$$

and eventually, by repeating the process the following is obtained:

$$\begin{bmatrix} 1.5 & -0.5 & 0 \\ -0.5 & 1 & -0.5 \\ 0 & -1 & 1 \end{bmatrix} \begin{Bmatrix} 1 \\ -1 \\ 1 \end{Bmatrix} = 2 \begin{Bmatrix} 1 \\ -1 \\ 1 \end{Bmatrix}$$

Hence $\lambda = 2$ and $\omega^2 = 2k/m$. λ is an *eigenvalue* of $[\mathbf{S}]$, and the associated value of $\{\mathbf{X}\}$ is the corresponding *eigenvector* of $[\mathbf{S}]$. The eigenvector gives the mode shape.

Thus the highest natural frequency is $1/2\pi \sqrt{(2k/m)}$ Hz, and the associated mode shape is $1:-1:1$. Thus if $X_1 = 1$, $X_2 = -1$ and $X_3 = 1$.

If the lowest natural frequency is required, it can be found from the lowest eigenvalue. This can be obtained directly by inverting $[\mathbf{S}]$ and premultiplying $[\mathbf{S}]\{\mathbf{X}\} = \lambda\{\mathbf{X}\}$ by $\lambda^{-1}[\mathbf{S}]^{-1}$.

Thus $[\mathbf{S}]^{-1}\{\mathbf{X}\} = \lambda^{-1}\{\mathbf{X}\}$. Iteration of this equation yields the largest value of λ^{-1} and hence the lowest natural frequency. A reasonable assumed mode for the first iteration would be

$$\begin{Bmatrix} 1 \\ 1 \\ 2 \end{Bmatrix}.$$

Alternatively, the lowest eigenvalue can be found from the flexibility matrix. The flexibility matrix is written in terms of the influence coefficients. The influence coefficient α_{pq} of a system is the deflection (or rotation) at the point p due to a unit force (or moment) applied at a point q. Thus, since the force each body applies is the product of its mass and acceleration.

$$X_1 = \alpha_{11}\, 2mX_1\omega^2 + \alpha_{12}\, 2mX_2\omega^2 + \alpha_{13} mX_3\omega^2,$$

$$X_2 = \alpha_{21}\, 2mX_1\omega^2 + \alpha_{22}\, 2mX_2\omega^2 + \alpha_{23} mX_3\omega^2,$$

and

$$X_3 = \alpha_{31}\,2mX_1\omega^2 + \alpha_{32}\,2mX_2\omega^2 + \alpha_{33}mX_3\omega^2.$$

or

$$\begin{bmatrix} 2\alpha_{11} & 2\alpha_{12} & \alpha_{13} \\ 2\alpha_{21} & 2\alpha_{22} & \alpha_{23} \\ 2\alpha_{31} & 2\alpha_{32} & \alpha_{33} \end{bmatrix} \begin{Bmatrix} X_1 \\ X_2 \\ X_3 \end{Bmatrix} = \frac{1}{m\omega^2} \begin{Bmatrix} X_1 \\ X_2 \\ X_3 \end{Bmatrix}.$$

The influence coefficients are calculated by applying a unit force or moment to each body in turn. Since the same unit force acts between the support and its point of application, the displacement of the point of application of the force is the sum of the extensions of the springs extended. The displacements of all points beyond the point of application of the force are the same.

Thus

$$\alpha_{11} = \alpha_{12} = \alpha_{13} = \alpha_{21} = \alpha_{31} = \frac{1}{2k},$$

$$\alpha_{22} = \alpha_{23} = \alpha_{32} = \frac{1}{2k} + \frac{1}{k} = \frac{3}{2k},$$

and

$$\alpha_{33} = \frac{1}{2k} + \frac{1}{k} + \frac{1}{k} = \frac{5}{2k}.$$

Iteration causes the eigenvalue $k/m\omega^2$ to converge to its highest value, and hence the lowest natural frequency is found. The other natural frequencies of the system can be found by applying the orthogonality relation between the principal modes of vibration.

3.2.1.1 Orthogonality of the principal modes of vibration

Consider a linear elastic system that has n degrees of freedom, n natural frequencies, and n principal modes.

The orthogonality relation between the principal modes of vibration for an n degree of freedom system is

$$\sum_{i=1}^{n} m_i A_i(r) A_i(s) = 0,$$

where $A_i(r)$ are the amplitudes corresponding to the rth mode, and $A_i(s)$ are the amplitudes corresponding to the sth mode.

This relationship is used to sweep unwanted modes from the system matrix, as illustrated in the following example.

Example 26

Consider the three degree of freedom model of a system shown.

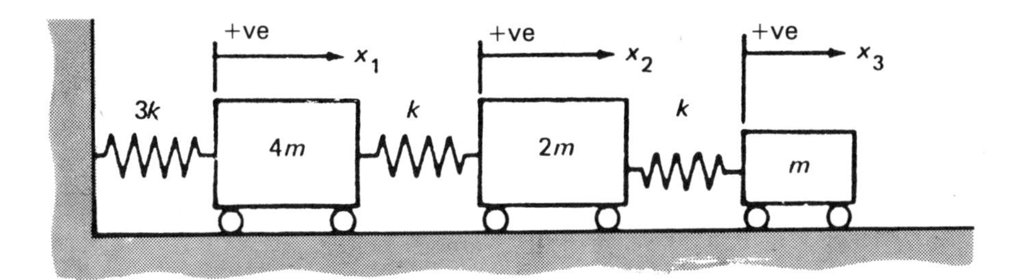

The equations of motion in terms of the influence coefficients are

$$X_1 = 4\alpha_{11}mX_1\omega^2 + 2\alpha_{12}mX_2\omega^2 + \alpha_{13}mX_3\omega^2,$$

$$X_2 = 4\alpha_{21}mX_1\omega^2 + 2\alpha_{22}mX_2\omega^2 + \alpha_{23}mX_3\omega^2,$$

and

$$X_3 = 4\alpha_{31}mX_1\omega^2 + 2\alpha_{32}mX_2\omega^2 + \alpha_{33}mX_3\omega^2.$$

That is,

$$\begin{Bmatrix} X_1 \\ X_2 \\ X_3 \end{Bmatrix} = \omega^2 m \begin{bmatrix} 4\alpha_{11} & 2\alpha_{12} & \alpha_{13} \\ 4\alpha_{21} & 2\alpha_{22} & \alpha_{23} \\ 4\alpha_{31} & 2\alpha_{32} & \alpha_{33} \end{bmatrix} \begin{Bmatrix} X_1 \\ X_2 \\ X_3 \end{Bmatrix}.$$

Now,

$$\alpha_{11} = \alpha_{12} = \alpha_{21} = \alpha_{13} = \alpha_{31} = \frac{1}{3k},$$

$$\alpha_{22} = \alpha_{32} = \alpha_{23} = \frac{4}{3k},$$

and

$$\alpha_{33} = \frac{7}{3k}.$$

Hence,

$$\begin{Bmatrix} X_1 \\ X_2 \\ X_3 \end{Bmatrix} = \frac{\omega^2 m}{3k} \begin{bmatrix} 4 & 2 & 1 \\ 4 & 8 & 4 \\ 4 & 8 & 7 \end{bmatrix} \begin{Bmatrix} X_1 \\ X_2 \\ X_3 \end{Bmatrix}.$$

To start the iteration a reasonable estimate for the first mode is

$$\begin{Bmatrix} 1 \\ 2 \\ 4 \end{Bmatrix};$$

this is inversely proportional to the mass ratio of the bodies.

Eventually iteration for the first mode gives

$$\begin{Bmatrix} 1.0 \\ 3.2 \\ 4.0 \end{Bmatrix} = \frac{14.4\ m\omega^2}{3k} \begin{Bmatrix} 1.0 \\ 3.18 \\ 4.0 \end{Bmatrix},$$

or $\omega_1 = 0.46\sqrt{(k/m)}$ rad/s.

To obtain the second principal mode, use the orthogonality relation to remove the first mode from the system matrix:

$$m_1 A_1 A_2 + m_2 B_1 B_2 + m_3 C_1 C_2 = 0.$$

Thus

$$4m(1.0)A_2 + 2m(3.18)B_2 + m(4.0)C_2 = 0,$$

or $A_2 = -1.59\ B_2 - C_2$, since the first mode is

$$\begin{Bmatrix} 1.0 \\ 3.18 \\ 4.0 \end{Bmatrix}$$

Hence, rounding 1.59 up to 1.6,

$$\begin{Bmatrix} A_2 \\ B_2 \\ C_2 \end{Bmatrix} = \begin{bmatrix} 0 & -1.6 & -1 \\ 0 & 1 & 0 \\ 0 & 0 & 1 \end{bmatrix} \begin{Bmatrix} A_2 \\ B_2 \\ C_2 \end{Bmatrix}.$$

When this sweeping matrix is combined with the original matrix equation, iteration makes convergence to the second mode take place because the first mode is swept out. Thus,

$$\begin{Bmatrix} X_1 \\ X_2 \\ X_3 \end{Bmatrix} = \frac{\omega^2 m}{3k} \begin{bmatrix} 4 & 2 & 1 \\ 4 & 8 & 4 \\ 4 & 8 & 7 \end{bmatrix} \begin{bmatrix} 0 & -1.6 & -1 \\ 0 & 1 & 0 \\ 0 & 0 & 1 \end{bmatrix} \begin{Bmatrix} X_1 \\ X_2 \\ X_3 \end{Bmatrix},$$

$$= \frac{\omega^2 m}{3k} \begin{bmatrix} 0 & -4.4 & -3 \\ 0 & 1.6 & 0 \\ 0 & 1.6 & 3 \end{bmatrix} \begin{Bmatrix} X_1 \\ X_2 \\ X_3 \end{Bmatrix}.$$

Now estimate the second mode as

$$\begin{Bmatrix} 1 \\ 0 \\ -1 \end{Bmatrix}$$

and iterate:

$$\begin{Bmatrix} 1 \\ 0 \\ -1 \end{Bmatrix} = \frac{\omega^2 m}{3k} \begin{bmatrix} 0 & -4.4 & -3 \\ 0 & 1.6 & 0 \\ 0 & 1.6 & 3 \end{bmatrix} \begin{Bmatrix} 1 \\ 0 \\ -1 \end{Bmatrix} = \frac{m\omega^2}{k} \begin{Bmatrix} 1 \\ 0 \\ -1 \end{Bmatrix}.$$

Hence $\omega_2 = \sqrt{(k/m)}$ rad/s, and the second mode was evidently estimated correctly as 1:0: − 1.

To obtain the third mode, write the orthogonality relation as

$$m_1A_2A_3 + m_2B_2B_3 + m_3C_2C_3 = 0,$$

and

$$m_1A_1A_3 + m_2B_1B_3 + m_3C_1C_3 = 0.$$

Substitute

$$A_1 = 1.0, \quad B_1 = 3.18, \quad C_1 = 4.0,$$

and

$$A_2 = 1.0, \quad B_2 = 0, \quad C_2 = -1.0,$$

as found above. Hence

$$\begin{Bmatrix} A_3 \\ B_3 \\ C_3 \end{Bmatrix} = \begin{bmatrix} 0 & 0 & 0.25 \\ 0 & 0 & -0.78 \\ 0 & 0 & 1 \end{bmatrix} \begin{Bmatrix} A_3 \\ B_3 \\ C_3 \end{Bmatrix}.$$

When this sweeping matrix is combined with the equation for the second mode the second mode is removed, so that it yields the third mode on iteration:

$$\begin{Bmatrix} X_1 \\ X_2 \\ X_3 \end{Bmatrix} = \frac{\omega^2 m}{3k} \begin{bmatrix} 0 & -4.4 & -3 \\ 0 & 1.6 & 0 \\ 0 & 1.6 & 3 \end{bmatrix} \begin{bmatrix} 0 & 0 & 0.25 \\ 0 & 0 & -0.78 \\ 0 & 0 & 1 \end{bmatrix} \begin{Bmatrix} X_1 \\ X_2 \\ X_3 \end{Bmatrix},$$

$$= \frac{\omega^2 m}{3k} \begin{bmatrix} 0 & 0 & 0.43 \\ 0 & 0 & -1.25 \\ 0 & 0 & 1.75 \end{bmatrix} \begin{Bmatrix} X_1 \\ X_2 \\ X_3 \end{Bmatrix},$$

or

$$\begin{Bmatrix} X_1 \\ X_2 \\ X_3 \end{Bmatrix} = 1.75 . \frac{\omega^2 m}{3k} \begin{bmatrix} 0 & 0 & 0.25 \\ 0 & 0 & -0.72 \\ 0 & 0 & 1 \end{bmatrix} \begin{Bmatrix} X_1 \\ X_2 \\ X_3 \end{Bmatrix}.$$

An estimate for the third mode shape now has to be made and the iteration procedure carried out once more. In this way the third mode eigenvector is found to be

$$\begin{Bmatrix} 0.25 \\ -0.72 \\ 1.0 \end{Bmatrix},$$

and $\omega_3 = 1.32\sqrt{(k/m)}$ rad/s.

The convergence for higher modes becomes more critical if impurities and rounding-off errors are introduced by using the sweeping matrices. It is well to check the highest mode by the inversion of the original matrix equation, which should be equal to the equation formulated in terms of the stiffness influence coefficients.

3.2.2 The Lagrange equation

Consideration of the energy in a dynamic system together with the use of the Lagrange equation is a very powerful method of analysis for certain physically complex systems.

It is an energy method which allows the equations of motion to be written in terms of any set of *generalized coordinates*. Generalized coordinates are a set of independent parameters which completely specify the system location and which are independent of any constraints. The fundamental form of Lagrange's equation can be written in terms of the generalized coordinates q_i as follows:

$$\frac{\mathrm{d}}{\mathrm{d}t}\left(\frac{\partial(T)}{\partial \dot{q}_i}\right) - \frac{\partial(T)}{\partial q_i} + \frac{\partial(V)}{\partial q_i} + \frac{\partial(DE)}{\partial \dot{q}_i} = Q_i,$$

where T is the total kinetic energy of the system, V is the total potential energy of the system, DE is the energy dissipation function when the damping linear (it is half the rate at which energy is dissipated so that for viscous damping $DE = \frac{1}{2}c\dot{x}^2$), Q_i is a generalized external force (or non-linear damping force) acting on the system, and q_i is a generalized coordinate that describes the position of the system.

The subscript i denotes n equations for an n degree of freedom system, so that the Lagrange equation yields as many equations of motion as there are degrees of freedom.

For a free conservative system Q_i and DE are both zero, so that

$$\frac{\mathrm{d}}{\mathrm{d}t}\left(\frac{\partial(T)}{\partial \dot{q}_i}\right) - \frac{\partial(T)}{\partial q_i} + \frac{\partial(V)}{\partial q_i} = 0.$$

The full derivation of the Lagrange equation can be found in *Vibration Theory and Applications* by W. T. Thomson (Allen & Unwin, 1989).

Example 27

A solid cylinder has a mass M and radius R. Pinned to the axis of the cylinder is an arm of length l which carries a bob of mass m. Obtain the natural frequency of free vibration of the bob. The cylinder is free to roll on the fixed horizontal surface shown.

The generalized coordinates are x_1 and x_2. They completely specify the position of the system and are independent of any constraints.

$$T = \tfrac{1}{2}M\dot{x}_1^2 + \tfrac{1}{2}(\tfrac{1}{2}MR^2)\dot{\theta}^2 + \tfrac{1}{2}m\dot{x}_2^2$$
$$= \tfrac{1}{2}M\dot{x}_1^2 + \tfrac{1}{2}(\tfrac{1}{2}M\dot{x}_1^2) + \tfrac{1}{2}m\dot{x}_2^2.$$
$$V = mgl(1 - \cos\phi) = (mgl/2)\phi^2 = (mg/2l)(x_2 - x_1)^2,$$

for small values of ϕ. Apply the Lagrange equation with $q_i = x_1$:

$$(\mathrm{d}/\mathrm{d}t)(\partial T/\partial \dot{x}_1) = M\ddot{x}_1 + \tfrac{1}{2}M\ddot{x}_1$$
$$\partial V/\partial x_1 = (mg/2l)(-2x_2 + 2x_1).$$

Hence $\frac{3}{2}M\ddot{x}_1 + (mg/l)(x_1 - x_2) = 0$ is an equation of motion.

Apply the Lagrange equation with $q_i = x_2$:

$$(\mathrm{d}/\mathrm{d}t)(\partial T/\partial \dot{x}_2) = m\ddot{x}_2$$
$$\partial V/\partial x_2 = (mg/2l)(2x_2 - 2x_1).$$

Hence

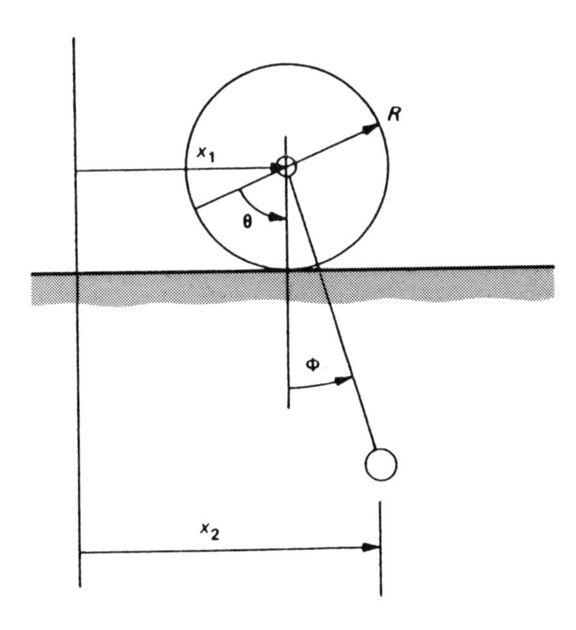

$$m\ddot{x}_2 + (mg/l)(x_2 - x_1) = 0 \text{ is an equation of motion.}$$

These equations of motion can be solved by assuming that $x_1 = X_1 \sin\omega t$ and $x_2 = X_2 \sin\omega t$. Then

$$X_1((mg/l) - (3M/2)\omega^2) + X_2(-mg/l) = 0,$$

and $\quad X_1(-mg/l) + X_2((mg/l) - m\omega^2) = 0.$

The frequency equation is therefore

$$(3M/2)\omega^4 - (g/l)\omega^2(m + (3M/2)) = 0.$$

Thus either $\omega = 0$, or $\omega = \sqrt{((1 + 2m/3M)g/l)}$ rad/s, and $X_1/X_2 = -2m/3M$.

Example 28

Find the equations of motion for free vibration of the triple pendulum shown overleaf.

$$T = \tfrac{1}{2}m_1L^2\dot{\theta}_1^2 + \tfrac{1}{2}m_2L^2\dot{\theta}_2^2 + \tfrac{1}{2}m_3L^2\dot{\theta}_3^2$$

$$V = m_1gL(1 - \cos\theta_1) + m_2gL(1 - \cos\theta_2) + m_3gL(1 - \cos\theta_3)$$
$$+ \tfrac{1}{2}k_1(a\sin\theta_2 - a\sin\theta_1)^2 + \tfrac{1}{2}k_2(a\sin\theta_3 - a\sin\theta_2)^2$$

Hence, by applying the Lagrange equation with $q_i = \theta_1, \theta_2, \theta_3$ in turn, the equations of motion are obtained. Assume θ is small so that $\sin\theta \simeq \theta$, and $\cos\theta \simeq 1 - \theta^2/2$.

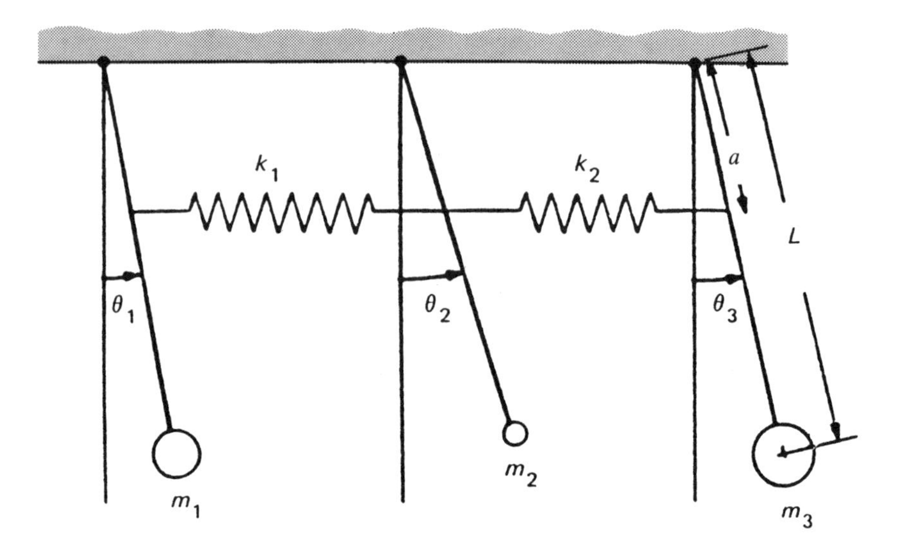

Example 29

To isolate a structure from the vibration generated by a machine, the machine is mounted on a large block. The block is supported on springs as shown. Find the equations which describe the motion of the block in the plane of the figure.

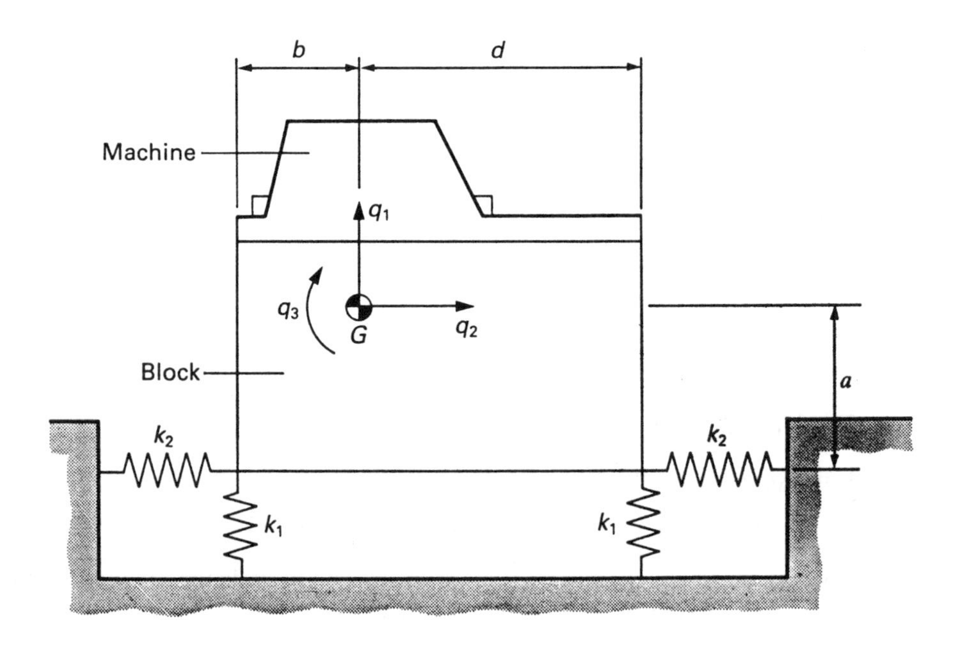

The coordinates used to describe the motion are q_1, q_2, and q_3. These are generalized coordinates because they completely specify the position of the system and are independent of any constraints. If the mass of the block and machine is M, and the total mass moment of inertia about G is I_G, then

$$T = \tfrac{1}{2}M\dot{q}_1^2 + \tfrac{1}{2}M\dot{q}_2^2 + \tfrac{1}{2}I_G\dot{q}_3^2, \text{ and}$$

$$\begin{aligned} V &= \text{strain energy stored in the springs} \\ &= \tfrac{1}{2}k_1(q_1 + bq_3)^2 + \tfrac{1}{2}k_1(q_1 - \mathrm{d}q_3)^2 \\ &\quad + \tfrac{1}{2}k_2(q_2 - aq_3)^2 + \tfrac{1}{2}k_2(q_2 - aq_3)^2. \end{aligned}$$

Now apply the Lagrange equation with $q_i = q_1$.

$$\frac{\partial T}{\partial q_1} = 0.$$

$$\frac{\partial T}{\partial \dot{q}_1} = M\dot{q}_1, \quad \text{so } \frac{\mathrm{d}}{\mathrm{d}t} \times \frac{\partial T}{\dot{q}_1} = M\ddot{q}_1,$$

and

$$\frac{\partial V}{\partial q_1} = k_1(q_1 + bq_3) + k_1(q_1 - \mathrm{d}q_3).$$

Thus the first equation of motion is

$$M\ddot{q}_1 + 2k_1q_1 + k_1(b - d)q_3 = 0.$$

Similarly by putting $q_i = q_2$ and $q_i = q_3$, the other equations of motion are obtained as

$$M\ddot{q}_2 + 2k_1q_2 - 2ak_2q_3 = 0,$$

and

$$I_G\ddot{q}_3 + k_1(b - d)q_1 - 2ak_2q_2 + ((b^2 + d^2)k_1 + 2a^2k_2)q_3 = 0.$$

The system therefore has three coordinate-coupled equations of motion. The natural frequencies can be found by substituting $q_i = A_i \sin \omega t$, and solving the resulting frequency equation. It is usually desirable to have all natural frequencies low so that the transmissibility is small throughout the range of frequencies excited.

3.2.3 Receptances

Some simplification in the analysis of multi-degree of freedom undamped dynamic systems can often be gained by using receptances, particularly if only the natural frequencies are required. If an *harmonic* force $F \sin \nu t$ acts at some point in a system so that the system responds at frequency ν, and the point of application of the force has a displacement $x = X \sin \nu t$, then if the equations of motion are linear, $x = \alpha F \sin \nu t$, where α, which is a function of the system parameters and ν but not a function of F, is known as the direct receptance at x. If the displacement is determined at some point other than that at which the force is applied, α is known as the transfer or cross receptance.

It can be seen that the frequency at which a receptance becomes infinite is a natural frequency of the system. Receptances can be written for rotational and translational coordinates in a system, that is the slope and deflection at a point.

Thus, if a body of mass m is subjected to a force $F \sin \nu t$ and the response of the body is $x = X \sin \nu t$,

$$F \sin \nu t = m\ddot{x} = m(-X\nu^2 \sin \nu t) = -m\nu^2 x.$$

Thus

$$x = -\frac{1}{m\nu^2} F \sin \nu t,$$

and

$$\alpha = -\frac{1}{m\nu^2}.$$

This is the direct receptance of a rigid body.

For a spring, $\alpha = 1/k$. This is the direct receptance of a spring.

In an undamped single degree of freedom model of a system, the equation of motion is

$$m\ddot{x} + kx = F \sin \nu t,$$

If $x = X \sin \nu t$, $\alpha = 1/(k - m\nu^2)$. This is the direct receptance of a single degree of freedom system.

In more complicated systems, it is necessary to be able to distinguish between direct and cross receptances and to specify the points at which the receptances are calculated. This is done by using subscripts. The first subscript indicates the coordinate at which the response is measured, and the second indicates that at which the force is applied. Thus α_{pq}, which is a cross receptance, is the response at p divided by the harmonic force applied at q, and α_{pp} and α_{qq} are direct receptances at p and q respectively.

Consider the two degree of freedom system shown in Fig. 3.12

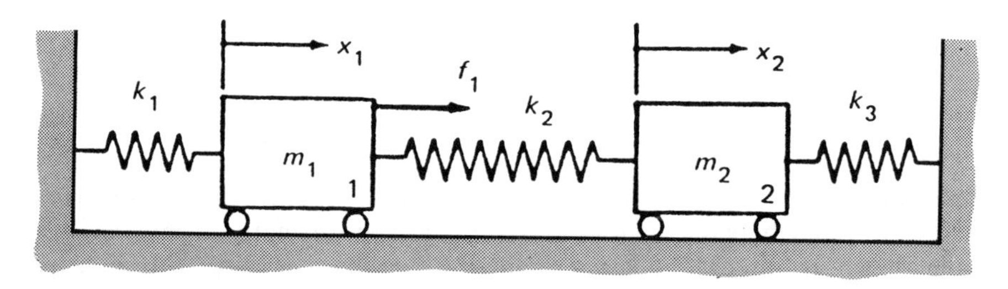

Fig. 3.12. Two degree of freedom system with forced excitation.

The equations of motion are

$$m_1\ddot{x}_1 + (k_1 + k_2)x_1 - k_2x_2 = f_1,$$

and

$$m_2\ddot{x}_2 + (k_2 + k_3)x_2 - k_2x_1 = 0.$$

Let $f_1 = F_1 \sin \nu t$, and assume that $x_1 = X_1 \sin \nu t$ and $x_2 = X_2 \sin \nu t$. Substituting into the equations of motion gives

$$(k_1 + k_2 - m_1\nu^2)X_1 + (-k_2)X_2 = F_1,$$

and

$$(-k_2)X_1 + (k_2 + k_3 - m_2\nu^2)X_2 = 0.$$

Thus

$$\alpha_{11} = \frac{X_1}{F_1} = \frac{k_2 + k_3 - m_2\nu^2}{\Delta},$$

where

$$\Delta = (k_1 + k_2 - m_1\nu^2)(k_2 + k_3 - m_2\nu^2) - k_2^2.$$

α_{11} is a direct receptance, and $\Delta = 0$ is the frequency equation.

Also the cross receptance

$$\alpha_{21} = \frac{X_2}{F_1} = \frac{k_2}{\Delta}.$$

This system has two more receptances, the responses due to f_2 applied to the second body. Thus α_{12} and α_{22} may be found. It is a fundamental property that $\alpha_{12} = \alpha_{21}$ (principle of reciprocity) so that symmetrical matrices result.

A general statement of the system response is

$$X_1 = \alpha_{11}F_1 + \alpha_{12}F_2,$$

and

$$X_2 = \alpha_{21}F_1 + \alpha_{22}F_2,$$

That is,

$$\begin{Bmatrix} X_1 \\ X_2 \end{Bmatrix} = \begin{bmatrix} \alpha_{11} & \alpha_{12} \\ \alpha_{21} & \alpha_{22} \end{bmatrix} \begin{Bmatrix} F_1 \\ F_2 \end{Bmatrix}.$$

Some simplification in the analysis of complex systems can be achieved by considering the complex system to be a number of simple systems (whose receptances are known) linked together by using conditions of compatibility and equilibrium. The method is to break the complex system down into subsystems and analyse each subsystem separately. Find each subsystem receptance at the point where it is connected to the adjacent subsystem, and 'join' all subsystems together, using the conditions of compatibility and equilibrium.

For example, to find the direct receptance γ_{11} of a dynamic system C at a single coordinate x_1 the system is considered as two subsystems A and B, as shown in Fig. 3.13.

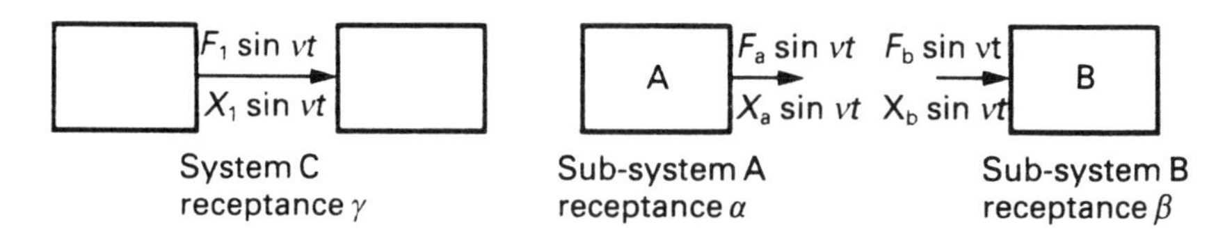

Fig. 3.13. Dynamic systems.

By definition

$$\gamma_{11} = \frac{X_1}{F_1}, \qquad \alpha_{11} = \frac{X_a}{F_a} \quad \text{and} \quad \beta_{11} = \frac{X_b}{F_b}.$$

Because the systems are connected.

$$X_a = X_b = X_1, \quad \text{(compatibility)}$$

and

$$F_1 = F_a + F_b, \quad \text{(equilibrium)}.$$

Hence

$$\frac{1}{\gamma_{11}} = \frac{1}{\alpha_{11}} + \frac{1}{\beta_{11}},$$

that is, the system receptance γ can be found from the receptances of the subsystems.

In a simple spring–body system, subsystems A and B are the spring and body respectively. Hence $\alpha_{11} = 1/k$ and $\beta_{11} = -1/mv^2$ and $1/\gamma_{11} = k - mv^2$. as above.

The frequency equation is $\alpha_{11} + \beta_{11} = 0$, because this condition makes $\gamma_{11} = \infty$.

Consider the undamped dynamic vibration absorber application shown in Fig. 3.14. The system is split into subsystems A and B.

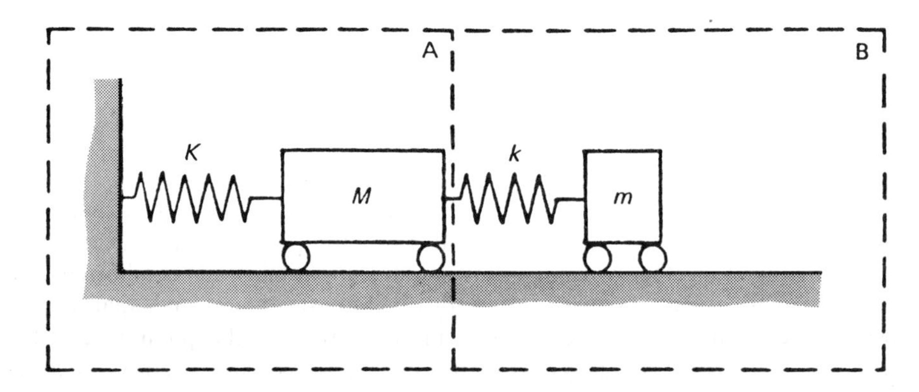

Fig. 3.14. Absorber sub systems.

For subsystem A,

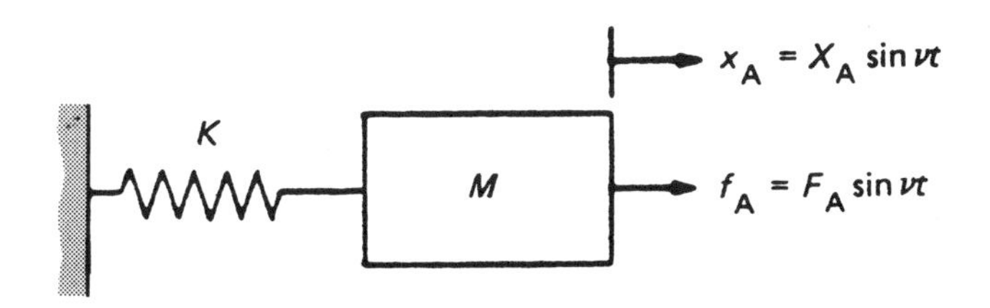

$$f_A = M\ddot{x}_A + Kx_A; \qquad \text{hence} \quad \alpha = 1/(K - M\nu^2).$$

For subsystem B,

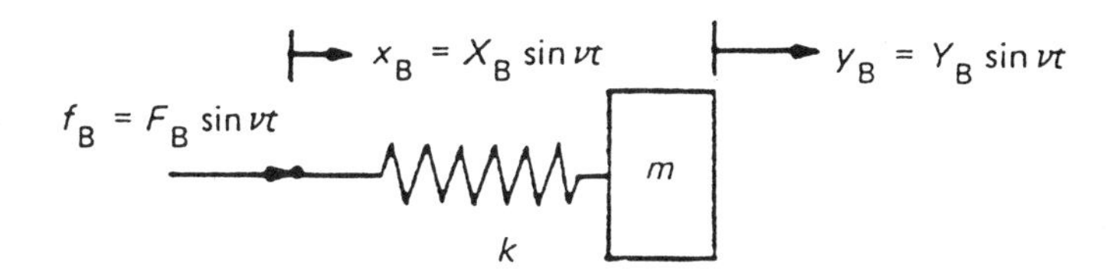

$$f_B = k(x_B - y_B) = m\ddot{y}_B = -m\nu^2 Y_B; \qquad \text{hence} \quad \beta = -(k - m\nu^2)/km\nu^2.$$

Thus the frequency equation $\alpha + \beta = 0$ gives

$$Mm\nu^4 - (mK + Mk + mk)\nu^2 + Kk = 0.$$

It is often convenient to solve the frequency equation $\alpha + \beta = 0$, or $\alpha = -\beta$, by a graphical method. In the case of the absorber, both α and $-\beta$ can be plotted as a function of ν, and the intersections give the natural frequencies Ω_1 and Ω_2 (Fig. 3.15).

It can be seen that the effect of adding different absorbers to the primary system can readily be determined without re-analysing the whole system. It is merely necessary to sketch the receptance of each absorber on Fig. 3.15 to find Ω_1 and Ω_2 for the complete system.

The receptance technique is particularly useful when it is required to investigate the effects of adding a dynamic system to an existing system, for example an extra floor, or an air-conditioning plant to a building. Once the receptance of the original system is known, it is only necessary to analyse the additional system, and then to include this in the original analysis. Furthermore, sometimes the receptances of dynamic systems are measured, and available only in graphical form. Example 30.

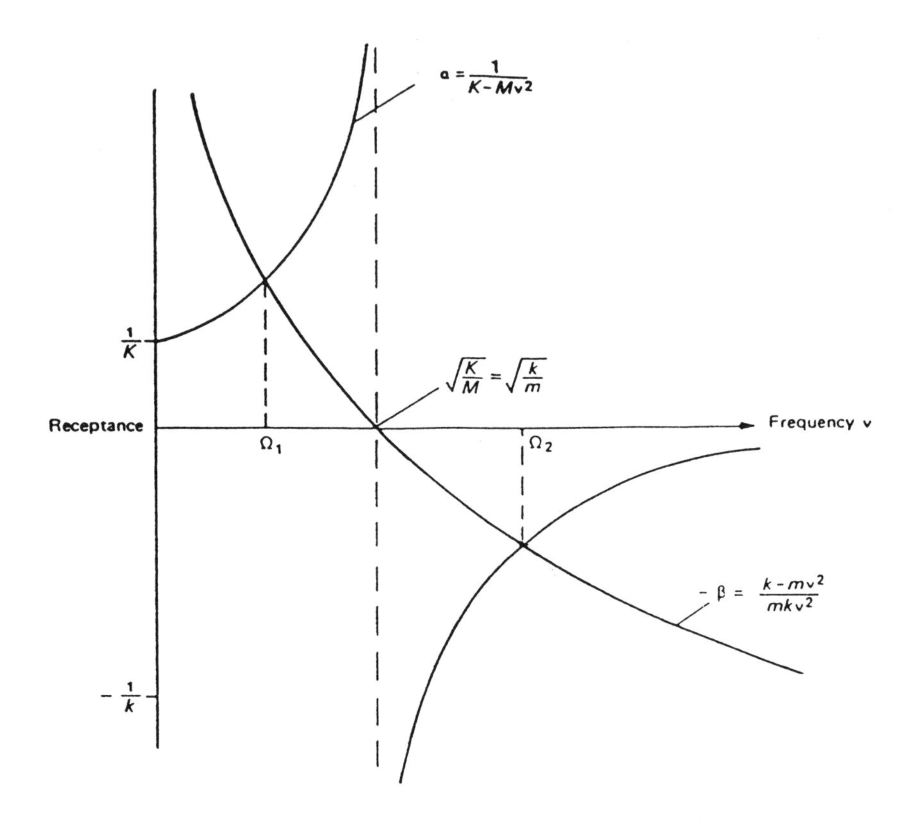

Fig. 3.15. Vibration absorber receptance.

Some subsystems, such as those shown in Fig. 3.16, are linked by two coordinates, for example deflection and slope at the common point.

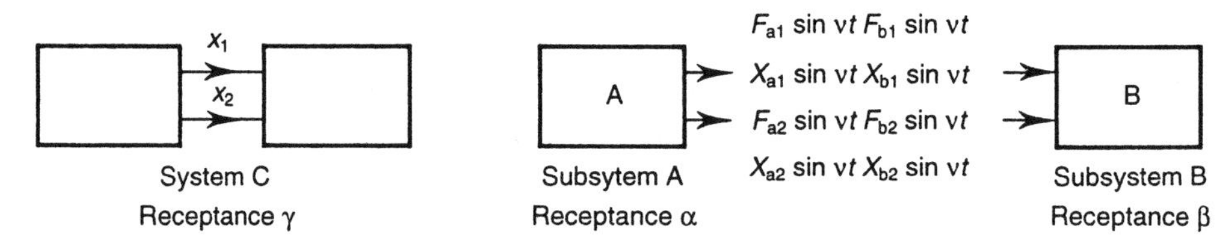

Fig. 3.16. Applied forces and system responses.

Now in this case,

$$X_{a1} = \alpha_{11}F_{a1} + \alpha_{12}F_{a2},$$

$$X_{a2} = \alpha_{21}F_{a1} + \alpha_{22}F_{a2},$$

$$X_{b1} = \beta_{11}F_{b1} + \beta_{12}F_{b2},$$

and

$$X_{b2} = \beta_{21}F_{b1} + \beta_{22}F_{b2}.$$

The applied forces or moments are $F_1 \sin vt$ and $F_2 \sin vt$ where

$$F_1 = F_{a1} + F_{b1},$$

and

$$F_2 = F_{a2} + F_{b2}.$$

Since the subsystems are linked

$$X_1 = X_{a1} = X_{b1},$$

and

$$X_2 = X_{a2} = X_{b2}.$$

Hence if excitation is applied at x_1 only, $F_2 = 0$ and

$$\gamma_{11} = \frac{X_1}{F_1} = \frac{\alpha_{11}(\beta_{11}\beta_{22} - \beta_{12}^2) + \beta_{11}(\alpha_{11}\alpha_{22} - \alpha_{12}^2)}{\Delta},$$

where

$$\Delta = (\alpha_{11} + \beta_{11})(\alpha_{22} + \beta_{22}) - (\alpha_{12} + \beta_{12}^2),$$

and

$$\gamma_{21} = \frac{X_2}{F_1} = \frac{\alpha_{12}(\beta_{11}\beta_{22} - \alpha_{12}\beta_{12}) - \beta_{12}(\alpha_{11}\alpha_{22} - \alpha_{12}\beta_{12})}{\Delta}.$$

If

$$F_1 = 0,$$

$$\gamma_{22} = \frac{X_2}{F_2} = \frac{\alpha_{22}(\beta_{11}\beta_{22} - \beta_{12}^2) - \beta_{22}(\alpha_{11}\alpha_{22} - \alpha_{12}^2)}{\Delta}.$$

Since $\Delta = 0$ is the frequency equation, the natural frequencies of the system C are given by

$$\begin{vmatrix} \alpha_{11} + \beta_{11} & \alpha_{12} + \beta_{12} \\ \alpha_{21} + \beta_{21} & \alpha_{22} + \beta_{22} \end{vmatrix} = 0.$$

This is an extremely useful method for finding the frequency equation of a system because only the receptances of the subsystems are required. The receptances of many dynamic systems have been published in *The Mechanics of Vibration* by R. E. D. Bishop & D. C. Johnson (CUP 1960/1979). By repeated application of this method, a system can be considered to consist of any number of subsystems. This technique is, therefore, ideally suited to a computer solution.

It should be appreciated that although the receptance technique is useful for writing the frequency equation, it does not simplify the solution of this equation.

Example 30

An engine with a flywheel is shown. The engine is to drive a propeller of inertia I through a shaft of torsional stiffness k as indicated. The receptance for torsional

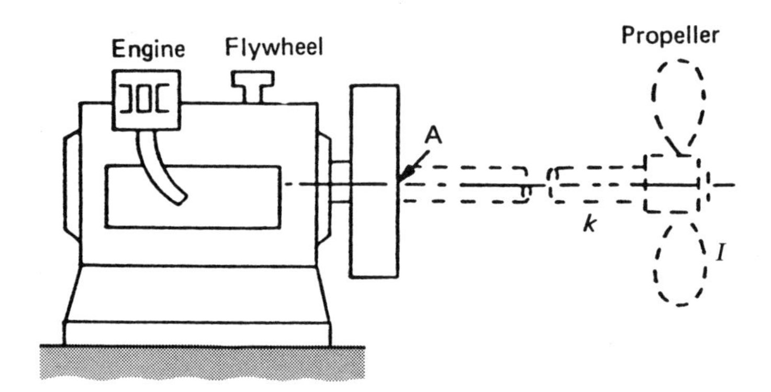

oscillation of the engine-flywheel system has been measured at point A over a limited frequency range which does not include any internal resonances of the system. The figure shows the receptance at A as a function of (frequency)^2.

Calculate the lowest non-zero natural frequency for the engine, flywheel and propeller system if $I = 0.9$ kg m^2 and $k = 300$ kN m/rad.

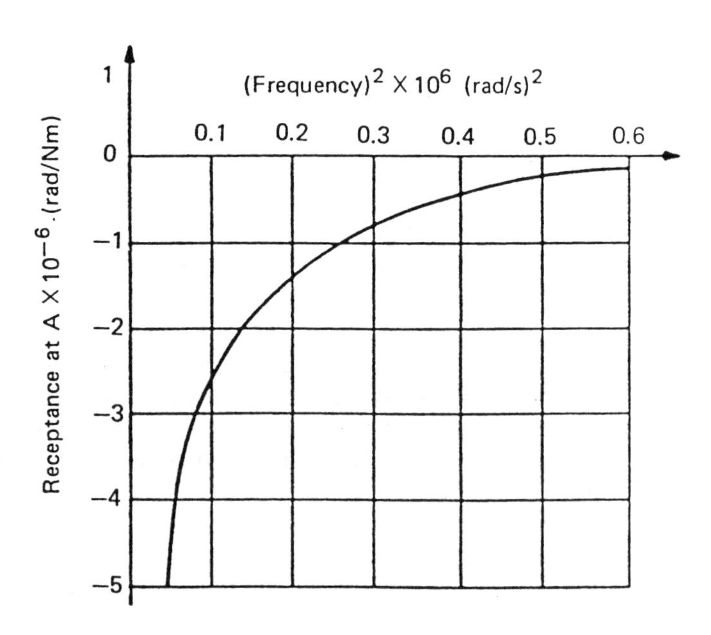

The engine – propeller system can be considered to be two subsystems, A and B:

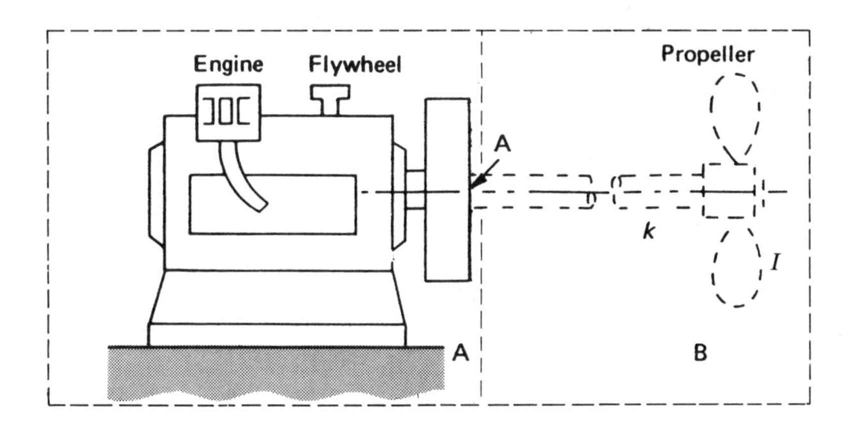

The frequency equation is $\alpha_{11} + \beta_{11} = 0$, where α_{11} is given in graphical form and β_{11} is found as follows. For the propeller system,

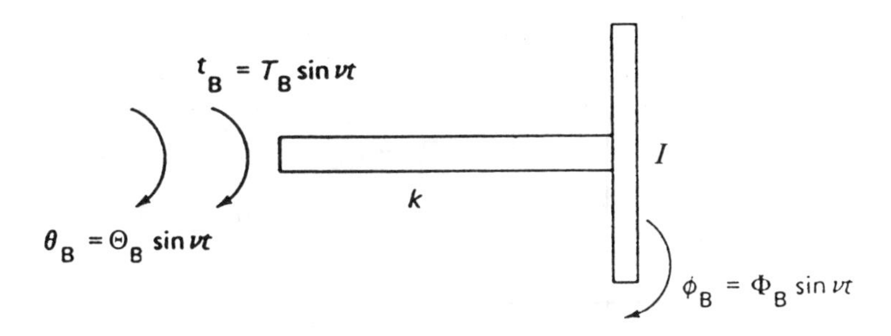

Now

$$t_B = k(\theta_B - \phi_B) = I\ddot{\phi}_B = -I\Phi_B \nu^2 \sin \nu t,$$

so

$$k(\Theta_B - \Phi_B) = -I\Phi_B \nu^2$$

and

$$\Theta_B = \Phi_B - \frac{I}{k}\Phi_B \nu^2 = \Phi_B\left(\frac{k - I\nu^2}{k}\right).$$

Thus

$$\beta_{11} = \frac{\Theta_B}{T_B} = -\Phi_B\left(\frac{k - Iv^2}{k}\right)\frac{1}{\Phi_B I v^2}$$

$$= -\left(\frac{k - Iv^2}{kIv^2}\right).$$

If α_{11} and $-\beta_{11}$ are plotted as functions of (frequency)^2, the intersection gives the value of the frequency which is a solution of $\alpha_{11} + \beta_{11} = 0$, that is, the natural frequency of free vibration is found. The table below can be calculated for $-\beta_{11}$ because $k = 300 \times 10^3$ Nm/rad, and $I = 0.9$ kg m^2.

v^2	Iv^2	$k - Iv^2$	kIv^2	$-\beta_{11}$
0.3×10^6	0.27×10^6	0.03×10^6	0.081×10^{12}	-0.37×10^{-6}
0.4×10^6	0.36×10^6	0.06×10^6	0.108×10^{12}	-0.55×10^{-6}
0.5×10^6	0.45×10^6	0.15×10^6	0.135×10^{12}	-1.11×10^{-6}

The receptance β_{11} can now be plotted against (frequency)^2 as below:

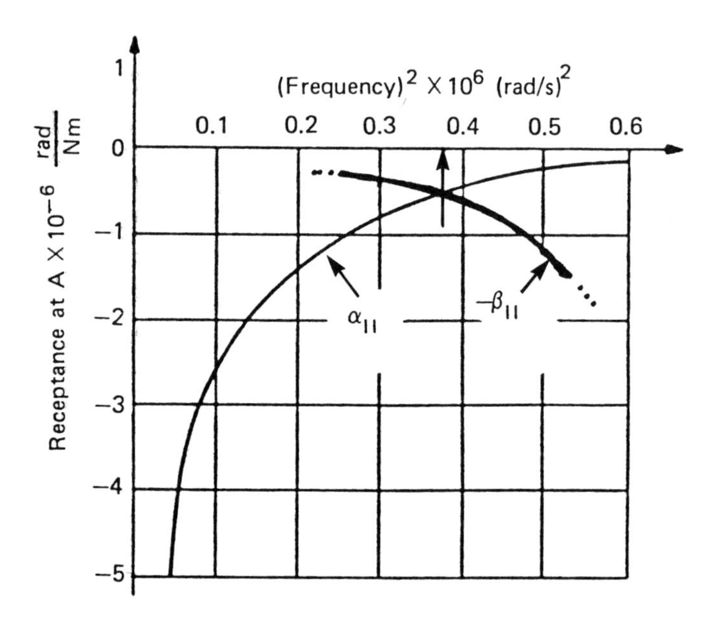

The intersection occurs at (frequency)^2 = 0.39×10^6 (rad/s)^2, that is, a frequency of 624 rad/s or 99.4 Hz. This is the natural frequency of the combined engine–propeller system. It can be seen that the effect of using different propellers with different k and I values is easily found, without having to re-analyse the whole engine–propeller system.

3.2.4 Impedance and Mobility

Impedance and mobility analysis techniques are frequently applied to systems and structures with many degrees of freedom. However, the method is best introduced by considering simple systems initially.

The impedance of a body is the ratio of the amplitude of the *harmonic* exciting force applied, to the amplitude of the resulting velocity. The mobility is the reciprocal of the impedance. It will be appreciated, therefore, that impedance and mobility analysis techniques are similar to those used in the receptance analysis of dynamic systems.

For a body of mass m subjected to an harmonic exciting force represented by $Fe^{j\nu t}$ the resulting motion is $x = Xe^{j\nu t}$, Thus

$$Fe^{j\nu t} = m\ddot{x} = -m\nu^2 Xe^{j\nu t},$$

and the receptance of the body,

$$\frac{X}{F} = -\frac{1}{m\nu^2}.$$

Now

$$\begin{aligned} Fe^{j\nu t} &= -m\nu^2 Xe^{j\nu t} \\ &= mj\nu(j\nu Xe^{j\nu t}) = mj\nu\, v, \end{aligned}$$

where v is the velocity of the body, and $v = Ve^{j\nu t}$.

Thus the impedance of a body of mass m is Z_m, where

$$Z_m = \frac{F}{V} = jm\nu,$$

and the mobility of a body of mass m is M_m, where

$$M_m = \frac{V}{F} = \frac{1}{jm\nu}.$$

Putting $s = j\nu$ so that $x = Xe^{st}$ gives

$$Z_m = ms,$$

$$M_m = \frac{1}{ms},$$

and

$$V = sX.$$

For a spring of stiffness k, $Fe^{j\nu t} = kXe^{j\nu t}$ and thus $Z_k = F/V = k/s$ and $M_k = s/k$, whereas for a viscous damper of coefficient c, $Z_c = c$ and $M_c = 1/c$.

If these elements of a dynamic system are combined so that the velocity is common to all elements, then the impedances may be added to give the system impedance, whereas if the force is common to all elements the mobilities may be added. This is demonstrated below by considering a spring–mass single degree of freedom system

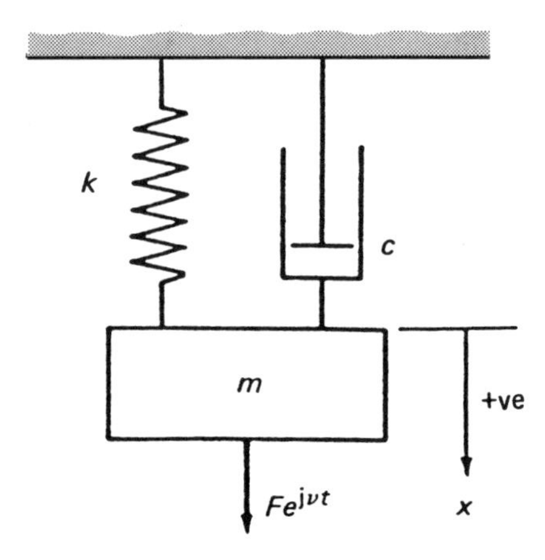

Fig. 3.17. Single degree of freedom system with elements connected in parallel.

with viscous damping, as shown in Fig. 3.17.

The velocity of the body is common to all elements, so that the force applied is the sum of the forces required for each element. The system impedance,

$$Z = \frac{F}{V} = \frac{F_m + F_k + F_c}{V}$$

$$= Z_m + Z_k + Z_c.$$

Hence

$$Z = ms + \frac{k}{s} + c.$$

That is

$$F = (ms^2 + cs + k)X,$$

or

$$F = (k - m\nu^2 + jc\nu)X.$$

Hence

$$X = \frac{F}{\sqrt{[(k - m\nu^2)^2 + (c\nu)^2]}}.$$

Thus when system elements are connected in parallel their impedances are added to give the system impedance.

In the system shown in Fig. 3.18, however, the force is common to all elements. In this case the force on the body is common to all elements so that the velocity at the driving point is the sum of the individual velocities. The system mobility,

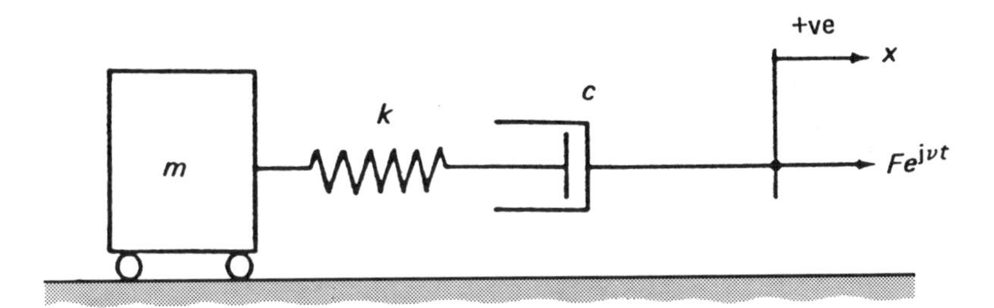

Fig. 3.18. Single degree of freedom system with elements connected in series.

$$M = \frac{V}{F} = \frac{V_m + V_k + V_c}{F},$$

$$= M_m + M_k + M_c,$$

$$= \frac{1}{ms} + \frac{s}{k} + \frac{1}{c}.$$

Thus when system elements are connected in series their mobilities are added to give the system mobility.

In the system shown in Fig. 3.19, the system comprises a spring and damper connected in series with a body connected in parallel.

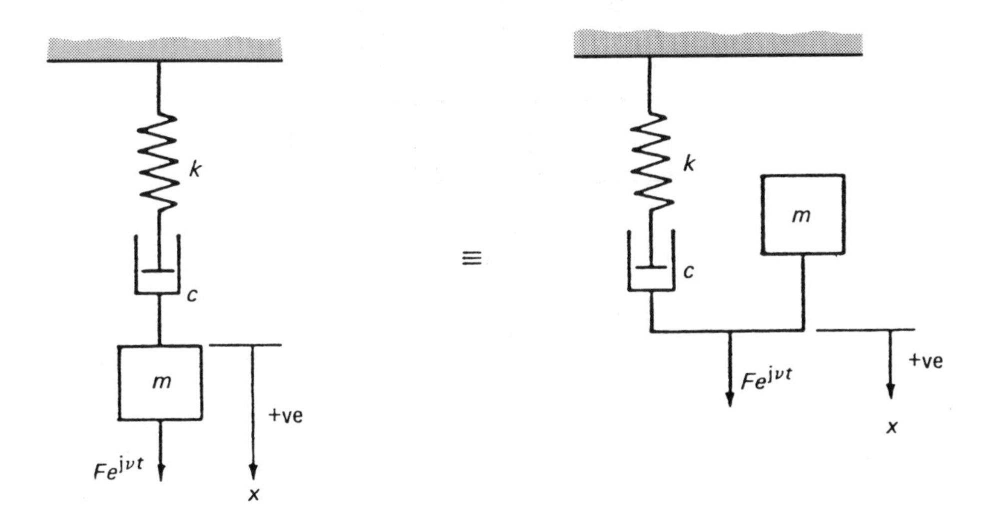

Fig. 3.19. Single degree of freedom system and impedance analysis model.

Thus the spring and damper mobilities can be added or the reciprocal of their impedances can be added. Hence the system driving point impedance Z is given by

$$Z = Z_m + \left[\frac{1}{Z_k} + \frac{1}{Z_c}\right]^{-1},$$

$$= ms + \left[\frac{1}{k/s} + \frac{1}{c}\right]^{-1}$$

$$= \frac{mcs^2 + mks + kc}{cs + k}.$$

Consider the system shown in Fig. 3.20. The spring k_1 and the body m_1 are connected in parallel with each other and are connected in series with the damper c_1.

Thus the driving point impedance Z is

$$Z = Z_{m_2} + Z_{k_2} + Z_{c_2} + Z_1$$

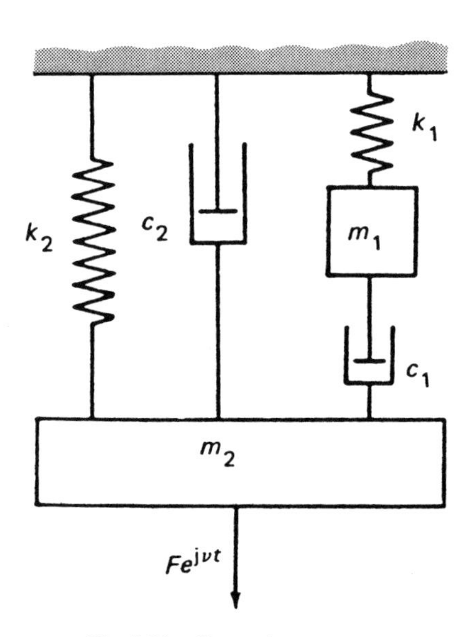

Fig. 3.20. Dynamic system.

where

$$Z_1 = \frac{1}{M_1},$$

$$M_1 = M_{c_1} + M_2, \quad M_2 = \frac{1}{Z_2},$$

and

$$Z_2 = Z_{k_1} + Z_{m_1}.$$

Thus

$$Z = Z_{m_2} + Z_{k_2} + Z_{c_2} + \cfrac{1}{\cfrac{1}{Z_{c_1}} + \cfrac{1}{Z_{k_1} + Z_{m_1}}}.$$

Hence

$$Z = \frac{m_1 m_2 s^4 + (m_1 c_2 + m_2 c_1 + m_1 c_1)s^3 + (m_1 k_1 + m_2 k_2 + c_1 c_2)s^2 + (c_1 k_2 + c_2 k_1 + c_1 k_1)s + k_1 k_2}{s(m_1 s^2 + c_1 s + k_1)}$$

The frequency equation is given when the impedance is made equal to zero or when the mobility is infinite. Thus the natural frequencies of the system can be found by putting $s = j\omega$ in the numerator above and setting it equal to zero.

To summarize, the mobility and impedance of individual elements in a dynamic system are calculated on the basis that the velocity is the relative velocity of the two ends of a spring or a damper, but the absolute velocity of the body. Individual impedances are added for elements or subsystems connected in parallel, and individual mobilities are added for elements or subsystems connected in series.

Example 31

Find the driving point impedance of the system shown in Fig. 3.6, and hence obtain the frequency equation.

The system of Fig. 3.6 can be redrawn as shown.

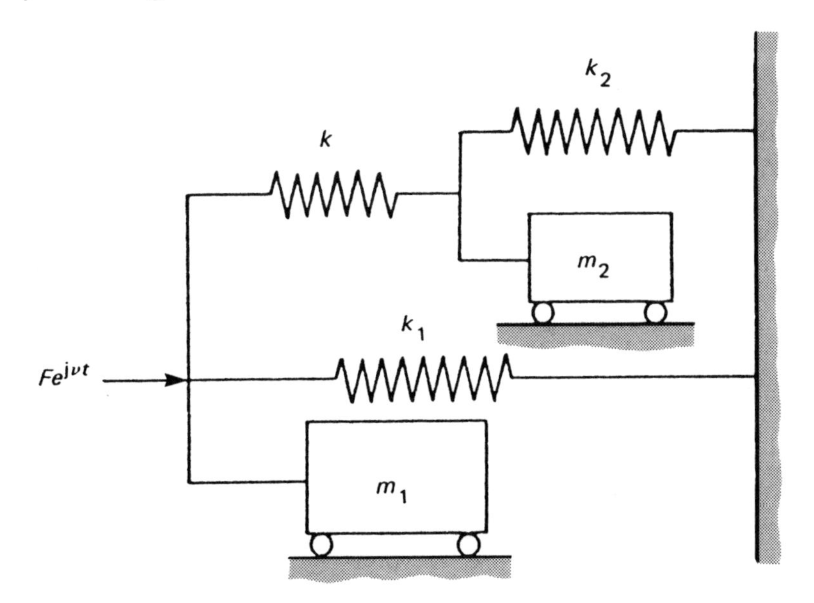

The driving point impedance is therefore

$$
\begin{aligned}
Z &= Z_{m_1} + Z_{k_1} + \frac{1}{\dfrac{1}{Z_k} + \dfrac{1}{Z_{m_2} + Z_{k_2}}}, \\
&= m_1 s + \frac{k_1}{s} + \frac{1}{\dfrac{1}{k/s} + \dfrac{1}{m_2 s + k_2/s}}, \\
&= \frac{(m_1 s^2 + k_1)(m_2 s^2 + k + k_2) + (m_2 s^2 + k_2)k}{s(m_2 s^2 + k + k_2)}
\end{aligned}
$$

The frequency equation is obtained by putting $Z = 0$ and $s = j\omega$, thus:

$$(k_1 - m_1\omega^2)(k + k_2 - m_2\omega^2) + k(k_2 - m_2\omega^2) = 0.$$

4

The vibration of systems with distributed mass and elasticity

Continuous systems such as beams, rods, cables and strings can be modelled by discrete mass and stiffness parameters and analysed as multi degree of freedom systems, but such a model is not sufficiently accurate for most purposes. Furthermore, mass and elasticity cannot always be separated in models of real systems. Thus mass and elasticity have to be considered as distributed parameters.

For the analysis of systems with distributed mass and elasticity it is necessary to assume an homogeneous, isotropic material which follows Hooke's law.

Generally free vibration is the sum of the *principal modes*. However, in the unlikely event of the elastic curve of the body in which motion is excited coinciding exactly with one of the principal modes, only that mode will be excited. In most continuous systems the rapid damping out of high-frequency modes often leads to the fundamental mode predominating.

4.1 WAVE MOTION

4.1.1 Transverse vibration of a string

Consider a vibrating flexible string of mass ρ per unit length and stretched under a tension T as shown in Fig. 4.1.

The FBDs of an element of length dx of the string assuming small deflections and slopes are given in Fig. 4.2. If the lateral deflection y is assumed to be small, the change in tension with deflection is negligible, and $\sin\theta \simeq \theta$.

The equation for motion in the y direction is

$$T\left(\theta + \frac{\partial\theta}{\partial x}\,dx\right) - T\theta = \rho\,dx\,\frac{\partial^2 y}{\partial t^2},$$

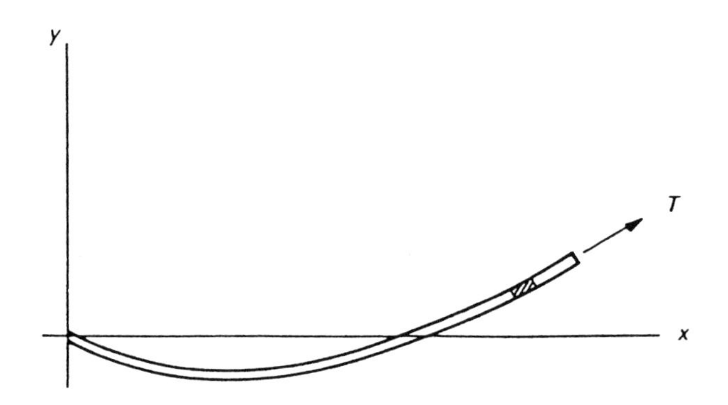

Fig. 4.1. Vibrating string.

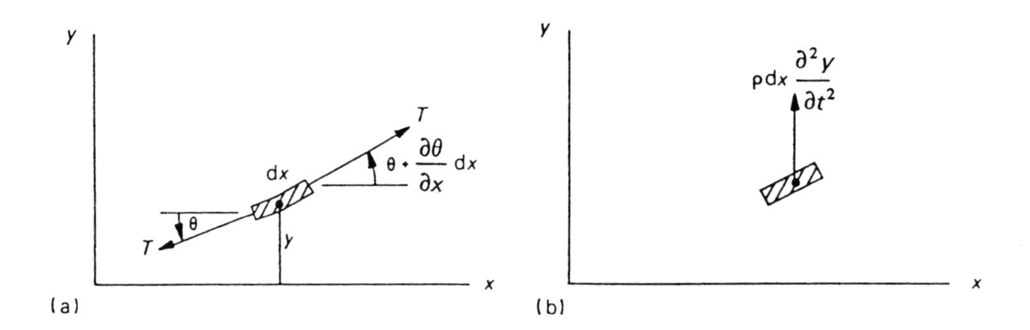

Fig. 4.2. (a) Applied forces; (b) effective force.

so

$$\partial\theta/\partial x = \frac{\rho}{T}\frac{\partial^2 y}{\partial t^2}.$$

Since

$$\theta = \partial y/\partial x.$$

$$\frac{\partial^2 y}{\partial x^2} = \frac{1}{c^2}\left(\frac{\partial^2 y}{\partial t^2}\right), \qquad \text{where} \quad c = \sqrt{(T/\rho)} \tag{4.1}$$

This is the *wave equation*, the solution of which is given in section 4.1.4. The velocity of propagation of waves along the string is c.

4.1.2 Longitudinal vibration of a thin uniform bar

Consider the longitudinal vibration of a thin uniform bar of cross-sectional area S, density ρ and modulus E under an axial force P, as shown in Fig. 4.3.

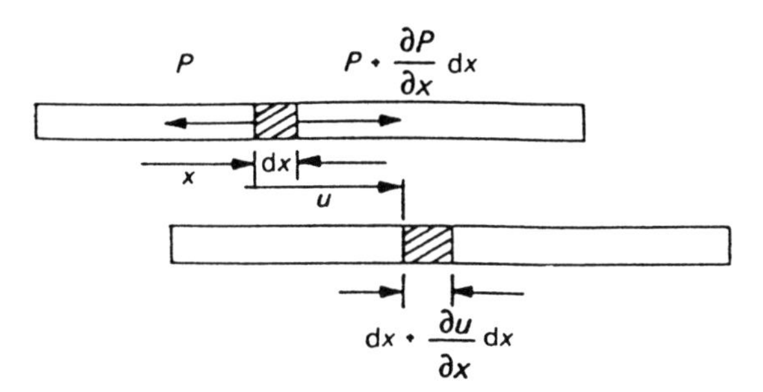

Fig. 4.3. Longitudinal beam vibration

From Fig. 4.3,

$$\frac{\partial P}{\partial x}\,\mathrm{d}x = \rho\, S\, \mathrm{d}x\, \frac{\partial^2 u}{\partial t^2}.$$

Now strain $\partial u/\partial x = P/SE$, so

$$\partial P/\partial x = SE(\partial^2 u/\partial x^2).$$

Thus

$$\partial^2 u/\partial t^2 = (E/\rho)(\partial^2 u/\partial x^2),$$

or

$$\partial^2 u/\partial x^2 = (1/c^2)(\partial^2 u/\partial t^2), \qquad \text{where} \quad c = \sqrt{(E/\rho)}. \tag{4.2}$$

This is the wave equation. The velocity of propagation of the displacement or stress wave in the bar is c.

4.1.3 Torsional vibration of a uniform shaft

In the uniform shaft shown in Fig. 4.4 the modulus of rigidity in torsion is G, the material density is ρ, and the polar second moment of area about the axis of twist is J. From Fig. 4.4,

$$\frac{\partial T}{\partial x}\,\mathrm{d}x = \rho\, J\, \mathrm{d}x\, \frac{\partial^2 \theta}{\partial t^2}.$$

The elastic equation is

$$T = GJ\,\frac{\partial \theta}{\partial x}.$$

Thus

$$\frac{\partial^2 \theta}{\partial t^2} = \left(\frac{G}{\rho}\right)\left(\frac{\partial^2 \theta}{\partial x^2}\right),$$

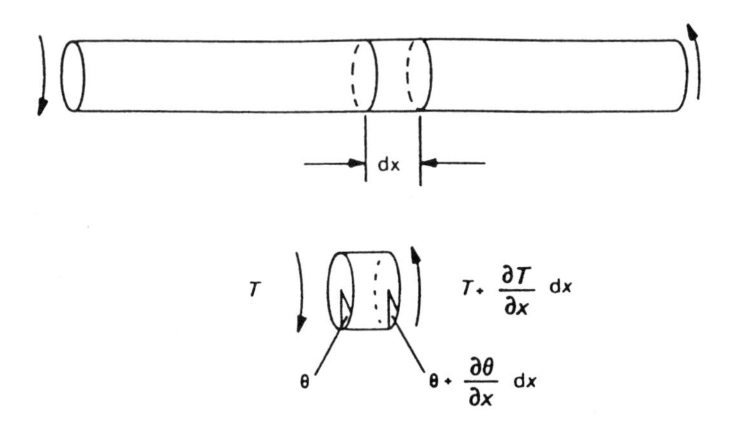

Fig. 4.4. Torsional vibration.

or

$$\frac{\partial^2 \theta}{\partial x^2} = \left(\frac{1}{c^2}\right)\left(\frac{\partial^2 \theta}{\partial t^2}\right), \qquad \text{where} \quad c = \sqrt{(G/\rho)}. \tag{4.3}$$

This is the wave equation. The velocity of propagation of the shear stress wave in the shaft is c.

4.1.4 Solution of the wave equation

The wave equation

$$\frac{\partial^2 y}{\partial x^2} = \left(\frac{1}{c^2}\right)\left(\frac{\partial^2 y}{\partial t^2}\right)$$

can be solved by the method of separation of variables and assuming a solution of the form

$$y(x, t) = F(x)G(t).$$

Substituting this solution into the wave equation gives

$$\frac{\partial^2 F(x)}{\partial x^2} G(t) = \frac{1}{c^2} \frac{\partial^2 G(t)}{\partial t^2} F(x),$$

that is,

$$\frac{1}{F(x)} \frac{\partial^2 F(x)}{\partial x^2} = \frac{1}{c^2} \frac{1}{G(t)} \frac{\partial^2 G(t)}{\partial t^2}.$$

The LHS is a function of x only, and the RHS is a function of t only, so partial derivatives are no longer required. Each side must be a constant, $-(\omega/c)^2$ say. (This quantity is chosen for convenience of solution.) Then

$$\frac{d^2F(x)}{dx^2} + \left(\frac{\omega}{c}\right)^2 F(x) = 0,$$

and

$$\frac{d^2G(t)}{dt^2} + \omega^2 G(t) = 0.$$

Hence

$$F(x) = A \sin\left(\frac{\omega}{c}\right) x + B \cos\left(\frac{\omega}{c}\right) x,$$

and

$$G(t) = C \sin \omega t + D \cos \omega t.$$

The constants A and B depend on the boundary conditions, and C and D on the initial conditions. The complete solution to the wave equation is therefore

$$u = \left(A \sin\left(\frac{\omega}{c}\right) x + B \cos\left(\frac{\omega}{c}\right)x\right)\left(C \sin \omega t + D \cos \omega t\right). \tag{4.4}$$

Example 32

Find the natural frequencies and mode shapes of longitudinal vibrations for a free–free beam with initial displacement zero.

Since the beam has free ends, $\partial u/\partial x = 0$ at $x = 0$ and $x = l$. Now

$$\frac{\partial u}{\partial x} = \left(A\left(\frac{\omega}{c}\right) \cos\left(\frac{\omega}{c}\right) x - B\left(\frac{\omega}{c}\right) \sin\left(\frac{\omega}{c}\right)x\right)\left(C \sin \omega t + D \cos \omega t\right).$$

Hence

$$\left(\frac{\partial u}{\partial x}\right)_{x=0} = A\left(\frac{\omega}{c}\right)(C \sin \omega t + D \cos \omega t) = 0, \qquad \text{so that} \quad A = 0,$$

and

$$\left(\frac{\partial u}{\partial x}\right)_{x=l} = \left(\frac{\omega}{c}\right)\left(-B \sin\left(\frac{\omega l}{c}\right)\right)\left(C \sin \omega t + D \cos \omega t\right) = 0.$$

Thus $\sin(\omega l/c) = 0$, since $B \neq 0$, and therefore

$$\frac{\omega l}{c} = \frac{\omega l}{\sqrt{(E/\rho)}} = \pi,\ 2\pi, \cdots n\pi \cdots$$

That is

$$\omega_n = \frac{n\pi}{l}\sqrt{\left(\frac{E}{\rho}\right)} \text{ rad/s},$$

where $\omega = c/\text{wavelength}$. These are the natural frequencies.

If the initial displacement is zero, $D = 0$ and

$$u = B' \cos\left(\frac{n\pi}{l}\right)x \times \sin\left(\frac{n\pi}{l}\right)\sqrt{\left(\frac{E}{\rho}\right)}\, t.$$

where $B' = B \times C$. Hence the mode shape is determined.

Example 33

A uniform vertical rod of length l and cross-section S is fixed at the upper end and is loaded with a body of mass M on the other. Show that the natural frequencies of longitudinal vibration are determined by

$$\omega l\sqrt{(\rho/E)} \tan \omega l\sqrt{(\rho/E)} = S\rho l/M$$

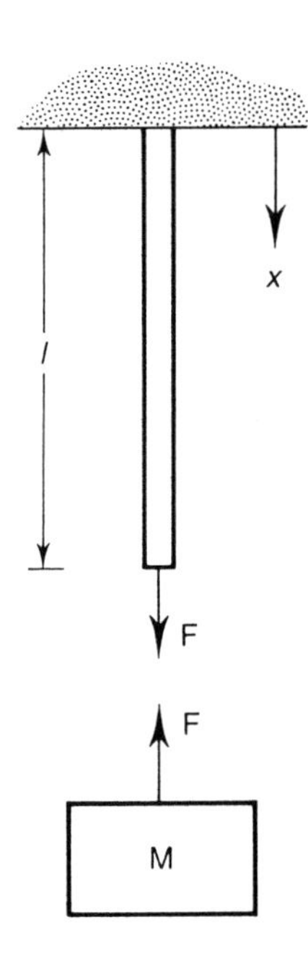

At

$$x = 0,\ u = 0, \qquad \text{and at} \quad x = l,\ F = SE\,(\partial u/\partial x).$$

Also

$$F = SE\,(\partial u/\partial x) = -M(\partial^2 u/\partial t^2).$$

The general solution is

$$u = (A \sin(\omega/c)x + B \cos(\omega/c)x)(C \sin \omega t + D \cos \omega t).$$

Now,

$$u_{x=0} = 0, \quad \text{so} \quad B = 0,$$

thus

$$u = (A \sin(\omega/c)x)(C \sin \omega t + D \cos \omega t),$$

$$(\partial u/\partial x)_{x=l} = (A(\omega/c) \cos(\omega l/c))(C \sin \omega t + D \cos \omega t),$$

and

$$(\partial^2 u/\partial t^2)_{x=l} = (-A\omega^2 \sin(\omega l/c))(C \sin \omega t + D \cos \omega t),$$

so

$$F = SEA\ (\omega/c) \cos(\omega l/c)(C \sin \omega t + D \cos \omega t)$$
$$= MA\omega^2 \sin(\omega l/c)(C \sin \omega t + D \cos \omega t).$$

Hence $(\omega l/c) \tan(\omega l/c) = SlE/Mc^2$, and

$$\omega l\sqrt{(\rho/E)} \tan \omega l\sqrt{(\rho/E)} = S\rho l/M, \qquad \text{since} \quad c^2 = E/\rho.$$

4.2 TRANSVERSE VIBRATION

4.2.1 Transverse vibration of a uniform beam

The transverse or lateral vibration of a thin uniform beam is another vibration problem in which both elasticity and mass are distributed. Consider the moments and forces acting on the element of the beam shown in Fig. 4.5. The beam has a cross-sectional area A, flexural rigidity EI, and material of density ρ.

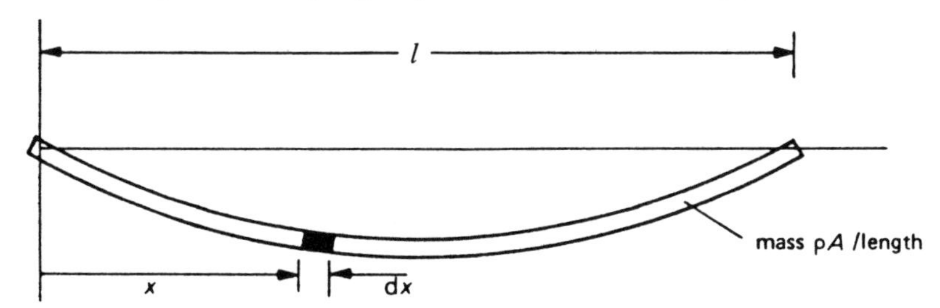

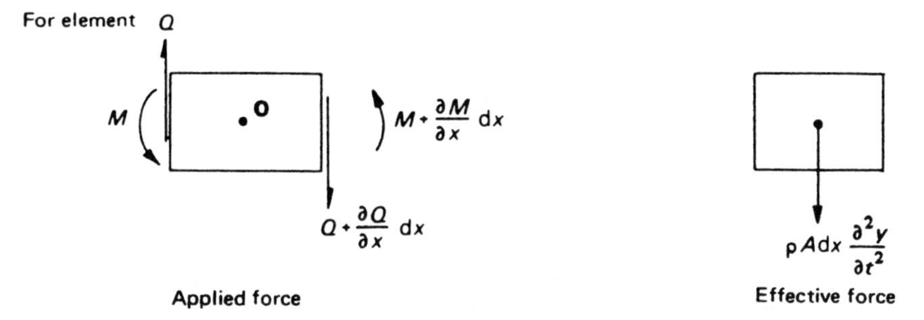

Fig. 4.5. Transverse beam vibration.

Then for the element, neglecting rotary inertia and shear of the element, taking moments about O gives

$$M + Q\,\frac{\mathrm{d}x}{2} + Q\,\frac{\mathrm{d}x}{2} + \frac{\partial Q}{\partial x}\,\mathrm{d}x\,\frac{\mathrm{d}x}{2} = M + \frac{\partial M}{\partial x}\,\mathrm{d}x.$$

That is,

$$Q = \partial M/\partial x. \tag{4.5}$$

Summing forces in the y direction gives

$$\frac{\partial Q}{\partial x}\,\mathrm{d}x = \rho A\,\mathrm{d}x\,\frac{\partial^2 y}{\partial t^2}.$$

Hence

$$\frac{\partial^2 M}{\partial x^2} = \rho A\,\frac{\partial^2 y}{\partial t^2}.$$

Now EI is a constant for a prismatical beam, so

$$M = -EI\,\frac{\partial^2 y}{\partial x^2} \quad \text{and} \quad \frac{\partial^2 M}{\partial x^2} = -EI\,\frac{\partial^4 y}{\partial x^4}.$$

Thus

$$\frac{\partial^4 y}{\partial x^4} + \left(\frac{\rho A}{EI}\right)\frac{\partial^2 y}{\partial t^2} = 0. \tag{4.6}$$

This is the general equation for the transverse vibration of a uniform beam.

When a beam performs a normal mode of vibration the deflection at any point of the beam varies harmonically with time, and can be written

$$y = X\,(B_1 \sin \omega t + B_2 \cos \omega t),$$

where X is a function of x which defines the beam shape of the normal mode of vibration. Hence

$$\frac{\mathrm{d}^4 X}{\mathrm{d}x^4} = \left(\frac{\rho A}{EI}\right)\omega^2 X = \lambda^4 X, \tag{4.7}$$

where

$\lambda^4 = \rho A \omega^2/EI.$ This is the *beam equation.*

The general solution to the beam equation (4.7) is

$$X = C_1 \cos \lambda x + C_2 \sin \lambda x + C_3 \cosh \lambda x + C_4 \sinh \lambda x,$$

where the constants $C_{1,2,3,4}$ are determined from the boundary conditions.

For example, consider the transverse vibration of a thin prismatical beam of length l, simply supported at each end. The deflection and bending moment are therefore zero at each end, so that the boundary conditions are $X = 0$ and $\mathrm{d}^2X/\mathrm{d}x^2 = 0$ at $x = 0$ and $x = l$.

Substituting these boundary conditions into the general solution above gives at

$$x = 0, \quad X = 0; \qquad \text{thus} \quad 0 = C_1 + C_3,$$

and at

$$x = 0, \quad \frac{\mathrm{d}^2X}{\mathrm{d}x^2} = 0; \qquad \text{thus} \quad 0 = C_1 - C_3.$$

That is,

$$C_1 = C_3 = 0 \qquad \text{and} \qquad X = C_2 \sin \lambda x + C_4 \sinh \lambda x.$$

Now at

$$x = l, X = 0 \qquad \text{so that} \qquad 0 = C_2 \sin \lambda l + C_4 \sinh \lambda l,$$

and at

$$x = l, \quad \frac{\mathrm{d}^2X}{\mathrm{d}x^2} = 0, \qquad \text{so that} \qquad 0 = C_2 \sin \lambda l - C_4 \sinh \lambda l.$$

That is,

$$C_2 \sin \lambda l = C_4 \sinh \lambda l = 0.$$

Since

$$\lambda l \neq 0, \quad \sinh \lambda l \neq 0 \qquad \text{and therefore} \qquad C_4 = 0.$$

Also $C_2 \sin \lambda l = 0$. Since $C_2 \neq 0$, otherwise $X = 0$ for all x, then $\sin \lambda l = 0$. Hence $X = C_2 \sin \lambda x$, and the solutions to $\sin \lambda l = 0$ give the natural frequencies. These are

$$\lambda = 0, \frac{\pi}{l}, \frac{2\pi}{l}, \frac{3\pi}{l} \ldots$$

so that

$$\omega = 0, \quad \left(\frac{\pi}{l}\right)^2 \sqrt{\left(\frac{EI}{A\rho}\right)}, \quad \left(\frac{2\pi}{l}\right)^2 \sqrt{\left(\frac{EI}{A\rho}\right)}, \quad \left(\frac{3\pi}{l}\right)^2 \sqrt{\left(\frac{EI}{A\rho}\right)}, \ldots \text{rad/s}.$$

$\lambda = 0$, $\omega = 0$ is a trivial solution because the beam is at rest. The lowest or first natural frequency is therefore $\omega_1 = (\pi/l)^2\sqrt{(EI/A\rho)}$ rad/s, and the corresponding mode shape is $X = C_2 \sin \pi x/l$. This is the first mode. $\omega_2 = (2\pi/l)^2\sqrt{(EI/A\rho)}$ rad/s is the second natural frequency, and the second mode is $X = C_2 \sin 2\pi x/l$, and so on. The mode shapes are drawn in Fig. 4.6.

These sinusoidal vibrations can be superimposed so that any initial conditions can be represented. Other end conditions give frequency equations with the solution

$$\omega = \frac{\alpha}{l^2} \sqrt{\left(\frac{\mathrm{EI}}{\mathrm{A}\rho}\right)} \text{ rad/s},$$

where the values of α are given in Table 4.1.

1st mode shape, one half-wave:

$$y = C_2 \sin \pi \left(\frac{x}{l}\right)(B_1 \sin \omega_1 t + B_2 \cos \omega_1 t); \qquad \omega_1 = \left(\frac{\pi}{l}\right)^2 \sqrt{\left(\frac{\mathrm{EI}}{\mathrm{A}\rho}\right)} \text{ rad/s.}$$

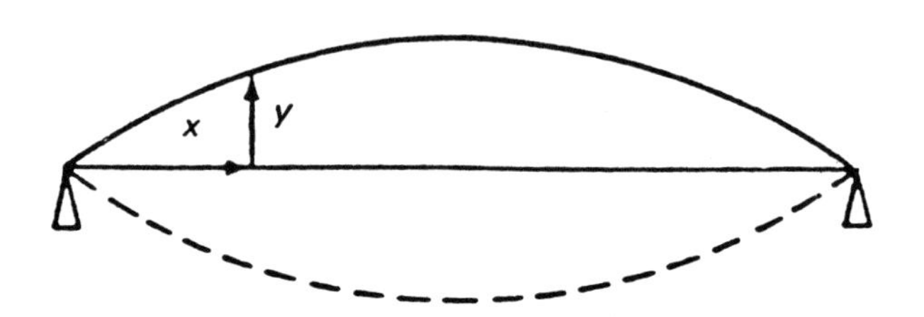

2nd mode shape, two half-waves:

$$y = C_2 \sin 2\pi \left(\frac{x}{l}\right)(B_1 \sin \omega_2 t + B_2 \cos \omega_2 t); \qquad \omega_2 = \left(\frac{2\pi}{l}\right)^2 \sqrt{\left(\frac{EI}{A\rho}\right)} \text{ rad/s.}$$

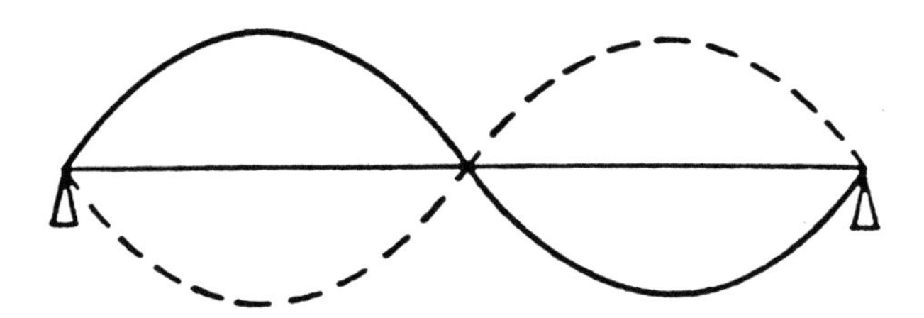

3rd mode shape, three half-waves:

$$y = C_2 \sin 3\pi \left(\frac{x}{l}\right)(B_1 \sin \omega_3 t + B_2 \cos \omega_3 t); \qquad \omega_3 = \left(\frac{3\pi}{l}\right)^2 \sqrt{\left(\frac{EI}{A\rho}\right)} \text{ rad/s.}$$

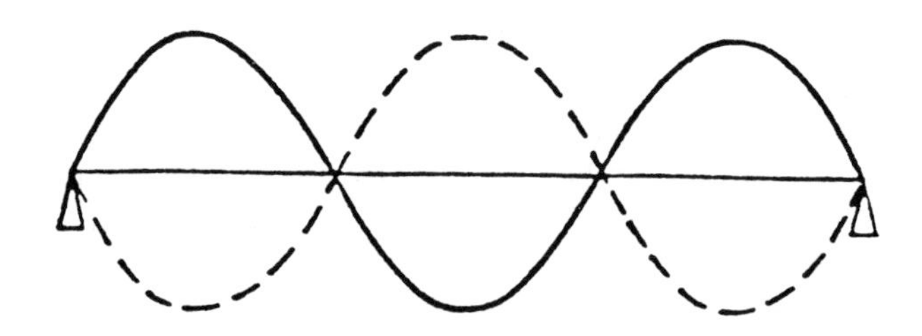

Fig. 4.6. Transverse beam vibration mode shapes and frequencies.

Table 4.1

End conditions	Frequency equation	1st mode	2nd mode	3rd mode	4th mode	5th mode
Clamped-free	$\cos \lambda l \cosh \lambda l = -1$	3.52	22.4	61.7	121.0	199.9
Pinned–pinned	$\sin \lambda l = 0$	9.87	39.5	88.9	157.9	246.8
Clamped–pinned	$\tan \lambda l = \tan \lambda l$	15.4	50.0	104.0	178.3	272.0
Clamped–clamped or Free–free	$\cos \lambda l \cosh \lambda l = 1$	22.4	61.7	121.0	199.9	298.6

The natural frequencies and mode shapes of a wide range of beams and structures are given in *Formulas for Natural Frequency and Mode Shape* by R. D. Blevins (Van Nostrand, 1979).

4.2.2 The whirling of shafts

An important application of the theory for transverse beam vibration is to the whirling of shafts. If the speed of rotation of a shaft is increased, certain speeds will be reached at which violent instability occurs. These are the critical speeds of whirling. Since the loading on the shaft is due to centrifugal effects the equation of motion is exactly the same as for transverse beam vibration. The centrifugal effects occur because it is impossible to make the centre of mass of any section coincide exactly with the axis of rotation, because of a lack of homogeneity in the material and other practical difficulties.

Example 34

A uniform steel shaft which is carried in long bearings at each end has an effective unsupported length of 3 m. Calculate the first two whirling speeds.

Take $I/A = 0.1 \times 10^{-3} \mathrm{m}^2$, $E = 200\ \mathrm{GN/m^2}$, and $\rho = 8000\ \mathrm{kg/m^3}$.

Since the shaft is supported in long bearings, it can be considered to be 'built in' at each end so that, from Table 4.1,

$$\omega = \frac{\alpha}{l^2} \sqrt{\left(\frac{EI}{A\rho}\right)} \text{ rad/s},$$

where $\alpha_1 = 22.4$ and $\alpha_2 = 61.7$. For the shaft,

$$\sqrt{\left(\frac{\mathrm{EI}}{\mathrm{A}\rho}\right)} = \sqrt{\left(\frac{200 \times 10^9 \times 0.1 \times 10^{-3}}{8000}\right)} = 50 \text{ m}^2/\text{s},$$

so that the first two whirling speeds are:

$$\omega_1 = \frac{22.4}{9} 50 = 124.4 \text{ rad/s},$$

so

$$f_1 = \frac{\omega_1}{2\pi} = \frac{124.4}{2\pi} = 19.8 \text{ cycle/s, and } N_1 = 1188 \text{ rev/min,}$$

and

$$N_2 = \frac{61.7}{22.4} 1188 = 3272 \text{ rev/min.}$$

Rotating this shaft at speeds at or near to the above will excite severe resonance vibration.

4.2.3 Rotary inertia and shear effects

When a beam is subjected to lateral vibration so that the depth of the beam is a significant proportion of the distance between two adjacent nodes, rotary inertia of beam elements and transverse shear deformation arising from the severe contortions of the beam during vibration make significant contributions to the lateral deflection. Therefore rotary inertia and shear effects must be taken into account in the analysis of high-frequency vibration of all beams, and in all analyses of deep beams.

The moment equation can be modified to take into account rotary inertia by adding a term $\rho I \, \partial^3 y/(\partial x \, \partial t^2)$, so that the beam equation becomes

$$EI \frac{\partial^4 y}{\partial x^4} - \rho I \frac{\partial^3 y}{\partial x \, \partial t^2} + \rho A \frac{\partial^2 y}{\partial t^2} = 0.$$

Shear deformation effects can be included by adding a term

$$\frac{EI\rho}{kg} \frac{\partial^4 y}{\partial x^2 \, \partial t^2},$$

where k is a constant whose value depends on the cross-section of the beam. Generally, k is about 0.85. The beam equation then becomes

$$EI \frac{\partial^4 y}{\partial x^4} - \frac{EI\rho}{kg} \frac{\partial^4 y}{\partial x^2 \, \partial t^2} + \rho A \frac{\partial^2 y}{\partial t^2} = 0.$$

Solutions to these equations are available, which generally lead to a frequency a few percent more accurate than the solution to the simple beam equation. However, in most cases the modelling errors exceed this. In general, the correction due to shear is larger than the correction due to rotary inertia.

4.2.4 The effect of axial loading

Beams are often subjected to an axial load, and this can have a significant effect on the lateral vibration of the beam. If an axial tension T exists, which is assumed to be constant for small amplitude beam vibrations, the moment equation can be modified by including a term $T \, \partial^2 y/\partial x^2$, so that the beam equation becomes

$$EI \frac{\partial^4 y}{\partial x^4} - T \frac{\partial^2 y}{\partial x^2} + \rho A \frac{\partial^2 y}{\partial t^2} = 0.$$

Tension in a beam will increase its stiffness and therefore increase its natural frequencies; compression will reduce these quantities.

Example 35

Find the first three natural frequencies of a steel bar 3 cm in diameter, which is simply supported at each end, and has a length of 1.5 m. Take $\rho = 7780\ \text{kg/m}^3$ and $E = 208\ \text{GN/m}^2$.

For the bar,

$$\sqrt{\left(\frac{EI}{A\rho}\right)} = \sqrt{\frac{208 \times 10^9 \cdot \pi(0.03)^4/64}{\pi(0.03/2)^2\ 7780}}\ \text{m/s}^2 = 38.8\ \text{m/s}^2.$$

Thus

$$\omega_1 = \frac{\pi^2}{1.5^2}\, 38.8 = 170.2\ \text{rad/s} \qquad \text{and} \qquad f_1 = 27.1\ \text{Hz}.$$

Hence

$$f_2 = 27.1 \times 4 = 108.4\ \text{Hz},$$

and

$$f_3 = 27.1 \times 9 = 243.8\ \text{Hz}.$$

If the beam is subjected to an axial tension T, the modified equation of motion leads to the following expression for the natural frequencies:

$$\omega_n^2 = \left(\frac{n\pi}{l}\right)^2 \frac{T}{A\rho} + \left(\frac{n\pi}{l}\right)^4 \frac{EI}{A\rho}.$$

For the case when $T = 1000$ the correction to ω_1^2 is ω_c^2, where

$$\omega_c^2 = \left(\frac{\pi}{1.5}\right)^2 \left(\frac{1000}{\pi(0.03/2)^2\ 7780}\right) = 795\ (\text{rad/s})^2.$$

That is, $f_c = 4.5$ Hz. Hence $f_1 = \sqrt{(4.5^2 + 27.1^2)} = 27.5$ Hz.

4.2.5 Transverse vibration of a beam with discrete bodies

In those cases where it is required to find the lowest frequency of transverse vibration of a beam which carries discrete bodies, Dunkerley's method may be used. This is a simple analytical technique which enables a wide range of vibration problems to be solved using a hand calculator. Dunkerley's method uses the following equation:

$$\frac{1}{\omega_1^2} \simeq \frac{1}{P_1^{\,2}} + \frac{1}{P_2^{\,2}} + \frac{1}{P_3^{\,2}} + \frac{1}{P_4^{\,2}} + \dots,$$

where ω_1 is the lowest natural frequency of a system and P_1, P_2, P_3,... are the frequencies of each body acting alone.

This equation may be obtained for a two degree freedom system by writing the equations of motion in terms of the influence coefficients as follows:

$$y_1 = \alpha_{11} m_1 \omega^2 y_1 + \alpha_{12} m_2 \omega^2 y_2,$$

and

$$y_2 = \alpha_{21} m_1 \omega^2 y_1 + \alpha_{22} m_2 \omega^2 y_2.$$

The frequency equation is given by

$$\begin{vmatrix} \alpha_{11} m_1 \omega^2 - 1 & \alpha_{12} m_2 \omega^2 \\ \alpha_{21} m_1 \omega^2 & \alpha_{22} m_2 \omega^2 - 1 \end{vmatrix} = 0.$$

By expanding this determinant, and solving the resulting quadratic equation, it is found that

$$\omega_{1,2}{}^2 = \frac{(\alpha_{11} m_1 + \alpha_{22} m_2) \pm \sqrt{[(\alpha_{11} m_1 + \alpha_{22} m_2)^2 - 4(\alpha_{11}\alpha_{22} - \alpha_{21}\alpha_{12})]}}{2(\alpha_{11}\alpha_{22} - \alpha_{21}\alpha_{12})}.$$

Hence it can be shown that

$$\frac{1}{\omega_1^2} + \frac{1}{\omega_2^2} = \alpha_{11} m_1 + \alpha_{22} m_2.$$

Now P_1 is the natural frequency of body 1 acting alone, hence

$$P_1{}^2 = \frac{k_1}{m_1} = \frac{1}{\alpha_{11} m_1}.$$

Similarly

$$P_2{}^2 = \frac{1}{\alpha_{22} m_2}.$$

Thus

$$\frac{1}{\omega_1{}^2} + \frac{1}{\omega_2{}^2} = \frac{1}{P_1{}^2} + \frac{1}{P_2{}^2}.$$

A similar relationship can be derived for systems with more than two degrees of freedom.

If $\omega_2 \gg \omega_1$, the left-hand side is approximately $1/\omega_1^2$, hence

$$\frac{1}{\omega_1{}^2} \simeq \frac{1}{P_1{}^2} + \frac{1}{P_2{}^2}.$$

Example 36

A steel shaft ($\rho = 8000$ kg/m^3, $E = 210$ GN/m^2) 0.055 m diameter, running in self-aligning bearings 1.25 m apart, carries a rotor of mass 70 kg, 0.4 m from one bearing. Estimate the lowest critical speed.

For the shaft alone

$$P_1 = (\pi/l)^2 \sqrt{(EI/A\rho)} \text{ rad/s} = 141 \text{ rad/s} = 1350 \text{ rev/min}.$$

This is the lowest critical speed for the shaft without the rotor. For the rotor alone, neglecting the mass of the shaft,

$$P_2 = \sqrt{(k/m)} \text{ rad/s}$$

$$k = 3EIl/(x^2(l - x)^2)$$

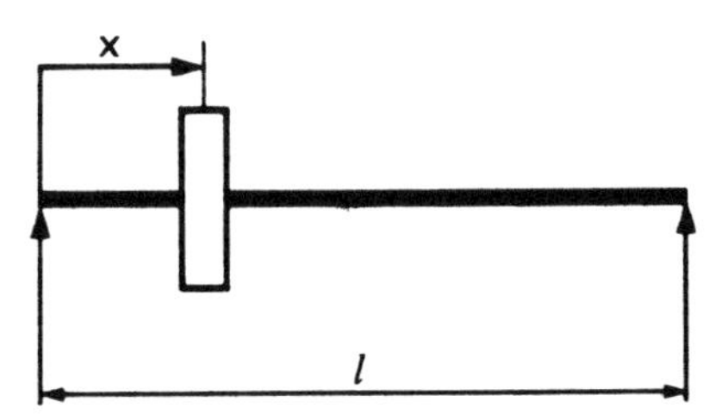

where $x = 0.4$ m and $l = 1.25$ m.

Thus

$$k = 3.07 \text{ MN/m}$$

and

$$P_2 = 209 \text{ rad/s} = 2000 \text{ rev/min}.$$

Now

$$1/N_1^{\,2} = 1/1350^2 + 1/2000^2, \qquad \text{hence} \quad N_1 = 1135 \text{ rev/min}$$

4.26 Receptance analysis

Many dynamic systems can be considered to consist of a number of beams fastened together. Thus if the receptances of each beam are known, the frequency equation of the system can easily be found by carrying out a subsystem analysis (Section 3.2.3). The required receptances can be found by inserting the appropriate boundary conditions in the general solution to the beam equation.

It will be appreciated that this method of analysis is ideal for computer solutions because of its repetitive nature.

For example, consider a beam which is pinned at one end ($x = 0$) and free at the other end ($x = l$). This type of beam is not commonly used in practice, but it is useful for analysis purposes. With an harmonic moment of amplitude M applied to the pinned end,

at $x = 0$, $X = 0$ (zero deflection) and

$$\frac{d^2X}{dx^2} = \frac{M}{EI} \qquad \text{(bending moment } M\text{)}$$

and at $x = l$,

$$\frac{d^2X}{dx^2} = 0 \qquad \text{(zero bending moment)}$$

and

$$\frac{d^3X}{dx^3} = 0 \qquad \text{(zero shear force).}$$

Now, in general,

$$X = C_1 \cos \lambda x + C_2 \sin \lambda x + C_3 \cosh \lambda x + C_4 \sinh \lambda x.$$

Thus applying these boundary conditions,

$$0 = C_1 + C_3 \quad \text{and} \quad \frac{M}{EI} = -C_1\lambda^2 + C_3\lambda^2.$$

Also

$$0 = -C_1\lambda^2 \cos \lambda l - C_2\lambda^2 \sin \lambda l + C_3\lambda^2 \cosh \lambda l + C_4\lambda^2 \sinh \lambda l.$$

and

$$0 = C_1\lambda^3 \sin \lambda l - C_2\lambda^3 \cos \lambda l + C_3\lambda^3 \sinh \lambda l + C_4\lambda^3 \cosh \lambda l.$$

By solving these four equations $C_{1,2,3,4}$ can be found and substituted into the general solution. It is found that the receptance moments/slope at the pinned end is

$$\frac{(1 + \cos \lambda l \cosh \lambda l)}{EI\lambda\ (\cos \lambda l \sinh \lambda l - \sin \lambda l \cosh \lambda l)}.$$

and at the free end is

$$\frac{2 \cos \lambda l \cosh \lambda l}{EI\lambda\ (\cos \lambda l \sinh \lambda l - \sin \lambda l \cosh \lambda l)}.$$

The frequency equation is given by

$$\cos \lambda l \sinh \lambda l - \sin \lambda l \cosh \lambda l = 0.$$

That is, $\tan \lambda l = \tanh \lambda l$.

Moment/deflection receptances can also be found.

By inserting the appropriate boundary conditions into the general solution, the receptance due to an harmonic moment applied at the free end, and harmonic forces applied to either end, can be deduced. Receptances for beams with all end conditions are tabulated in *The Mechanics of Vibration* by R. E. D. Bishop & D. C. Johnson (CUP, 1960/79), thereby greatly increasing the ease of applying this technique.

Example 37

A hinged beam structure is modelled by the array shown below:

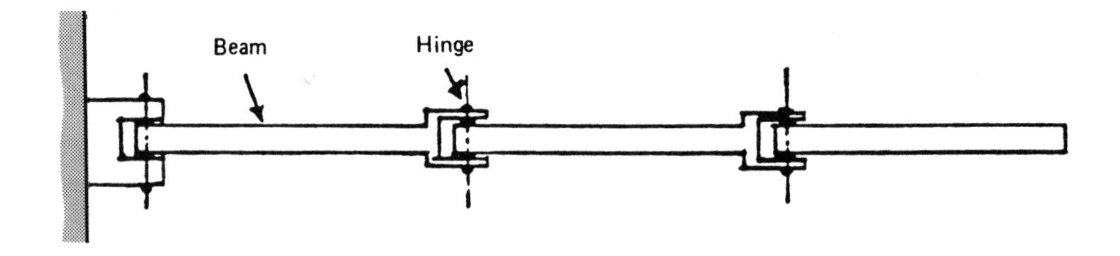

The hinges are pivots with torsional stiffness k_T and their mass is negligible. All hinges and beams are the same.

It is required to find the natural frequencies of free vibration of the array, so that the excitation of these frequencies, and therefore resonance, can be avoided.

Since all the beams are identical, the receptance technique is relevant for finding the frequency equation. This is because the receptances of each subsystem are the same, which leads to some simplification in the analysis.

There are two approaches:

(i) to split the array into subsystems comprising torsional springs and beams, or
(ii) to split the array into subsystems comprising spring-beam assemblies.

This approach results in a smaller number of subsystems.

Considering the first approach, and only the first element of the array, the subsystems could be either

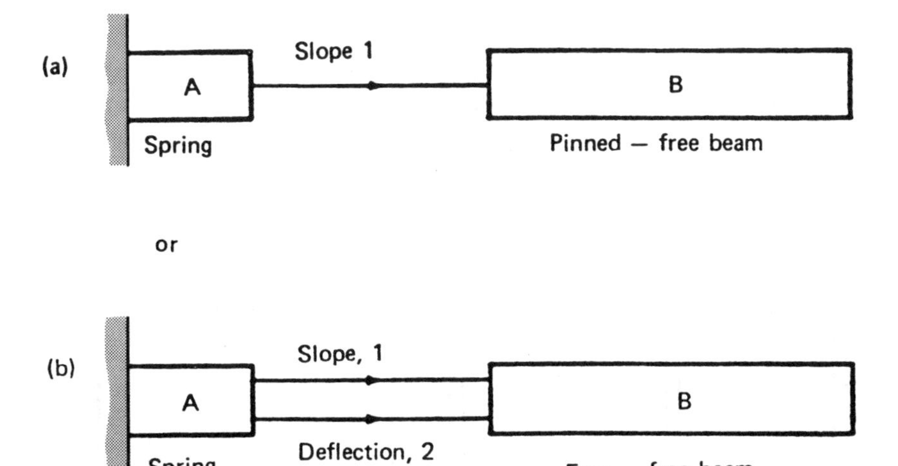

For (a) the frequency equation is $\alpha_{11} + \beta_{11} = 0$, whereas for (b) the frequency equation is

$$\begin{vmatrix} \alpha_{11} + \beta_{11} & \alpha_{12} + \beta_{12} \\ \alpha_{21} + \beta_{21} & \alpha_{22} + \beta_{22} \end{vmatrix} = 0,$$

where α_{11} is the moment/slope receptance for A, β_{11} is the moment/slope receptance for B, β_{12} is the moment/deflection receptance for B, β_{22} is the force/deflection receptance for B, and so on.

For (a), either calculating the beam receptances as above, or obtaining them from tables, the frequency equation is

$$\frac{1}{k_{\mathrm{T}}} + \frac{\cos \lambda l \cosh \lambda l + 1}{EI\lambda(\cos \lambda l \sinh \lambda l - \sin \lambda l \cosh \lambda l)} = 0$$

where

$$\lambda = 4\sqrt{\left(\frac{A\rho\omega^2}{EI}\right)}.$$

For (b), the frequency equation is

$$\begin{vmatrix} \dfrac{1}{k_T} + \dfrac{\cos \lambda l \sinh \lambda l + \sin \lambda l \cosh \lambda l}{EI\lambda(\cos \lambda l \cosh \lambda l - 1)} & \dfrac{-\sin \lambda l \sinh \lambda l}{EI\lambda^2(\cos \lambda l \cosh \lambda l - 1)} \\ \dfrac{-\sin \lambda l \sinh \lambda l}{EI\lambda^2(\cos \lambda l \cosh \lambda l - 1)} & \dfrac{-(\cos \lambda l \sinh \lambda l - \sin \lambda l \cosh \lambda l)}{EI\lambda^3(\cos \lambda l \cosh \lambda l - 1)} \end{vmatrix} = 0$$

which reduces to the equation given by method (a).

The frequency equation has to be solved after inserting the structural parameters, to yield the natural frequencies of the structure.

For the whole array it is preferable to use approach (ii), because this results in a smaller number of subsystems than (i), with a consequent simplification of the frequency equation. However, it will be necessary to calculate the receptances of the spring pinned–free beam if approach (ii) is adopted.

The analysis of structures such as frameworks can also be accomplished by the receptance technique, by dividing the framework to be analysed into beam substructures. For example, if the in-plane natural frequencies of a portal frame are required, it can be divided into three substructures coupled by the conditions of compatibility

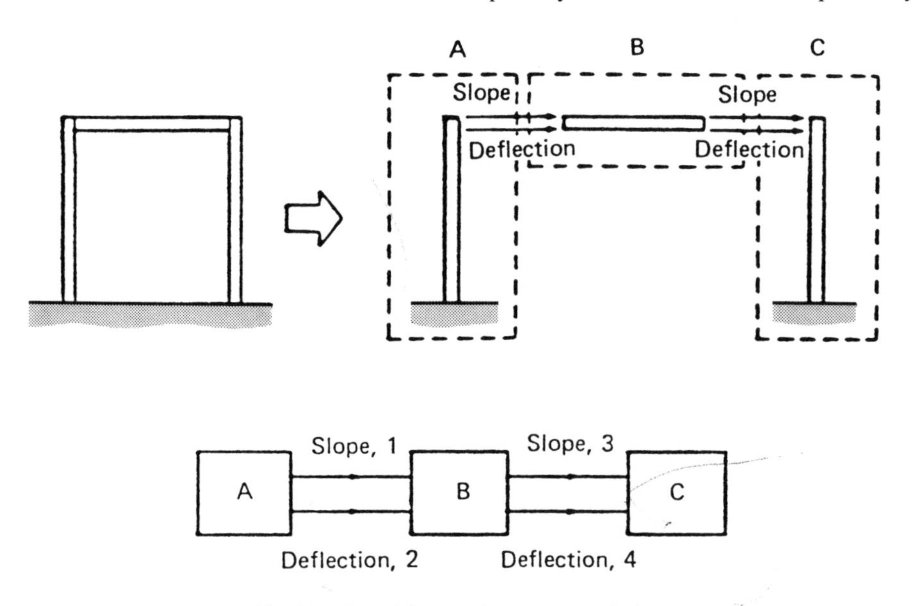

Fig. 4.7. Portal frame substructure analysis.

and equilibrium, as shown in Fig. 4.7.

Substructures A and C are cantilever beams undergoing transverse vibration, whereas B is a free-free beam undergoing transverse vibration. Beam B is assumed rigid in the horizontal direction, and the longitudinal deflection of beams A and C is assumed to be negligible.

Because the horizontal member B has no coupling between its horizontal and flexural motion $\beta_{12} = \beta_{14} = \beta_{23} = \beta_{34} = 0$, so that the frequency equation becomes

$$\begin{vmatrix} \alpha_{11}+\beta_{11} & \alpha_{12} & \beta_{13} & 0 \\ \alpha_{21} & \alpha_{22}+\beta_{22} & 0 & \beta_{24} \\ \beta_{31} & 0 & \gamma_{33}+\beta_{33} & \beta_{34} \\ 0 & \beta_{42} & \gamma_{43} & \gamma_{44}+\beta_{44} \end{vmatrix} = 0.$$

4.3 THE ANALYSIS OF CONTINUOUS SYSTEMS BY RAYLEIGH'S ENERGY METHOD

Energy methods can be used in the analysis of continuous systems if the mode shape of the system is known; this is necessary for evaluating the kinetic and potential energies. It has been shown by Rayleigh that the lowest natural frequency of such systems can be fairly accurately found by assuming any reasonable deflection curve for the shape of the first mode. Naturally the curve must be compatible with the end conditions of the system, and since any deviation from the true mode shape puts additional constraints on the system, the frequency determined by Rayleigh's method is never less than the exact frequency. Generally the static deflection curve of the system gives a frequency within a few per cent of the exact value. The frequency of vibration is found by considering the conservation of energy in the system; the natural frequency is determined by dividing the expression for potential energy in the system by the expression for kinetic energy.

4.3.1 The vibration of systems with heavy springs

The mass of the spring elements can have a considerable effect on the frequency of vibration of those systems in which relatively heavy springs are used, such as in valve mechanisms and in some rotational systems.

Consider the translational system shown in Fig. 4.8, where a rigid body of mass M is connected to a fixed frame by a spring of mass m, length l, and stiffness k. The body moves in the x direction only. If the dynamic deflected shape of the spring is assumed to be the same as the static shape under an applied axial force, the velocity of the spring element is $\dot{y} = (y/l)\dot{x}$, and the mass of the element is $(m/l)\mathrm{d}y$.

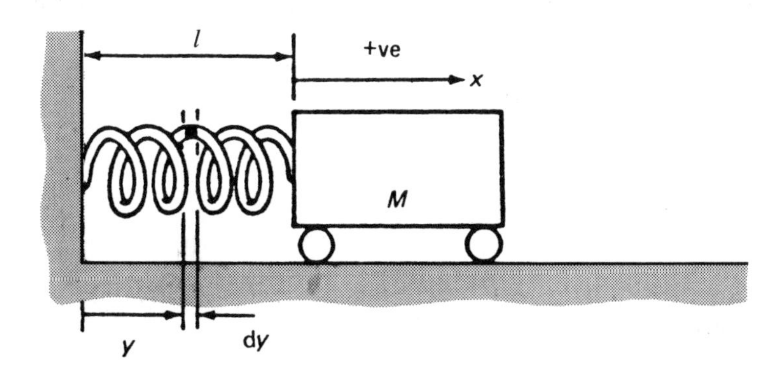

Fig. 4.8. Single degree of freedom system with heavy spring.

Thus

$$T = \frac{1}{2}M\dot{x}^2 + \int_0^l \frac{1}{2}\left(\frac{m}{l}\right)\left[\frac{y}{l}\dot{x}\right]^2 dy$$

$$= \frac{1}{2}\left(M + \frac{m}{3}\right)\dot{x}^2$$

and

$$V = \tfrac{1}{2}kx^2.$$

Assuming simple harmonic motion and putting $T_{max} = V_{max}$ gives the frequency of free vibration as

$$f = \frac{1}{2\pi}\sqrt{\left(\frac{k}{M + (m/3)}\right)}\ \text{Hz}.$$

That is, if the system is to be modelled with a massless spring, one third of the actual spring mass must be added to the mass of the body in the frequency calculation.

Alternatively, $\dfrac{d}{dt}(T + V) = 0$ can be used for finding the frequency of oscillation.

4.3.2 Transverse vibration of a beam

For the beam shown in Fig. 4.9, if m is the mass/length and y is the amplitude of the assumed deflection curve, then

$$T_{max} = \tfrac{1}{2}\int \dot{y}_{max}^2\ dm = \tfrac{1}{2}\omega^2 \int y^2\ dm,$$

where ω is the natural circular frequency of the beam.

The strain energy of the beam is the work done on the beam which is stored as elastic energy. If the bending moment is M and the slope of the elastic curve is θ,

$$V = \tfrac{1}{2}\int M\ d\theta.$$

Usually the deflection of beams is small so that the following relationships can be assumed to hold:

$$\theta = \frac{dy}{dx} \quad \text{and} \quad R\ d\theta = dx,$$

thus

$$\frac{1}{R} = \frac{d\theta}{dx} = \frac{d^2y}{dx^2}.$$

From beam theory, $M/I = E/R$, where R is the radius of curvature and EI is the flexural rigidity. Thus

$$V = \tfrac{1}{2}\int \frac{M}{R}\ dx = \tfrac{1}{2}\int EI\left(\frac{d^2y}{dx^2}\right)^2 dx.$$

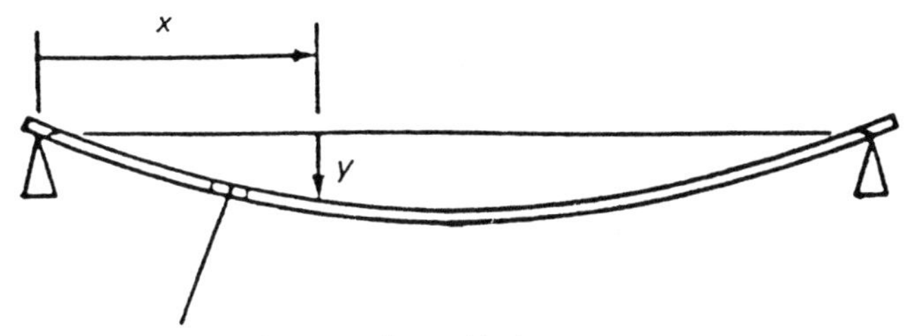

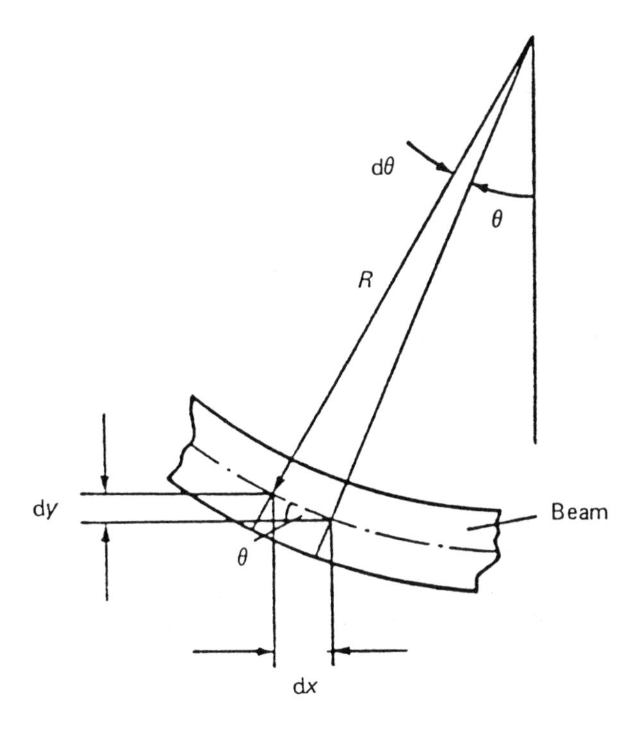

Fig. 4.9. Beam deflection.

Since

$$T_{\max} = V_{\max};$$

$$\omega^2 = \frac{\int EI\left(\frac{d^2y}{dx^2}\right)^2 dx}{\int y^2 \, dm}.$$

This expression gives the lowest natural frequency of transverse vibration of a beam. It can be seen that to analyse the transverse vibration of a particular beam by this method requires y to be known as a function of x. For this the static deflected shape or a part sinusoid can be assumed, provided the shape is compatible with the beam boundary conditions.

Example 38

A simply supported beam of length l and mass m_2 carries a body of mass m_1 at its mid-point. Find the lowest natural frequency of transverse vibration.

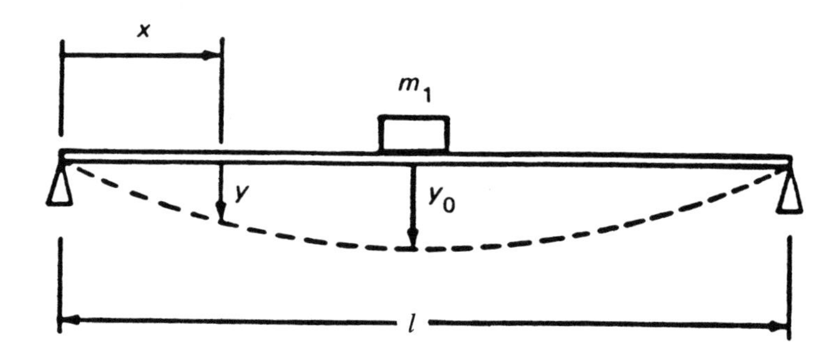

The boundary conditions are $y = 0$ and $d^2y/dx^2 = 0$ at $x = 0$ and $x = l$. These conditions are satisfied by assuming that the shape of the vibrating beam can be represented by a half sine wave. A polynomial expression can be derived for the deflected shape, but the sinusoid is usually easier to manipulate.

$y = y_0 \sin(\pi x/l)$ is a convenient expression for the beam shape, which agrees with the boundary conditions. Now

$$\dot{y} = \dot{y}_0 \sin\left(\frac{\pi x}{l}\right) \quad \text{and} \quad \frac{d^2y}{dx^2} = -y_0\left(\frac{\pi}{l}\right)^2 \sin\left(\frac{\pi x}{l}\right).$$

Hence

$$\int_0^l EI\left(\frac{d^2y}{dx^2}\right)^2 dx = \int_0^l EI{y_0}^2\left(\frac{\pi}{l}\right)^4 \sin^2\left(\frac{\pi x}{l}\right) dx$$

$$= EI{y_0}^2\left(\frac{\pi}{l}\right)^4 \frac{l}{2},$$

and

$$\int y^2\, dm = \int_0^l {y_0}^2 \sin^2\left(\frac{\pi x}{l}\right) \frac{m}{l}\, dx + y_0^2\, m_1$$

$$= y_0^2\left(m_1 + \frac{m_2}{2}\right). \qquad \text{Thus}$$

$$\omega^2 = \frac{EI(\pi/l)^4 \; l/2}{(m_1 + m_2/2)} \text{ (rad/s)}^2.$$

If $\quad m_2 = 0, \qquad \omega^2 = \dfrac{EI}{2}\dfrac{\pi^4}{l^3 m_1} = 48.7\,\dfrac{EI}{m_1 l^3}.$

The exact solution is 48 $EI/m_1 l^3$, so the Rayleigh method solution is 1.4% high.

Alternatively the Dunkerley method can be used. Here,

$$P_1{}^2 = \frac{48\;EI}{m_1 l^3} \qquad \text{and} \qquad P_2{}^2 = \frac{EI\;\pi^4}{m_2 l^3}.$$

Thus

$$\frac{1}{\omega^2} = \frac{m_1 l^3}{48\;EI} + \frac{m_2 l^3}{\pi^4 EI}.$$

Hence

$$\omega^2 = \frac{EI\left(\dfrac{\pi}{l}\right)^4 \dfrac{l}{2}}{\left(1.015 m_1 + \dfrac{m_2}{2}\right)} \text{ (rad/s)}^2.$$

which is very close to the value determined by the Rayleigh method.

Example 39

Find the lowest natural frequency of transverse vibration of a cantilever of mass m, which has rigid body of mass M attached at its free end.

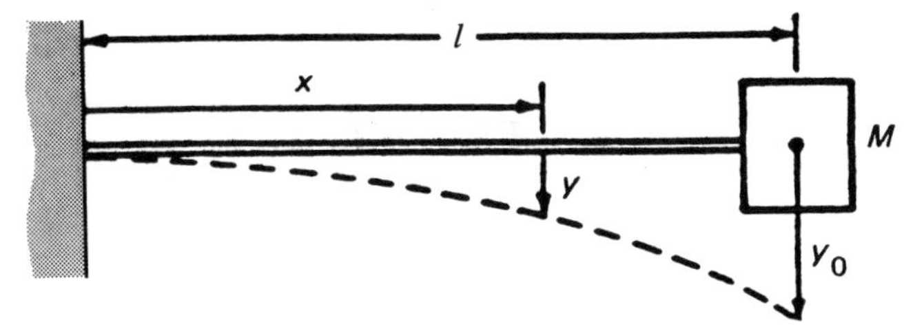

The static deflection curve is $y = (y_0/2l^3)(3lx^2 - x^3)$. Alternatively $y = y_0(1 - \cos \pi x/2l)$ could be assumed. Hence

$$\int_0^l EI\left(\frac{d^2y}{dx^2}\right)^2 dx = EI\int_0^l \left(\frac{y_0}{2l^3}\right)^2 (6l - 6x)^2\, dx = \frac{3EI}{l^3}\, y_0^2,$$

and

$$\int y^2\, dm = \int_0^l y^2 \frac{m}{l}\, dx + y_0{}^2 M,$$

$$= \int_0^l \frac{y_0^2}{4l^6} \frac{m}{l} (3lx^2 - x^3)^2 \,\mathrm{d}x + y_0^2 M,$$

$$= y_0^2 \left\{ M + \frac{33}{140} m \right\}.$$

Thus

$$\omega^2 = \left\{ \frac{3 \; EI}{\left(M + \frac{33}{140} m \right) l^3} \right\} (\text{rad/s})^2.$$

Example 40

A pin-ended strut of length l has a vertical axial load P applied. Determine the frequency of free transverse vibration of the strut, and the maximum value of P for stability. The strut has a mass m and a second moment of area I, and is made from material with modulus of elasticity E.

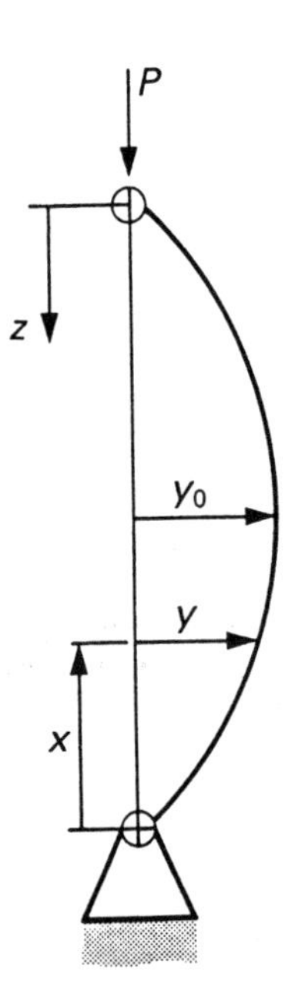

The deflected shape can be expressed by

$$y = y_0 \sin \pi \left(\frac{x}{l} \right),$$

since this function satisfies the boundary conditions of zero deflection and bending moment at $x = 0$ and $x = l$. Now

$$V_{\max} = \frac{1}{2}\int EI\left(\frac{d^2y}{dx^2}\right)^2 dx - Pz,$$

where

$$\frac{1}{2}\int EI\left(\frac{d^2y}{dx^2}\right)^2 dx = \frac{1}{2}\int_0^l EI\left(\frac{\pi}{l}\right)^4 y_0{}^2 \sin^2 \pi\left(\frac{x}{l}\right) dx$$

$$= \frac{EI}{4}\frac{\pi^4}{l^3} y_0{}^2,$$

and

$$z = \int_0^l \left(\sqrt{\left[1 + \left(\frac{dy}{dx}\right)^2\right]} - 1\right) dx,$$

$$= \int_0^l \frac{1}{2}\left(\frac{dy}{dx}\right)^2 dx,$$

$$= \frac{1}{2}\int_0^l y_0{}^2 \left(\frac{\pi}{l}\right)^2 \cos^2 \pi\left(\frac{x}{l}\right) dx,$$

$$= \frac{y_0{}^2}{4}\frac{\pi^2}{l}.$$

Thus

$$V_{\max} = \left(\frac{EI}{4}\frac{\pi^4}{l^3} - \frac{P}{4}\frac{\pi^2}{l}\right) y_0{}^2.$$

Now,

$$T_{\max} = \frac{1}{2}\int y^2 \, dm = \frac{1}{2}\int_0^l y^2 \frac{m}{l} dx,$$

$$= \frac{1}{2}\int_0^l y_0{}^2 \sin^2 \pi \frac{x}{l}\frac{m}{l} dx = \frac{m}{4} y_0{}^2.$$

Thus

$$\omega^2 = \frac{\left(\dfrac{EI}{4}\dfrac{\pi^4}{l^3} - \dfrac{P}{4}\dfrac{\pi^2}{l}\right)}{\dfrac{m}{4}} \ (\text{rad/s})^2,$$

and

$$f = \frac{1}{2}\sqrt{\left(\frac{EI(\pi/l)^2 - P}{ml}\right)} \text{ Hz}.$$

For stability $f > 0$, hence $EI(\pi/l)^2 > P$. That is, $P_{max} = EI(\pi/l)^2$. $EI\pi^2/l^2$ is known as the *Euler buckling load.*

Example 41

Part of an industrial plant incorporates a horizontal length of uniform pipe, which is rigidly embedded at one end and is effectively free at the other. Considering the pipe as a cantilever, derive an expression for the frequency of the first mode of transverse vibration using Rayleigh's method.

Calculate this frequency, given the following data for the pipe:

Modulus elasticity	200 GN/m^2
Second moment of area about bending axis	0.02 m^4
Mass	6×10^4 kg
Length	30m
Outside diameter	1 m

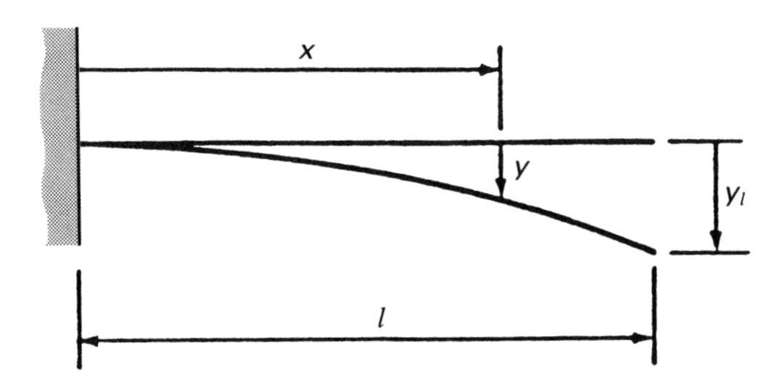

For a cantilever, assume

$$y = y_l\left(1 - \cos\frac{\pi x}{2l}\right).$$

This is compatible with zero deflection and slope when $x = 0$, and zero shear force and bending moment when $x = l$. Thus

$$\frac{d^2y}{dx^2} = y_l\left(\frac{\pi}{2l}\right)^2 \cos\frac{\pi x}{2l}.$$

Now

$$\int_0^l EI\left(\frac{d^2y}{dx^2}\right)^2 dx = EI\int_0^l y_l^2\left(\frac{\pi}{2l}\right)^4 \cos^2\frac{\pi x}{2l}\, dx$$

$$= EIy_l^2\left(\frac{\pi}{2l}\right)^4 \frac{l}{2},$$

and

$$\int_0^l y^2\, \mathrm{d}m = \int_0^l y_l^2\left(1 - \cos\frac{\pi x}{2l}\right)^2 \frac{m}{l}\, \mathrm{d}x$$

$$= y_l^2\, m\left(\frac{3}{2} - \frac{4}{\pi}\right).$$

Hence, assuming the structure to be conservative, that is the total energy remains constant throughout the vibration cycle,

$$\omega^2 = \frac{EIy_l^2\left(\dfrac{\pi}{2l}\right)^4 \dfrac{l}{2}}{y_l^2\, m\left(\dfrac{3}{2} - \dfrac{4}{\pi}\right)}$$

$$= \frac{EI}{ml^3}\, 13.4$$

Thus

$$\omega = 3.66\sqrt{\frac{EI}{ml^3}}\ \text{rad/s} \quad \text{and} \quad f = \frac{3.66}{2\pi}\sqrt{\frac{EI}{ml^3}}\ \text{Hz}.$$

In this case

$$\frac{EI}{ml^3} = \frac{200 \times 10^9 \times 0.02}{6 \times 10^4 \times 30^3} \Big/ \text{s}^2.$$

Hence

$$\omega = 5.75\ \text{rad/s} \quad \text{and} \quad f = 0.92\ \text{Hz}.$$

4.3.3 Wind or current excited oscillation

A structure exposed to a fluid stream is subjected to an harmonically varying force in a direction perpendicular to the stream. This is because of eddy or vortex shedding on alternate sides of the structure on the leeward side. Tall structures such as masts, bridges, and chimneys are susceptible to excitation from steady winds blowing across them. The Strouhal number relates the excitation frequency f_s, to the velocity of fluid flow v(m/s) and the hydraulic mean diameter D (m) of the structure as follows:

$$\text{Strouhal number} = \frac{f_s D}{v}.$$

If the frequency f_s is close to the natural frequency of the structure, resonance may occur.

For a structure,

$$D = \frac{4 \times \text{area of cross-section}}{\text{circumference}},$$

so that for a chimney of circular cross-section and diameter d,

$$D = \frac{4(\pi/4)\ d^2}{\pi d} = d,$$

and for a building of rectangular cross-section $a \times b$,

$$D = \frac{4ab}{2(a+b)} = \frac{2ab}{(a+b)}.$$

Experimental evidence suggests a value of 0.2–0.24 for the Strouhal number for most flow rates and wind speeds encountered. This value is valid for Reynolds numbers in the range $3 \times 10^5 - 3.5 \times 10^6$.

Example 42

For constructing a tanker terminal in a river estuary a number of cylindrical concrete piles were sunk into the river bed and left free standing. Each pile was 1 m diameter and protruded 20 m out of the river bed. The density of the concrete was 2400 kg/m^3 and the modulus of elasticity 14×10^6 kN/m^2. Estimate the velocity of the water flowing past a pile which will cause it to vibrate transversely to the direction of the current, assuming a pile to be a cantilever and taking a value for the Strouhal number

$$\frac{f_s D}{v} = 0.22,$$

where f_s is the frequency of flexural vibrations of a pile, D is the diameter, and v is the velocity of the current.

Consider the pile to be a cantilever of mass m, diameter D, and length l; then the deflection y at a distance x from the root can be taken to be, $y = y_l(1 - \cos \pi x/2l)$, where y_l is the deflection at the free end.

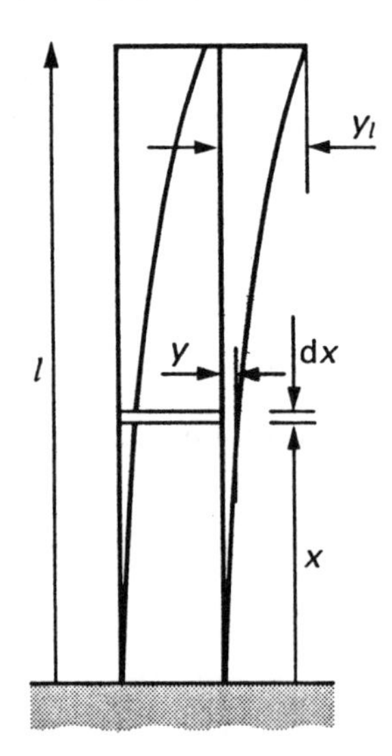

Thus

$$\frac{d^2y}{dx^2} = y_l\left(\frac{\pi}{2l}\right)^2 \cos\left(\frac{\pi x}{2l}\right).$$

$$\int_0^l EI\left(\frac{d^2y}{dx^2}\right)^2 dx = EI\int_0^l y_l^{\,2}\left(\frac{\pi}{2l}\right)^4 \cos^2\left(\frac{\pi x}{2l}\right) dx$$

$$= EIy_l^{\,2}\left(\frac{\pi}{2l}\right)^4 \frac{l}{2}$$

$$\int y^2\, dm = \int_0^l y_l^{\,2}\left(1 - \cos\left(\frac{\pi x}{2l}\right)\right)^2\left(\frac{m}{l}\right) dx$$

$$= y_l^{\,2}\,\frac{m}{l}\left(\frac{3}{2} - \frac{4}{\pi}\right) l$$

Hence

$$\omega^2 = \frac{EIy_l^{\,2}\left(\frac{\pi}{2l}\right)^4 \frac{l}{2}}{y_l^{\,2}\,\frac{m}{l}\left(\frac{3}{2} - \frac{4}{\pi}\right) l}.$$

Substituting numerical values gives $\omega = 5.53$ rad/s, that is, $f = 0.88$ Hz. When $f_s = 0.88$ Hz resonance occurs; that is, when

$$v = \frac{f_s D}{0.22} = \frac{0.88}{0.22} = 4 \text{ m/s}.$$

4.4 THE STABILITY OF VIBRATING SYSTEMS

If a system is to vibrate about an equilibrium position, it must be stable about that position. That is, if the system is disturbed when in an equilibrium position, the elastic forces must be such that the system vibrates about the equilibrium position. Thus the expression for ω^2 must be positive if a real value of the frequency of vibration about the equilibrium position is to exist, and hence the potential energy of a stable system must also be positive.

The principle of minimum potential energy can be used to test the stability of systems which are conservative. Thus a system will be stable at an equilibrium position if the potential energy of the system is a minimum at that position. This requires that

$$\frac{dV}{dq} = 0 \quad \text{and} \quad \frac{d^2V}{dq^2} > 0,$$

where q is an independent or generalized coordinate. Hence the necessary conditions for vibration to take place are found, and the position about which the vibration occurs is determined.

4.5 THE FINITE ELEMENT METHOD

Many structures, such as a ship hull or engine crankcase, are too complicated to be analysed by classical techniques, so that an approximate method has to be used. It can be seen from the receptance analysis of complicated structures that breaking a dynamic structure down into a large number of substructures is a useful analytical technique, provided that sufficient computational facilities are available to solve the resulting equations. The finite element method of analysis extends this method to the consideration of continuous structures as a number of elements, connected to each other by conditions of compatibility and equilibrium. Complicated structures can thus be modelled as the aggregate of simpler structures.

The principal advantage of the finite element method is its generality; it can be used to calculate the natural frequencies and mode shapes of any linear elastic system. However, it is a numerical technique which requires a fairly large computer, and care has to be taken over the sensitivity of the computer output to small changes in input.

For beam type systems the finite element method is similar to the lumped mass method, because the system is considered to be a number of rigid mass elements of finite size connected by massless springs. The infinite number of degrees of freedom associated with a continuous system can thereby be reduced to a finite number of degrees of freedom, which can be examined individually.

The finite element method therefore consists of dividing the dynamic system into a series of elements by imaginary lines, and connecting the elements only at the intersections of these lines. These intersections are called nodes. It is unfortunate that the word node has been widely accepted for these intersections; this meaning should not be confused with the zero vibration regions referred to in vibration analysis. The stresses and strains in each element are then defined in terms of the displacements and forces at the nodes, and the mass of the elements is lumped at the nodes. A series of equations are thus produced for the displacement of the nodes and hence the system. By solving these equations the stresses, strains, natural frequencies, and mode shapes of the system can be determined. The accuracy of the finite element method is greatest in the lower modes, and increases as the number of elements in the model increases. The finite element method of analysis is considered in *Techniques of Finite Elements* by B. Irons & S. Ahmad (Ellis Horwood, 1980).

5

Automatic control systems

Automatic control systems, wherein a variable quantity is made to conform to a predetermined level, have been in use for several centuries; but, as with much technological development, the most rapid advances have taken place during recent years. Feedforward, or open loop systems, merely control the input, such as in the case of a machine tool cutter which simply follows a given guide or pattern. An early example of an open loop system is the Jacquard loom of 1801, in which a set of punched cards programmed the patterns to be woven by the loom, and no information from the process or results was used to correct the loom operation. Feedback, or closed loop, systems feed back information from the process to control the operation of the machine. One of the earliest closed loop systems was that used by the Romans to maintain water levels in their aqueducts by means of floating valves. Later, windmills were the spawning ground of several control systems, for example the sails were automatically kept into the wind by means of a fantail (1745), centrifugal governors were used to control the speed of the millstones (1783), and the speed of rotation of the sails was automatically controlled by roller reefing (1789). In the late eighteenth century centrifugal governors were also being used to control the speed of steam engines by regulating the steam supply. These devices provided much closer control than manual systems could, and they were cheaper to operate, so that the overall efficiency of the machine increased. This led to demands for even better control systems which operated within narrower margins so that the efficiency was further increased.

In principle, many variables can be controlled by humans, but in practice this may be impossible, difficult, costly, or undesirable because of the need for continuous operation regardless of environment, large forces, and a fast response. The human reaction time of about 0.3 seconds is too slow for many applications. Further examples of control systems are liquid level control by ball valve, temperature control by thermostat, and surface control such as a ship's rudder or aircraft flaps by hydraulic servo. However, all types of control system can be modelled for analysis purposes, irrespective of the operating mechanism. The first theoretical analysis of a control system was published by Maxwell in the nineteenth century: this theory was soon generalized and followed by a large number of contributions to explain and improve the operation of control systems. In the early twentieth century the rapid development of automatic control systems took place because of the need to position guns and ships quickly and accurately, which led to the development of servomechanisms in the mid-1930s. In the 1950s the potential of multiple loop systems was investigated, and the introduction of computers opened the way for much greater complexity in control systems. Computer control is usually applied to industrial problems in one of three ways: supervisory control which continually adjusts the plant to optimum operation conditions, direct digital control, and hierarchy control which integrates the plants operation at every level from management decisions through to valve settings. For example, the new Waterloo rail terminal has a building management system which monitors and controls air conditioning, lighting, escalators and travelaters.

Computer control is aided by the ability to measure and convert into electrical signals a wide range of system parameters such as temperature, pressure, speed, level, weight, flow, conductivity, and thickness. A fast rate of progress has been maintained to the present day, particularly since the introduction of microprocessors, so that only an introduction to automatic control systems can be attempted in a text of this length.

The essential feature of an automatic control system is the existence of a feedback loop to give good performance. This is a *closed* loop system; if the measured output is not compared with the input the loop is *open*. Usually it is required to apply a specific input to a system and for some other part of the system to respond in the desired way. The error between the actual response and the ideal response is detected and fed back to the input to modify it so that the error is reduced, as shown in Fig. 5.1. The output of a device represented by a block in a block diagram cannot affect the input to that device unless a specific feedback loop is provided.

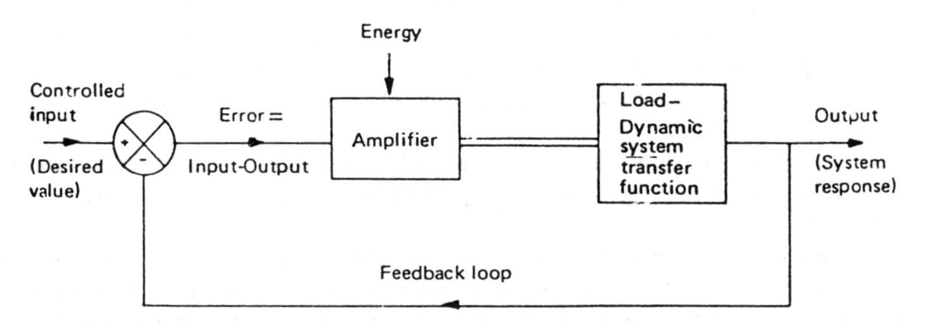

Fig. 5.1. Control system block diagram.

In general both input and output vary with time, and the control system can be mechanical, pneumatic, hydraulic and electrical in operation, or any combination of these or other power sources. The system should be absolutely stable so that if excited it will settle to some steady value, and it should be accurate in the steady state.

This concept can be illustrated by considering a simple dynamic system of the type previously considered, as shown in Fig. 5.2.

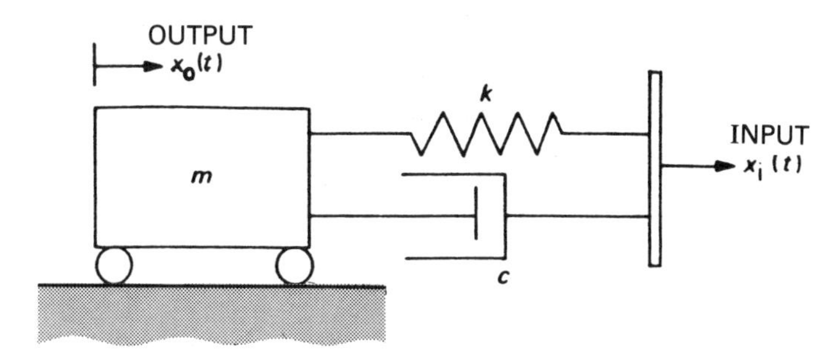

Fig. 5.2. Dynamic system.

The system comprises a body of mass m connected to an input controlled platform by a spring and viscous damper. A specific input $x_i(t)$ is applied to this platform, and, as a control system, the response of the body or output $x_o(t)$ should be identical to the input.

Considering the forces acting on the body, the equation of motion is

$$m\ddot{x}_o = k(x_i - x_o) + c(\dot{x}_i - \dot{x}_o). \tag{5.1}$$

Equations of motion of this type have been solved for an harmonic input (equation 2.18). For a general solution irrespective of input it is convenient to use the D-operator. Thus (5.1) becomes

$$m\mathrm{D}^2x_o = k(x_i - x_o) + c\mathrm{D}(x_i - x_o) = (k + c\mathrm{D})(x_i - x_o).$$

It should be noted that although using the D-operator is a neat and compact form of writing the equation it does not help with the solution of the response problem.

Now the force F on the body is $m\mathrm{D}^2x_o$, so $F = m\mathrm{D}^2x_o$, or

$$F\left(\frac{1}{m\mathrm{D}^2}\right) = x_o.$$

The *transfer function* of a system is the function by which the input is multiplied to give the output, so that since F is the input to the body and x_o the output, $(1/m\mathrm{D}^2)$ is the transfer function (TF) of the body.

This is shown in block diagram form in Fig. 5.3.

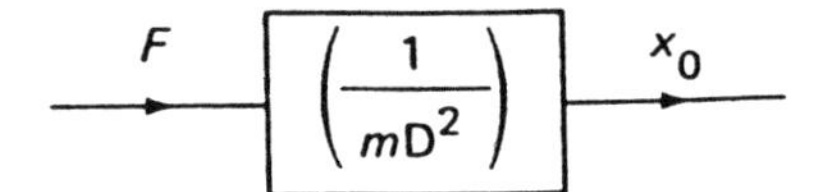

Fig. 5.3. Rigid body block diagram.

For the spring/damper unit $F = (k + c\mathrm{D})(x_i - x_o)$ as shown in Fig. 5.4.

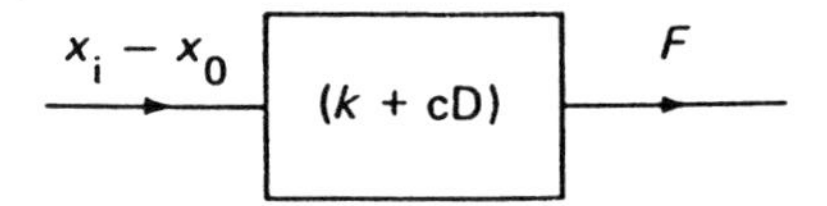

Fig. 5.4. Spring/damper block diagram.

Because the input to the spring/damper unit is $(x_i - x_o)$ and the output is F, the TF is $(k + c\mathrm{D})$. These systems can be combined as shown in Fig 5.5.

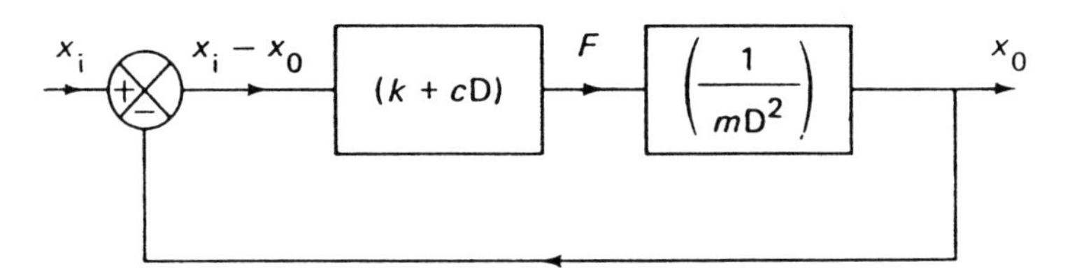

Fig. 5.5. Block diagram.

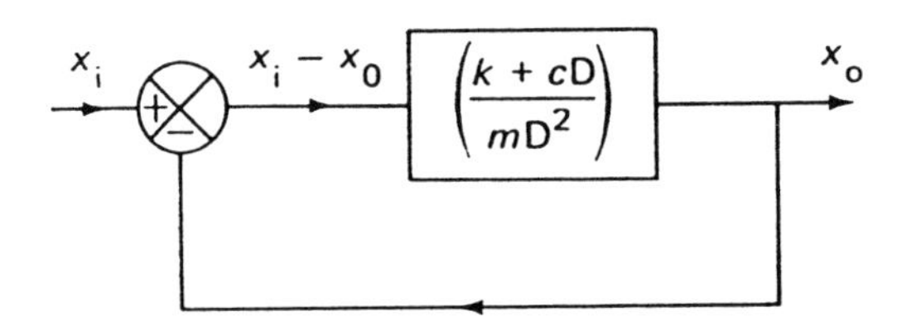

Fig. 5.6. System block diagram.

Fig. 5.6 is the conventional unity feedback loop form. Essentially, the spring/damper acts as an error-sensing device and generates a restoring force related to that error. Since

$$(x_i - x_o)\left(\frac{k + c\mathrm{D}}{m\mathrm{D}^2}\right) = x_o,$$

$$x_o = \left(\frac{cD + k}{mD^2 + cD + k}\right)x_i,$$

or

$$\frac{x_o}{x_i} = \left(\frac{cD + k}{mD^2 + cD + k}\right).$$

This is the TF of the dynamic system with feedback, that is, it is the closed loop TF.

The system response, or output for a sustained harmonic input motion $x_i = X_i \cos vt$, has already been discussed, see section 2.3.2. If the output x_o is assumed to be $X_o \cos (vt - \phi)$, then substitution into the closed loop transfer function (x_o/x_i) gives

$$- mv^2 X_o \cos (vt - \phi) - cvX_o \sin (vt - \phi) + kX_o \cos (vt - \phi)$$
$$= - cvX_i \sin vt + kX_i \cos vt.$$

This equation can be solved by using the phasor technique (Fig. 5.7).

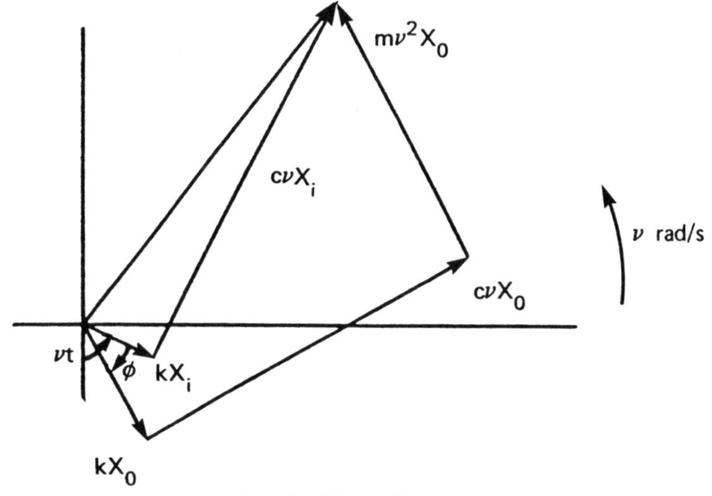

Fig. 5.7. Phasor diagram.

From the phasor diagram,

$$(kX_i)^2 + (cvX_i)^2 = (kX_o - mv^2X_o)^2 + (cvX_o)^2.$$

Hence

$$X_i\sqrt{(k^2 + (cv)^2)} = X_o\sqrt{((k - mv^2)^2 + (cv)^2)}.$$

This equation gives the amplitude of the sustained oscillation. However, there is also an initial transient vibration which is given by the complementary function; although this is a damped oscillation which dies away with time, it is often important in control systems, particularly if accurate positioning is required in a short time.

In the study of control system dynamics, the response of the system to a range of types of input must be considered; for example the impulse, step, and ramp shown in Fig. 5.8. Impulse excitation has been discussed in section 2.3.6.

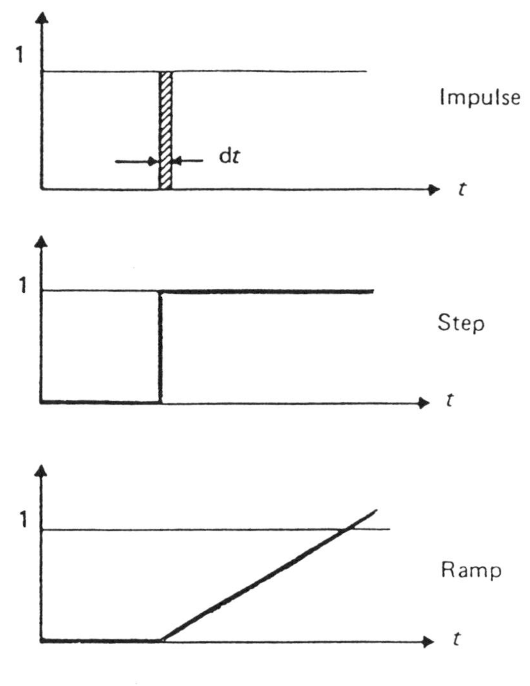

Fig. 5.8. System inputs.

For the system considered, if the input x_i is given a step change so that $x_i = X$,

$$x_o = \left(\frac{k + cD}{mD^2 + cD + k}\right)x_i$$

$$= \left(1 + \frac{c}{k}D + \frac{m}{k}D^2\right)^{-1}\left(1 + \frac{c}{k}D\right)X$$

$$= \left(1 - \frac{c}{k}D + \cdots\right)\left(X\right)$$

$$= X.$$

That is, in the steady state $x_o = X$, so that there is no error between the input and the output. However, the complementary function is given by $(mD^2 + cD + k)x_o = 0$, so $x_o = Ae^{-\zeta\omega t} \sin (\omega\sqrt{(1 - \zeta^2)}t + \phi)$.

That is, x_o may also contain an initial damped oscillation, so that the system response to a step input would be as shown in Fig. 5.9.

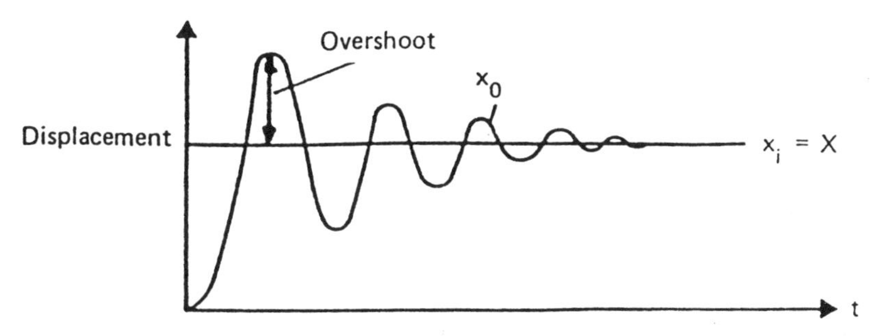

Fig. 5.9. Step input response.

The amount by which x_o exceeds x_i on the first oscillation is called the first *overshoot*. In a stable system, that is one in which the transient dies away, this will be the maximum overshoot.

If the input is a ramp, $x_i = \beta t$,

so that

$$x_o = \left(1 - \frac{c}{k}\,\mathrm{D} + \cdots\right)\left(1 + \frac{c}{k}\,\mathrm{D}\right)(\beta t)$$

$$= \left(1 - \frac{c}{k}\,\mathrm{D} + \cdots\right)\left(\beta t + \frac{c}{k}\,\beta\right)$$

$$= \beta t + \frac{c}{k}\,\beta - \frac{c}{k}\,\beta = \beta t.$$

That is, there is no steady-state error. However, the initial transient exists, so that the system output response is as shown in Fig. 5.10.

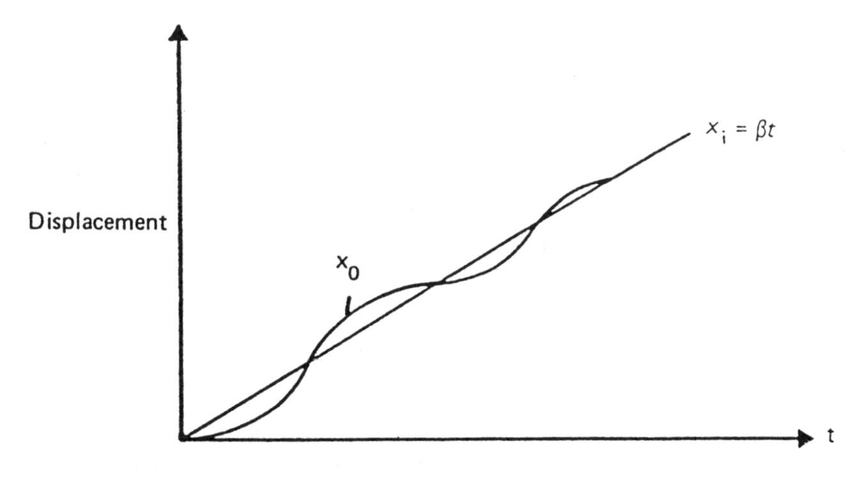

Fig. 5.10. Ramp input response.

There are many needs and applications for systems that will move a remote device to some prescribed position in an acceptable and controlled way. The input for this movement may be a manual action such as turning a knob or moving a lever, or it may come from the motion or signal from some other device. Some systems have relatively low power at the input in comparison with that dissipated at the output; essentially, therefore, they are error-actuated power amplifiers, and they are usually referred to as servomechanisms, or just servos.

5.1 THE SIMPLE HYDRAULIC SERVO

Hydraulic remote position control (RPC) systems are used extensively; they generally rely on some form of supply control valve feeding actuators, pumps, and hydraulic motors. The type of control valve used can vary, but the most common is the spool valve which is shown diagrammatically in Fig. 5.11.

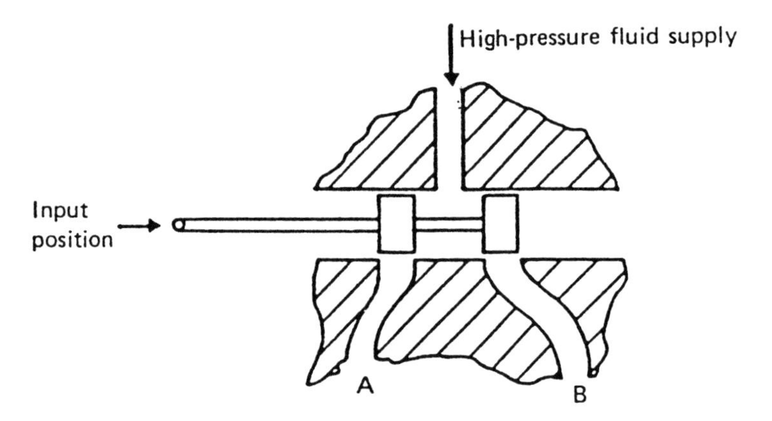

Fig. 5.11. Simple spool valve.

The input position of the spool may be controlled by direct mechanical linkage, or by electrical means. When the spool is centrally situated, as shown, both ports are closed and channels A and B are cut off. When the spool is moved to the right, B is connected to the high-pressure supply and when the spool is moved to the left, A is connected to the high-pressure supply. For small valve openings, it is usually assumed that the rate of flow of fluid is proportional to the valve opening, and that the fluid is incompressible. Some improvement in performance can be obtained by using the three-seal spool valve shown in Fig. 5.12.

5.1.1 Open loop hydraulic servo

Fig. 5.13 shows an open loop hydraulic servo. The input controls the position of the spool valve which directs fluid under pressure through a port to one end of the working cylinder. The working piston is connected to the controlled element so that

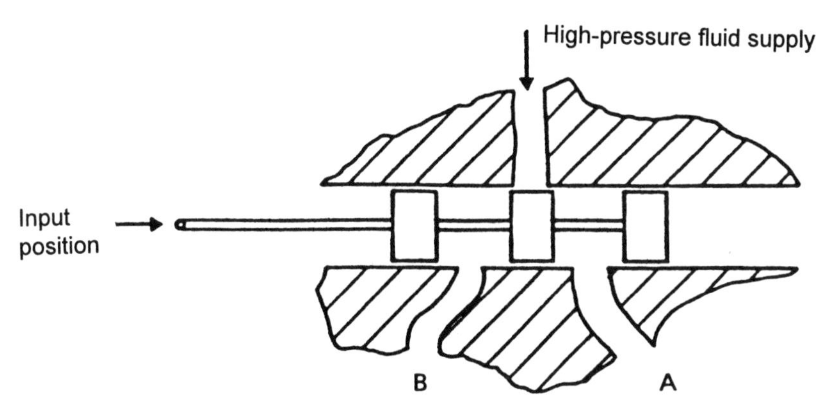

Fig. 5.12. Three-seal spool valve.

its position may be considered to be the output.

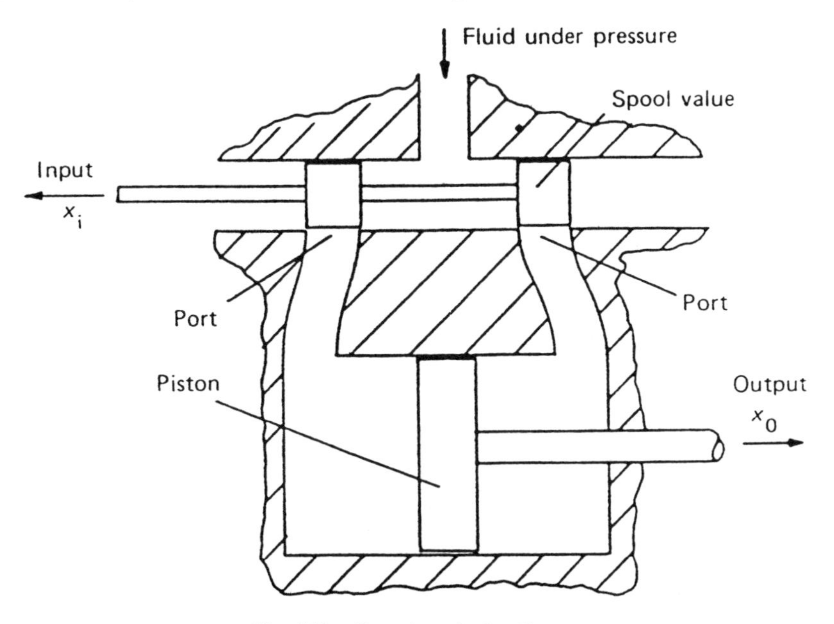

Fig. 5.13. Open loop hydraulic servo.

If the area of the piston is A and the port coefficient is b, the flow equation gives

$$Q = A\mathrm{D}x_{\mathrm{o}} = bx_{\mathrm{i}}$$

where Q is the rate of fluid flow.

Thus, $x_o = \left(\frac{b}{A}\right)\left(\frac{1}{D}\right)x_i$,

that is, the system integrates the input.
If a step input is applied,

at $\quad t < 0, \quad x_i = x_o = 0,$

and at $t \geqslant 0, \quad x_i = X.$

Thus when $t \geqslant 0$,

$$x_o = \left(\frac{b}{A}\right)\left(\frac{1}{D}\right)X = \left(\frac{b}{A}\right)Xt,$$

so that the output increases with the time – as shown in Fig. 5.14.

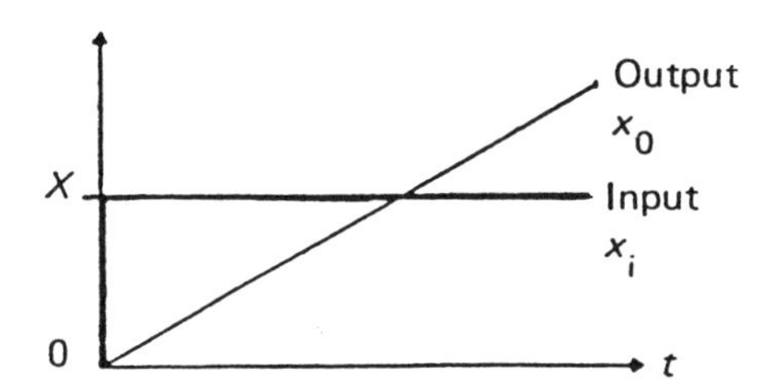

Fig. 5.14. Servo output.

This type of response is usually undesirable in a control system, so the system is improved by applying feedback to close the control loop.

5.1.2 Closed loop hydraulic servo

Because of the limitations of the open loop servo performance, feedback of the output is usually added, as shown in Fig. 5.15. As the output reaches the desired value it acts to move the spool valve and thereby close the port.

For the spool valve displacement y, from the figure,

$$\frac{x_i + x_o}{l + m} = \frac{x_i - y}{l},$$

so that

$$y = x_i - (l/(l + m))(x_o + x_i).$$

The flow equation gives $by = ADx_o$. Thus if y is eliminated,

$$ADx_o = -(bl/(l + m))x_o + (bm/(l + m))x_i.$$

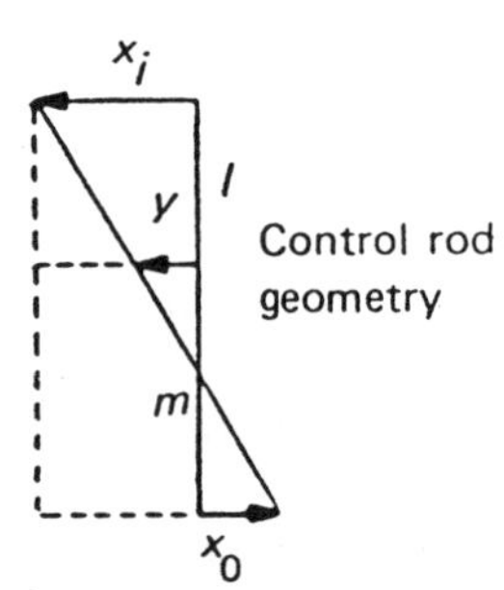

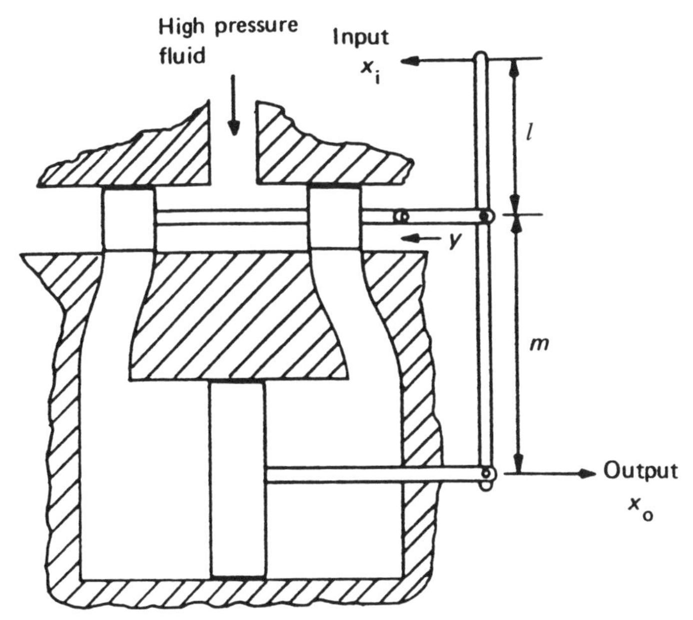

Fig. 5.15. Closed loop hydraulic servo.

that is,

$$[1 + ((l + m)/bl)A\mathrm{D}]x_o = (m/l)x_i,$$

or

$$[1 + T\mathrm{D}]x_o = (m/l)x_i. \tag{5.2}$$

where $T = ((l + m)/bl)A$ and has dimensions of time.

T is known as the *time constant* of the system. Equation (5.2) is the equation of motion, and it relates the output to the input. It is sometimes convenient to consider the error ε between input and output where $\varepsilon = x_i - x_o$.

Thus from (5.2), $(1 + T\mathrm{D})x_o = (m/l)(\varepsilon + x_o)$, or

$$X_o = (1 - (m/l) + T\mathrm{D})^{-1}(m/l)\varepsilon.$$

Hence a block diagram can be drawn to represent the system, as shown in Fig. 5.16.

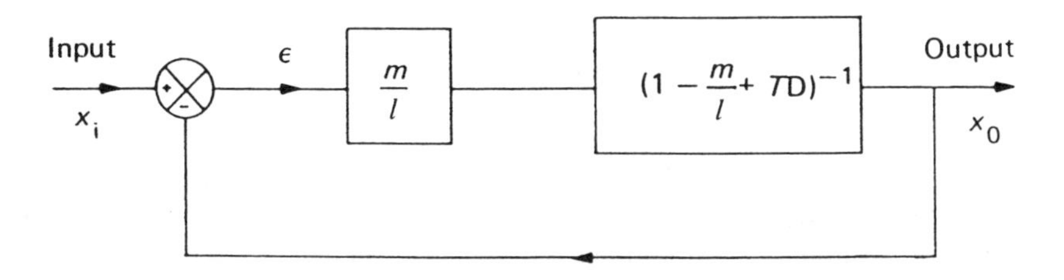

Fig. 5.16. Servo block diagram.

It should be noted that in the idealized hydraulic servo considered, many factors have been neglected such as friction, slackness in connections, non-linearity in the spool valve, pressure losses, compressibility of the fluid, and flexibility of the metal parts. In a real system some of these factors may have to be taken into account if they have a significant effect on the system response.

Response to a step input

A step input is when x_i is suddenly increased to a value, say X, and when $t < 0$ $x_i = 0$, and when $t \geqslant 0$, $x_i = X$.

From equation (5.2), for the output x_o the P.I. = $(m/l)X$ and C.F. = $Be^{-t/T}$. Thus

$$x_o = (m/l)X + Be^{-t/T}.$$

Since $x_o = 0$ when $t = 0$, $B = -(m/l)X$, hence

$$x_o = (m/l)X[1 - e^{-t/T}].$$

That is, x_o increases exponentially with time to a value $(m/l)X$; the system time constant T has considerable effect on this response (see Fig. 5.17).

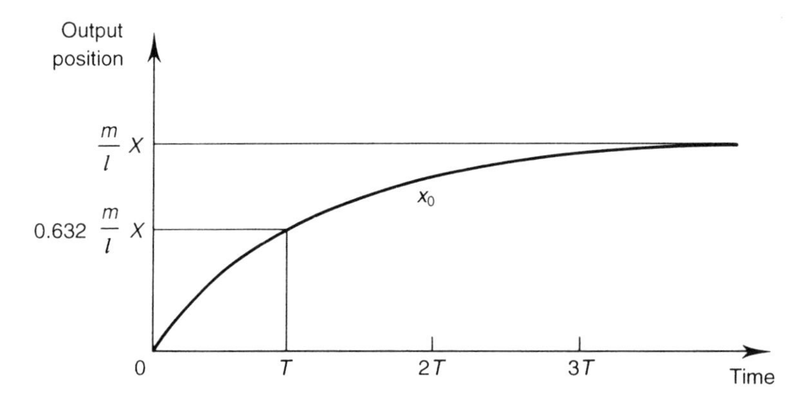

Fig. 5.17. Step input response.

After time T, the time constant of the servo,

$$x_o = \left(\frac{m}{l}\right)X(1 - e^{-1}) = \left(\frac{m}{l}\right)X(0.632),$$

that is, the output reaches 63.2% of its final value at time T. It can also be shown that the output reaches 86.5% of its final after a time $2T$, and 95% after a time $3T$. The factor (m/l) amplifies or attenuates X to give the final x_o value reached. This may be compared with the servo without feedback when $x_o = (b/A)Xt$, and the output steadily increases with time.

Response to a ramp input

A ramp input is an input which increases uniformly with time. Thus $x_i = \beta t$ is a ramp input, where β is a constant.

From equation (5.2), for the output x_o,

$$\text{P.I.} = (m/l)\beta t - (m/l)T\beta \text{ and C.F.} = Be^{-t/T}.$$

Thus

$$x_o = Be^{-t/T} + (m/l)\beta t - (m/l)T\beta.$$

If $x_o = 0$ when $t = 0$, $B = (m/l)T\beta$. Hence

$$x_o = (m/l)T\beta(e^{-t/T} - 1) + (m/l)\beta t.$$

For the case when $m = l$,

$$x_o = T\beta(e^{-t/T} - 1) + \beta t.$$

Thus x_o increases exponentially with time towards a value βt ($= x_i$). The error between the output and the input is $T\beta(e^{-t/T} - 1)$.

Even when $t = \infty$, this error is $- T\beta$. This is known as a *steady-state error* because it persists after steady conditions have been attained (see Fig. 5.18).

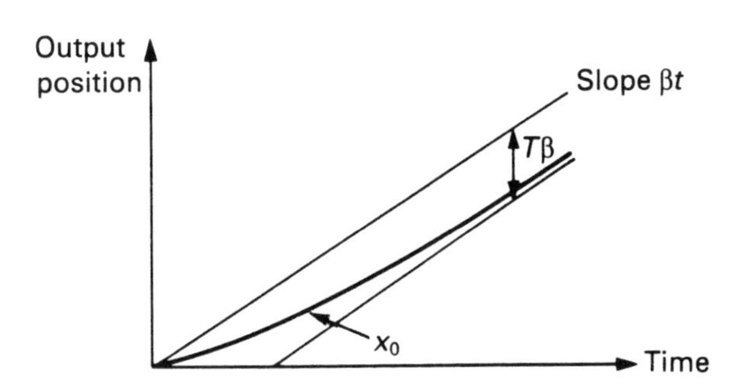

Fig. 5.18. Ramp input response.

Response to a sinusoidal input

For a sinusoidal input $x_i = X_i \cos vt$.

For x_o the C.F. is an exponential decay, and this can usually be neglected. A P.I. can be assumed, $x_o = X_o \cos (vt - \phi)$.

Substituting this solution into equation (5.2) gives

$$X_o \cos (vt - \phi) - vTX_o \sin (vt - \phi) = (m/l)X_i \cos vt.$$

This equation can be solved by drawing a phasor diagram as shown in Fig. 5.19.

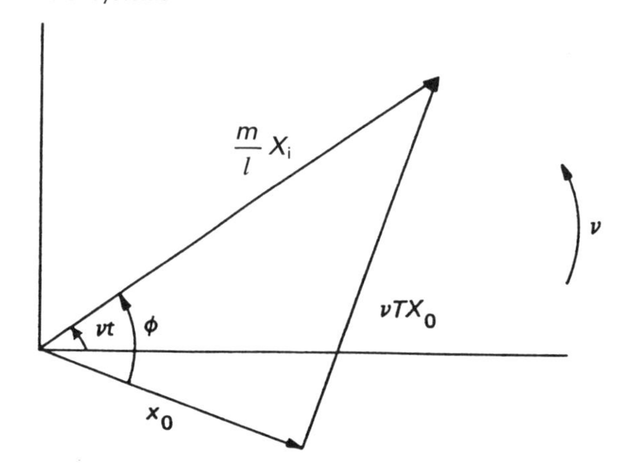

Fig. 5.19. Sinusoidal input response.

Thus

$$(m/l)^2X_i^2 = (1 + \nu^2T^2)X_o^2$$

or

$$X_o/X_i = (m/l)/\sqrt{(1 + \nu^2T^2)} \quad \text{and} \quad \phi = \tan^{-1}\nu T. \tag{5.3}$$

It is often desirable to show graphically how the steady-state response to a sinusoidal input varies with frequency. There are several methods of doing this; one is to draw a Bode diagram. The Bode diagram is a plot of gain (X_o/X_i) against frequency. The magnitude of (X_o/X_i) is plotted on a log scale, usually dB.

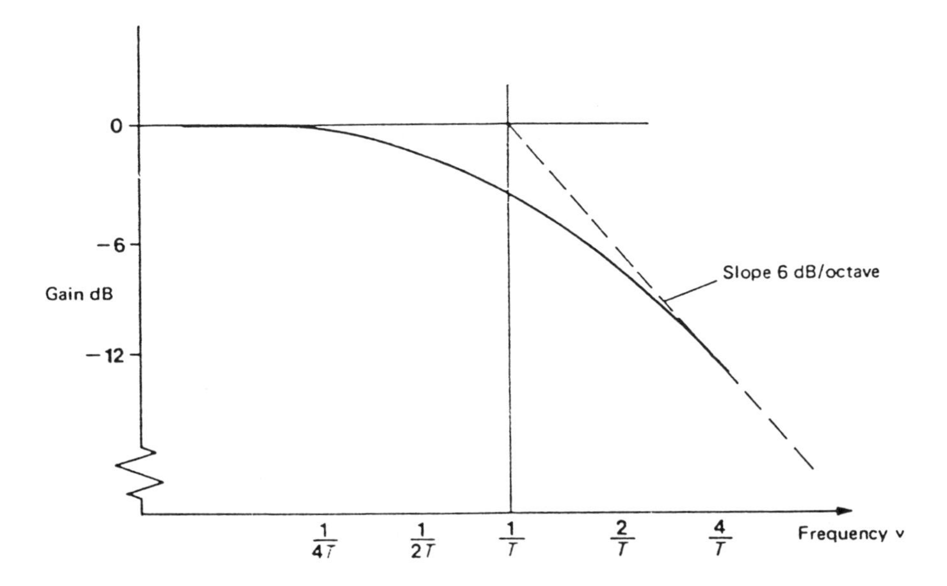

Fig. 5.20. Bode diagram.

The gain (or attenuation) in dB = 20 $\log_{10}(X_o/X_i)$, so for equation (5.3)

$$\text{Gain (dB)} = 20\log_{10}(m/l) - 10\log_{10}[1 + v^2T^2].$$

Fig. 5.20 shows the diagram for the case $m = l$.

Further discussion of the Bode analysis technique is given in section 5.7.2.

An alternative diagram is the harmonic response locus which can be drawn by plotting gain and phase as a polar diagram. Since

$$X_i = (l/m)X_o\sqrt{(1 + v^2T^2)} = [(lX_o/m)^2 + (lX_ovT/m)^2]^{1/2},$$

the locus is as shown in Fig. 5.21.

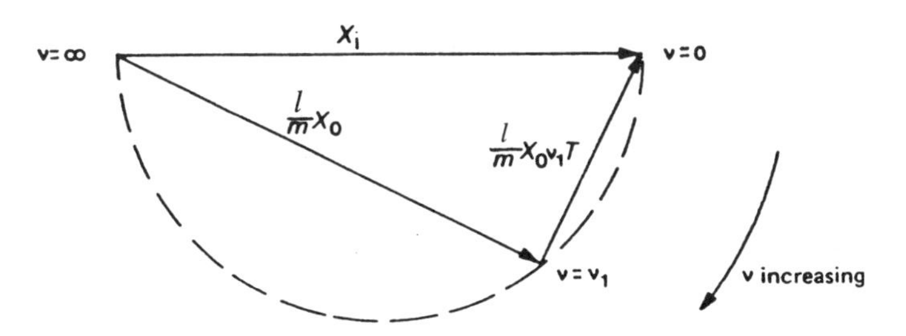

Fig. 5.21. Harmonic response locus.

5.2 MODIFICATIONS TO THE SIMPLE HYDRAULIC SERVO

5.2.1. Derivative control

An intelligent human controller would take account not only of the instantaneous value of the input but also its rate of change, or the rate of change of error. This would help to avoid overshooting and would improve the system response. The simple servo can be modified to act in a similar way, as shown in Fig. 5.22.

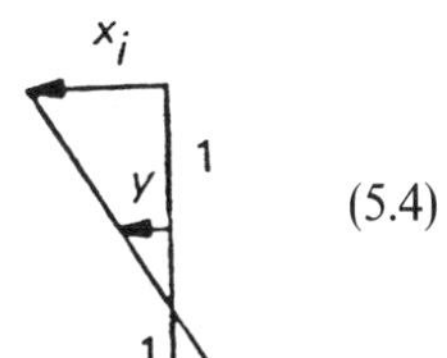

From the control rod geometry, the valve displacement

$$y = (x_i - v)/2. \qquad (5.4)$$

The flow equation gives

$$by = ADx_o \qquad (5.5)$$

and force balance gives

$$k(rx_o - v) - cDv = 0. \qquad (5.6)$$

It is required to eliminate v and y and to give x_o as $f(x_i)$. From (5.6),

$$v = (kr/(k + cD))x_o.$$

Substitute in (5.4):

$$y = (x_i/2) - (kr/(k + cD))(x_o/2)$$

and in (5.5)

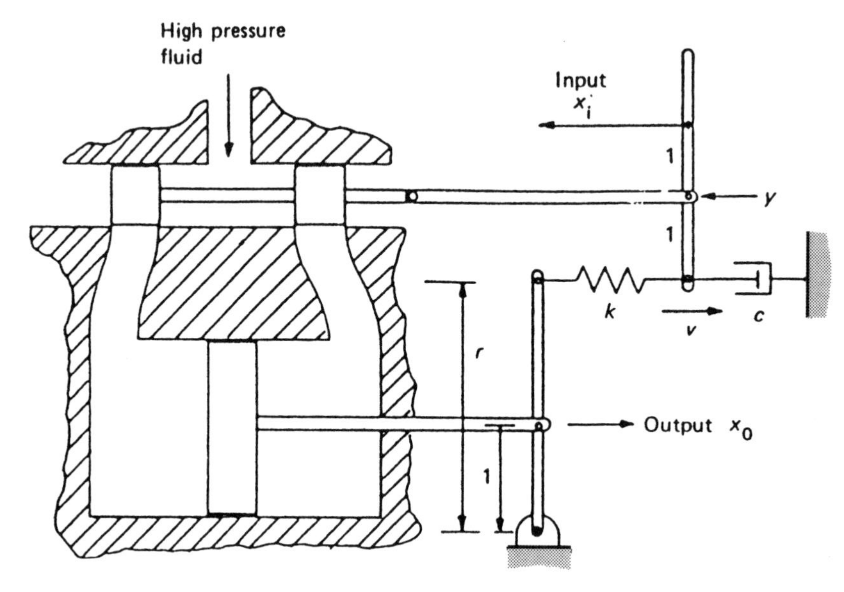

Fig. 5.22. Hydraulic servo with derivative control.

$$AD x_o = bx_i/2 - (b/2)(kr/(k + cD))x_o,$$

that is

$$(2AcD^2 + 2kAD + bkr)x_o = (bk + bcD)x_i$$

or

$$((2Ac/bkr)D^2 + (2A/br)D + 1)x_o = (1 + c/kD)x_i/r. \tag{5.7}$$

Comparing equation (5.7) with equation (5.2) which was derived for a similar system without a velocity element, it can be seen that both sides of the equation of motion have gone up by one order. Thus when x_i is changed the valve setting y is not reduced by the full value corresponding to the change in x_o, because the faster x_i and x_o change, the less y is reduced.

From equation (5.7),

$$x_o = \left(1 + \frac{2A}{br}D + \frac{2Ac}{bkr}D^2\right)^{-1}\left(1 + \frac{c}{k}D\right)\frac{x_i}{r},$$

so for a step input, $x_i = X$,

$$x_o = \left(1 - \frac{2A}{br}D - \cdots\right)\left(\frac{X}{r}\right) = \frac{X}{r}.$$

The C.F. of x_o is a damped oscillation or decay, so the response to the step input is as shown in Fig. 5.23.

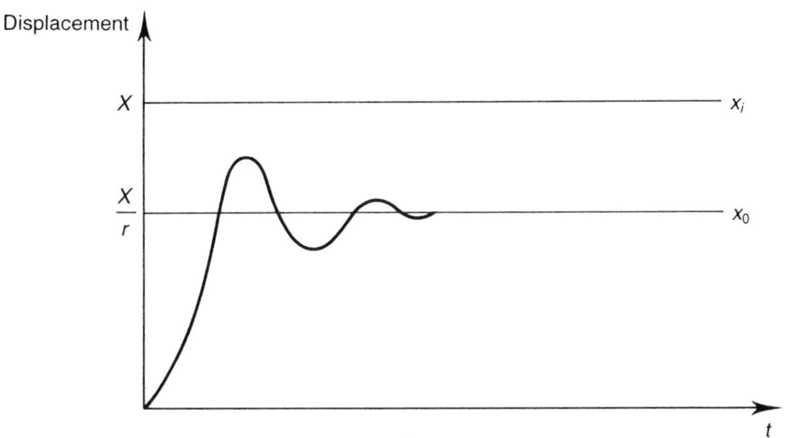

Fig. 5.23. Step input response.

For a ramp input, $x_i = \beta t$, hence,

$$x_o = \left(1 - \frac{2A}{br}\,\mathrm{D} - \cdots\right)\left(\frac{\beta t}{r} + \frac{c}{kr}\beta\right)$$

$$= \frac{\beta t}{r} + \frac{c\beta}{kr} - \frac{2A\beta}{br\,r},$$

so that a steady-state error exists. The initial oscillation is given by the C.F. as above.

The steady-state response to a sinusoidal input can be found by drawing a phasor diagram. If $x_i = X_i \cos vt$, $x_o = X_o \cos(vt - \phi)$, then from Fig. 5.24.

$$X_o = \frac{\sqrt{(1 + (cv/k)^2)}}{\sqrt{((r - (2Ac/bk)v^2)^2 + (2Av/b)^2)}}\, X_i.$$

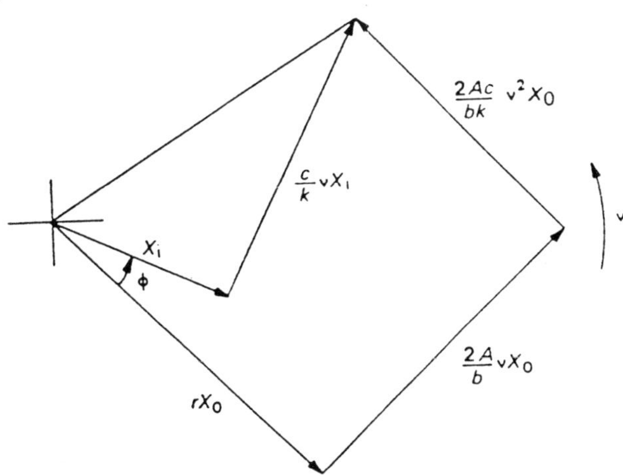

Fig. 5.24. Phasor diagram.

At low frequencies $\nu \simeq 0$, so $X_o \to X_i/r$ and $\phi \to 0$.
At high frequencies $\nu \simeq \infty$, so $X_o \to 0$ and $\phi \to \pi/2$.
A detailed comparison of the simple system performance with that for a system with derivative control added depends on the system parameters; but, in general derivate control allows a faster transient response, with overshoot if too fast.

5.2.2 Integral control

Both the simple system and that with derivative control will give steady-state errors when the input is a ramp, $x_i = \beta t$. Real systems, in which fewer assumptions are made than here, can give steady-state errors with other kinds of input. To eliminate these errors a correction can be added to the system which is proportional to the time integral of the error: such a system is shown in Fig. 5.25. It can be seen that compared to the system with derivation action, Fig. 5.22, k and c have been interchanged.

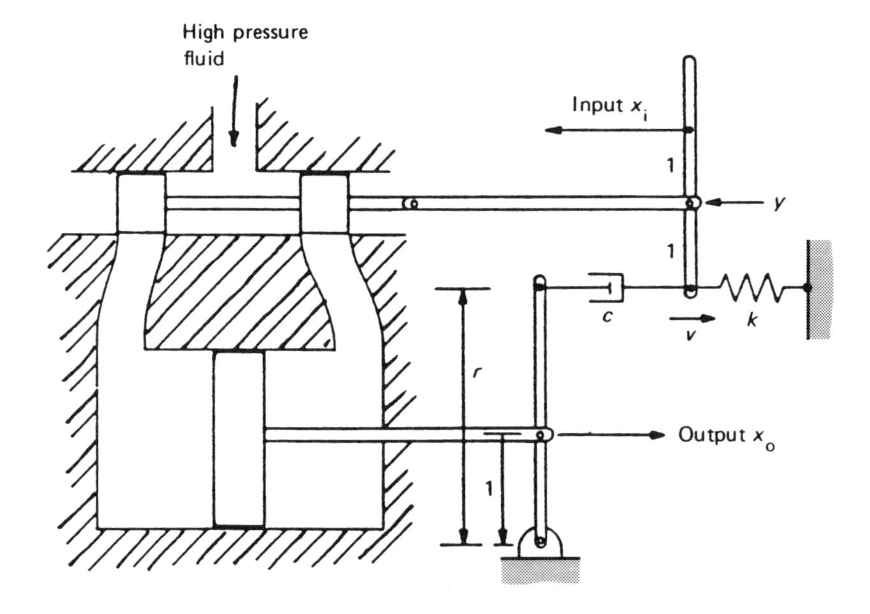

Fig. 5.25. Hydraulic servo with integral control.

Equations (5.4) and (5.5) apply, but force balance gives $c\mathrm{D}(rx_o - v) - kv = 0$ and thus

$$\left(1 + \frac{2Ac}{2Ak + bcr}\mathrm{D}\right)x_o = \left(\frac{bc}{2Ak + bcr} + \frac{bk}{2Ak + bcr}\frac{1}{\mathrm{D}}\right)x_i \tag{5.8}$$

Comparing equation (5.8) with equation (5.2) shows an extra term proportional to the integral of x_i. The response to a step or sinusoidal input can be found from equation (5.8).

Example 43
A servomechanism incorporating an hydraulic relay with displacement feedback through a dashpot and spring assembly is shown below.

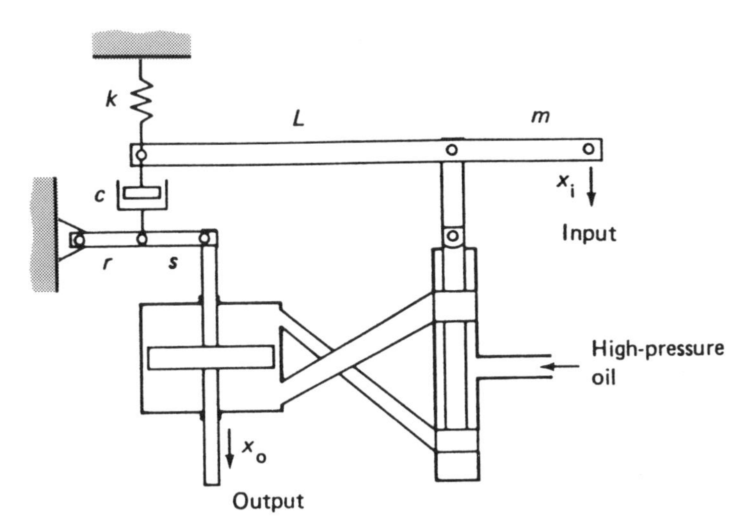

The velocity of the output ram (Dx_o) is equal to K times the movement of the pilot valve from its neutral position. Given that the inertia of all parts and the effect of friction (other than at the dashpot c) may be neglected, determine the equation of motion of the system.

For the control rod,

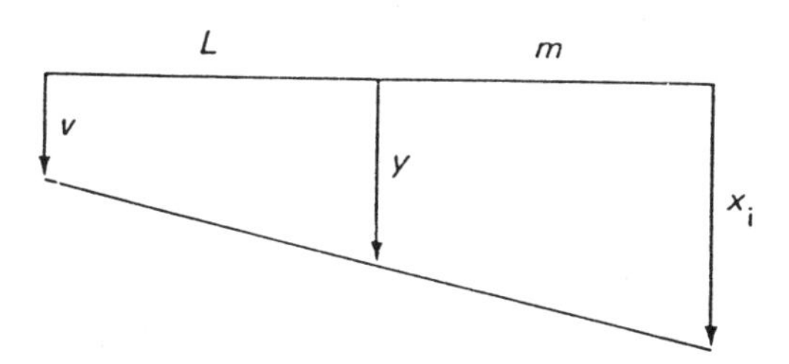

where v is the spring displacement and y is the movement of the spool valve.

Thus

$$\frac{y-v}{L}=\frac{x_i-v}{L+m},$$

and

$$y=\left(\frac{L}{L+m}\right)x_i-\left(\frac{m}{L+m}\right)v.$$

Force balance for the spring and dashpot gives

$$kv = c\mathrm{D}\left(\left(\frac{r}{r+s}\right)x_o - v\right),$$

so that $(k + c\mathrm{D})v = \left(\dfrac{r}{r+s}\right)c\mathrm{D}x_o$.

The flow equation gives $\mathrm{D}x_o = Ky$.
Substituting for y and v gives

$$\mathrm{D}x_o = K\left[\left(\frac{L}{L+m}\right)x_i - \left(\frac{m}{L+m}\right)\left(\frac{r}{r+s}\right)\left(\frac{c\mathrm{D}}{k+c\mathrm{D}}\right)x_o\right],$$

which can be rearranged to give the equation of motion as

$$\left[\mathrm{D}^2 + \mathrm{D}\left(\frac{k}{c} + \frac{Kmr}{(L+m)(r+s)}\right)\right]x_o = \left(\frac{KL}{L+m}\right)\left(\frac{k+c\mathrm{D}}{c}\right)x_i.$$

Example 44

The figure opposite shows an hydraulic servo, in which it can be assumed that

(a) all friction (except that intended to be present in the dashpot which has a viscous damping coefficient c) is negligible,
(b) the oil is incompressible,
(c) the inertia forces are negligible, and
(d) the volume flow per unit time through either port in the valve is b times the displacement of the valve spool from its neutral position.

Derive the transfer function relating the output displacement z to the input displacement x, this being measured from the position corresponding to zero oil flow. Find the value of z as a function of time following a sudden change in x of magnitude X.

The flow equation gives $by = A\mathrm{D}z$, where y is the displacement of the spool valve. If the spring extension is p,

$$2(y + p) = (x + p)$$

so

$$y = \frac{x - p}{2}$$

Thus

$$b\left(\frac{x-p}{2}\right) = A\mathrm{D}z$$

For equilibrium,

$$kp = c\mathrm{D}(nz - p)$$

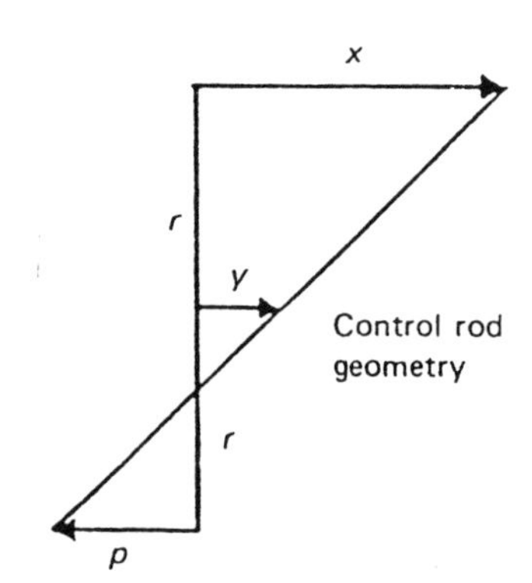

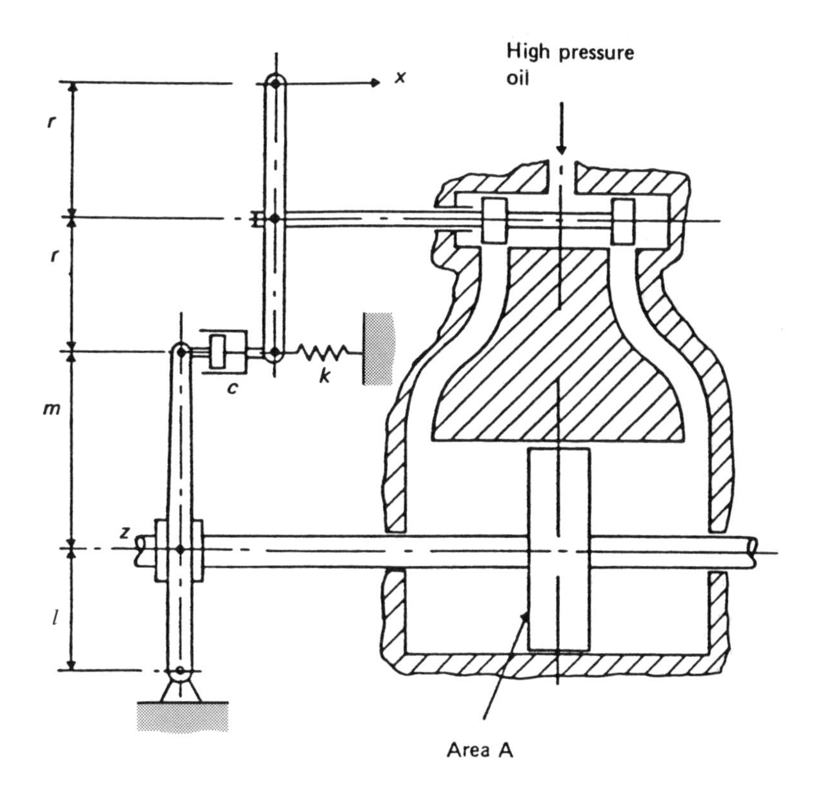

where

$$n = \frac{l + m}{l}.$$

Since

$$p = x - \frac{2A}{b}\mathrm{D}z,$$

$$cnDz - c\mathrm{D}x + \frac{2Ac}{b}\mathrm{D}^2 z = kx - \frac{2kA}{b}\mathrm{D}z,$$

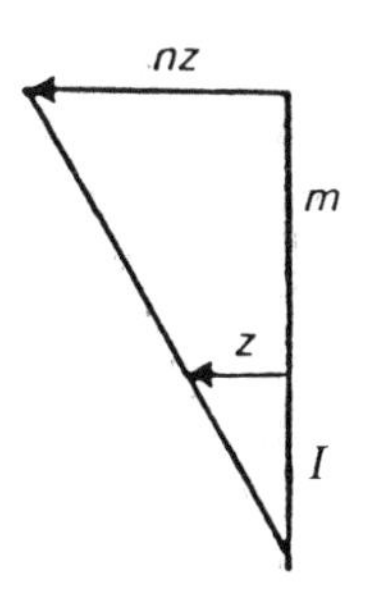

so

$$z = \left[\frac{k + c\mathrm{D}}{cn\mathrm{D} + (2Ak/b)\mathrm{D} + (2Ac/b)\mathrm{D}^2}\right]x,$$

or the transfer function

$$\frac{z}{x} = \left[\frac{b(k + c\mathrm{D})}{\mathrm{D}(cnb + 2Ak + 2Ac\mathrm{D})}\right]$$

If x changes by a step X,

$$\text{P.I. is } \left(\frac{bk}{(cnb + 2Ak)\mathrm{D}(1 + \cdots)}\right)X = \left(\frac{bk}{cnb + 2Ak}\right)Xt,$$

$$\text{C.F. is } P + Q \exp\left[-\left(\frac{cnb + 2Ak}{2Ac}\right)t\right],$$

so

$$z = P + Q\exp\left[-\left(\frac{cnb + 2Ak}{2Ac}\right)\right]t + \left(\frac{bk}{cnb + 2Ak}\right)Xt.$$

Now when $t = 0, z = 0,$ so $P + Q = 0$. And when $t = 0,$ $p = 0,$ so $\mathrm{D}z = (b/2A)X$. Thus

$$\frac{bX}{2A} = \mathrm{D}z = 0 + P\left(\frac{cnb + 2Ak}{2Ac}\right) + \left(\frac{bk}{cnb + 2Ak}\right)X$$

so

$$-P\left(\frac{cnb + 2Ak}{2Ac}\right) = b\left(\frac{k}{cnb + 2Ak} - \frac{1}{2A}\right)X$$

or

$$P = \frac{b^2c^2nX}{(cnb + 2Ak)^2}$$

Thus,

$$z = \left(\frac{bkX}{cnb + 2Ak}\right)t + \frac{b^2c^2nX}{(cnb + 2Ak)^2}\left\{1 - \exp\left[-\left(\frac{cnb + 2Ak}{2Ac}\right)t\right]\right\}$$

Example 45

In the hydraulic position control servo shown, the desired input position is converted to a voltage V_i and compared with the output position voltage V_o. Any error is applied to a power amplifier and position transducer to move the spool valve controlling the oil flow. The oil is delivered to the cylinder so that the piston acts on the load as required; a position transducer generates the signal V_o as shown.

With the parameters given in the figure, find the relationship between the output position and the input position,

$$\frac{X_o}{X_i}$$

If an input of 5 cm/s is applied, find the error between the input and output after 2 s; neglect transient effects. Also calculate the required coefficient of the output displacement transducer for this error to be a minimum, and the size of this minimum.

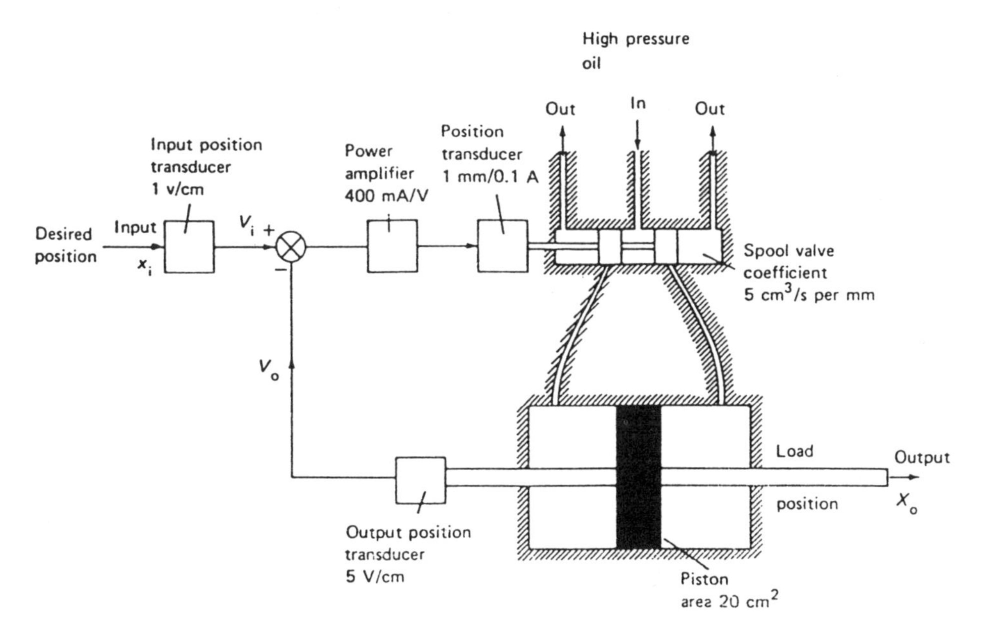

$$\text{Spool valve opening} = (V_i - V_o)\,400\frac{1}{100}\ \text{mm}.$$

Oil flow

$$= 20\ DX_o\ \text{cm}^3$$
$$= 5(V_i - V_o)4\ \text{cm}^3.$$

But

$$V_o = 5X_o \text{ and } V_i = X_i,$$

so

$$20\ DX_o = 5(X_i - 5X_o)4$$

or

$$\frac{X_o}{X_i} = \frac{0.2}{1 + 0.2D}$$

Input is 5 cm/s, so $X_i = 5t$ cm.

In the steady state

$$X_o = \left(\frac{0.2}{1 + 0.2\mathrm{D}}\right)X_i$$
$$= 0.2[1 - 0.2\mathrm{D} + \cdots]5t$$
$$= t - 0.2.$$

Hence steady-state error $= X_i - X_o$

$$= 5t - (t - 0.2)$$

$$= 4t + 0.2.$$

After $2s$, ss error $= +8.2$cm.
For minimum error, let output position coefficient be α V/cm.

Then

$$V_o = \alpha X_o$$

and

$$20\,DX_o = 5(X_i - \alpha X_o)4.$$

Hence

$$DX_o = X_i - \alpha X_o.$$

Thus

$$\frac{X_o}{X_i} = \frac{1}{D + \alpha},$$

and steady-state output $X_o = \dfrac{1}{\alpha}\left(1 - \dfrac{D}{\alpha} + \cdots\right)5t$

$$= \frac{5}{\alpha}t - \frac{5}{\alpha^2}.$$

When $t = 2$, the steady-state error is

$$\varepsilon = 5 \times 2 - \frac{5}{\alpha} \times 2 + \frac{5}{\alpha^2}.$$

for minimum ε, $d\varepsilon/d\alpha = 0$.

That is,

$$d\varepsilon/d\alpha = 0 = \frac{10}{\alpha^2} - \frac{10}{\alpha^3},$$

so $\alpha = 1$.
Therefore minimum steady-state error $= 10 - 10 + 5 = +5$ cm.

5.3 THE ELECTRIC POSITION SERVOMECHANISM

Electric position servos are widely used in control systems; an example of turntable position control has already been considered in Chapter 1.

Open loop systems are available, but in practice the operator often has to act as the feedback to ensure correct positioning. Usually, therefore, only the closed loop servo is considered.

5.3.1 The basic closed loop servo

The elements of the basic closed loop position servo are shown in block diagram form in Fig. 5.26. The position of the load is compared to the desired position, thereby creating an error signal. The amplifier and motor respond to the error signal and act on the load position.

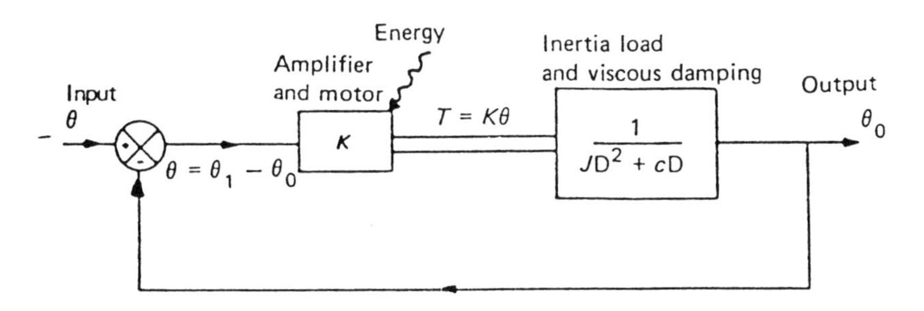

Fig. 5.26. Closed loop position servo.

The error signal θ is the difference between the desired position (θ_i) and the actual position (θ_o) that is, $\theta = \theta_i - \theta_o$. The amplifier and motor are such that the torque T produced by the motor is $K\theta$. For the load,

$$T = (JD^2 + cD)\theta_o,$$

since it is considered to comprise an inertia load with viscous damping. Thus

$$K(\theta_i - \theta_o) = (JD^2 + cD)\theta_o$$

or

$$(JD^2 + cD + K)\theta_o = K\theta_i.$$

This is the same equation of motion as found for forced damped vibration, and therefore the solution is the same for a given input.

Response of servo to a step input $\theta_i = X$.
For θ_o, C.F. $= Ae^{-\zeta\omega t} \sin(\omega\sqrt{(1-\zeta^2)}t + \phi)$,
where $\zeta = c/c_c$ and $\omega = \sqrt{(K/J)}$ rad/s, and

$$\text{P.I.} = [1 + (c/K)D + (J/K)D^2]^{-1}X$$

$$= [1 - (c/K)D + \cdots]X$$

$$= X.$$

Thus

$$\theta_o = Ae^{\zeta\omega t} \sin(\omega\sqrt{(1-\zeta^2)}t + \phi) + X. \tag{5.9}$$

The response is shown in Fig. 5.27.

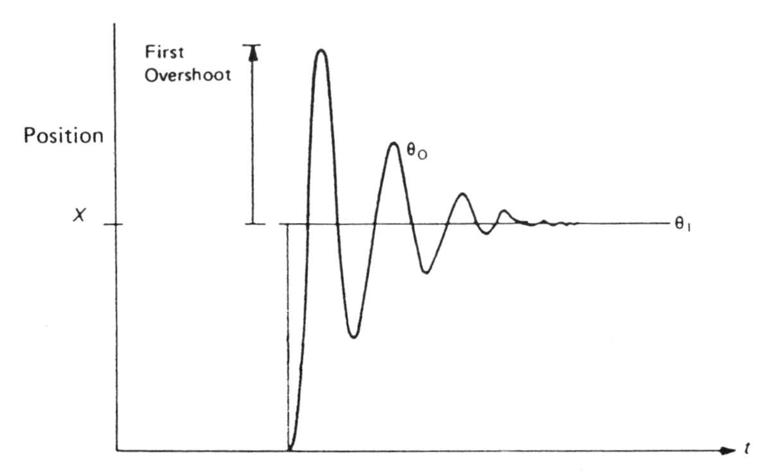

Fig. 5.27. Step input response.

Clearly the fastest response is when the damping is low. Unfortunately this is accompanied by a large first overshoot and a poor transient response, that is, the oscillations take a long time to die away. Increasing the damping slows the response but reduces the overshoot and improves the transient. Critical damping gives the quickest response without overshoot.

To find the overshoot, $d\theta_o/dt$ is found and made zero and hence the times to overshoot are obtained. Substituting these times into (5.9) gives θ_o. See Example 46. There is no steady state error.

Response of servo to a ramp input $\theta_i = \beta t$

For θ_o, C.F. as above (same system), but

$$\begin{aligned} \text{P.I.} = (1 + (c/K)\mathrm{D} + (J/K)\mathrm{D}^2)^{-1}\beta t &= \beta t - (c/K)\beta \\ &= \theta_i - (c/K)\beta. \end{aligned}$$

Hence there is a steady-state error $(c/K)\beta$.

Thus high damping to achieve a good transient response also results in a large steady-state error to a ramp input.

Response of servo to an harmonic input $\theta_i = A \sin vt$

The analysis is exactly as for the forced damped vibration system, so

$$\theta_o = \frac{KA}{\sqrt{((K - Jv^2)^2 + c^2v^2)}} \sin(vt - \phi), \text{ where } \phi = \tan^{-1}\left(\frac{cv}{K - Jv^2}\right).$$

Hence there are gain and phase errors which are functions of the system parameters and the frequency of excitation.

Because of the shortcomings of the basic system, it is often modified to provide an improved response.

Example 46

In the closed loop system shown in Fig. 5.26 the input is given a step increase θ_i. Find the output, θ_o, as a function of time, assuming that both θ_o and $d\theta_o/dt$ are zero at $t = 0$. Hence find the time to the first overshoot and the corresponding magnitude of θ_o.

From equation (5.9),

$$\theta_o = \theta_i + e^{-\zeta\omega t}[A \sin \omega\sqrt{(1-\zeta^2)}t + B \cos \omega\sqrt{(1-\zeta^2)}t].$$

Now, when $\theta_o = 0$, $t = 0$, so $\theta_i = -B$, and when $d\theta_o/dt = 0$, $t = 0$, so $0 = -\zeta\omega B + \omega\sqrt{(1-\zeta^2)}A$

and

$$A = -\frac{\zeta}{\sqrt{(1-\zeta^2)}}\theta_i.$$

Hence

$$\theta_o = \theta_i - \theta_i e^{-\zeta\omega t}\left[\frac{\zeta}{\sqrt{(1-\zeta^2)}} \sin \omega\sqrt{(1-\zeta^2)}t + \cos \omega\sqrt{(1-\zeta^2)}t\right]$$

that is,

$$\theta_o = \theta_i\left[1 - \frac{\zeta}{\sqrt{(1-\zeta^2)}} e^{-\zeta\omega t}[\sin (\omega\sqrt{(1-\zeta^2)}t + \phi)]\right]$$

where

$$\tan \phi = \frac{\sqrt{(1-\zeta^2)}}{\zeta}.$$

Now at overshoot, time is given by $d\theta_o/dt = 0$.

That is,

$$0 = +\zeta\omega e^{-\zeta\omega t}\frac{\theta_i}{\sqrt{(1-\zeta^2)}} \sin (\omega\sqrt{(1-\zeta^2)}t + \phi)$$

$$-\frac{\theta_i}{\sqrt{(1-\zeta^2)}}e^{-\zeta\omega t}\omega\sqrt{(1-\zeta^2)}\,[\cos(\omega\sqrt{(1-\zeta^2)}t + \phi)].$$

Hence

$$\frac{\sqrt{(1-\zeta^2)}}{\zeta} = \tan (\omega\sqrt{(1-\zeta^2)}t + \phi)$$

$$= \tan\left(\omega\sqrt{(1-\zeta^2)}t + \tan^{-1}\frac{\sqrt{(1-\zeta^2)}}{\zeta}\right)$$

$$\text{or } \omega\sqrt{(1-\zeta^2)}t = 0, \pi, 2\pi \ldots$$

so that the time to the first overshoot is

$$\frac{\pi}{\omega\sqrt{(1-\zeta^2)}}.$$

This is one half of the period of the damped oscillation.

At this time,

$$\theta_o = \theta_i - \frac{\theta_i}{\sqrt{(1-\zeta^2)}} \exp\left(\frac{-\zeta\omega\pi}{\omega\sqrt{(1-\zeta^2)}}\right) \sin(\pi + \phi)$$

$$= \theta_i - \frac{\theta_i}{\sqrt{(1-\zeta^2)}} \exp\left(\frac{-\zeta\pi}{\sqrt{(1-\zeta^2)}}\right)(-\sin\phi)$$

$$= \theta_i + \frac{\theta_i}{\sqrt{(1-\zeta^2)}} \exp\left(\frac{-\zeta\pi}{\sqrt{(1-\zeta^2)}}\right)\sqrt{(1-\zeta^2)}$$

Thus

$$\theta_o = \theta_i\left[1 + \exp\left(-\frac{\zeta\pi}{\sqrt{(1-\zeta^2)}}\right)\right]$$

Now if in a given servo, $\omega = 5$ rad/s, $\zeta = 0.1$ and $\theta_i = 10$ deg.

Time to first overshoot is

$$\frac{\pi}{5\sqrt{(1-0.1^2)}} = 0.63 \text{ s}$$

and

$$\theta_o = 10\left[1 + \exp\left(-\frac{0.1\pi}{\sqrt{(1-0.1^2)}}\right)\right] = 17.3°.$$

Increasing ζ reduces θ_o but increases the time to the first overshoot.

Example 47

The angular position θ_o of a turntable is controlled by a closed loop servomechanism which is critically damped by viscous friction. The moment of inertia of the turntable is 400 kg m^2, and the motor torque is 3.6 kNm/rad of misalignment between θ_o and the desired position θ_i. Given that $\dot{\theta}_i = 10$ rev/min, find the steady-state position error.

The equation of motion is

$$(JD^2 + cD + K)\theta_o = K\theta_i$$

so that

$$\theta_o = \left[1 + \frac{c}{K}\,\mathrm{D} + \frac{J}{K}\,\mathrm{D}^2\right]^{-1}\theta_i.$$

If $\theta_i = \beta t$, $\quad \theta_o = \beta t - (c/K)\beta$,

so the steady-state error is $\dfrac{c}{K}\,\beta$.

Now $(c/K)\beta = 2\sqrt{(J/K)}\beta$, since damping is critical and $c_c = 2\sqrt{(JK)}$.

Hence

$$\theta_{SS} = 2\sqrt{\left(\frac{400}{3600}\right)\left(\frac{2\pi 10}{60}\right)\left(\frac{180}{\pi}\right)} = 40^\circ.$$

Response of servo to sudden loads

Servomechanisms are sometimes subjected to sudden loads due to shocks and impulses, such as those arising from gust loads and impacts. Fig. 5.28 shows a servo with a sudden load T_L applied.

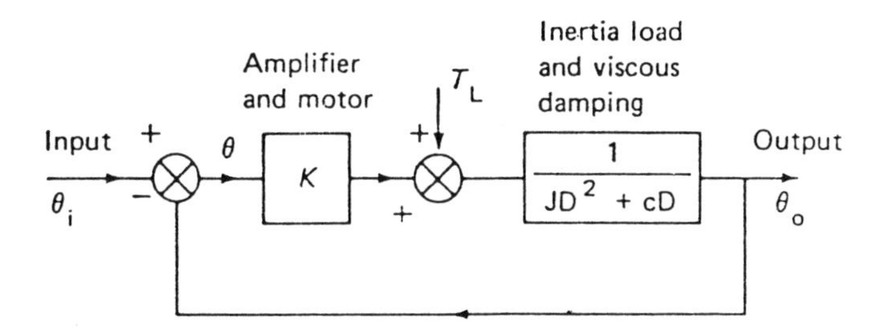

Fig. 5.28. Servo with external load, T_L.

In this system,

$$K(\theta_i - \theta_o) + T_L = (J\mathrm{D}^2 + c\mathrm{D})\theta_o,$$

so that

$$(J\mathrm{D}^2 + c\mathrm{D} + K)\theta_o = T_L + K\theta_i.$$

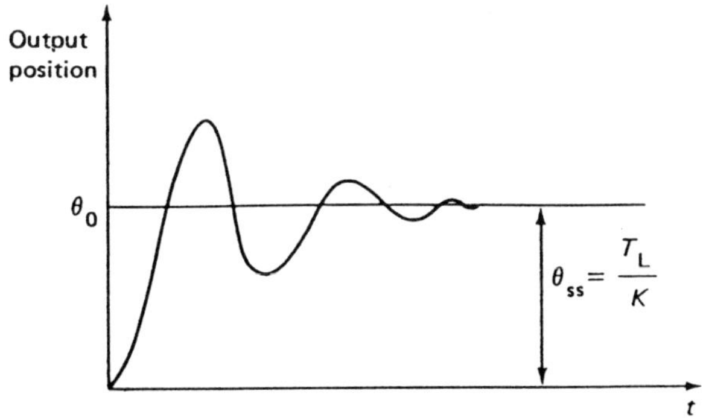

Fig. 5.29. Servo response to sudden load.

If the input θ_i is zero, the response to the sudden load T_L gives a steady-state error T_L/K.
The complementary function gives the transient response so that the output position is as shown in Fig. 5.29.

Example 48
In a simple servo mechanism, $J = 0.1$ kg m^2, $K = 20$ N m/rad error, and $c = \frac{1}{2}c_c$. Given that a step input is applied,
(i) find the frequency of the transient oscillation of the output, and the time to rise to the first maximum;
(ii) find the position lag if an output of 10 rev/min is applied; and
(iii) calculate the steady state error when a load of 1 Nm is applied.

(i) $$\omega = \sqrt{\left(\frac{K}{J}\right)} = \sqrt{\left(\frac{2.0}{0.1}\right)} = 14.1 \text{ rad/s},$$

and

$$\zeta = 0.5$$

so

$$\omega_v = \omega\sqrt{(1 - \zeta^2)} = 14.1\sqrt{[1 - (0.5)^2]} = 12.3 \text{ rad/s},$$

and frequency of oscillation, $f = 12.3/2\pi = 1.95$ Hz.

The time to the first maximum $= (\frac{1}{2})(1/f) = 0.256$ s.

(ii) $$\text{Steady-state error} = \frac{c}{K}\beta$$

$$= \frac{\sqrt{(KJ)}}{K}\beta = \frac{\beta}{\omega}$$

$$= \frac{2\pi 10}{60 \times 14.1} = 0.074 \text{ rad} = 4.2°$$

(iii) $$\text{Steady-state error} = \frac{F}{K} = \frac{1}{20} = 0.05 \text{ rad} = 2.9°$$

Example 49
The block diagram of a satellite tracking system is shown

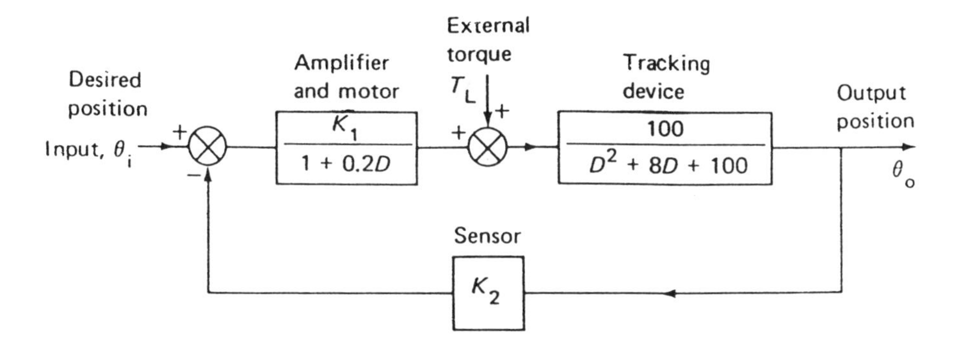

The transfer functions of the amplifier plus motor, tracking device and sensor are as shown. The system is subjected to the desired position input, and also an external torque T_L which is due to wind loads acting on the structure.

(i) When $T_L = 0$, find the closed loop transfer function, θ_o/θ_i,
(ii) Find K_1 and K_2 if
  (a) there is to be no steady-state error with a step position input, and
  (b) when a ramp position input of 2 rad/s is applied the steady-state error is to be 0.056 rad.
(iii) A sudden steady wind can be approximated to a suddenly applied torque $T_L = 2$ N m. Find the magnitude of the sustained displacement.

(i) When $T_L = 0$, the equation of motion is

$$(\theta_i - K_2\theta_o)\left(\frac{K_1}{1 + 0.2D}\right)\left(\frac{100}{D^2 + 8D + 100}\right) = \theta_o,$$

so

$$(\theta_i - K_2\theta_o) = \theta_o\left(\frac{(D^2 + 8D + 100)(1 + 0.2D)}{100K_1}\right),$$

and

$$\theta_i = \theta_o\left(\frac{(D^2 + 8D + 100)(1 + 0.2D) + 100K_1K_2}{100K_1}\right).$$

Hence CLTF (closed loop transfer function),

$$\frac{\theta_o}{\theta_i} = \left(\frac{100K_1}{(D^2 + 8D + 100)(1 + 0.2D) + 100K_1K_2}\right)$$

$$= \left(\frac{100K_1}{100 + 100K_1K_2 + 28D + 2.6D^2 + 0.2D^3}\right).$$

(ii) With a step input $\theta_i = X$,

$$\theta_o = \frac{100K_1}{100 + 100K_1K_2}(1 - \cdots \mathrm{D} \cdots)\mathrm{X}$$

so that the steady-state error θ_{SS} is given by

$$\theta_{SS} = X\left(\frac{100K_1}{100 + 100K_1K_2} - 1\right)$$

For θ_{SS} to be zero, $K_1 = 1 + K_1K_2$.

With a ramp input $\theta_i = \beta t$,

$$\theta_o = \frac{K_1}{1 + K_1K_2}\left(1 - \frac{28\mathrm{D}}{100 + 100K_1K_2} \cdots \mathrm{D}^2 \cdots\right)\beta t$$

$$= \frac{K_1}{1 + K_1K_2}\left(\beta t - \left(\frac{28}{100 + 100K_1K_2}\right)\beta\right).$$

Now $K_1 = 1 + K_1K_2$, so

$$\theta_o = \beta t - \frac{28}{100K_1}\beta.$$

When $\beta = 2$ rad/s, require θ_{SS} to be 0.056 rad.

That is

$$\frac{28 \times 2}{100 \times K_1} = 0.056,$$

so that

$K_1 = 10$ and hence $K_2 = 0.9$

(iii) With a sudden load T_L applied the equation of motion is

$$\left[-K_2\theta_o\left(\frac{K_1}{1 + 0.2\mathrm{D}}\right) + T_L\right]\left[\frac{100}{\mathrm{D}^2 + 8\mathrm{D} + 100}\right] = \theta_o.$$

That is

$$\theta_o\left(\frac{\mathrm{D}^2 + 8\mathrm{D} + 100}{100} + \frac{\mathrm{K}_1K_2}{1 + 0.2\mathrm{D}}\right) = T_L.$$

Now $K_1K_2 = 9$, so

$$\theta_o\left(\frac{(\mathrm{D}^2 + 8\mathrm{D} + 100)(1 + 0.2\mathrm{D}) + 900}{100(1 + 0.2\mathrm{D})}\right) = T_L,$$

or

$$\theta_o = \left(\frac{100(1 + 0.2D)}{1000 + 28D + 2.6D^2 + 0.2D^3}\right)T_L$$

$$= \frac{100}{1000}(1 + 0.2D)(1 - \frac{28}{1000}D \cdots)T_L.$$

When $T_L = 2$, the sustained displacement is $(100/1000) \times 2 = 0.2$ rad.

5.3.2 Servo with negative output velocity feedback

If viscous damping is not great enough, the damping effect can be simulated by providing a signal proportional to $D\theta_o$. This is shown in Fig. 5.30.

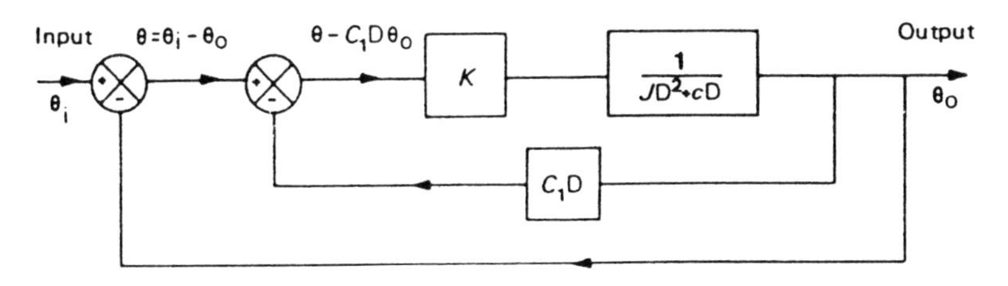

Fig. 5.30. Servo with output velocity feedback.

For this system

$$K(\theta_i - \theta_o - C_1 D\theta_o) = (JD^2 + cD)\theta_o,$$

that is

$$(JD^2 + (c + KC_1)D + K)\theta_o = K\theta_i.$$

Thus the value of C_1 can control the total damping, and the step and transient responses can be improved at the expense of the steady-state error with a ramp input.

However, if $C_1 D\theta_o$ is added to θ instead of subtracted, the opposite is true, but there is a danger of negative damping occurring with the associated unstable response

Example 50

A certain servomechanism is required to control the angular position θ_o of a rotatable load of moment of inertia J. The rotation of the load is subject to a viscous friction torque c per unit angular velocity. The mechanism has velocity feed back such that the motor torque is

$$K\left(\varepsilon - C_1 \frac{d\theta_o}{dt}\right)$$

where ε is the error between the desired angular position input θ_i and the output position θ_o, and K and C_1 are constants.

(i) Draw a block diagram of the system and establish the equation of motion.
(ii) Find the steady-state positional error when the input signal has constant velocity given by

$$\frac{d\theta_i}{dt} = \beta.$$

(iii) When a step input is applied the response overshoot must not exceed 10%. Find the minimum value of the damping ratio required, and express this ratio in terms of the given parameters.

(i) The block diagram is as below:

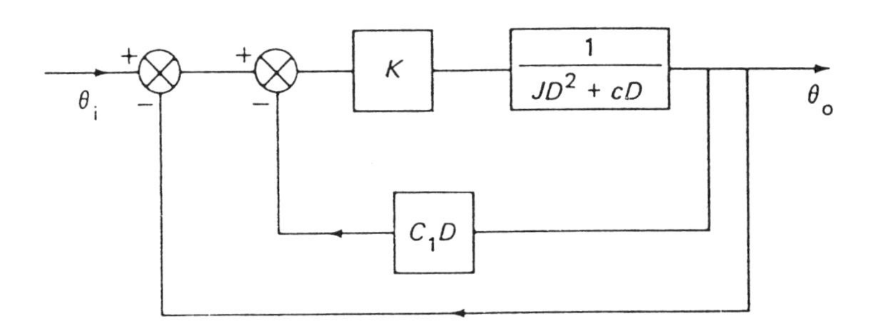

Thus

$$[\theta_i - \theta_o - C_1 D\theta_o]K = [JD^2 + cD.]\theta_o,$$

and the equation of motion is

$$JD^2\theta_o + (c + KC_1)D\theta_o + K\theta_o = K\theta_i.$$

(ii) Input $\theta_i = \beta t$.

$$\theta_o = \left[1 + \left(\frac{c + KC_1}{K}\right)D + \frac{JD^2}{K}\right]^{-1} \beta t$$

$$= \beta t - \left(\frac{c + KC_1}{K}\right)\beta.$$

$$\text{steady-state positional error} = \left(\frac{c + KC_1}{K}\right)\beta.$$

(iii) Damping ratio $\zeta = \frac{\text{`}c\text{'}}{c_c} = \frac{c + KC_1}{2\sqrt{(KJ)}}$.

At first overshoot,

$$t = \frac{\pi}{\omega\sqrt{(1 - \zeta^2)}} \qquad \text{and } \theta_o = 1.1\theta_i.$$

Thus

$$0.1 = \exp[-\zeta\pi/\sqrt{(1-\zeta^2)}],$$

so that

$$2.3 = \frac{\zeta\pi}{\sqrt{(1-\zeta^2)}}$$

Hence

$$\zeta = 0.59 = \frac{c + KC_1}{2\sqrt{(KJ)}}.$$

Example 51

A linear remote position control system with negative output feedback consists of a potentiometer giving 8 V/rad error to an amplifier, and a motor which applies a torque of 3 N m/V to the load. The load has an inertia of 6 kg m^2 and viscous friction of 12 N m s/rad.

(i) Draw a block diagram for the system and derive its equation of motion.
(ii) Calculate the maximum overshoot in the output response to a step input of 2 rad, and
(iii) Given that a tachogenerator is employed to provide negative output velocity feedback, derive the new equation of motion and calculate the velocity feedback coefficient needed to give critical damping.

(i) The block diagram for the system is

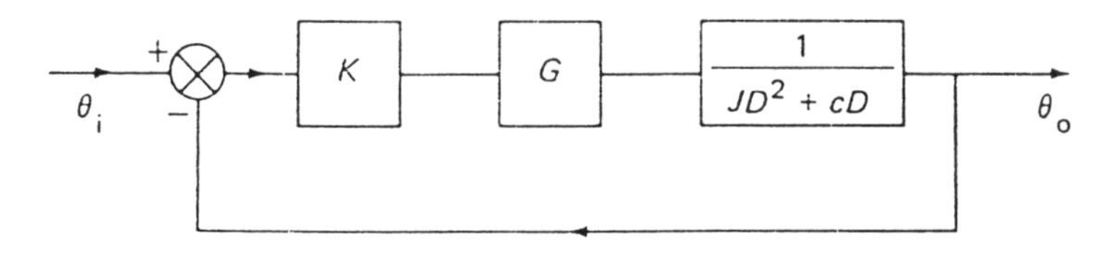

Now $(\theta_i - \theta_o)KG = (JD^2 + cD)\theta_o$,

So the equation of motion is

$$(JD^2 + cD + KG)\theta_o = KG\theta_i.$$

where
$K = 8$ V/rad error,

$G = 3$ Nm/V,

$J = 6$ kg m^2, and

$c = 12$ N m s/rad,

(ii) With a step input, overshoot is $\theta_i e^{-\zeta\pi/\sqrt{(1-\zeta^2)}}$,

where

$$\zeta = \frac{c}{2\sqrt{(KGJ)}} = \frac{12}{2\sqrt{(8 \times 3 \times 6)}} = \frac{1}{2}.$$

Thus

$$e^{-1/2\pi/\sqrt{[1-(1/2)^2]}} = e^{-1.813} = 0.163 \text{ rad}$$

and overshoot = 2 × 0.163 rad = 0.326 rad, for a 2 rad input.

(iii) With output velocity feedback, the block diagram is

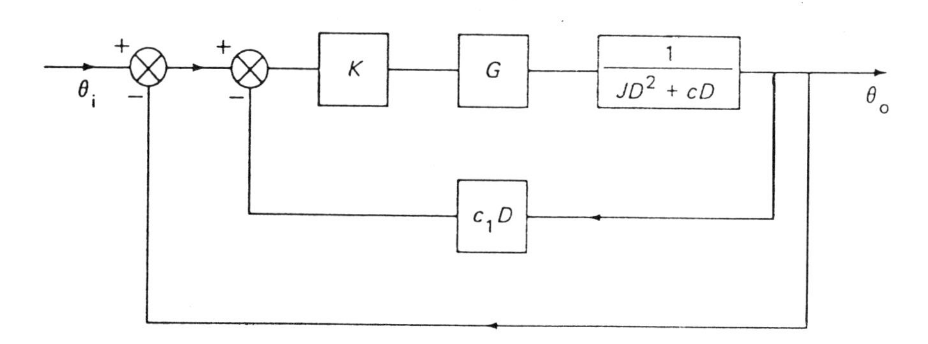

So that

$$[\theta_i - \theta_o - c_1 D\theta_o]KG = [JD^2 + cD]\theta_o$$

and the equation of motion is

$$[JD^2 + (c + c_1 KG)D + KG]\theta_o = KG\theta_i.$$

For critical damping

$$c + c_1 KG = 2\sqrt{(KGJ)},$$

that is

$$12 + (c_1 \times 8 \times 3) = 2\sqrt{(8 \times 3 \times 6)}.$$

Hence

$$c_1 = 0.5 \text{ (N m s/rad)/(Nm/rad)},$$

or

$$c_1 = 0.5 \text{ s}.$$

5.3.3 Servo with derivative of error control

An alternative modification to the simple feedback servo is to add derivative of error control in which the signal is fed forward.

This system is shown in Fig. 5.31.

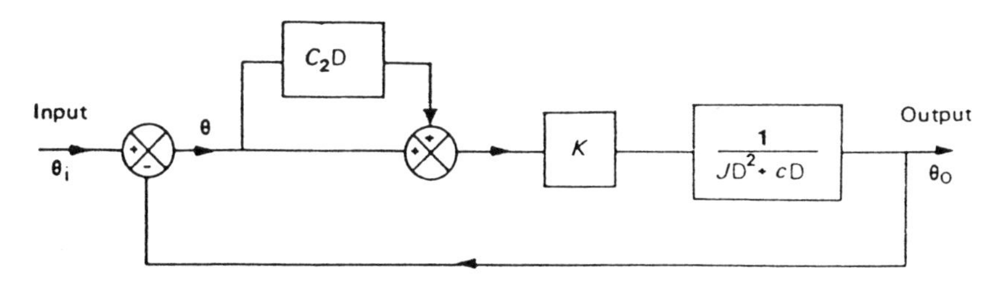

Fig. 5.31. Servo with derivative of error control.

For this system

$$K(\theta_i - \theta_o + C_2D(\theta_i - \theta_o)) = (JD^2 + cD)\theta_o,$$

that is

$$(JD^2 + (c + KC_2)D + K)\theta_o = K(1 + C_2D)\theta_i.$$

Because the damping term is increased the transient response dies away more quickly.

For the response to a ramp input $\theta_i = \beta t$,

$$\text{P.I.} = \frac{K(1 + C_2D)\beta t}{K(1 + (c/K + C_2)D + \cdots)}$$

$$= (1 - (c/K + C_2)D + \cdots)(\beta t + C_2\beta),$$

$$= \beta t - (c/K)\beta;$$

that is, $(c/K)\beta$ is the steady-state error due to the damping; this has not been affected by derivative control. Thus derivative control is a method for temporarily increasing the total damping term and hence the rate of convergence of the transient response, without increasing the physical damping or the steady-state error.

5.3.4 Servo with integral of error control

A further alternative is to feed forward an integral of the error signal as shown in Fig. 5.32.

For this system

$$K(\theta_i - \theta_o + (K_1/D(\theta_i - \theta_o)) = (JD^2 + cD)\theta_o$$

that is,

$$(JD^2 + cD + K + KK_1/D)\theta_o = K(1 + K_1/D)\theta_i$$

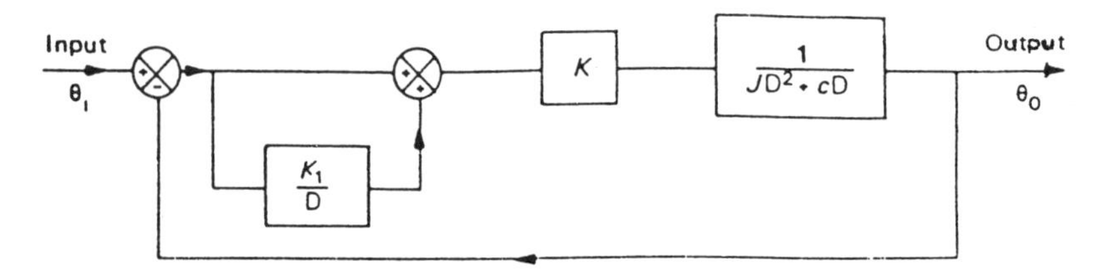

Fig. 5.32. Servo with integral of error control.

or

$$(JD^3 + cD^2 + KD + KK_1)\theta_o = K(D + K_1)\theta_i.$$

For the response to a ramp input $\theta_i = \beta t$,

$$\text{P.I.} = \frac{K(D + K_1)}{K(K_1 + D + \cdots)}\beta t = (1 + D/K_1 + \cdots)^{-1}(K_1\beta t + \beta)/K_1 = \beta t,$$

that is, there is no steady-state error. Although this is a very desirable quality, it is possible for this system to become unstable. This can be demonstrated by considering the response to an harmonic input.

The left-hand-side of the equation of motion is third order, but the response to a sinusoidal input can be found by the phasor technique.

If $\theta_i = A \cos \nu t$ and $\theta_o = B \cos (\nu t - \phi)$ in the steady state, then the equation of motion can be represented by Fig. 5.33.

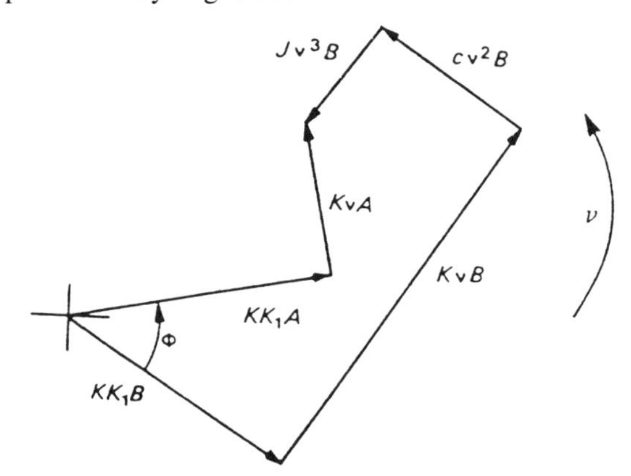

Fig. 5.33. Phasor diagram.

$$\text{Thus} \quad \theta_o = \frac{A\sqrt{((K\nu)^2 + (KK_1)^2)}}{\sqrt{((K\nu - J\nu^3)^2 + (KK_1 - c\nu^2)^2)}}\cos(\nu t - \phi).$$

If the denominator is zero, B is infinite and the system is unstable. For the denominator to be zero,

$$Kv = Jv^3 \text{ and } KK_1 = cv^2.$$

Thus

$$c = JK_1 \text{ and } v^2 = K/J.$$

This is a critical relationship for a stable response.

The performance of the main forms of the electric position servo are compared in Table 5.1 overleaf.

Example 52

In the position control servo shown in block diagram form below, the amplifier voltage gain is 100 and the error detector produces 2 V per degree of error.

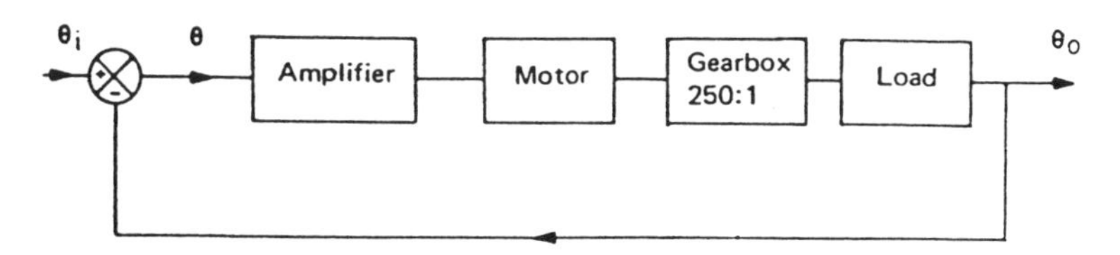

The motor produces a torque of 0.1 N m per 100 V applied. The inertia of the load is 0.1 kg m^2, and the viscous damping is 10 N m s/rad. Find the damping ratio of the system.

Given that the input speed is 5 rad/s find the steady state error.

$$\text{Motor torque} = 0.1/100 \times 2 \times 57.3 \times 100 \text{ Nm/rad error}$$

$$= 11.46 \text{ Nm/rad error.}$$

Torque applied to load = 250 × 11.46 = 2865 Nm/rad error.

For damping ratio,

$$\zeta = c/c_c = 10/2\sqrt{(2865 \times 0.1)} = 0.295.$$

For $\theta_i = 5$ rad/s,

$$\text{steady-state error} = 10/2865 \times 5 \times 57.3° = 1°.$$

Example 53

An engine drives a load of constant inertia which is subjected to external moments which vary unpredictably. To limit the speed changes resulting from load changes, a governor is fitted. This measures the engine speed and changes the moment exerted by the engine to try to keep the speed constant. The changes are related by the equation

$$(1 + TD)m = -ks,$$

where T is a time constant,
m is the change in engine moment from an initial steady value,
k is a constant, and
s is the change in speed from an initial steady value.

Table 5.1 Comparison of main forms of electric position servo

Input		Basic system	Basic system with output velocity feedback	Basic system with derivative of error	Basic system with integral of error
Step	Response	Exponential decay oscillation, rate of decay increasing →			Perhaps unstable
	s.s. error	—— NIL ——			
Ramp	Response	Exponential decay oscillation, rate of decay increasing →			Perhaps unstable
	s.s. error	Proportional to $D\theta_i$	Larger than $D\theta_i$	Proportional to $D\theta_i$	NIL
Harmonic	Response	Gain and phase errors proportional to system parameters and frequency of excitation.			Perhaps unstable

Given that the load moment increases by a step function of magnitude m_L and then remains constant, derive an expression for the consequent speed change as a function of time. If the inertia of the engine and load is 4.0 kg m^2, T is 0.35 s, and k is such that the damping of the closed loop is 0.6 of critical, find the final drop in speed following an increase of load, m_L, of 11 N m.

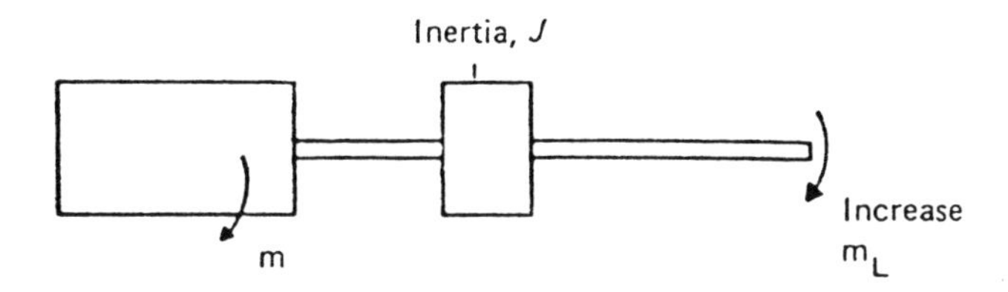

Since s is the *change* in speed,

$$m - m_L = JDs.$$

Now

$$(1 + TD)m = -ks.$$

so

$$m = \frac{-ks}{1 + TD} = JDs + m_L.$$

For a step change in load m_L, $Dm_L = 0$, hence

$$JTD^2s + JDs + ks = -m_L.$$

Solution comprises P.I. $-m_L/k$ and C.F., so

$$s = -\frac{m_L}{k} + e^{-\zeta\omega t}(A \sin \omega t + B \cos \omega t).$$

Substituting initial conditions $s = 0$ and $\dot{s} = 0$ at $t = 0$, gives

$$A = \frac{\zeta m_L}{k} \quad \text{and} \quad B = \frac{m_L}{k},$$

thus

$$s = \frac{m_L}{k}[-1 + e^{-\zeta\omega t}(\zeta \sin \omega t + \cos \omega t)].$$

The steady-state error is $-\dfrac{m_L}{k}$.

From the equation of motion, $c_c = \sqrt{(kJT)}$, so

$$c = 1.2\sqrt{(kJT)} = J,$$

hence

$$k = \frac{J}{1.44T} = \frac{4}{1.44 \times 0.35} = 7.936 \text{ kg m}^2\text{/s},$$

$$\text{Thus steady-state error} = -\frac{11}{7.936} = -1.386 \text{ rad/s}.$$

Example 54

A linear servomechanism consists of a proportional controller which supplies a torque equal to K times the error between the input and output positions, and drives a rotational load of mass moment of inertia J and viscous damping coefficient (less than critical) c.

Draw a block diagram for the mechanism and write down its equation of motion.

During tests on a mechanism for which $J = 5$ kg m^2 it was found that

(i) an external torque of 100 Nm applied to the output shaft gave a steady-state error of 0.2 rad, and
(ii) a constant velocity input of 3 rad/s produced a steady state error of 0.3 rad. Given that a step displacement of 10° is applied to the mechanism when it is at rest, find the magnitude of the first overshoot.

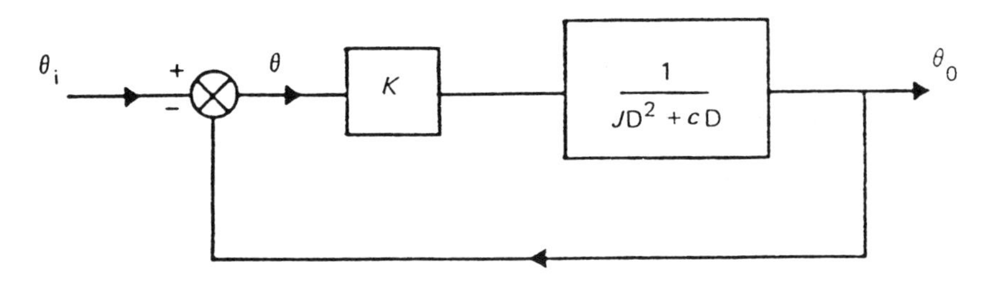

The equation of motion is

$$JD^2\theta_o + cD\theta_o + K\theta_o = K\theta_i.$$

(i) $K = \frac{100}{0.2} = 500$ Nm/rad.

(ii) $0.3 = \frac{c}{500} \times 3,$ so $c = 50$ N m s/rad.

Since $c_c = 2\sqrt{(kJ)} = 2\sqrt{(500 \times 5)} = 100$ N m s/rad,

$$\zeta = \frac{c}{c_c} = 0.5.$$

At first overshoot, from Example 46,

$$\theta_o = X\left[1 + \exp\left(-\frac{\zeta\pi}{\sqrt{(1-\zeta^2)}}\right)\right].$$

Now,

$$-\frac{\zeta\pi}{\sqrt{(1-\zeta^2)}} = -\frac{0.5\pi}{0.866} = -1.82 \qquad \text{and} \qquad e^{-1.82} = 0.162.$$

Thus $\theta_o = 10[1.162] = 11.62°$, so the magnitude of the first overshoot is 1.6°.

Example 55

The angular position θ_o of a turntable is controlled by a servomechanism which has positive acceleration feedback. The moment of inertia of the turntable about its axis of rotation is 5 kg m^2 and the effective viscous damping coefficient is 10 N m s/rad.

The motor torque is $20(\varepsilon + k\ddot{\theta}_o)$ N m, where ε is the error, in radians, between the position of the input, θ_i, and the turntable position, θ_o, and $k\,\ddot{\theta}_o$ is the feedback signal proportional to the acceleration of the turntable.

(i) Draw a block diagram for the servomechanism and derive its equation of motion.
(ii) If a step input is applied, determine the steady-state error and the value of k required for the maximum output overshoot to be 5% of the steady state value.
(iii) Determine the steady-state position error if an external torque of 5 N m is suddenly applied to the turntable.

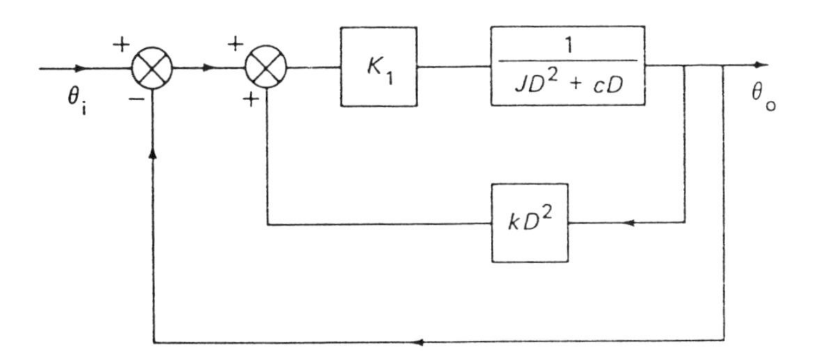

(i) The equation of motion is given by

$$K_1(\theta_i - \theta_o + kD^2\theta_o) = (JD^2 + cD)\theta_o.$$

Rearranging gives

$$(JD^2 + cD - kK_1D^2 + K_1)\theta_o = K_1\theta_i$$

That is,

$$[(J - kK_1)D^2 + cD + K_1]\theta_o = K_1\theta_i.$$

(ii) A step input is applied, so let $\theta_i = X$
Then P.I. is

$$\theta_o = \left[\left(\frac{J - kK_1}{K_1}\right)D^2 + \frac{c}{K_1}D + 1\right]^{-1} X$$

$$= \left[1 - \frac{c}{K_1}D + \cdots\right]X = X,$$

That is, since $\theta_o = \theta_i$, steady-state error is zero

For a step input, the complete solution is the C.F. and the P.I. Thus

$$\theta_o = \theta_i - \frac{\theta_i}{\sqrt{(1-\zeta^2)}} e^{-\zeta\omega t} \sin(\omega\sqrt{(1-\zeta^2)}t + \phi),$$

where $\phi = \tan^{-1}\dfrac{\sqrt{(1-\zeta^2)}}{\zeta}$

At the first overshoot, $D\theta_o = 0$, so

$$\zeta \sin(\omega\sqrt{(1-\zeta^2)}t + \phi) = \sqrt{(1-\zeta^2)} \cos(\omega\sqrt{(1-\zeta^2)}t + \phi)$$

or

$$\tan(\omega\sqrt{(1-\zeta^2)}t + \phi) = \tan\phi,$$

so that

$$\omega\sqrt{(1-\zeta^2)}t = 0, \pi, \cdots$$

and

$$t = 0, \quad \frac{\pi}{\omega\sqrt{(1-\zeta^2)}}, \ldots .$$

Now at the first overshoot

$$t = \frac{\pi}{\omega\sqrt{(1-\zeta^2)}} \quad \text{and} \quad \theta_o = 1.05\theta$$

so

$$0.05 = e^{-\zeta\pi/\sqrt{(1-\zeta^2)}}$$

Thus

$$3 = \zeta\pi/\sqrt{(1-\zeta^2)},$$

and hence $\zeta = 0.69$.

From the equation of motion

$$\zeta = \frac{c}{c_c} = \frac{c}{2\sqrt{[K_1(J - kK_1)]}},$$

where $c = 10$ N m s/rad,

$K_1 = 20$ N m/rad,

and $J = 5$ kg m^2.

Thus

$$0.69 = \frac{10}{2\sqrt{[20(5 - 20K)]}}$$

and

$$k = 0.119 \text{ s}^2/\text{rad}.$$

(iii) With a steady external torque T_L applied the equation of motion is

$$K_1(\theta_i - \theta_o + kD^2\theta_o) + T_L = (JD^2 + cD)\theta_o.$$

When $\theta_i = 0$,

$$[(J - kK_1)D^2 + cD + K_1]\theta_o = T_L,$$

so $\theta_{o_{\text{steady state}}} = \dfrac{T_L}{K_1}$.

With $T_L = 5$ N m and $K_1 = 20$ N m/rad,

$$\theta_{o_{\text{steady state}}} = \frac{5}{20} = 0.25 \text{ rad}.$$

Example 56

A linear servomechanism consists of a proportional controller which supplies a torque equal to K_p times the error between the input and output positions of the rotational load. The load has a mass moment of inertia J and viscous damping coefficient (less than critical) c. For the given mechanism $J = 0.5$ kg m^2, $c = 1$ N m s/rad and $K_p = 32$ N m/rad error.

(i) Draw a block diagram for the mechanism and write down its equation of motion.
(ii) Derive the closed loop transfer function and calculate the steady-state error if the input is increased at a constant rate of 2 rad/s.
(iii) Given that a derivative of error controller, with a gain constant K_D is added in parallel with K_p, sketch the new block diagram, calculate the steady-state error with a ramp input of 2 rad/s, explain the effect on the transient response, and define the value of K_D which will give critical damping.
(iv) When an integral of error controller, with a gain constant K_i is added in parallel with K_p and K_D, show that the steady-state error is zero with a ramp input, and find the range of values K_i can have if the mechanism is to have a stable response.

(i)

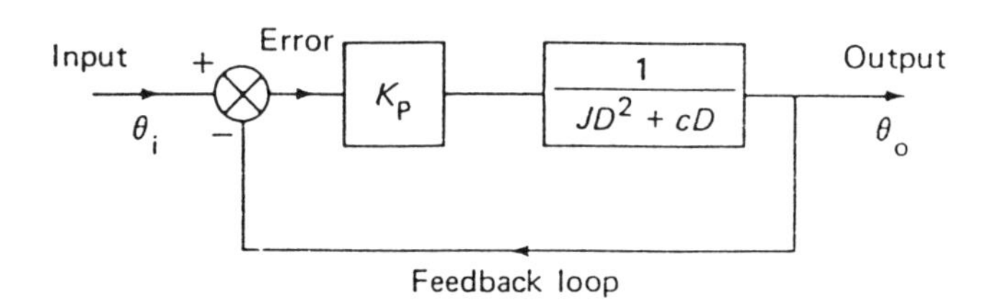

The equation of motion is given by

$$K_p(\theta_i - \theta_o) = (JD^2 + cD)\theta_o,$$

that is

$$(JD^2 + cD + K_p)\theta_o = K_p\theta_i$$

(ii) $\text{CLTF} = \dfrac{\theta_o}{\theta_i} = \dfrac{K_p}{JD^2 + cD + K_p}$

The system is given a ramp input of $\theta_i = \beta t$ where $\beta = 2$ rad/s, so

$$\theta_o = \left[1 + \frac{c}{K_p} D + \frac{J}{K_p} D^2\right]^{-1} \theta_i,$$

$$= \left[1 - \frac{c}{K_p} D + \cdots\right] \beta t$$

$$= \beta t - \frac{c}{K_p}\beta.$$

Hence steady-state error is a lag $\dfrac{c}{K_p}\beta$, that is a lag of $(\frac{1}{32}) \times 2 = \frac{1}{16}$ rad.

(iii)

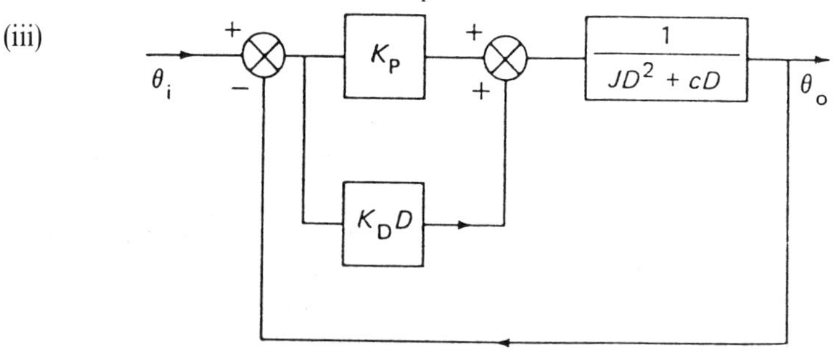

Equation of motion is given by

$$(\theta_i - \theta_o)(K_p + K_D D) = (JD^2 + cD)\theta_o,$$

that is

$$(JD^2 + (c + K_D)D + K_p)\theta_o = (k_p + K_D D)\theta_i,$$

With a ramp input $\theta_i = \beta t$,

$$\theta_o = \left[1 + \left(\frac{c + K_D}{K_P}\right)D + \frac{J}{K_P}D^2\right]^{-1}\left[1 + \frac{K_D}{K_P}D\right]\beta t$$

$$= \left[1 - \left(\frac{c + K_D}{K_P}\right)D + \cdots\right]\left[\beta t + \frac{K_D}{K_P}\beta\right]$$

$$= \beta t + \frac{K_D}{K_P}\beta - \left(\frac{c + K_D}{K_P}\right)\beta$$

$$= \beta t - \frac{c}{K_P}\beta.$$

Hence steady-state error is unchanged at $\dfrac{c}{K_P}\beta = \frac{1}{16}$ rad, but transient dies away more quickly because damping term increased from c to $(c + K_D)$.

For critical damping, require

$$c + K_D = 2\sqrt{(K_D J)},$$

so

$$1 + K_D = 2\sqrt{(32 \times \tfrac{1}{2})} = 8$$

or

$$K_D = 7 \text{ N m s/rad}.$$

(iv)

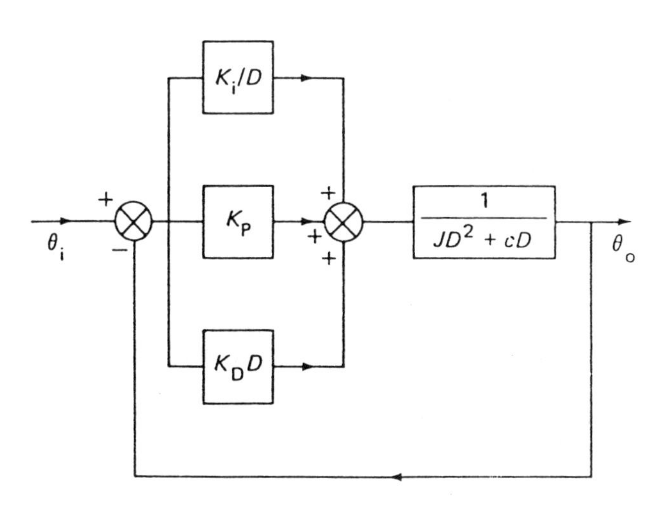

Equation of motion is given by

$$\left[\frac{K_i}{D} + K_P + K_D D\right]\left[\theta_i - \theta_o\right] = \left[JD^2 + cD\right]\theta_o,$$

that is,

$$[JD^3 + (c + K_D)D^2 + K_P D + K_i]\theta o = [K_D D^2 + K_P D + K_i]\theta_i,$$

With a ramp input $\theta_i = \beta t$,

$$\theta_o = \left[1 + \frac{K_P}{K_i}D + \left(\frac{c + K_D}{K_i}\right)D^2 + \frac{J}{K_i}D^3\right]^{-1}\left[\frac{K_D D^2}{K_i} + \frac{K_P}{K_i}D + 1\right]\beta t$$

$$= \left[1 - \frac{K_P}{K_i}D + \cdots\right]\left[\frac{K_P}{K_i}\beta + \beta t\right],$$

$$= \frac{K_P}{K_i}\beta + \beta t - \frac{K_P}{K_i}\beta t = \beta t.$$

Since $\theta_o = \theta_i$, there is no steady-state error.

For stable response limit substitute $D = j\omega$ in the frequency equation

$$JD^3 + (c + K_D)D^2 + K_P D + K_i = 0.$$

Hence

$$-Jj\omega^3 - (c + K_D)\omega^2 + K_P j\omega + K_i = 0.$$

Equating the real parts gives

$$-(c + K_D)\omega^2 + K_i = 0 \quad \text{or} \quad \omega^2 = \frac{K_i}{c + K_D},$$

and equating the imaginary parts gives

$$-J\omega^3 + K_P\omega = 0 \text{ i.e. } \omega = 0 \text{ or } \omega^2 = \frac{K_P}{J}.$$

Thus either $K_i = 0$ $(\omega = 0)$

or

$$K_i = \left(\frac{c + K_D}{J}\right)K_P \quad \left(\omega^2 = \frac{K_P}{J}\right).$$

Hence the stability limits are given by $K_i = 0$ and

$$K_i = \left(\frac{1 + 7}{0.5}\right)32 = 512,$$

so that for a stable response $512 > K_i > 0$.

Example 57

A tension control device such as those used in the paper industry is shown below. It is desired to maintain a constant tension in the paper as it is wound up. To increase the tension F in the paper, the tension control level is lowered: this increases the torque T_m applied by the motor to the wind-up roll. The change in torque provided by the motor is t_m where

$$t_m = \frac{K_m Z}{1 + \tau D}$$

where K_m is the motor constant, Z is the movement of the tension control lever at its mid-point, and τ is a time constant.

If the tension control lever is given a small downward step movement X at its free end, derive an expression for the resulting change in paper tension and sketch this tension as a function of time.

Obtain the steady-state value of this change and evaluate this quantity if $k = 20$ kN/m, $R = 0.4$m, $K_m = 10$ kN and $X = 10$ mm.

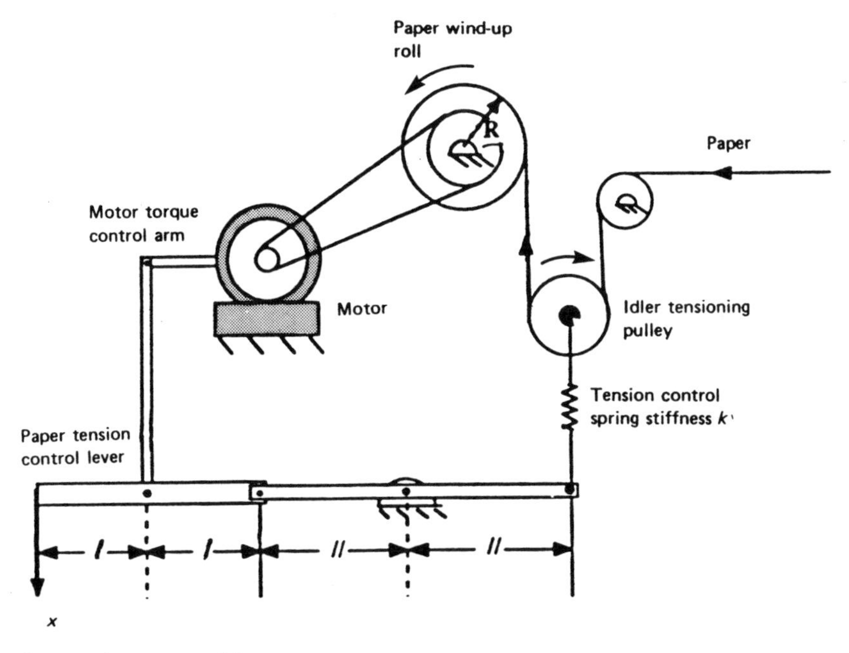

For the tension control lever,

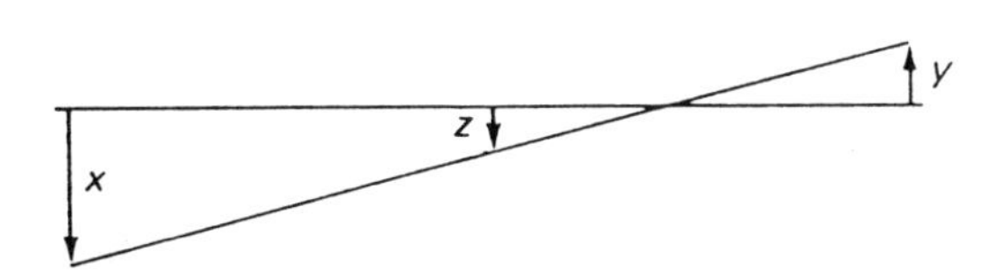

$$x + y = 2(z + y),$$

$$\text{so } z = \frac{x - y}{2}.$$

For the wind-up roll, $T_m = FR$, and if f is the change in paper tension, $t_m = Rf$.
For the tension control spring $ky = 2f$, so $y = 2f/k$.
The change in motor torque t_m following a movement x is

$$t_m = \frac{K_m z}{1 + \tau D} = fR,$$

so

$$fR = \left(\frac{K_m}{1 + \tau D}\right)\left(\frac{x - y}{2}\right),$$

$$= \left(\frac{K_m}{1 + \tau D}\right)\left(x - \frac{2f}{k}\right) \times \frac{1}{2}.$$

That is,

$$f\left[\frac{2R}{K_m}(1 + \tau D) + \frac{2}{k}\right] = x$$

or

$$f = \left[\frac{kK_m}{2kR(1 + \tau D) + 2K_m}\right]x.$$

For a step change $x = X$,

$$\text{C.F. is } [2kR(1 + \tau D) + 2K_m]f = 0$$

or

$$f\left[1 + \left(\frac{kR}{K_m + kR}\right)\tau D\right] = 0.$$

Thus

$$f = A \exp\left[-t \bigg/ \left(\frac{kR\tau}{K_m + kR}\right)\right].$$

P.I. is

$$\left[\frac{kK_m/2(K_m + kR)}{(1 + 2kR\tau D/2(K_m + kR)}\right]X,$$

$$= \left[1 - \cdots D\right]\left(\frac{kK_m}{2(kR + K_m)}\right)X$$

$$= \left(\frac{kK_m}{2(kR + K_m)}\right)X.$$

The complete solution is therefore

$$f = A\exp\left[-t\Bigg/\left(\frac{kR\tau}{K_m + kR}\right)\right] + \left(\frac{kK_m}{2(kR + K_m)}\right)X.$$

Now, if $t = 0, f = 0$,

$$f = \left(\frac{kK_m}{2(kR + K_m)}\right)X\left[1 - \exp\left[-t\Bigg/\left(\frac{kR\tau}{kR + K_m}\right)\right]\right],$$

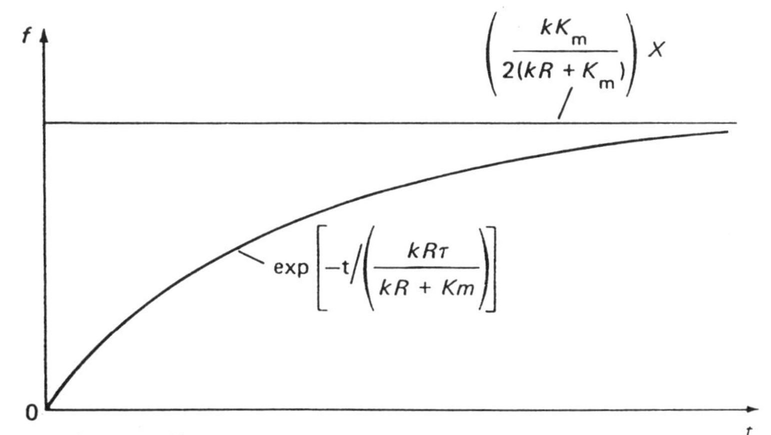

Thus steady state f is

$$\left[\frac{kK_m}{2(kR + K_m)}\right]X.$$

Substituting the given data gives

$$f_{ss} = \left[\frac{20 \times 10^3 \times 10 \times 10^3}{2[(20 \times 10^3 \times 0.4) + (10 \times 10^3)]}\right]0.01\text{N},$$

$$= 55.6 \text{ N}.$$

5.4 THE LAPLACE TRANSFORMATION

In the solution of differential equations of motion describing the dynamic behaviour of mechanical systems, d/dt is commonly represented by the D operator, as above. However, in general control system analysis the Laplace transformation has tended to replace other forms of differential equation representation. This is because with time-dependent functions, it is often very convenient to represent transfer functions in terms of the Laplace transform, which is a linear transformation from a time-dependent function to a frequency-dependent function. The basic harmonic nature of the response of dynamic systems means that a frequency representation is often much simpler than the equivalent time-dependent form. The Laplace transform is defined as

$$\mathscr{L}[f(t)] = F(s) = \underset{t \to \infty}{\mathrm{Lt}} \int_0^{\infty} f(t)\mathrm{e}^{-st}\,\mathrm{d}t,$$

where $F(s)$ is the Laplace transform,
$f(t)$ is the time function, and
s is a complex variable.

The use of different notations does not usually cause any confusion, because the initial conditions are assumed to be zero, so that the differential equation is transformed into the Laplace domain by replacing $\mathrm{d}/\mathrm{d}t$ by s, $\mathrm{d}^2/\mathrm{d}t^2$ by s^2, and so on. s is the Laplace operator $a + \mathrm{j}b$, where s is a complex variable, a and b are real variables, and $\mathrm{j} = \sqrt{-1}$.

Some common Laplace transforms used in control system analysis are given in the table below.

For a full description of the Laplace transformation, see for example, Engineering Science Data Unit Item No. 69025, *Solution of ordinary linear differential equations by the Laplace transform method*, or any advanced engineering mathematics text.

	Time function $f(t)$	Laplace transform $F(s)$
Unit impulse	$\delta(t)$	1
Delayed impulse	$\delta(t - T)$	e^{-Ts}
Unit step	$u(t)$ or 1	$\frac{1}{s}$
Delayed step	$u(t - T)$	$\frac{1}{s}\mathrm{e}^{-Ts}$
Rectangular pulse	$u(t) - u(t - T)$	$\frac{1}{s}(1 - \mathrm{e}^{-Ts})$
Unit ramp	t	$\frac{1}{s^2}$
Polynomial	t^n	$\frac{n!}{s^{n+1}}$
Exponential	$\mathrm{e}^{-\alpha t}$	$\frac{1}{s + \alpha}$
Sine wave	$\sin \omega t$	$\frac{\omega}{s^2 + \omega^2}$

Sine wave	$\sin(\omega t + \phi)$	$\dfrac{s \sin \phi + \omega \cos \phi}{s^2 + \omega^2}$
Cosine wave	$\cos \omega t$	$\dfrac{s}{s^2 + \omega^2}$
Damped sine wave	$e^{-\alpha t} \sin \omega t$	$\dfrac{\omega}{(s + \alpha)^2 + \omega^2}$
Damped cosine wave	$e^{-\alpha t} \cos \omega t$	$\dfrac{s + \alpha}{(s + \alpha)^2 + \omega^2}$

Analysing the hydraulic closed loop servo by the Laplace transformation method gives the equation of motion, from equation (5.2), as

$$(1 + Ts)x_o = \frac{m}{l} x_i.$$

For a step input X, $\quad x_i = \dfrac{X}{s}$,

so

$$x_o = \frac{(m/l)X}{s(1 + Ts)} = \frac{-(m/l)XT}{(1 + Ts)} + \frac{(m/l)X}{s}$$

$$= \left(\frac{m}{l}\right) X \left(\frac{1}{s} - \frac{T}{1 + Ts}\right),$$

so that

$$X_o = \left(\frac{m}{l}\right) X[1 - e^{-t/T}],$$

as before.

The *final value theorem* (FVT), which states that

$$\underset{t \to \infty}{\text{Lt}}\ f(t) = \underset{s \to 0}{\text{Lt}}\ [s \times F(s)]$$

can be used to determine the steady-state error, so that for a ramp input βt, if $m/l = 1$,

$$x_i = \frac{\beta}{s^2} \quad \text{and} \quad x_o = \frac{\beta}{s^2(1 + Ts)},$$

so that the error,

$$x_o - x_i = \frac{-\beta Ts}{s^2(1 + Ts)}.$$

The steady-state error $= (x_o - x_i)_{t \to \infty}$

$$= \underset{s \to 0}{\text{Lt}}\ s\left(\frac{-\beta Ts}{s^2(1 + Ts)}\right) = \underset{s \to 0}{\text{Lt}}\left(\frac{-\beta T}{1 + Ts}\right).$$

$$= -\beta T,$$

as before, see section 5.3.3.

In the case of the electric closed loop position servo, the equation of motion is, from section 5.5.2,

$$(Js^2 + cs + K)\theta_o = K\theta_i,$$

so that if a ramp input βt is applied,

$$\theta_i = \frac{\beta}{s^2} \quad \text{and} \quad \left(\frac{J}{K}s^2 + \frac{c}{K}s + 1\right)\theta_o = \theta_i.$$

Hence the error,

$$\theta_o - \theta_i = \theta_i\left(\frac{-\dfrac{J}{K}s^2 - \dfrac{c}{K}s}{\dfrac{J}{K}s^2 + \dfrac{c}{K}s + 1}\right),$$

and the FV Theorem gives, for θ_{SS},

$$\theta_{SS} = (\theta_o - \theta_i)_{t \to \infty} = \underset{s \to 0}{\text{Lt}}\ s\frac{\beta}{s^2}\left(\frac{-\dfrac{J}{K}s^2 - \dfrac{c}{K}s}{\dfrac{J}{K}s^2 + \dfrac{c}{K}s + 1}\right)$$

$$= -\frac{c}{K}\beta,$$

as before.

5.5 SYSTEM TRANSFER FUNCTIONS

The block diagram of any linear closed loop system incorporating negative feedback and having one input and one output variable can be reduced to the form shown in Fig. 5.34.

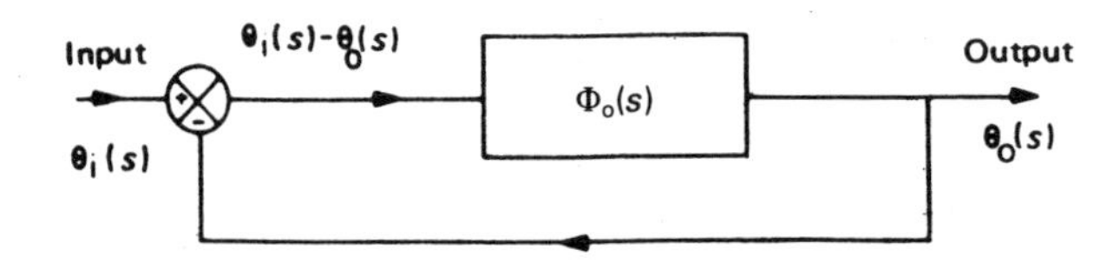

Fig. 5.34. Closed loop system block diagram.

The input and output variables have Laplace transforms $\theta_i(s)$ and $\theta_o(s)$ respectively, and the forward path of the system has a transfer function (*TF*) $\Phi_o(s)$. This is the open loop transfer function (OLTF) since it describes the behaviour of the system with the feedback loop open.

When the loop is closed, the input to $\Phi_o(s)$ is the error signal $\theta_i(s) - \theta_o(s)$ and thus

$$\theta_o(s) = \Phi_o(s)(\theta_i(s) - \theta_o(s))$$

that is

$$\theta_o(s)/\theta_i(s) = \Phi_o(s)/(1 + \Phi_o(s)).$$

This equation determines the overall behaviour of the system when the loop is closed and $\Phi_o(s)/(1 + \Phi_o(s))$ is accordingly known as the closed loop transfer function (CLTF) denoted by $\Phi_c(s)$.

Example 58

Find the CLTF for the system shown below in block form.

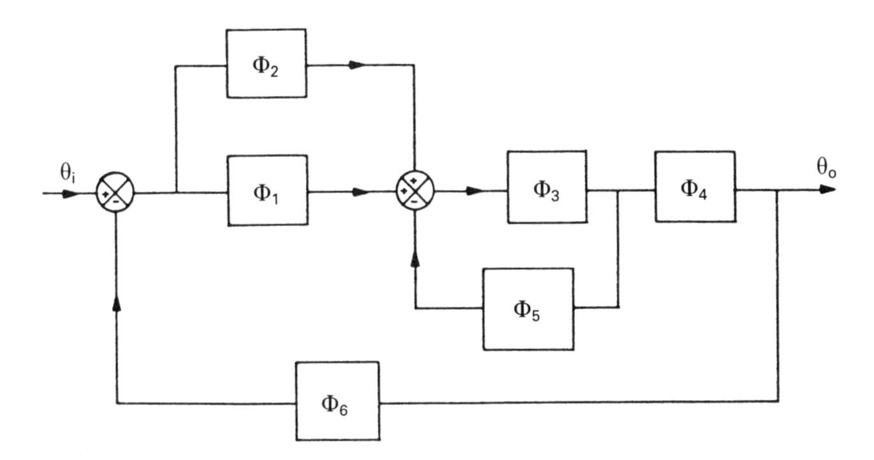

The signal leaving the first junction is

$$\theta_i - \Phi_6\theta_o.$$

The signal leaving the second junction is

$$[\theta_i - \Phi_6\theta_o](\Phi_1 + \Phi) - \Phi_5\theta_o/\Phi_4.$$

$$\text{Thus } [(\theta_i - \Phi_6\theta_o)(\Phi_1 + \Phi_2) - \Phi_5\theta_o/\Phi_4]\Phi_3\Phi_4 = \theta_o.$$

Hence the CLTF,

$$\frac{\theta_o}{\theta_1} = \frac{(\Phi_1 + \Phi_2)\Phi_3\Phi_4}{1 + \Phi_3\Phi_5 + \Phi_3\Phi_4\Phi_6(\Phi_1 + \Phi_2)}.$$

For an electric position servo used for controlling the angular position of a turntable the block diagram model is as shown in Fig. 5.35.

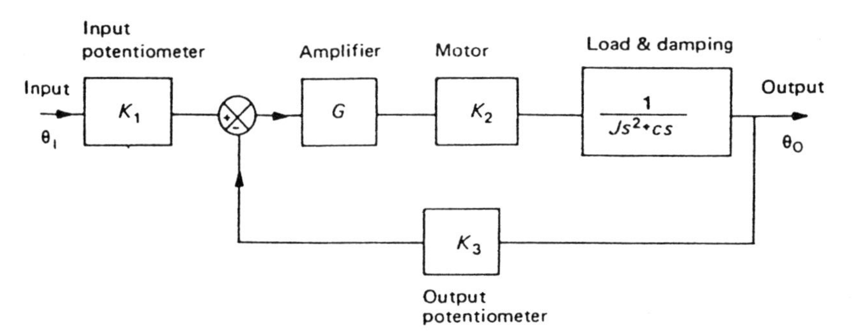

Fig. 5.35. Electric position servo.

This block diagram can be simplified as shown in Fig. 5.36.

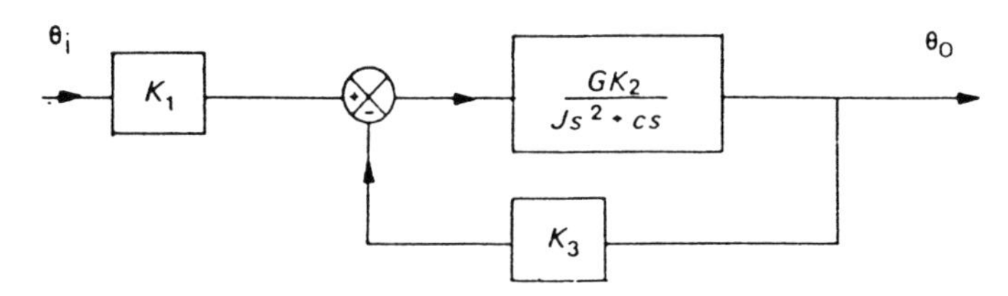

Fig. 5.36. Block diagram.

From Fig. 5.36,

$$\theta_o = (K_1\theta_i - K_3\theta_o)(GK_2/(Js^2 + cs)). \tag{5.10}$$

If the OLTF and CLTF are to be determined the block diagram is required in the form of Fig. 5.34. The OLTF,

$$\Phi_o(s) = \theta_o/(\theta_i - \theta_o)$$

and CLTF,

$$\Phi_c(s) = \theta_o/\theta_i.$$

From equation (5.10),

$$((Js^2 + cs)/(GK_2) + K_3)\theta_o = K_1\theta_i.$$

Hence

$$\Phi_c(s) = \frac{GK_1K_2}{Js^2 + cs + GK_2K_3}.$$

and

$$\Phi_o(s) = \frac{GK_1K_2}{Js^2 + cs + GK_2K_3 - GK_1K_2}.$$

It can be seen from the expression for $\Phi_c(s)$ that the frequency equation is

$$Js^2 + cs + GK_2K_3 = 0$$

or

$$s^2 + cs/J + GK_2K_3/J = 0,$$

since the values of s which satisfy this equation make $\Phi_c(s) = \infty$. These values can be denoted by p_1 and p_2 where

$$(s - p_1)(s - p_2) = 0,$$

$$p_1 = a + jb \qquad \text{and} \qquad p_2 = a - jb.$$

Now

$$s = -c/2J \pm j\sqrt{(GK_2K_3/J - (c/2J)^2)}$$

so

$$a = -c/2J \qquad \text{and} \qquad b = \sqrt{(GK_2K_3/J - (c/2J)^2)}.$$

These roots can be plotted on the s-plane as shown in Fig. 5.37, as G increases from zero.

For an oscillatory response $b > 0$, that is,

$$GK_2K_3 > \frac{c^2}{4J}.$$

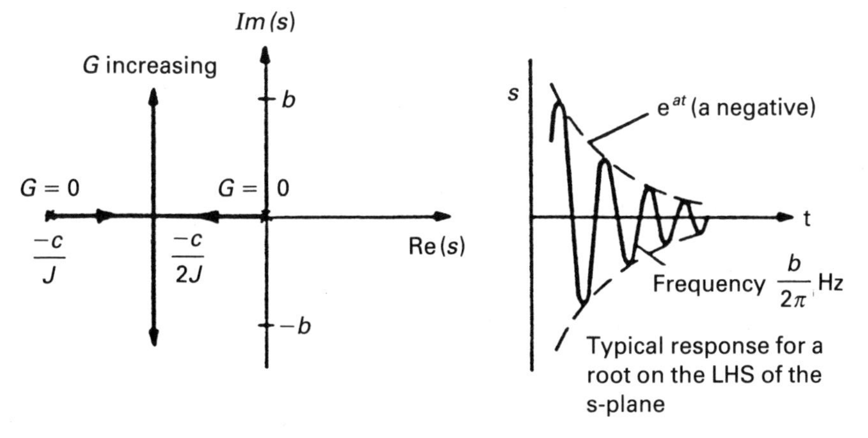

Fig. 5.37. s-plane.

The frequency equation of a system governs its response to a stimulus because its roots are the same as the roots of the complementary function. If the roots lie on the left-hand side of the s-plane the response dies away with time and the system is stable. If the roots lie on the right-hand side of the s-plane the response grows with time and the system is unstable.

If general any transfer function $\Phi(s)$ has the form

$$\Phi(s) = \frac{K(s^m + \alpha_1 s^{m-1} + \alpha_2 s^{m2} + \ldots \alpha_m)}{(s^n + \beta_1 s^{n-1} + \beta_2 s^{n-2} + \ldots \beta_n)} \text{ and } n \geqslant m$$

with a set of roots $s_1, s_2, s_3, \ldots, s_n$. Thus

$$\Phi(s) = \frac{K(s - z_1)(s - z_2) \cdots (s - z_m)}{(s - p_1)(s - p_2) \cdots (s - p_n)},$$

where the values of $s = p_1, p_2, \ldots, p_n$ are those which make $\Phi(s) = \infty$ and are called *poles*, and the values of $s = z_1, z_2, \ldots, z_m$ are those which make $\Phi(s) = 0$ and are called *zeros*. Hence the poles of $\Phi_c(s)$ are the natural frequencies of the system.

5.6 ROOT LOCUS

In most systems instability is intolerable. In dynamic systems a degree of damping is usually desired for safety, while in control systems a considerable margin of stability is essential, preferably with little or no oscillation. However, excessive damping is wasteful of energy.

Consider the diagram of the s-plane Figs 3.11 and 5.37; the right-hand side is a completely unstable region, the imaginary axis is neutral equilibrium, and the left-hand side is stable. In both dynamic and control systems, therefore, it is important to know where the roots of the frequency, or characteristic equation lie on this plane, and if necessary how to adjust the system to move them.

The idea of the root locus technique is to use a diagram such as the s-plane to see how the roots of the frequency equation vary as various system parameters are changed.

Consider the problem of wheel shimmy. This is an unstable oscillation of a castored wheel about the axis of the support pin. A plan view of a wheel which has its axle supported in a carrier is shown opposite in Fig. 5.38; the carrier is pinned so that it can rotate in a direction normal to the wheel axis, and lateral motion is resisted by a spring of stiffness k. The wheel moves with a speed V as shown.

If the mass of the wheel and carrier is m, and the moment of inertia of the wheel and carrier about an axis through G is I_G, then for small displacements x and ϕ the equation of motions are:

(ΣF_x) $\quad m(\ddot{x} + l_2\ddot{\phi}) + kx + F = 0$

and

(ΣM_G) $\quad kxl_2 - Fl_1 = I_G\ddot{\phi}.$

Also, for no side slip of the wheel,

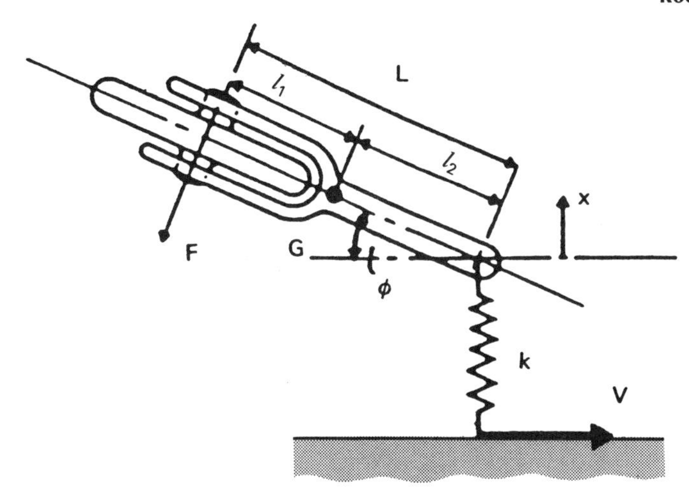

Fig. 5.38. Wheel shimmy model.

$$\dot{x} + L\dot{\phi} + V\phi = 0.$$

Now F can be eliminated, and if it is assumed that

$$x = X\mathrm{e}^{st} \text{ and } \phi = \Phi\mathrm{e}^{st},$$

where s is the Laplace operator $a + \mathrm{j}b$, then

$$Xs + \Phi(V + Ls) = 0$$

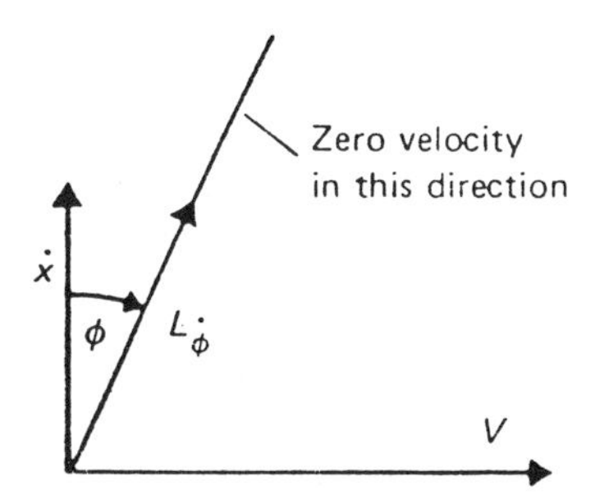

and

$$X(l_1ms^2 + kL) + \Phi(-I_Gs^2 + ml_1l_2s^2) = 0.$$

Hence the frequency equation is

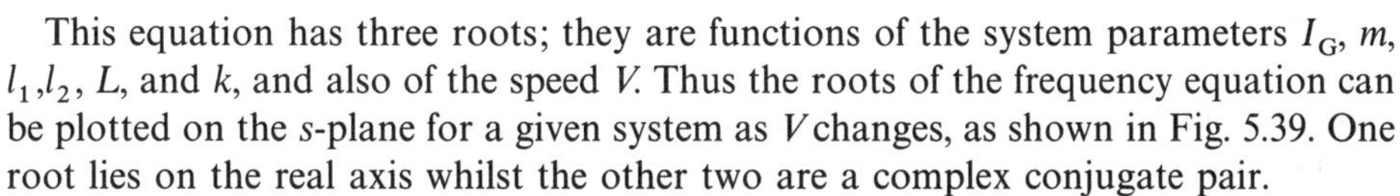

$$s^3 + s^2\left(\frac{l_1mV}{I_G + ml_1^2}\right) + s\left(\frac{kL^2}{I_G + ml_1^2}\right) + \left(\frac{LkV}{I_G + ml_1^2}\right) = 0.$$

This equation has three roots; they are functions of the system parameters I_G, m, l_1,l_2, L, and k, and also of the speed V. Thus the roots of the frequency equation can be plotted on the s-plane for a given system as V changes, as shown in Fig. 5.39. One root lies on the real axis whilst the other two are a complex conjugate pair.

Since the system becomes unstable when roots of the frequency equation appear on the right-hand side of the s-plane, it can be seen from Fig. 5.39 (overleaf) that a critical speed exists. For speeds in excess of V_{crit} unstable oscillation results from a system disturbance. This phenomenon is known as *wheel shimmy*.

Considerable labour is required to obtain the frequency equation of a dynamic system and to solve it repeatedly for a range of values of a certain parameter such as the speed V above, particularly if the frequency equation is of a high order. Although some relief can be obtained by using a computer, it is usually only necessary to sketch the locus of the roots of the frequency equation, so that a curve sketching technique has been developed.

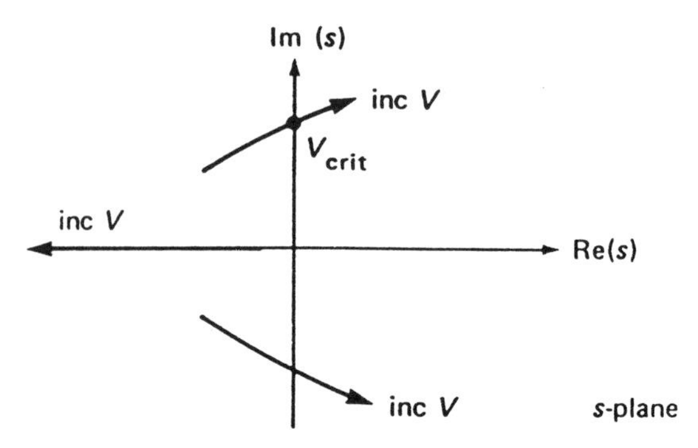

Fig. 5.39. Roots of frequency equation.

Consider the system shown in Fig. 5.34,

$$\Phi_c(s) = \theta_o/\theta_i = \Phi_o(s)/(1 + \Phi_o(s)).$$

Instead of writing $\Phi_c(s) = \infty$ for the condition of resonance it is more convenient to put $1 + \Phi_o(s) = 0$: this is the frequency equation. For a system to be stable all roots of the frequency equation must lie on the left-hand side of the imaginary axis when plotted on the s-plane.

If $\Phi_o(s) = K\Phi'_o(s)$ where

$$\Phi'_o(s) = \frac{(s - z_1)(s - z_2) \cdots (s - z_m)}{(s - p_1)(s - p_2) \cdots (s - p_n)},$$

the root locus can be sketched as K increases from zero to infinity and the conditions for instability determined. K is the overall system gain constant.

Techniques for the construction of a root locus for a known control system have been developed and can be summarized as a set of rules.

It is important to realize that the roots of the frequency equation are the same as the roots of the complementary function. Now, since the complementary function governs the system transient response to a disturbance, a stable response can be achieved only if the roots of the complementary function, and therefore also the roots of the frequency equation, lie on the left-hand side of the s-plane. This is why the roots of the frequency equation are so important in control system response and stability.

5.6.1 Rules for constructing root loci

Rule 1 Number of loci
The number of loci is equal to the degree of the characteristic equation.

Proof The number of characteristic roots is equal to the degree of the characteristic equation, and each root has its own locus.

Rule 2 Symmetry of loci
The root loci for a real characteristic equation are symmetrical with respect to the real axis.

Proof The complex roots of a real characteristic equation occur in conjugate pairs.

Rule 3 Poles of $\Phi_o'(s)$
The poles of $\Phi_o'(s)$ lie on the root loci and correspond to $K = 0$.

Proof Since $1 + \Phi_o(s) = 0$ and $\Phi_o(s) = K\Phi_o'(s)$, then $\Phi_o'(s) = -1/K$. Thus if $K = 0$, $\Phi_o'(s) = \infty$ and $(s - p_1)(s - p_2)\cdots = 0$; that is, when $K = 0$ the roots of the characteristic equation are the poles of $\Phi_o'(s)$.

Rule 4 Zeros of $\Phi_o'(s)$
The zeros of $\Phi_o'(s)$ lie on the root loci and correspond to $K = \pm\infty$.

Proof Since $1/K + \Phi_o'(s) = 0$ as $K \to \pm\infty$, $\Phi_o'(s) \to 0$; that is, $(s - z_1)(s - z_2)\cdots = 0$.

Rule 5 Asymptotes of root loci
If $\Phi_o'(s)$ has r more poles than zeros, the root loci are asymptotic to r straight lines making angles $(2N + 1)\pi/r$ with the real axis ($N = 0, 1, 2, \ldots, r - 1$), and also to r straight lines making angles $2N\pi/r$ with the real axis. The root loci approach the former asymptotes when $K \to +\infty$ and the latter when $K \to -\infty$.

Proof
$$\Phi_o'(s) = \frac{s^m + \alpha_1 s^{m-1} + \cdots + \alpha_m}{s^n + \beta_1 s^{n-1} + \cdots + \beta_n}$$

Since

$$1 + K\Phi_o'(s) = 0, \quad \text{if } r = n - m > 0$$

$$1/\Phi_o'(s) = s^r + (\beta_1 - \alpha_1)s^{r-1}\cdots = -K,$$

that is, $|K| \to \infty$ when $|s| \to \infty$. Furthermore, since the first two terms dominate the expression for $1/\Phi_o'(s)$ when $|s| \to \infty$, this equation is approximately

$$(s - a_o)^r = -K \begin{array}{ll} = |K|e^{j(2N+1)\pi} & \text{for } K > 0 \\ = |K|e^{j2N\pi} & \text{for } K < 0 \end{array}$$

where $a_o = -(\beta_1 - \alpha_1)/r$ and N is any integer.
Each of these equations has r distinct solutions given by

$$s - a_o = \begin{array}{ll} |K|^{1/r}e^{j(2N+1)\pi/r} & \text{for } K > 0 \\ |K|^{1/r}e^{j2N\pi/r} & \text{for } K < 0, \end{array}$$

where $N = 0, 1, 2, \ldots, r - 1$. Putting $s = a + jb$ and equating real and imaginary parts yields:

$$\left.\begin{array}{r} a - a_o = |K|^{1/r} \cos(2N + 1)\pi/r \\ b = |K|^{1/r} \sin(2N + 1)\pi/r \end{array}\right\} K > 0$$

and

$$\left.\begin{aligned} a - a_o &= |K|^{1/r} \cos 2N\pi/r \\ b &= |K|^{1/r} \sin 2N\pi/r \end{aligned}\right\} K < 0$$

By division

$$b = \begin{matrix} (a - a_o) \tan (2N + 1)\pi/r & K > 0 \\ (a - a_o) \tan 2N\pi/r & K < 0. \end{matrix}$$

These are the equations of the asymptotes of the root loci. Each of these equations represents a family of r straight lines in the s-plane. The angular inclinations of these lines with the real axis are $(2N + 1)\pi/r$ and $2N\pi/r$ when $K > 0$ and $K < 0$ respectively.

Rule 6 Point of intersection of asymptotes
Both sets of asymptotes intersect on the real axis at a point with abscissa

$$a_o = \left(\sum_1^n p_i - \sum_1^m z_i \right) \Big/ r$$

where p_i and z_i are the poles and zeros respectively of $\Phi_o'(s)$.

Proof From Rule 5 proof, in both cases, when $b = 0$, $a = a_o = -(\beta_1 + \alpha_1)/r$ for all values of K. All the asymptotes therefore intersect on the real axis at the point $(a_o, 0)$. The abscissa a_o has the value quoted above since it can be deduced from the expression for $\theta_o'(s)$ given that $\alpha_1 = -\sum_1^m z_i$ and $\beta_1 = -\sum_1^n p_i$ (algebraic rule of roots of equations).

Rule 7 Root loci on the real axis
If $\Phi_o'(s)$ has at least one real pole or zero, the whole of the real axis is occupied by root-loci: a segment of the real axis corresponds to positive or negative values of K according to whether an odd or even number of poles and zeros of $\Phi_o'(s)$ lie to its right.

Proof The arguments of the complex numbers represented by the vectors drawn from the conjugate complex poles or zeros of $\Phi_o'(s)$ to a point on the real axis cancel out in pairs.

The argument of a vector drawn *from* a real pole or zero lying to the left of a point on the real axis is zero, whilst the corresponding quantity for a real pole or zero lying to the right is π.

Now

$$\Phi_o'(s) = \frac{(s - z_1)(s - z_2) \cdots (s - z_m)}{(s - p_1)(s - p_2) \cdots (s - p_n)}$$

and

$$K\Phi_o'(s) = -1 = e^{j(2N+1)\pi} \text{ for } K > 0,$$

so

$$\arg K\Phi_o'(s) = (2N + 1)\pi \text{ for } K > 0.$$

Thus

$$\sum_{1}^{m} \arg(s - z_i) - \sum_{1}^{n} \arg(s - p_i) = \begin{matrix} (2N + 1)\pi \text{ for } K > 0 \\ 2N\pi \text{ for } K < 0 \end{matrix}$$

It follows that the first of these equations will be satisfied on the real axis only at those points having an odd number of poles and zeros to their right, and that the second equation will be satisfied only when this number is even. Zero is regarded as an even number.

Rule 8 Breakaway points
Breakaway points indicate the existence of multiple characteristic roots and occur at those values of s which satisfy $\mathrm{d}K/\mathrm{d}s = 0$.

Consider a characteristic equation which has a root of multiplicity $q(\geqslant 2)$ at $s = s_o$ when $K = K_o$. In the root locus for such an equation q loci will converge on the point $s = s_o$ as K increases towards K_o, and will then break away from this common point as K increases beyond K_o. Typical breakaway points are shown in Fig. 5.40.

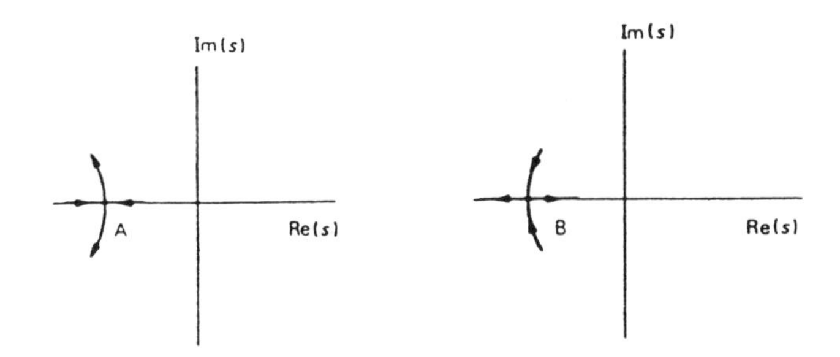

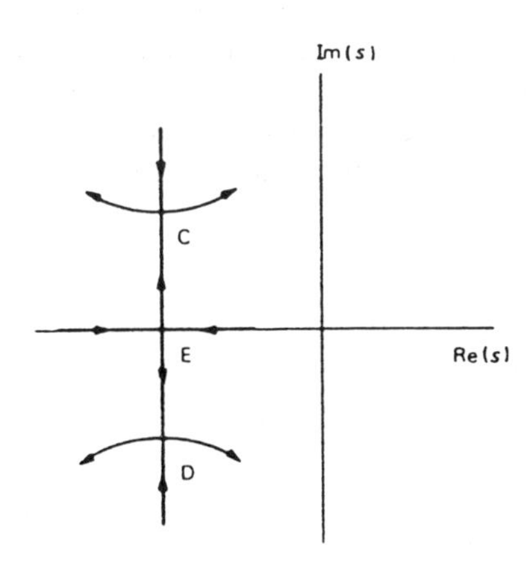

Fig. 5.40. Breakaway points.

A and B each represents two equal real roots, C and D each represents two equal complex roots, and E represents two equal real roots.

Proof The characteristic equation can be written

$$f(s,K) = P(s) + KQ(s) = 0.$$

Thus

$$df/ds = \partial f/\partial s + \partial f/\partial K \times dK/ds = 0$$

and consequently

$$\frac{dK}{ds} = -\frac{\partial f/\partial s}{Q(s)}.$$

Clearly

$$\partial f/\partial K = Q(s).$$

and since

$$f(s,K_o) = (s - s_o)^q g(s)$$

at a breakaway point $s = s_o$, where $g(s)$ is some function such that $g(s_o) \neq 0$, it is evident that

$$\partial f(s,K_o)/\partial s = q(s - s_o)^{q-1} g(s) + (s - s_o)^q g'(s)$$
$$= 0 \quad \text{at } s = s_o.$$

Hence

$$(dK/ds)_{s=s_o} = 0, \qquad \text{if } Q(s_o) \neq 0.$$

Rule 9 Intersections of root loci with the imaginary axis
The intersections of root loci with the imaginary axis can be found by calculating the values of K which result in the existence of imaginary characteristic roots. These values of K together with the corresponding imaginary roots can be found by writing $s = j\omega$ in the characteristic equation and equating the real and imaginary parts.

For example let

$$F(s) = s^3 + 2s^2 + 3s + K + 2 = 0$$

Putting $s = j\omega$ gives

$$-j\omega^3 - 2\omega^2 + 3j\omega + K + 2 = 0.$$

Equate real parts:

$$-2\omega^2 + K + 2 = 0, \qquad \text{so } K = 2\omega^2 - 2.$$

Equate imaginary parts:

$$-\omega^3 + 3\omega = 0, \qquad \text{so } \omega = 0 \quad \text{or} \quad \omega^2 = 3.$$

Hence $\omega^2 = 3$ and $K = 4$ at the intersection of the loci with the imaginary axis.

Rule 10 Slopes of root-loci at the complex poles and zeros of $\Phi_o'(s)$
The slope of a root locus at a complex pole or zero of $\Phi_o'(s)$ can be found by applying the equations

$$\sum_1^m \arg(s - z_i) - \sum_1^n \arg(s - p_i) = \begin{matrix} (2N + 1)\pi \text{ for } K > 0 \\ 2N\pi \text{ for } K < 0 \end{matrix}$$

to a point in the neighbourhood of the pole or zero.

For example, consider the complex pole P_1 shown in Fig. 5.41. where γ is the unknown slope of the locus at P_1.

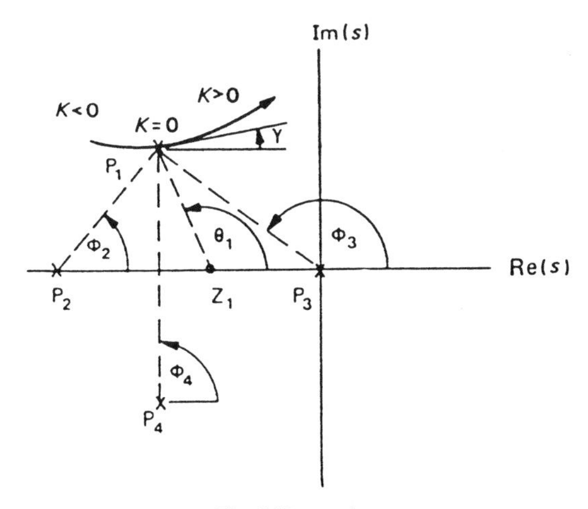

Fig. 5.41. *s*-plane

The arguments of the complex numbers represented by the vectors drawn from the other poles P_2, P_3 and P_4 and the zero Z_1 to a point on the root locus *near* P_1 can be considered to be the angles ϕ_2, ϕ_3, ϕ_4 and θ_1.

If the first of the above equations is applied to a point in the neighbourhood of P_1 for which $K > 0$,

$$\theta_1 - \phi_2 - \phi_3 - \phi_4 - \gamma = (2N + 1)\pi$$

where N is an integer. Hence γ.

The second equation may be applied to a point near P_1 for which $K < 0$.

Rule 11 Calculation of K on the root loci
The absolute magnitude of the value of K corresponding to any point s_r on a root locus can be found by measuring the lengths of the vectors drawn to s_r from the poles and zeros of $\Phi_o'(s)$ and evaluating

$$|K| = \frac{(s_r - p_1)(s_r - p_2)\cdots}{(s_r - z_1)(s_r - z_2)\cdots}$$

Summary

The rules for constructing root loci can be summarized as follows:

1. number of loci;
2. symmetry of loci;
3. poles;
4. zeros;
5. asymptotes;
6. intersection of asymptotes;
7. root loci on real axis;
8. breakaway points;
9. intersection with imaginary axis;
10. slopes of loci at complex poles and zeros;
11. calculation of K.

Poles are plotted as an 'x', zeros as an 'o'. It is not usually necessary to use all the rules for a root locus sketch.

Example 59

Sketch the root locus for a closed-loop control system with open loop transfer function given by

$$\frac{K}{s(s+1)(s^2+s+1)}$$

as K increases from $-\infty$ to $+\infty$. Hence determine the range of values K can have and the system be stable.

The rules are applied as follows:

1. 4 loci.
2. Loci are symmetrical about the real axis for real K.
3. Poles given by $s(s+1)(s^2+s+1)=0$ that is, $s = 0, -1, -\frac{1}{2} \pm j\sqrt{3}/2$. Root loci pass through poles when $K = 0$.
4. No zeros.
5. $r = 4$. Thus inclination of asymptotes are

$$\frac{\pi}{4}, \frac{3\pi}{4}, \frac{5\pi}{4}, \frac{7\pi}{4} \text{ for } K \to +\infty \text{, and}$$

$$0, \frac{2\pi}{4}, \frac{4\pi}{4}, \frac{6\pi}{4} \text{ for } K \to -\infty.$$

6. Asymptotes intersect at

$$\frac{0 - 1 - \frac{1}{2} + j\sqrt{3}/2 - \frac{1}{2} - j\sqrt{3}/2 - 0}{4} = -\frac{1}{2}.$$

7. Whole of real axis occupied.
8. $dK/ds = 0$ gives $4s^3 + 6s^2 + 4s + 1 = 0$.
   Breakaway points are $-\frac{1}{2}, -\frac{1}{2} \pm j\frac{1}{2}$. ($-\frac{1}{2}$ found by trial and error).
   When $s = -\frac{1}{2}$, $K = \frac{3}{16}$, and when $s = -\frac{1}{2} \pm j\frac{1}{2}$, $K = \frac{1}{4}$.

9. Since $1 + \Phi_o(s) = 0$,

$$s(s+1)(s^2+s+1) + K = 0. \text{ Thus}$$

in $\quad s^4 + 2s^3 + 2s^2 + s + K = 0 \quad$ put $s = j\omega$, to give

$$\omega^4 - 2j\omega^3 - 2\omega^2 + j\omega + K = 0.$$

Equate real part: $\qquad \omega^4 - 2\omega^2 + K = 0.$

and imaginary part: $-2\omega^3 + \omega = 0$,

that is, $\omega = 0$ and $K = 0$,

or $\omega^2 = \frac{1}{2}$ and $K = \frac{3}{4}$.

This is sufficient information to sketch the root locus, as shown in the diagram below.

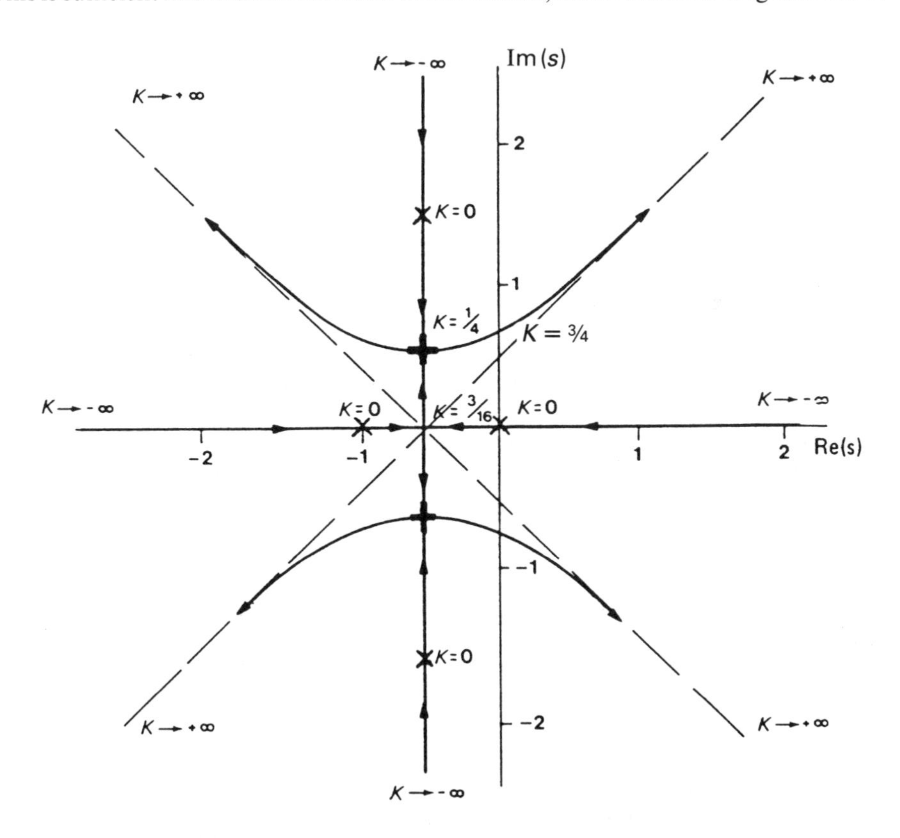

As K increases from $\to\infty$ to zero the four roots move along the 0, $\pi/2$, π and $3\pi/2$ asymptotes to the poles ($K = 0$). As K increases from zero the loci on the real axis meet at a breakaway point ($K = \frac{3}{16}$) and move to meet the other two loci at the other breakaway points ($K = \frac{1}{4}$). As K increases further the four loci move off towards the

$\pi/4$, $3\pi/4$, $5\pi/4$, and $7\pi/4$ asymptotes. The loci cross the imaginary axis when $K = \frac{3}{4}$. Therefore the range of values K can have for the system to be stable is > 0 and $< \frac{3}{4}$.

It can be seen that as K increases from $\frac{1}{4}$ to $\frac{3}{4}$, the frequency of oscillation of the response increases slightly and the rate of decay of oscillation decreases.

For practical reasons, root loci are often only drawn for positive values of the system gain K.

Example 60

Sketch the root loci of the equation

$$s^4 + 5s^3 + 8s^2 + (6 + K)s + 2K = 0$$

if K can have positive real values.

Rewrite equation as: $$1 + \frac{K(s + 2)}{s(s + 3)(s^2 + 2s + 2)} = 0,$$

so

$$\Phi_o'(s) = \frac{s + 2}{s(s + 3)(s^2 + 2s + 2)}.$$

To draw the root loci apply the rules:

1. Four loci
2. Loci are symmetrical about the real axis for real K.
3. Poles are given by $s(s + 3)(s^2 + 2s + 2) = 0$;
   that is $s = 0, -3, -1 \pm j$. Root loci pass through poles when $K = 0$.
4. Zeros given by $s + 2 = 0$, that is, $s = -2$.
   Root loci pass through zeros when $K = \infty$.
5. $r = 3$. Thus inclination of asymptotes are $\pi/3$, π, $5\pi/3$ for $K \to \infty$.
6. Asymptotes intersect at

$$\frac{(0 - 3 - 1 + j - 1 - j) - (-2)}{3} = -1.$$

7. Real axis occupied as below:

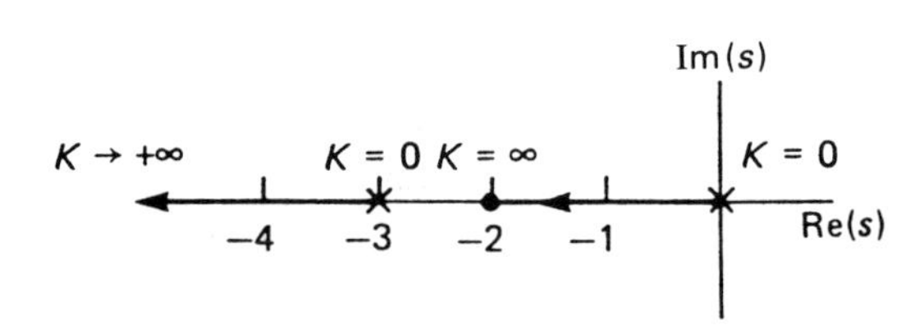

8. No breakaway points because two loci on real axis and only two other loci which are symmetrical with respect to the real axis.

9. In $s^4 + 5s^3 + 8s^2 + (6 + K)s + 2K = 0$ put $s = j\omega$.
   Hence $K = 5\sqrt{13} - 11 = 7$.
10. The slope, γ, of the tangent to locus at the pole $-1 + j$ is given by

$$45^\circ - 30^\circ - 90^\circ - 135^\circ - \gamma = (2N + 1)180^\circ.$$

   Hence $\gamma = -30^\circ$.

The root locus is as shown in the diagram below.

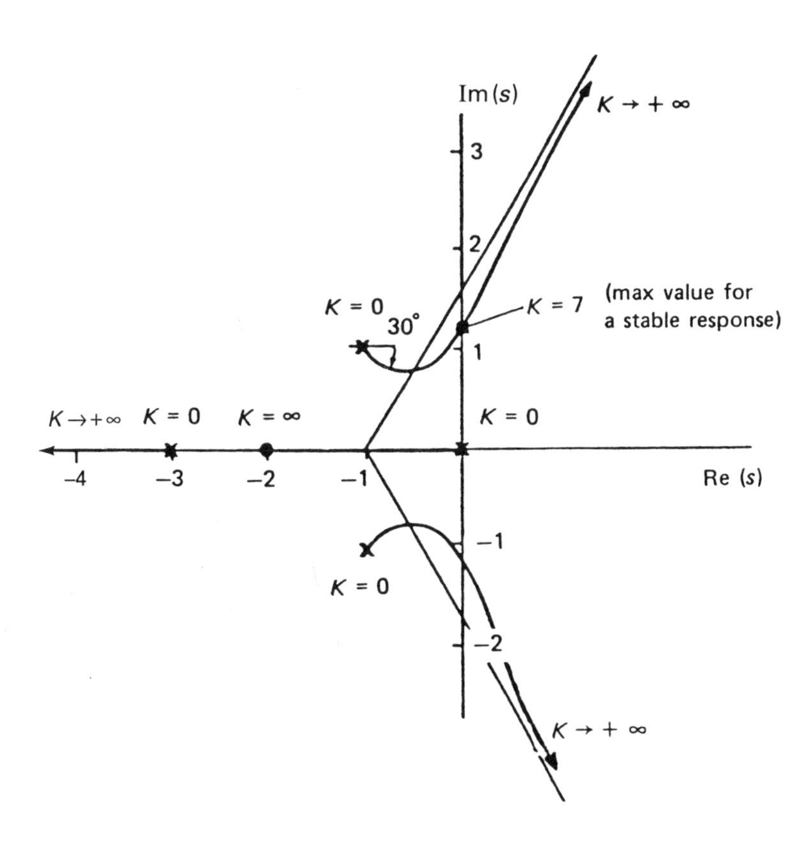

Example 61

An electric position servo is designed to move a load on a production line. When the load is in the correct position an electric limit switch applies a step input to the servo. All the components of the servo are standard items, so that the only variable parameter in the system is the gain constant K. From an analysis of the servo the transfer function has been found and the root locus plotted: the figure shows part of the result.

It is required that the positioning operation be completed in two seconds; determine a suitable value of K, and sketch the form of the servo response to be expected from a step input.

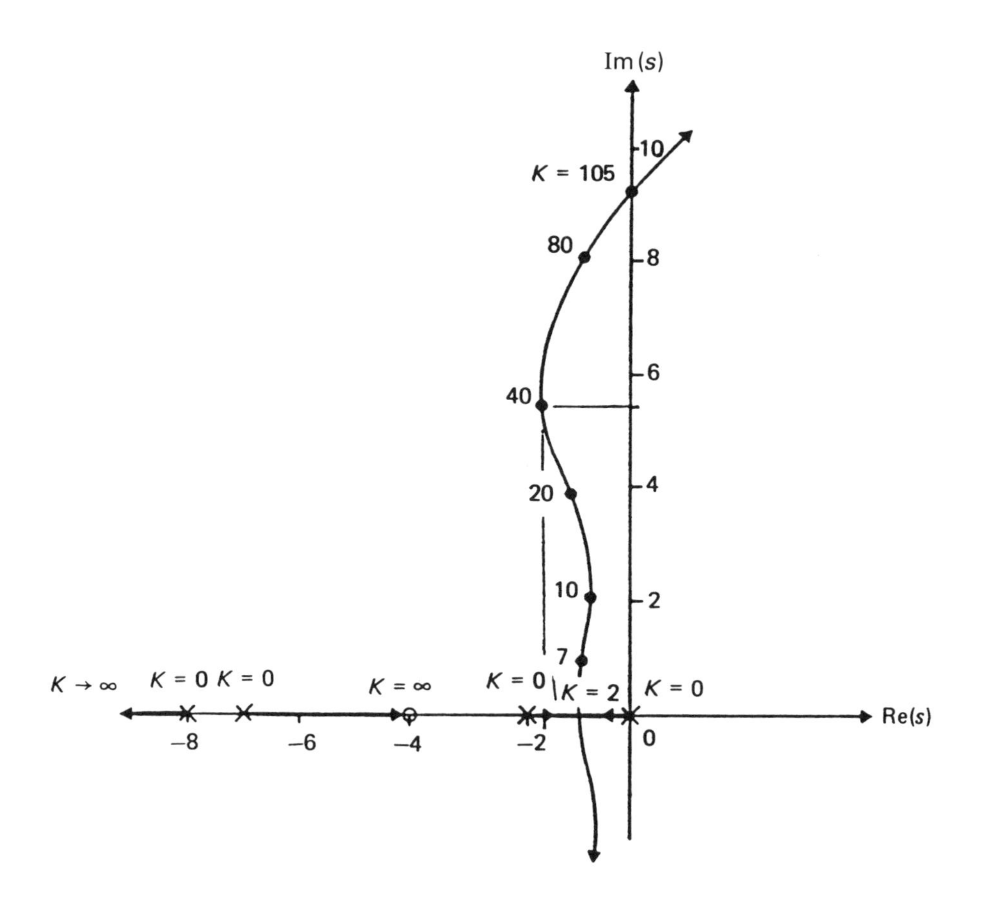

Because there are four loci and therefore four roots to the frequency equation, the response is the sum of four exponential decays. When $K = 2$, two are given by e^{-t}, the other two roots decay at approximately e^{-6t} and e^{-8t} which are fast and stable responses. Thus it is the e^{-t} response which limits the performance. After two seconds, the position is only 87% of the input step because e^{-2} is 0.13: this is not good enough, so that K must be increased to a value which gives a faster decay. However, the response will then be oscillatory. For example, if $K = 40$, the root locus plot indicates the decay to be $e^{-1.7t}$ and the frequency of oscillation to be 5.5 rad/s. When $t = 2$, $e^{-3.4} = 0.03$ so that the position is 97% of the input step, and further, the oscillation assists the positioning because $\cos 5.5t$ is 0.004 when $t = 2$. That is, the output is nearly 99% of the input when $t = 2s$. These responses are shown below: they should be compared with those for other values of K.

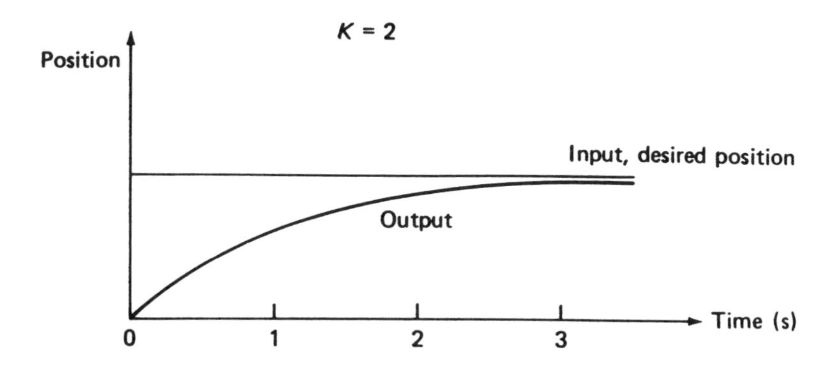

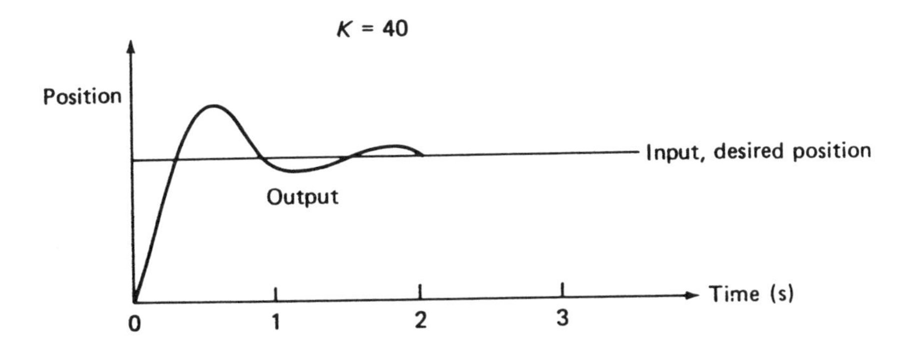

Damping ratio determination

It is worth noting that for a frequency equation

$$s^2 + 2\zeta\omega s + \omega^2 = 0.$$

$$s = -\zeta\omega \pm j\omega\sqrt{(1 - \zeta^2)}.$$

These roots can be plotted on the s-plane as shown in Fig. 5.42.

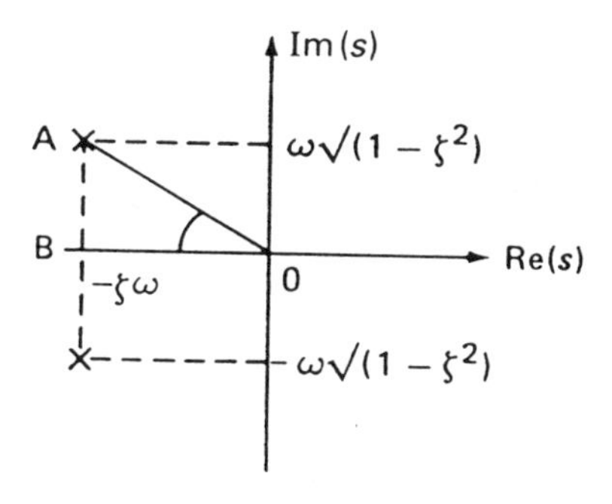

Fig. 5.42. s-plane.

So that OA $= \omega$, and cos BOA $= \zeta$.

That is, the damping ratio pertaining to a particular root can be obtained from the RL diagram, by finding the cosine of the angle between the real axis and the line joining the origin to the root.

5.6.2 The Routh–Hurwitz criterion

The Routh–Hurwitz (RH) criterion is a method for determining whether or not the frequency equation of a dynamic system has any roots containing positive real parts. If positive real parts exist in any root the system will be unstable, because any disturbance will produce a response which grows with time.

The frequency equation of a dynamic system can be written as

$$a_m s^m + a_{m-1} s^{m-1} + a_{m-2} s^{m-2} + \cdots + a_2 s^2 + a_1 s + a_o = 0.$$

For there to be no roots with positive real parts and thus the system be stable two conditions must be fulfilled:

1. All of the coefficients of the frequency equation must have the same sign, and
2. Each member of the sequence of determinants $R_1 R_2, \ldots R_{m-1}$ defined below must be positive:

$$R_1 = a_1$$

$$R_2 = \begin{vmatrix} a_1 & a_0 \\ a_3 & a_2 \end{vmatrix}$$

$$R_3 = \begin{vmatrix} a_1 & a_0 & 0 \\ a_3 & a_2 & a_1 \\ a_5 & a_4 & a_3 \end{vmatrix} \quad \ldots \text{ etc.}$$

Note. This arrangement of coefficients can be remembered as follows:

$$\begin{array}{cc|c|c|cccccc}
R_1 & a_1 & a_0 & 0 & 0 & 0 & 0 & 0 & 0 & \cdots \\
\cline{1-2}
R_2 & a_3 & a_2 & a_1 & a_0 & 0 & 0 & 0 & 0 & \cdots \\
\cline{1-3}
R_3 & a_5 & a_4 & a_3 & a_2 & a_1 & a_0 & 0 & 0 & \cdots \\
\cline{1-4}
 & a_7 & a_6 & a_5 & a_4 & a_3 & a_2 & a_1 & a_0 & \cdots \\
 & \vdots & \vdots & \vdots & \vdots & \vdots & \vdots & \vdots & \vdots &
\end{array}$$

When a_{m-1} first appears in the diagonal the process is stopped, and the array is completed by putting zeros in any spaces. The last determinant is thus always $(m - 1)$ square. Any coefficient absent in a particular frequency equation is replaced by a zero.

Example 62

A system has the following frequency equation,

$$7s^4 + 3s^3 + 8s^2 + 5s + 9 = 0.$$

Use the RH criterion to determine whether the system is stable or not.

Since all the coefficients of s are positive, condition (1) is satisfied.
The determinants for condition (2) are

$$R_1 = +5$$

$$R_2 = \begin{vmatrix} 5 & 9 \\ 3 & 8 \end{vmatrix} = 40 - 27 = +13$$

$$R_3 = \begin{vmatrix} 5 & 9 & 0 \\ 3 & 8 & 5 \\ 0 & 7 & 3 \end{vmatrix} = -136$$

Since R_1 and R_2 are positive and R_3 is negative, condition (2) fails and the system is unstable.

Note that this is the only information about the system obtained from the RH criterion. However, it is quick and easy to apply this criterion, at least in the first instance before carrying out a root locus analysis if this result is satisfactory.

Example 63
Apply the RH criterion to the system considered in Example 59 and determine the maximum value of K for stability

From Example 59,

$$\text{OLTF} = \frac{K}{s(s+1)(s^2+s+1)},$$

so that the frequency equation is

$$s^4 + 2s^3 + 2s^2 + s + K = 0.$$

Applying the RH criterion, condition (1) is satisfied provided that $K > 0$.
The determinants from condition (2) are

$$R_1 = +1$$

$$R_2 = \begin{vmatrix} 1 & K \\ 2 & 2 \end{vmatrix} = 2 - 2K$$

$$R_3 = \begin{vmatrix} 1 & K & 0 \\ 2 & 2 & 1 \\ 0 & 1 & 2 \end{vmatrix} = 3 - 4K$$

Thus R_1 is positive, but for R_2 to be positive, $K < 1$, and for R_3 to be positive $K < \frac{3}{4}$. Therefore the range of values K can have for the system to be stable is between 0 and $\frac{3}{4}$. This result was also obtained by applying Rule 9 of the root locus analysis. Example 59.

Example 64
A control system with overall gain K has an open loop transfer function

$$\frac{K}{(s+1)(s+3)(s+6)}.$$

(i) Draw the root locus diagram as K increases to infinity.
(ii) Find the range of values K can have for the system to be stable, and the corresponding frequencies of oscillation.
(iii) Confirm this range of values of K by using the Routh–Hurwitz criterion, and
(iv) If K is adjusted to be 100, find the damping ratio of the system response.

(i) To draw the root locus diagram, apply the rules.

1. There are three loci
2. The locus is symmetrical with respect to the real axis.
3. There are poles at $s = -1$, $s = -3$ and $s = -6$.
   These points correspond to $K = 0$.
4. There are no zeros.
5. $r = 3$, so the inclinations of the asymptotes are

$$\frac{\pi}{3}, \frac{3\pi}{3}, \frac{5\pi}{3}, \ldots \quad K \to +\infty$$

and
$$0, \frac{2\pi}{3}, \frac{4\pi}{3}, \ldots \quad K \to -\infty$$

6. The asymptotes intersect at the point given by

$$\left(\frac{0-1-3-6}{3}, 0\right), \text{ that is } \left(-\frac{10}{3}, 0\right),$$

7. For the real axis,

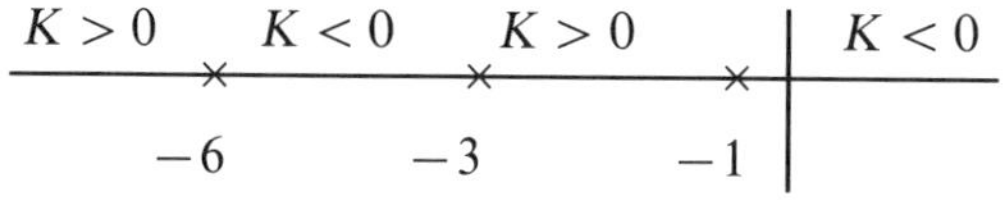

$K > 0$	$K < 0$	$K > 0$	$K < 0$
	-6	-3	-1

8. The breakaway points are when $\frac{dK}{ds} = 0$.

Now

$$-K = (s+1)(s+3)(s+6)$$
$$= s^3 + 10s^2 + 27s + 18$$

So putting $dK/ds = 0$ gives $3s^2 + 20s + 27 = 0$

Hence $s = -\dfrac{20 \pm \sqrt{(400 - 27 \times 4 \times 3)}}{6}$

$$= -4.75 \text{ or } -1.85.$$

The corresponding values of K are -8 and $+4.1$ respectively.

9. The intersection with the imaginary axis is when $s = j\omega$. That is

$$-j\omega^3 - 10\omega^2 + 27j\omega + 18 + K = 0.$$

Equating the real parts gives $-10\omega^2 + 18 + K + 0$ and equating the imaginary parts gives $-\omega^3 + 27\omega = 0$. That is, $\omega = 0$ or $\omega^2 = 27$. When $\omega = 0$, $K = -18$

and when $\omega^2 = 27$, $K = 252$. The root locus may now be drawn:

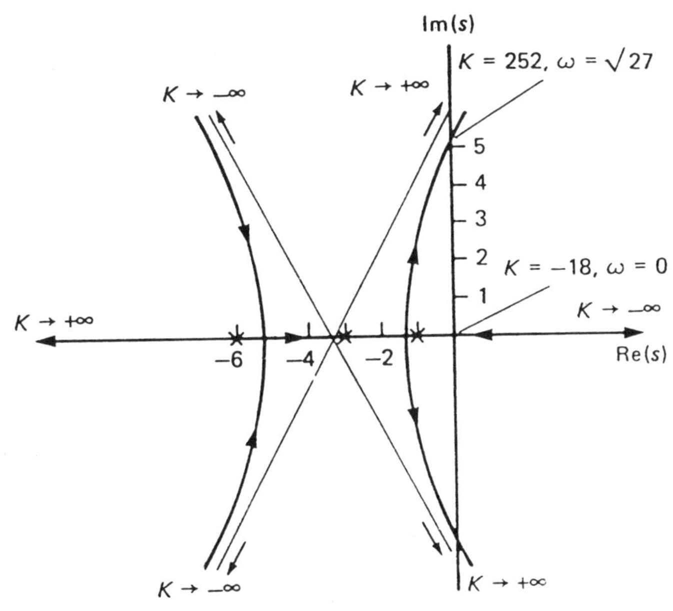

(ii) From the diagram, or by rule 9, for a stable system $252 > K > -18$. Each root then lies on the left-hand side of the s-plane.
When $K = 252$, $\omega = \sqrt{27} = 5.2$ rad/s, and when $K = -18$, $\omega = 0$.

(iii) For Routh–Hurwitz, the equation is

$$s^3 + 10s^2 + 27s + 18 + K = 0.$$

Rule 1 is satisfied if $K > -18$.
For rule 2,

$$R_1 = 27,$$

$$R_2 = \begin{vmatrix} 27 & 18 + K \\ 1 & 10 \end{vmatrix}$$

$$= 270 - (18 + K).$$

R_2 is positive provided $K < 252$.

Hence $252 > K > -18$, as before

(iv) If $K = 100$, the frequency equation is

$$s^3 + 10s^2 + 27s + 118 = 0.$$

By trail and error, $s \simeq -8.5$ (must be less than -6). This root represents a non-oscillatory exponential decay. Dividing the frequency equation by $(s + 8.5)$ gives

$$s^2 + 1.5s + 14.25 = 0.$$

Hence $s = -0.75 \pm 3.7\text{j}$.

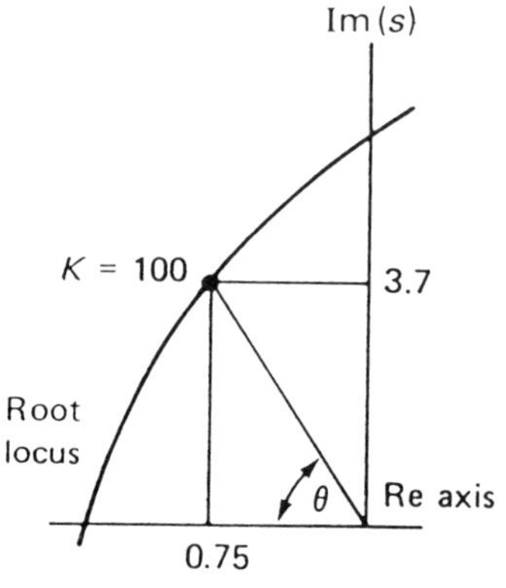

These roots are oscillatory with frequency 3.7 rad/s.

$$\text{Damping ratio } \zeta = \cos\theta = \frac{0.75}{\sqrt{(0.75^2 + 3.7^2)}}$$

$$= 0.2$$

Example 65

In a simple unity feedback control system the elements have individual functions as given in the figure.

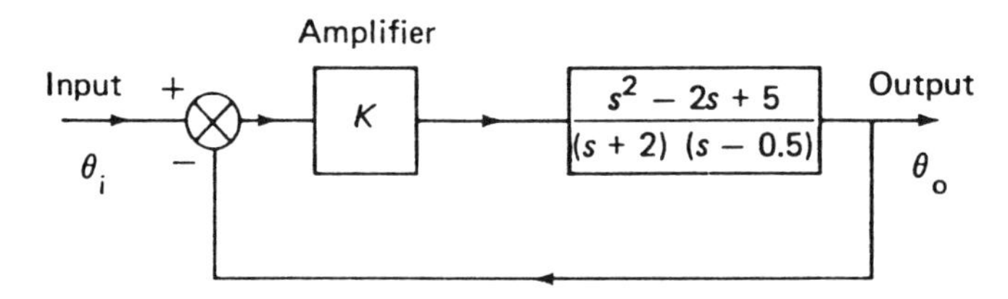

(i) Sketch the root locus diagram of this system for all positive values of the amplifier gain K, and determine the range of values for K for stability.
(ii) Confirm this range of values by applying the Routh–Hurwitz criterion, and
(iii) Find the damping ratio of the system response when $K = 0.3$

(i) $\text{OLTF} = \dfrac{K(s^2 - 2s + 5)}{(s + 2)(s - 0.5)}$

To sketch the root locus diagram, apply the rules

1. There are two loci.
2. The locus is symmetrical with respect to the real axis.
3. There are poles at $s = -2$ and $s = 0.5$.
   These points correspond to $K = 0$.
4. There are zeros at $s = \dfrac{2 \pm \sqrt{(4 - 20)}}{2} = 1 \pm 2\text{j}$.
   These points correspond to $K = \pm\infty$.
8. The breakaway points are when $\mathrm{d}K/\mathrm{d}s = 0$.

   Now

$$K = -\frac{(s + 2)(s - 0.5)}{(s^2 - 2s + 5)},$$

so $\dfrac{\mathrm{d}K}{\mathrm{d}s} = 0$ gives

$$(s^2 + 1.5s - 1)(2s - 2) = (s^2 - 2s + 5)(2s + 1.5).$$

Thus,

$$3.5s^2 - 12s - 5.5 = 0,$$

so that

$$s = 3.84 \text{ or } -0.41.$$

When

$$s = 3.84,\ K = -9.44, \text{ and}$$

when

$$s = -0.41,\ K = +0.24.$$

9. The intersection with the imaginary axis is when $s = j\omega$, That is,

$$K(-\omega^2 - 2j\omega + 5) = -(-\omega^2 + 1.5j\omega - 1).$$

Equate real parts: $-\omega^2 K + 5K = \omega^2 + 1$.
Equate imaginary parts: $-2K\omega = -1.5\omega$.

So that either $\omega = 0$ and $K = 0.2$,

or $\omega^2 = 11/7$ and $K = 0.75$.

Therefore for stability, $0.75 > K > 0.2$. The root locus can now be sketched.

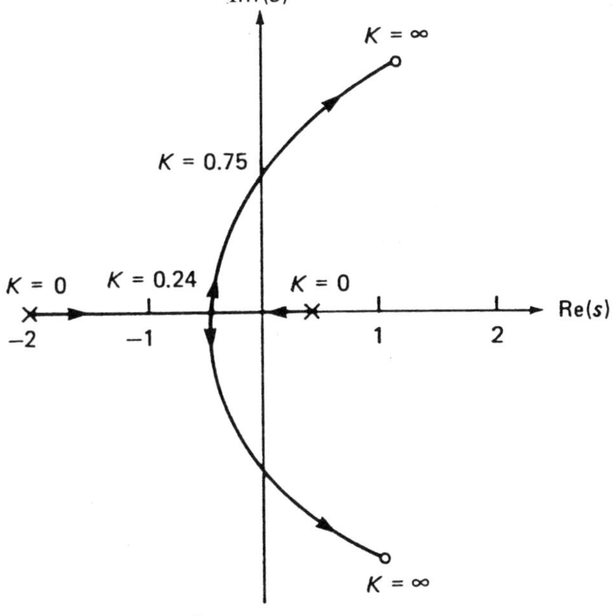

(ii) $K(s^2 - 2s + 5) = -(s^2 + 1.5s - 1)$,

so that

$$(K + 1)s^2 + (1.5 - 2K)s + (5K - 1) = 0$$

Require +ve coefficients, so

$$K > -1, \quad 2K < 1.5, \quad \text{and } 5K > 1.$$

That is,

$$K < 0.75 \text{ and } K > 0.2.$$

Determinants require

$$R_1 = 1.5 - 2K > 0, \text{ that is } K < 0.75,$$

and

$$R_2 = (1.5 - 2K)(K + 1) > 0, \text{ that is } K < 0.75 \quad \text{or} > -1.$$

Thus $\quad 0.75 > K > 0.2$ for stability.

(iii) $K\left[\dfrac{s^2 - 2s + 5}{(s + 2)(s - 0.5)}\right] = -1,$

so if $K = 0.3$.

$$1.3\, s^2 + 0.9\, s + 0.5 = 0.$$

or $\quad s = -0.346 \pm j0.515.$

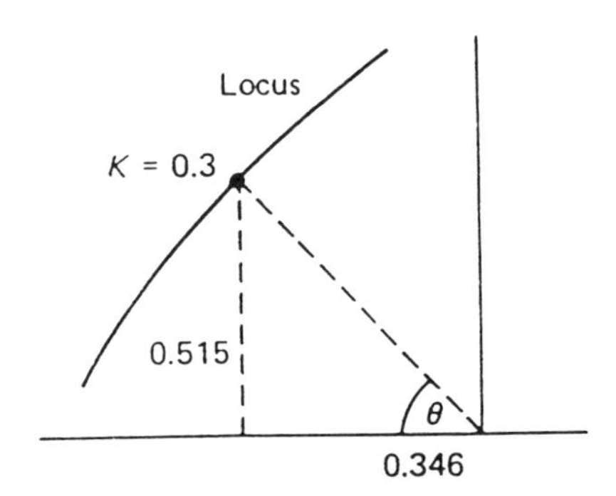

$$\theta = \tan^{-1}\frac{0.515}{0.346}$$

$$= 56^\circ$$

Damping ratio $\zeta = \cos 56^\circ = 0.558.$

Example 66

A control system with overall gain K has an open loop transfer function given by

$$\phi_o(s) = \frac{K}{s(1 + s)(1 + 2s)}.$$

(i) Draw the root locus diagram as K increases to ∞.

(ii) Find K_{max}, which is the maximum value K can have for the system to be stable, and the corresponding frequency of oscillation.

(iii) Confirm the value of K_{max} by using the Routh–Hurwitz criterion.

(iv) The system is modified by adding to the open loop derivative action with a time constant of 3.75 s. Write down the open loop transfer function of the modified system and comment on the stability. Also find the system oscillation frequency and damping factor when $K = K_{max}$.

To draw the root locus diagram, apply the rules.

1. There are three loci.
2. The loci are symmetrical with respect to the real axis.
3. There are poles at $s = 0$, $s = -1$ and $s = -\frac{1}{2}$.
   These points correspond to $K = 0$.

4. There are no zeros.
5. $r = 3$, so the inclinations of the asymptotes are

$$\frac{\pi}{3}, \frac{3\pi}{3}, \frac{5\pi}{3}, \ldots \; K \to +\infty,$$

and $$0, \frac{2\pi}{3}, \frac{4\pi}{3} \ldots \; K \to -\infty.$$

6. The asymptotes intersect at the point given by

$$\left(\frac{0 - \frac{1}{2} - 1 - 0}{3}, 0\right), \text{ that is } \left(-\frac{1}{2}, 0\right).$$

7. For the real axis:

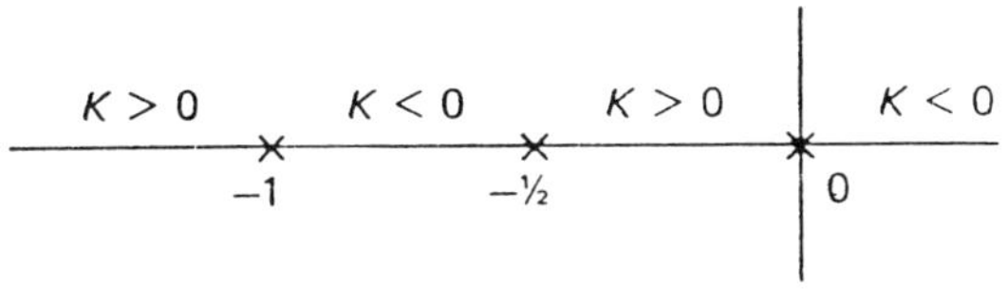

$K > 0$ $K < 0$ $K > 0$ $K < 0$

-1 $-\frac{1}{2}$ 0

8. The breakaway points are when $\dfrac{dK}{ds} = 0$.

Now

$$-K = 2s^3 + 3s^2 + s,$$

so

$$\frac{dK}{ds} = 6s^2 + 6s + 1, \quad \text{and when } \frac{dK}{ds} = 0,$$

$$s = \frac{-6 \pm \sqrt{(36 - 24)}}{12} = -0.2 \text{ or } 0.8.$$

When

$$s = -0.2, \quad K = -0.1,$$

and when

$$s = -0.8, \quad K = -0.12.$$

9. The intersection with the imaginary axis is when $s = j\omega$. That is

$$-2j\omega^3 - 3\omega^2 + j\omega + K = 0.$$

Equating the real parts gives $-3\omega^2 + K = 0$, and equating the imaginary parts gives $-2\omega^3 + \omega = 0$. That is $\omega = 0$ or $\omega^2 = \frac{1}{2}$; when $\omega = 0$, $K = 0$ and when $\omega^2 = \frac{1}{2}$, $K = \frac{3}{2}$.

The root locus may now be drawn:

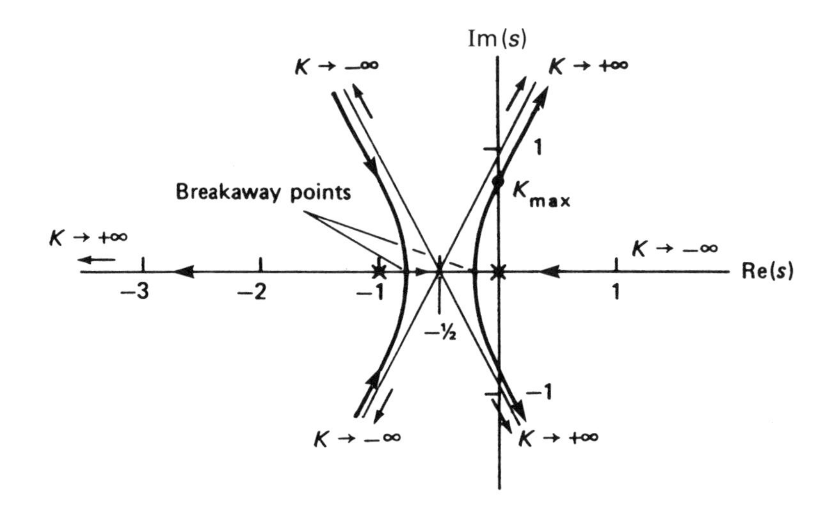

(ii) Maximum K, K_{max} for the system to be stable is given when $s = j\omega$. From rule 9,

$$K_{max} = \tfrac{3}{2} \text{ and } \omega^2 = \tfrac{1}{2}.$$

That is, the frequency of oscillation = 0.707 rad/s = 0.11 Hz.

(iii) For Routh–Hurwitz, the equation is

$$2s^3 + 3s^2 + s + K = 0,$$

that is $a_3 = 2$, $a_2 = 3$, $a_1 = 1$ and $a_0 = K$.

Condition 1 requires $K > 0$.
For condition 2,

$$R_1 = 1$$

$$R_2 = \begin{vmatrix} 1 & K \\ 2 & 3 \end{vmatrix} = 3 - 2K, \text{ that is } K \not> \tfrac{3}{2}$$

$$R_3 = \begin{vmatrix} 1 & K & 0 \\ 2 & 3 & 1 \\ 0 & 0 & 2 \end{vmatrix} = 1.6 - 2.2K, \text{ that is } K \not> \tfrac{3}{2}$$

Hence for a stable response $\tfrac{3}{2} > K > 0$.

(iv) The OLTF of the modified system is

$$\Phi_0(s) = \frac{K(1 + 3\tfrac{3}{4}s)}{s(1 + s)(1 + 2s)}$$

The addition of a zero makes this system stable for all values of K.

When $K = K_{max} = \frac{3}{2}$,

$$\Phi_0(s) = \frac{3(1 + \frac{15}{4}s)}{2s(1 + s)(1 + 2s)}$$

The frequency equation is therefore

$$2s(1 + s)(1 + 2s) + 3(1 + \tfrac{15}{4}s) = 0.$$

That is

$$4s^3 + 6s^2 + \tfrac{53}{4}s + 3 = 0.$$

By trial and error, one root is when $s = -\frac{1}{4}$.
Dividing $(s + \frac{1}{4})$ into the frequency equation gives $(4s^2 + 5s + 12)$. The roots of $4s^2 + 5s + 12 = 0$ are given by

$$s = \frac{-5 \pm \sqrt{(25 - 4 \times 4 \times 12)}}{8} = -\frac{5}{8} \pm \mathrm{j}\,1.62.$$

Thus the frequency of oscillation is 1.62 rad/s or 0.26 Hz, and the damping factor is $\frac{5}{8}$.

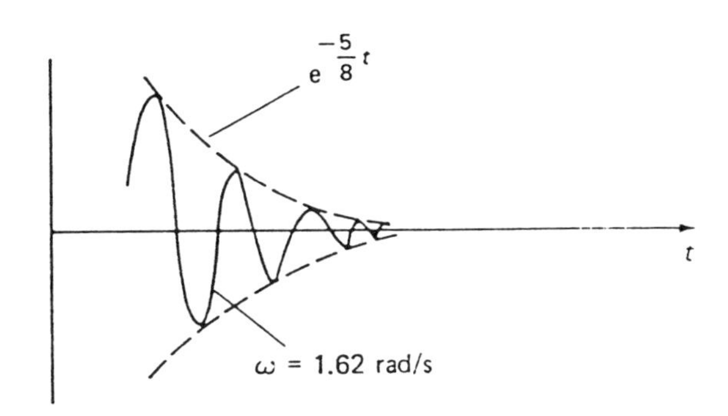

Example 67
A position control system is shown below in block diagram form.

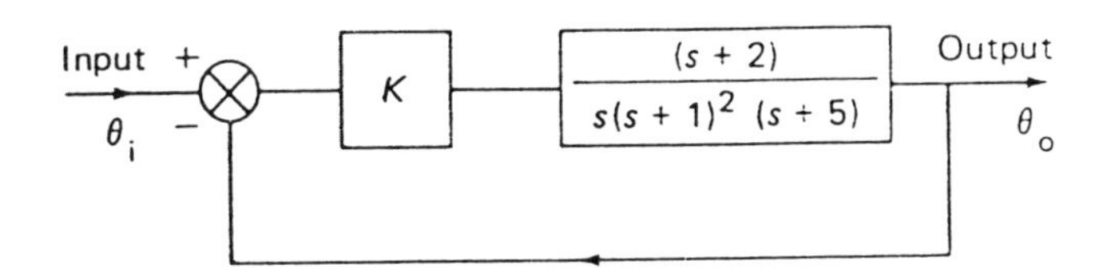

(i) Use the Routh–Hurwitz criterion to find the maximum value of gain K for stability
(ii) Because this value of K is too low for the application, a compensating subsystem with transfer function $(s + 1)/(s + 10)$ is to be inserted in the forward feed of the system.
Sketch the root locus diagram of the modified system for all values of the gain K, and find the range of values K can have for stability.
Given that the damped natural frequency of the system response is 1 rad/s, estimate the damping ratio.

(i) The frequency equation is

$$1 + \frac{K(s + 2)}{s(s + 1)^2(s + 5)} = 0.$$

That is,

$$s^4 + 7s^3 + 11s^2 + (5 + K)s + 2K = 0.$$

For the Routh–Hurwitz criterion, condition 1 requires $K > 0$.
Also for condition 2,

$$R_1 = 5 + K$$

$$R_2 = \begin{vmatrix} 5 + K & 2K \\ 7 & 11 \end{vmatrix}$$

$$= 55 - 3K, \text{ so } +\text{ve for } K < \tfrac{55}{3},$$

and

$$R_3 = \begin{vmatrix} 5 + K & 2K & 0 \\ 7 & 11 & 5 + K \\ 0 & 1 & 7 \end{vmatrix}$$

$$= (5 + K)[77 - 5 - K] - 7[14K],$$

$$= 360 - 5K + 72K - K^2 - 98K.$$

Limit when $R_3 = 0$, i.e. when

$$K^2 + 31K - 360 = 0,$$

that is

$$(K - 9)(K + 40) = 0$$

or

$$K = 9 \text{ or } -40.$$

Thus maximum value of K is 9.

(ii) Modified system block diagram:

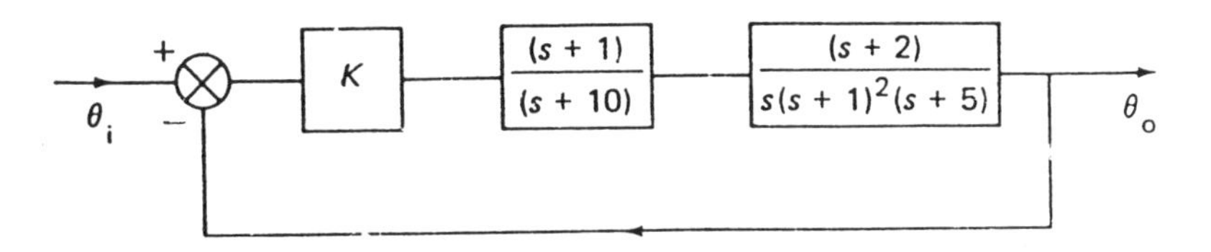

That is,

$$\theta_i \quad \frac{K(s+2)}{s(s+1)(s+5)(s+10)} \quad \theta_o$$

To draw the root locus diagram, apply the rules.

1. There are four loci.
2. The locus is symmetrical with respect to the real axis.
3. There are poles at $s = 0, -1, -5, -10$ $(K = 0)$.
4. There is a zero at $s = -2$ $(K = \pm\infty)$.
5. $r = 3$, so the inclination of the asymptotes are

$$\frac{\pi}{3}, \frac{3\pi}{3}, \frac{5\pi}{3}, \ldots K \rightarrow +\infty,$$

and $$0, \frac{2\pi}{3}, \frac{4\pi}{3} \ldots K \rightarrow -\infty.$$

6. The asymptotes intersect at the point given by

$$\left(\frac{0-1-5-10-(-2)}{3}, 0\right), \text{ that is } \left(-\frac{14}{3}, 0\right).$$

7. For the real axis;

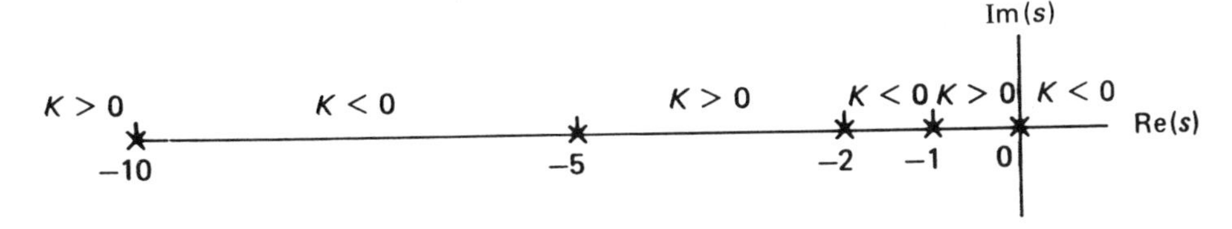

8. The breakaway points are when $\dfrac{\mathrm{d}K}{\mathrm{d}s} = 0.$

   Now

$$1 + \frac{K(s+2)}{s(s+1)(s+5)(s+10)} = 0,$$

so $$K = -\frac{s^4 + 16s^3 + 65s^2 + 50s}{s+2}.$$

$$\frac{\mathrm{d}K}{\mathrm{d}s} = 0 \text{ gives}$$

$$3s^4 + 40s^3 + 161s^2 + 260s + 100 = 0.$$

   Trial and error gives (approximately) $s = -0.55$ and $K = 4$.

   and $s = -7.5$.

9. The intersection with the imaginary axis is when $s = \mathrm{j}\omega$. That is,

$$\omega^4 - 16\mathrm{j}\omega^3 - 65\omega^2 + 50\mathrm{j}\omega + K\mathrm{j}\omega + K.2 = 0.$$

   Equate real parts: $\omega^4 - 65\omega^2 + 2K = 0.$

   Equate Imaginary parts: $-16\omega^3 + 50\omega^2 + K\omega = 0.$

   so that $\omega = 0$ and $K = 0$,

   or $\omega = 6$ and $K = 526$.

   Thus max K for stability $= 526$.
   The root locus can now be drawn as shown opposite.

   For stability, $526 > K > 0$.
   For the damping ratio at 1 rad/s,

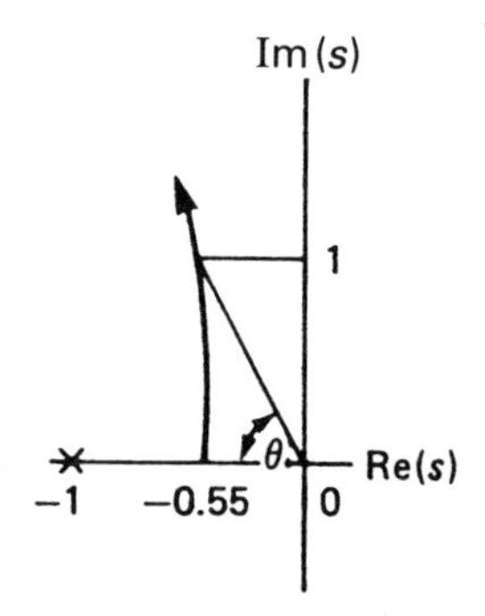

$$\zeta = \cos\theta$$

$$\simeq \frac{0.55}{\sqrt{(1^2 + 0.55^2)}}$$

$$\simeq 0.48.$$

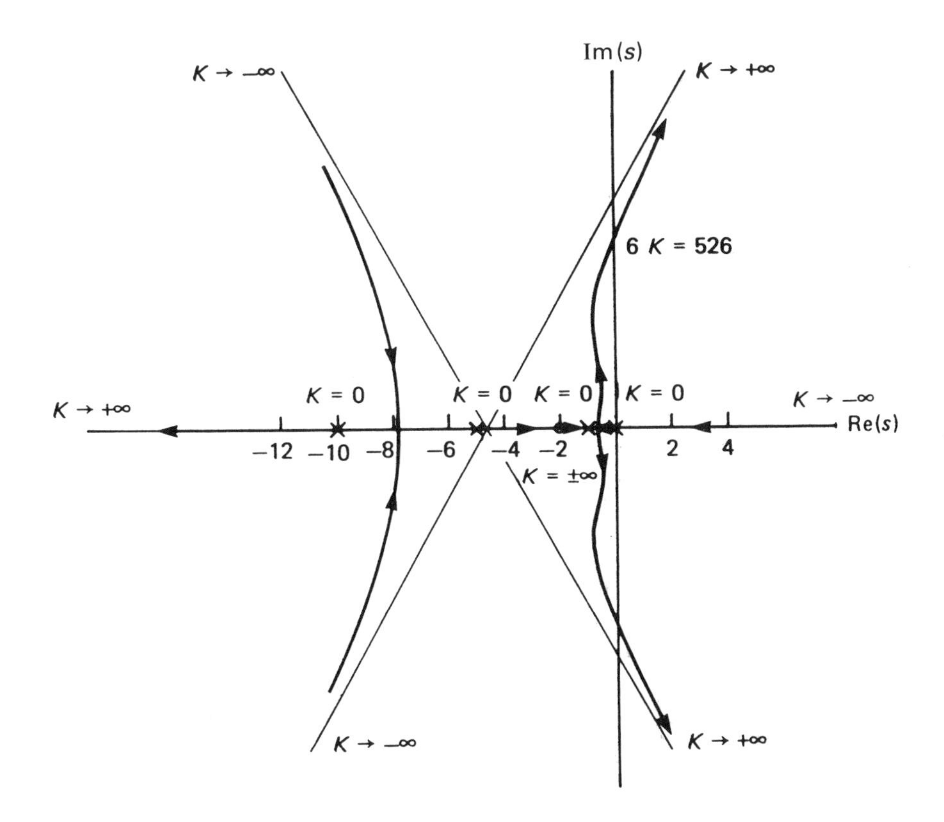

5.7 CONTROL SYSTEM FREQUENCY RESPONSE

Frequency response methods provide a convenient means for investigating the dynamic behaviour of control systems. By frequency response is meant the response of a system to an harmonic input of the form $x_i = X_i \cos \omega t$.

5.7.1 The Nyquist criterion

The conditions for limiting stability were found by the root locus method by plotting $1 + \Phi_o(s) = 0$ and finding the value of K for $s = j\omega$. An alternative method is to plot $\Phi_o(s)$ with $s = j\omega$, i.e. sinusoidal forcing, and find when this quantity is equal to -1. Thus the closed loop stability is determined from the open loop response. This is the basis of the Nyquist criterion. The Nyquist criterion is an important method for studying linear feedback systems since it is expressed only in terms of the open loop transfer function.

The open loop transfer function can be plotted on a Nyquist diagram; an example is shown in Fig. 5.43. If the loop formed by the open loop transfer function as the frequency increases from $-\infty$ to $+\infty$ *encloses* the point $\Phi_0(s) = (-1, 0)$ the system is unstable. The proximity of the loop to that point is a measure of how stable the system is.

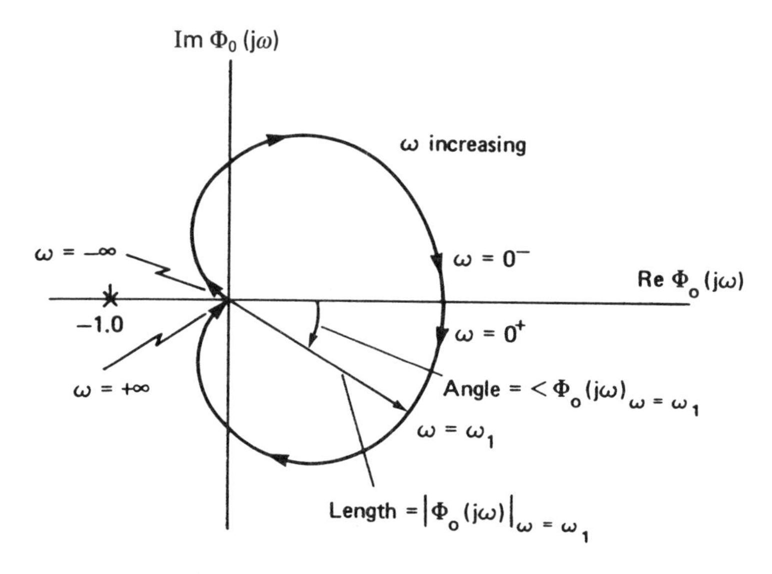

Fig. 5.43. Nyquist diagram. Stable system.

All points to the right of a contour, as it is traversed in a clockwise direction, are said to be enclosed by it (Fig. 5.44). Because the contour is symmetrical about the real axis, it is only necessary to calculate the contour as ω increases from $-\infty$ to zero.

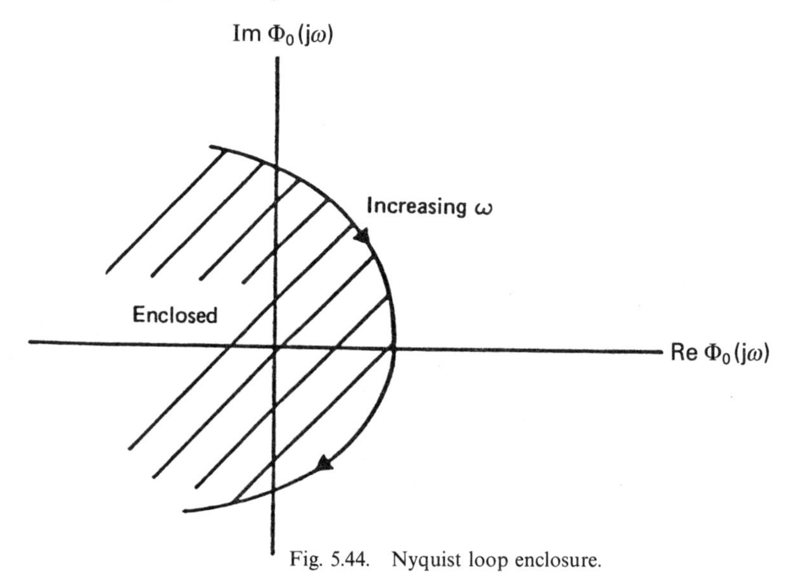

Fig. 5.44. Nyquist loop enclosure.

Example 68

Draw the Nyquist diagram for the system with an open loop transfer function

$$\frac{K}{(s + a)(s + b)}$$

and hence determine whether the closed loop system will be stable.

$$\Phi_o(s) = \frac{K}{(s + a)(s + b)}. \quad \text{Substitute } s = j\omega,$$

So,

$$\Phi_o(j\omega) = \frac{K}{(a + j\omega)(b + j\omega)} = \frac{K}{(ab - \omega^2) + j\omega(a + b)}$$

Rationalize and hence

$$\Phi_o(j\omega) = \frac{K((ab - \omega^2) - j\omega(a + b))}{((ab - \omega^2)^2 + \omega^2(a + b)^2)}.$$

Thus

$$\text{Re } \Phi_o(j\omega) = \frac{K(ab - \omega^2)((a - \omega^2)^2 + \omega^2(a + b)^2)}{((ab - \omega^2)^2 + \omega^2(a + b)^2)},$$

and

$$\text{Im } \Phi_o(j\omega) = \frac{-K\omega(a + b)}{((ab - \omega^2)^2 + \omega^2(a + b)^2)}.$$

Hence the following table can be deduced:

ω	$\text{Re}\Phi_o(j\omega)$	$\text{Im}\Phi_o(j\omega)$
$-\infty$	0	0
− ve large	− ve	+ ve
− ve small	+ ve	+ ve
0^-	+ ve	0

The loop is symmetrical for positive values of ω.

From this table the Nyquist diagram can be sketched, as shown.

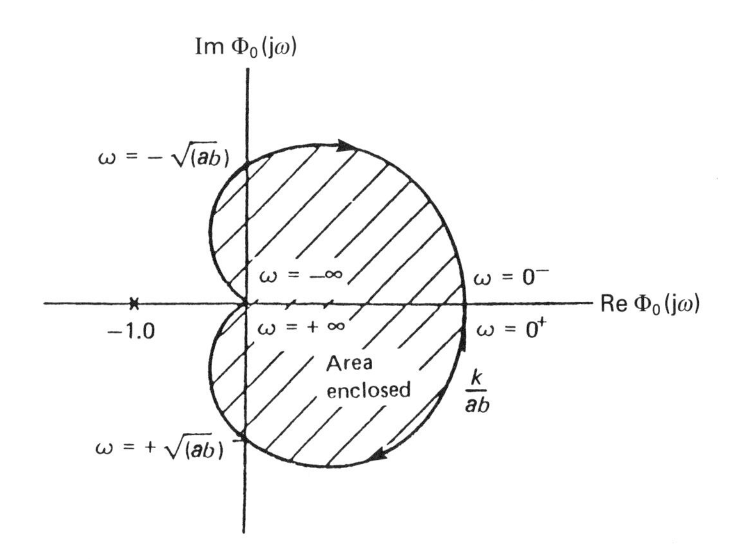

It can be seen that increasing K merely increases the size of the loop, and it will never pass through (− 1,0). Thus the system is always stable.

Example 69

Consider the closed loop system with an open loop transfer function given by

$$\frac{K}{s^3 + 6^2 + 11s + 6}.$$

Determine the maximum value of K for a stable response.

Put $s = j\omega$, so that

$$\Phi_o(j\omega) = \frac{K}{(6 - 6\omega^2) + j\omega(11 - \omega^2)}.$$

Rationalize and split into real and imaginary parts.
Thus

$$\mathrm{Re}\Phi_o(j\omega) = \frac{6K(1 - \omega^2)}{\omega^6 + 14\omega^4 + 49\omega^2 + 36},$$

and

$$\mathrm{Im}\Phi_o(j\omega) = \frac{-K\omega(11 - \omega^2)}{\omega^6 + 14\omega^4 + 49\omega^2 + 36}.$$

Hence the following tåble.

ω	$\text{Re}\Phi_0(j\omega)$	$\text{Im}\Phi_0(j\omega)$
$-\infty$	0	0
− ve large	− ve small	− ve small
$-\sqrt{11}$	− ve	0
− ve small	− ve	+ ve
− 1	0	+ ve
0	$K/6$	0

From this table the Nyquist diagram can be sketched as shown:

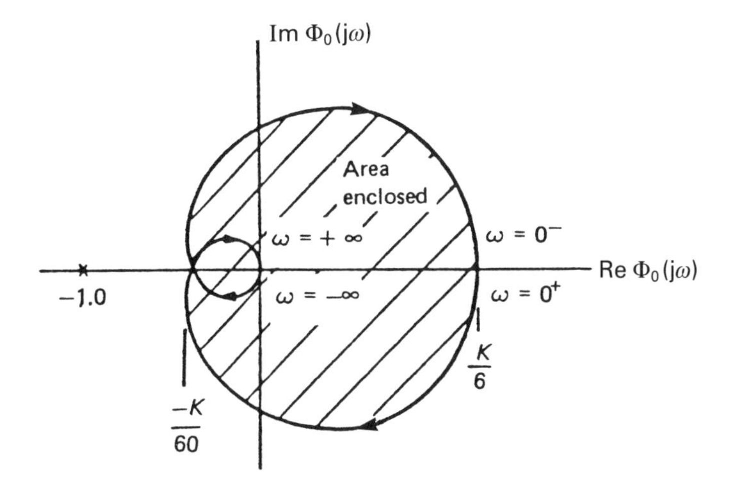

When $\omega = -\sqrt{11}$, $\text{Im}\Phi_0(j\omega) = 0$
and

$$\text{Re}\Phi_0(j\omega) = \frac{6K(1 - 11)}{11^3 + 14.11^2 + 49.11 + 36} = -\frac{K}{60},$$

so that K can be increased to 60 before instability occurs.

Alternatively, the maximum value of K for stability can be found by putting $\text{Re}\Phi_0(j\omega) = -1$ when $\text{Im}\Phi_0(j\omega) = 0$. In this case $-1 = -K/60$, and $K = 60$, as before.

When $K = 60$, the curve passes through $(-1,0)$. This point is not enclosed.

The closeness of a contour to the $(-1, 0)$ point can be expressed in terms of the *gain margin* and the *phase margin*, Fig. 5.45, so that in the above example the gain margin is $60/K$. Hence the Nyquist criterion gives information on 'absolute' stability and 'relative' stability. 'Relative' stability is used to indicate the degree of stability of a system, and is associated with the nearness of the open loop frequency response plot to the (− 1, 0) point. The quantitative measures 'gain margin' and 'phase margin' determine the degree of stability.

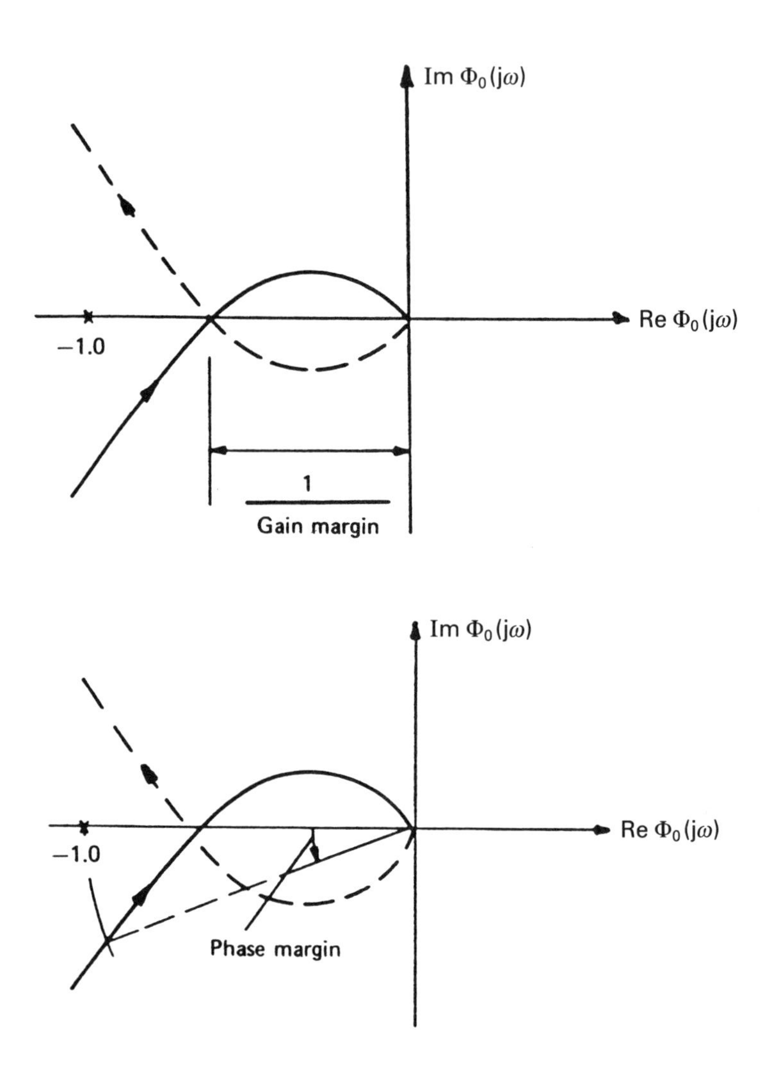

Fig. 5.45. Gain and phase margins.

The Nyquist method is also useful for assessing system performance based on experimental measurements, as shown in the following example.

Example 70
The following experimental results were obtained from an open loop frequency response test of a control system:

ω rad/s	4	5	6	8	10
Gain	0.66	0.48	0.36	0.23	0.15
Phase angle	– 134°	– 143°	– 152°	– 167°	– 180°

Plot the locus of the transfer function, and measure the gain and phase margins.

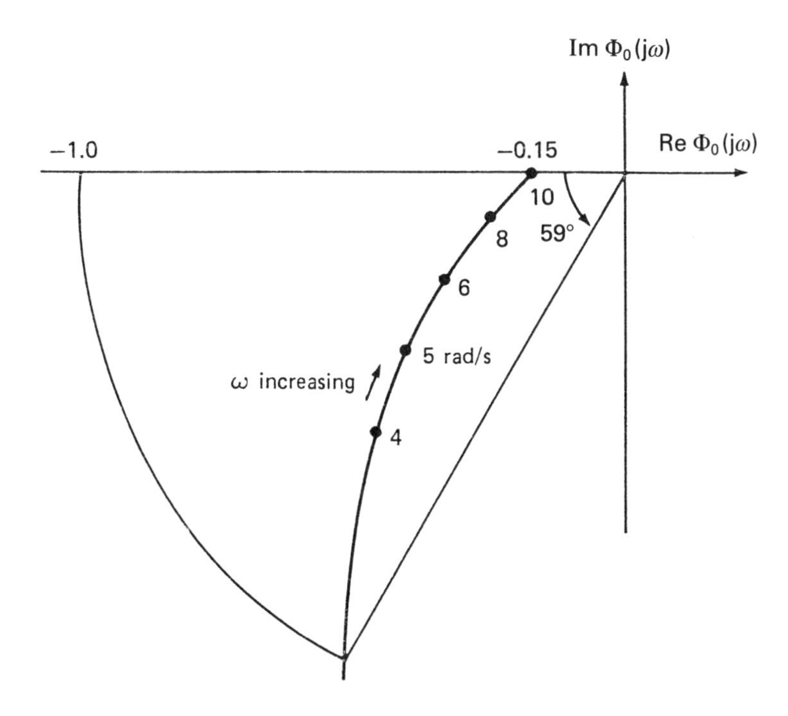

Hence the gain margin = 1/0.15 = 6.7, and the phase margin = 59°.

It can be seen that the Nyquist diagram can be drawn directly from sinusoidal steady-state measurements on the components that make up the open loop transfer function. This is very useful for determining system stability characteristics when transfer functions of the loop components are not available in analytic form, or when physical systems are to be tested and evaluated experimentally.

The Nyquist technique is also useful when the input is periodic but not simple harmonic; in this case the input may be considered to comprise a series of harmonic components, as in the Fourier analysis technique.

The effect on the Nyquist diagram of adding poles to the open loop transfer function is shown in Fig. 5.46. It can be seen that if three or more poles exist, the system can become unstable. However, the effect of adding derivative action to the system, that is, a zero to the open loop transfer function, is shown in Fig. 5.47 opposite. In this case, stability can be restored to a system having a transfer function with three poles.

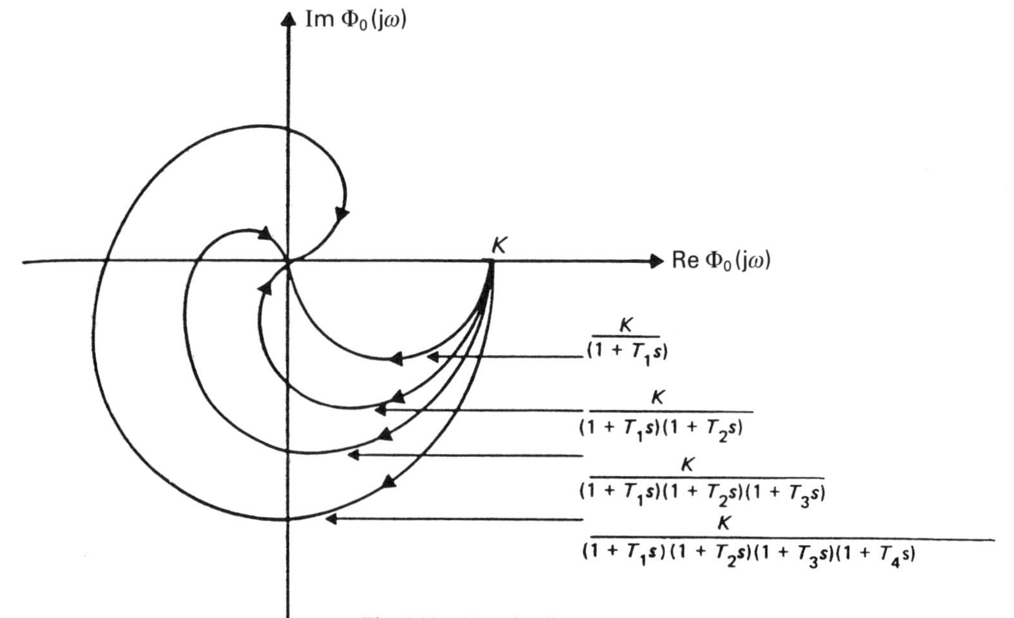

Fig. 5.46. Nyquist diagram.

Example 71

A control system has an open loop transfer function

$$\Phi_0(s) = \frac{K}{s(s+2)(s+3)}.$$

Sketch the Nyquist frequency response diagram, and determine whether the closed loop system will be stable when $K = 20$.

The system is to be modified by adding derivative action of time constant one second. What will be the effect on the stability of the closed loop system?

$$\Phi_0(s) = \frac{K}{s(s+2)(s+3)} = \frac{K}{s^3 + 5s^2 + 6s}.$$

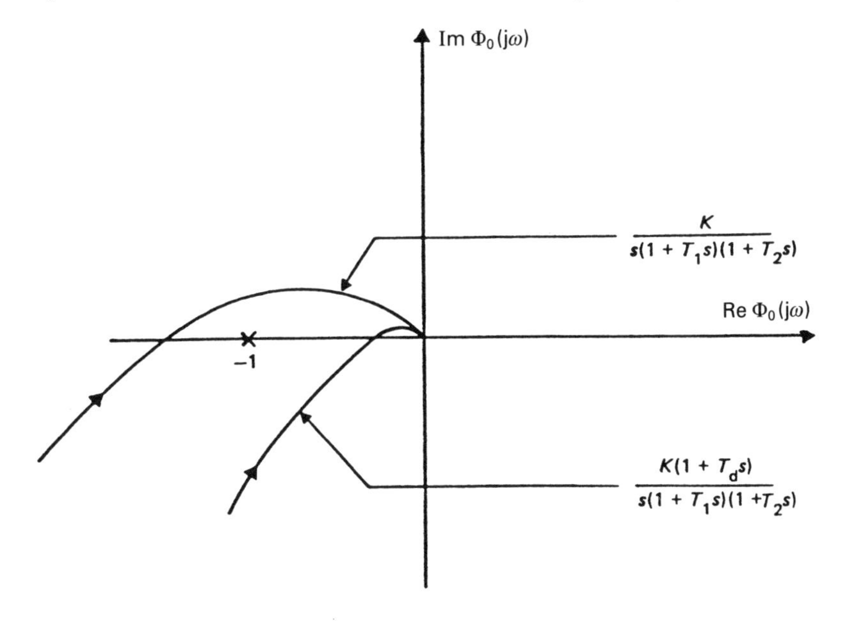

Fig. 5.47. Derivative action effect.

Hence

$$\Phi_0(\mathrm{j}\omega) = \frac{K}{-5\omega^2 + \mathrm{j}\omega(6-\omega^2)}.$$

Rationalize and split into real and imaginary parts, so that

$$\mathrm{Re}\ \Phi_0(\mathrm{j}\omega) = \frac{-5\omega^2 K}{(-5\omega^2)^2 + \omega^2(6-\omega^2)^2},$$

and

$$\mathrm{Im}\ \Phi_0(\mathrm{j}\omega) = \frac{-\omega(6-\omega^2)}{(-5\omega^2)^2 + \omega^2(6-\omega^2)^2}.$$

The following table can now be deduced:

ω	Re $\Phi_0(\mathrm{j}\omega)$	Im $\Phi_0(\mathrm{j}\omega)$
$-\infty$	0	0
− ve large	− ve	− ve
− ve small	− ve	+ ve
0^-	$-5K/36$	∞

Also, when $\omega^2 = 6$ or 0, $\text{Im } \Phi_0(j\omega) = 0$.

If $\omega = 0$

$$\text{Re } \Phi_0(j\omega) = 0,$$

and if $\omega^2 = 6$,

$$\text{Re } \Phi_0(j\omega) = -\frac{K}{30}.$$

The Nyquist diagram can now be drawn:

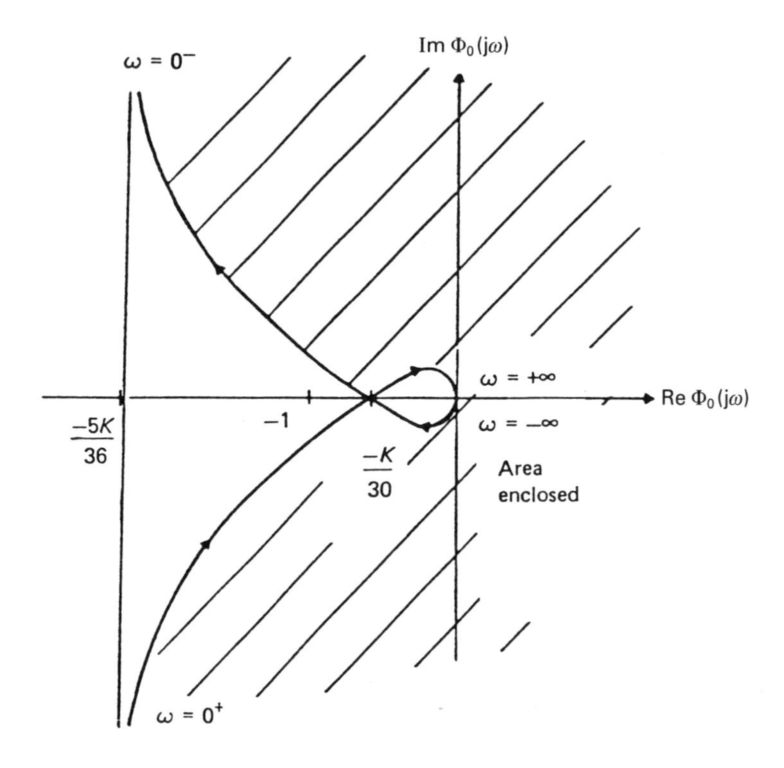

When $K = 20$ the $(-1, 0)$ point is not enclosed, so the system will be stable.

However, although the gain margin is reasonable, the phase margin is small, so that derivate action should be considered. If derivative action with a time constant of one second is added, the OLTF becomes

$$\Phi_0(s) = \frac{K(s+1)}{s(s+2)(s+3)}.$$

Thus

$$\Phi_0(j\omega) = \frac{K(j\omega + 1)}{j\omega(-\omega^2 + 5j\omega + 6)},$$

so that

$$\text{Re } \Phi_0(j\omega) = \frac{-K(\omega^4 - \omega^2)}{\omega^6 + 13\omega^4 + 36\omega^2},$$

and

$$\text{Im } \Phi_0(j\omega) = \frac{-K(4\omega^3 + 6\omega)}{\omega^6 + 13\omega^4 + 36\omega^2}.$$

The following table can now be deduced:

ω	Re $\Phi_0(j\omega)$	Im $\Phi_0(j\omega)$
$-\infty$	0	0
$-$ ve large	$-$ ve small	+ ve small
$-$ ve small	$-$ ve large	+ ve large
-1	0	$K/5$
0^-	$K/36$	$+\infty$

Hence the Nyquist diagram can be plotted as shown below.

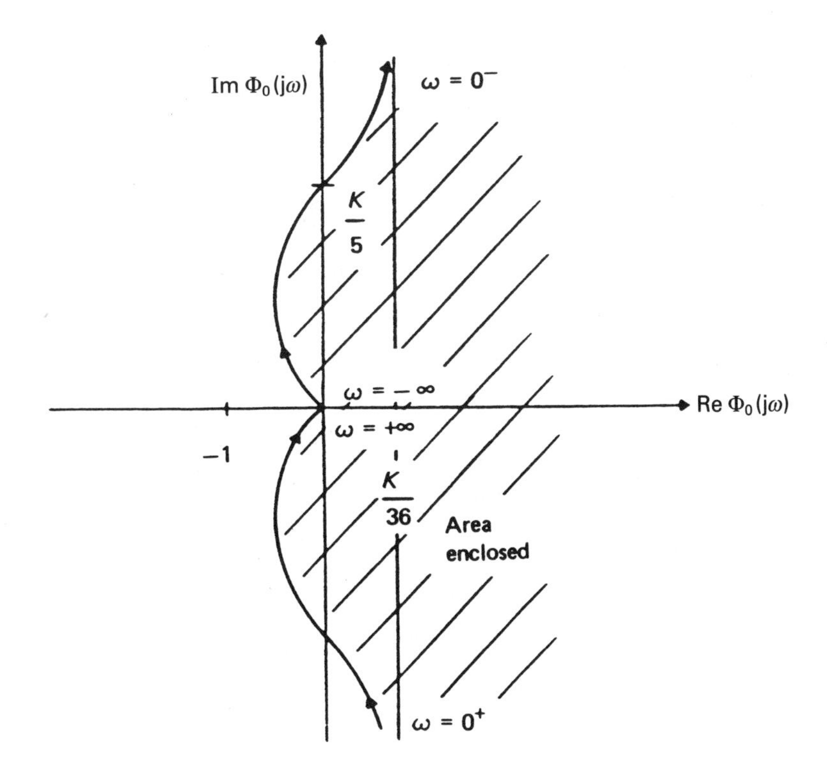

The modified system is thus stable for all values of K, with a much improved phase margin, which gives a better transient performance.

Example 72

A process is controlled by adding negative feedback through an integral action controller as shown:

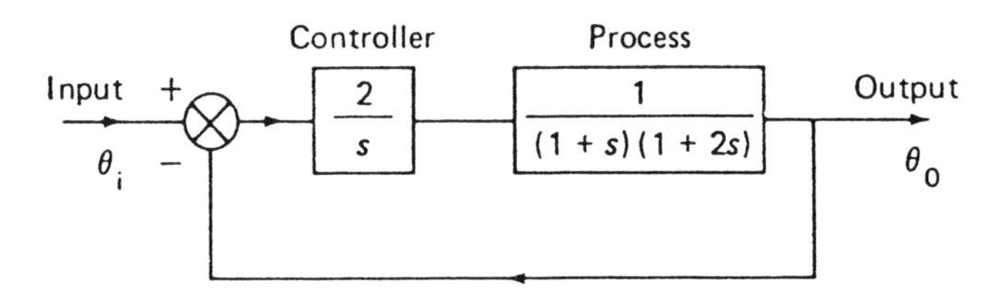

(i) Sketch the Nyquist frequency response diagram for the open loop transfer function of the system, and thus determine whether the system is stable or not.

(ii) The system is modified by adding derivative action so that the controller transfer function becomes

$$\left(\frac{2+s}{s}\right)$$

Write down the open loop transfer function of the modified system and hence sketch the Nyquist frequency response diagram. Comment on the stability of the modified system and calculate the gain margin.

(i) $$\Phi_0(s) = \frac{2}{s(1+s)(1+2s)} = \frac{2}{3s^2 + s + 2s^3}$$

Put $s = \mathrm{j}\omega$ for harmonic response, and rationalize:

$$\Phi_0(\mathrm{j}\omega) = \frac{2}{-3\omega^2 + \mathrm{j}(\omega - 2\omega^3)}\left[\frac{-3\omega^2 - \mathrm{j}(\omega - 2\omega^3)}{-3\omega^2 - \mathrm{j}(\omega - 2\omega^2)}\right].$$

So

$$\mathrm{Re}\ \Phi_0(\mathrm{j}\omega) = \frac{-6\omega^2}{4\omega^6 + 5\omega^4 + \omega^2},$$

and

$$\mathrm{Im}\ \Phi_0(\mathrm{j}\omega) = \frac{-2(\omega - 2\omega^3)}{4\omega^6 + 5\omega^4 + \omega^2}.$$

Hence the following table:

ω	Re $\Phi_0(j\omega)$	Im $\Phi_0(j\omega)$
$-\infty$	0	0
− ve large	− ve	− ve
− ve small	− ve	+ ve
0^-	-6	∞
$-\sqrt{\frac{1}{2}}$	$-\frac{4}{3}$	0

The Nyquist diagram can now be drawn.

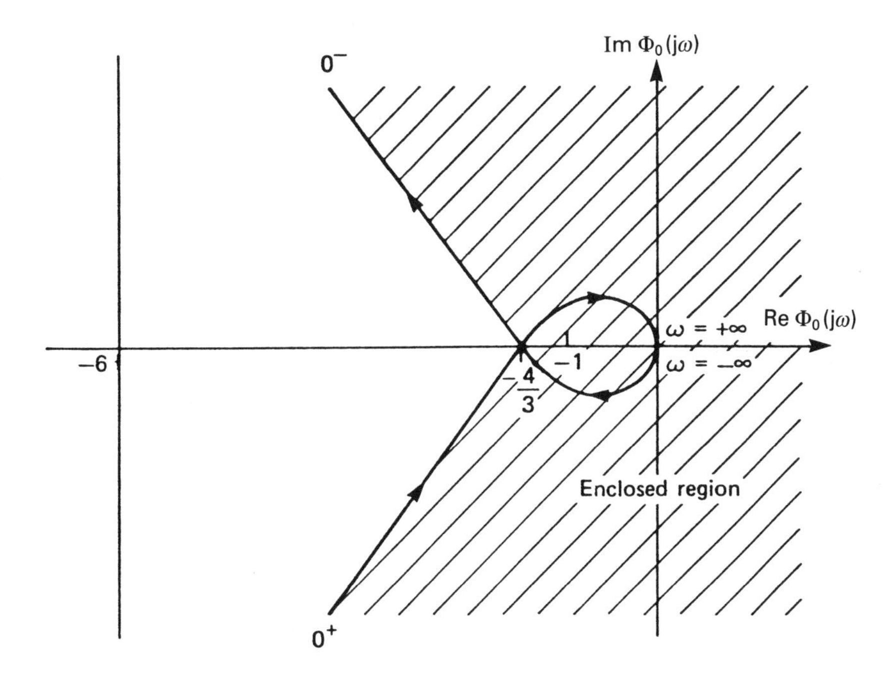

The system is unstable because the point (− 1,0) is enclosed.

(ii) For the modified system

$$\Phi_0(s) = \frac{(2 + s)}{2s^3 + 3s^2 + s}.$$

Putting $s = j\omega$ and rationalizing as before gives

$$\Phi_0(j\omega) = \frac{2 + j\omega}{-3\omega^2 + j(\omega - 2\omega^3)}\left[\frac{-3\omega^2 - j(\omega - 2\omega^3)}{-3\omega^2 - j(\omega - 2\omega^3)}\right],$$

So

$$\text{Re } \Phi_0(\text{j}\omega) = \frac{-5\omega^2 - 2\omega^4}{4\omega^6 + 5\omega^4 + \omega^2},$$

and $$\text{Im } \Phi_0(\text{j}\omega) = \frac{-2\omega + \omega^3}{4\omega^6 + 5\omega^4 + \omega^2}.$$

Hence the following table:

ω	Re $\Phi_0(\text{j}\omega)$	Im $\Phi_0(\text{j}\omega)$
$-\infty$	0	0
$-$ ve large	$-$ ve	$-$ ve
$-$ ve small	$-$ ve	+ ve
0^-	-5	∞
$-\sqrt{\frac{5}{2}}$	0	+ ve
$-\sqrt{2}$	$-\frac{1}{3}$	0

The Nyquist diagram can now be drawn.

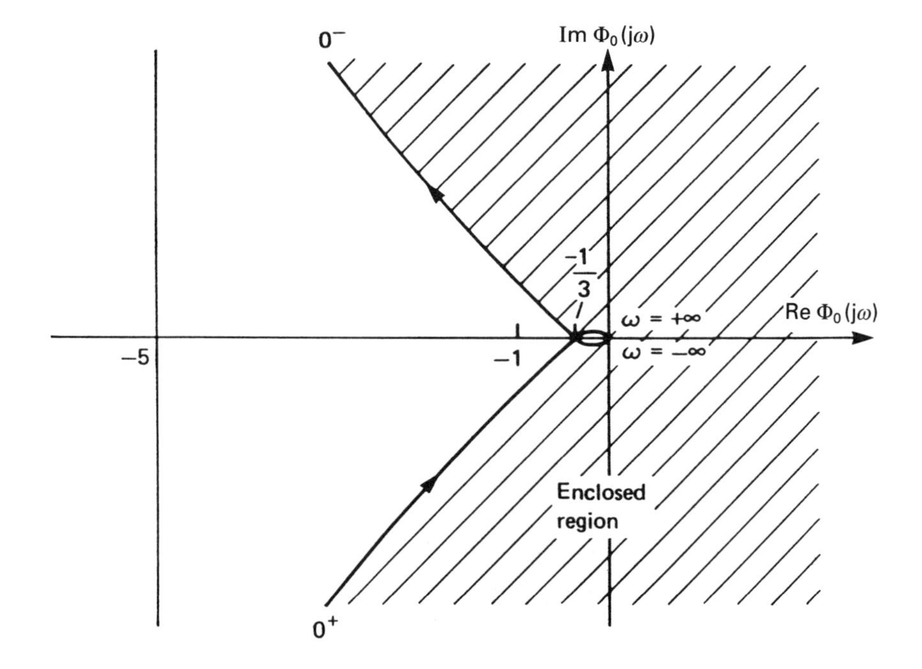

The system is now stable because the point (– 1,0) is not enclosed.
The intercept on the real axis is $-\frac{1}{3}$ so the gain margin = 3.

Example 73
A process is controlled by adding negative feedback through an integral action controller as shown below

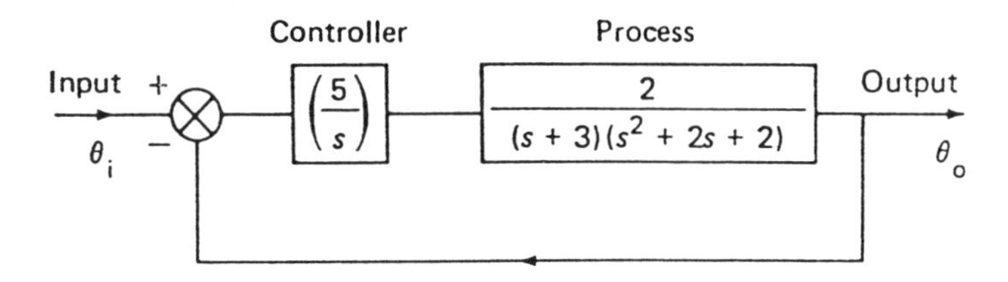

(i) Sketch the Nyquist frequency response diagram for the open loop transfer function (OLTF) of the system, and comment on the stability.
(ii) The system is modified by adding proportional action so that the controller transfer function becomes $5(s + 1)/s$.
Sketch the Nyquist frequency response diagram for the OLTF of the modified system, comment on the stability and calculate the gain margin.

(i) $$\text{OLTF} = \frac{10}{s(s+3)(s^2+2s+2)} = \frac{10}{s^4+5s^3+8s^2+6s}.$$

Put $s = j\omega$, and $\Phi_o(j\omega)$ becomes

$$\frac{10}{\omega^4 - 5j\omega^3 - 8\omega^2 + 6j\omega} = \frac{10}{(\omega^4 - 8\omega^2) + j(6\omega - 5\omega^3)}.$$

Rationalize to give

$$\text{Re}\ \Phi_0(j\omega) = \frac{10\omega^2(\omega^2 - 8)}{\omega^8 + 9\omega^6 + 4\omega^4 + 36\omega^2},$$

and

$$\text{Im}\ \Phi_0(j\omega) = \frac{-10\omega(6 - 5\omega^2)}{\omega^8 + 9\omega^6 + 4\omega^4 + 36\omega^2}.$$

Hence the following table:

ω	Re $\Phi_0(j\omega)$	Im $\Phi_0(j\omega)$
$-\infty$	0	0
− ve large	+ ve	− ve
$-\sqrt{8}$	0	− 0.104
$-\sqrt{\frac{6}{5}}$	− 1.23	0
− ve small	− ve	+ ve
0^-	− 2.2	∞

Thus the Nyquist diagram is as below. Because $(-1, 0)$ is enclosed, the system is unstable.

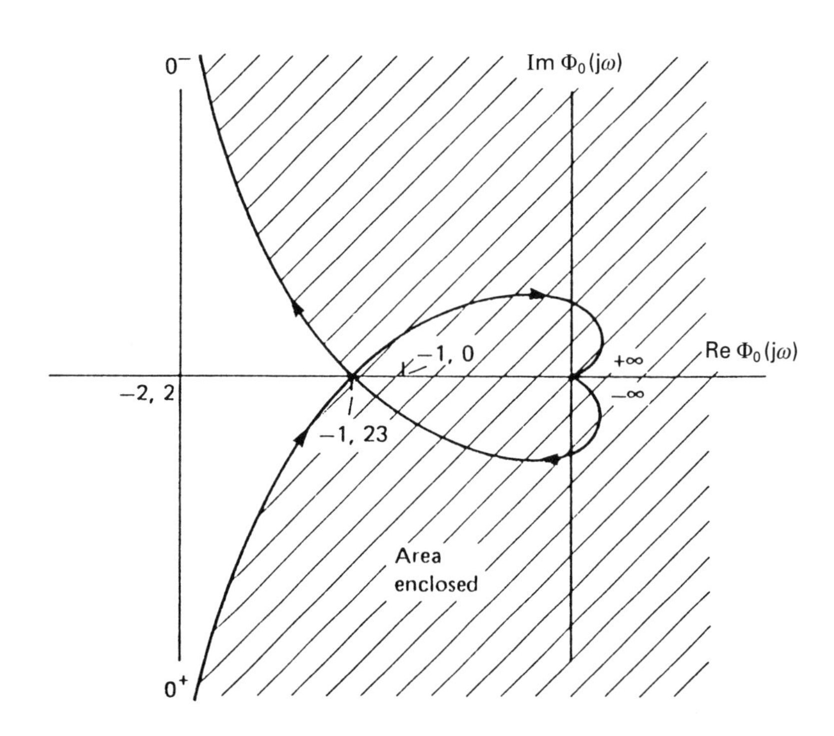

(ii) With proportional control added,

$$\text{OLTF} = \frac{10(s+1)}{s(s+3)(s^2+2s+2)}.$$

Putting $s = j\omega$ and rationalizing gives

$$\text{Re } \Phi_0(j\omega) = \frac{10(\omega^4 - 8\omega^2) + 10\omega(6\omega - 5\omega^2)}{\omega^8 + 9\omega^6 + 4\omega^4 + 36\omega^2}$$

and

$$\text{Im } \Phi_0(j\omega) = \frac{10\omega(\omega^4 - 8\omega^2) - 10(6\omega - 5\omega^3)}{\omega^8 + 9\omega^6 + 4\omega^4 + 36\omega^2}.$$

so

$$\text{Re } \Phi_0(j\omega) = \frac{-40\omega^4 - 20\omega^2}{\omega^8 + 9\omega^6 + 4\omega^4 + 36\omega^2},$$

and

$$\text{Im } \Phi_0(j\omega) = \frac{10\omega^5 - 30\omega^3 - 60\omega}{\omega^8 + 9\omega^6 + 4\omega^4 + 36\omega^2}.$$

Hence the following table:

ω	Re $\Phi_0(j\omega)$	Im $\Phi_0(j\omega)$
$-\infty$	0	0
– ve large	– ve	– ve
– 4.37	– 0.63	0
– ve small	– ve	+ ve
0^-	– 0.56	∞

The Nyquist diagram is shown overleaf.

The system is now stable because (– 1,0) is not enclosed.
Gain Margin $= \frac{1}{0.63} = 1.59$, with satisfactory phase margin.

5.7.2 Bode analysis

Bode analysis consists of plotting two graphs: the magnitude of $\Phi_0(s)$ with $s = j\omega$, and the phase angle of $\Phi_0(s)$ with $s = j\omega$, both plotted as a function of the frequency ω. Log scales are usually used for the frequency axis and for the magnitude of $\Phi_0(j\omega)$.

The magnitude, $|\Phi_0(j\omega)|$ of the transfer function $\Phi_0(j\omega)$ for any value of ω is plotted on a log scale in decibel (dB) units, where

$$\text{dB} = 20 \log_{10}|\Phi_0(j\omega)|.$$

Thus the magnitude plot of a frequency response function expressible as a product of more than one term can be obtained by adding the individual dB magnitude plots for each product term. Thus the Bode magnitude plot for

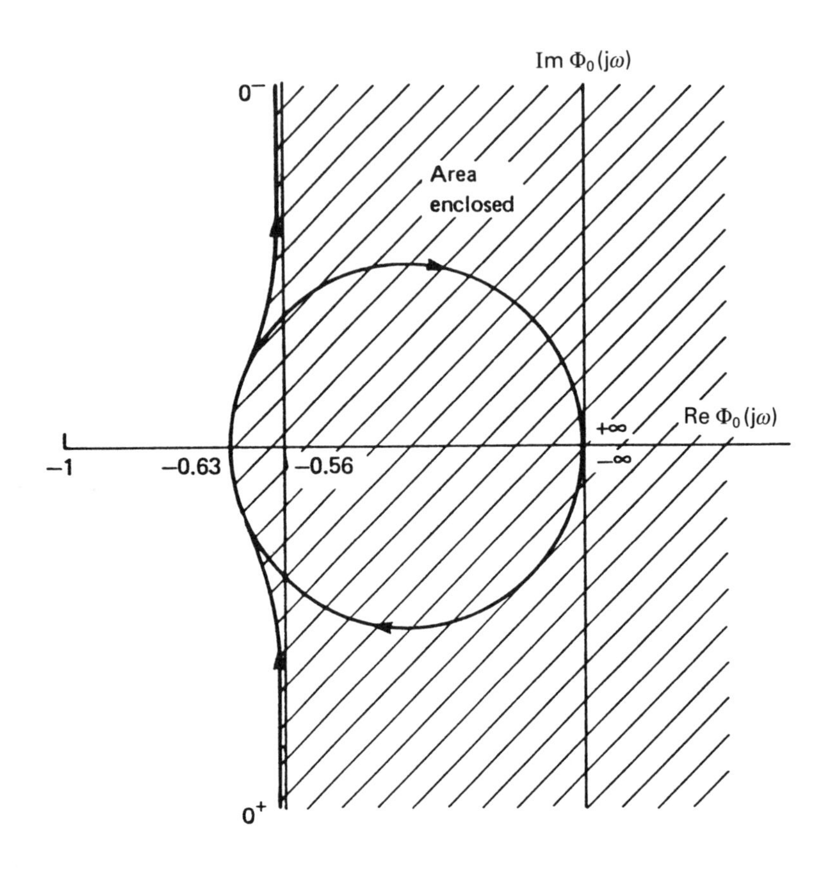

$$\Phi_0(j\omega) = \frac{100(1 + 0.1j\omega)}{(1 + j\omega)}$$

is obtained by adding the Bode magnitude plots for 100, $(1 + 0.1j\omega)$ and $1/(1 + j\omega)$.

The dB magnitude against log ω plot is the Bode magnitude plot, and the phase angle against log ω plot is the Bode phase angle plot.

Bode analysis techniques can be investigated by considering Bode diagrams of some simple frequency response functions.

The gain constant k has a magnitude $|k|$ and a phase angle of 0 deg if k is positive, and $-180°$ if k is negative. This is true for all values of ω so that the Bode diagrams, or plots are as in Fig. 5.48 opposite.

The frequency response function, or sinusoidal transfer function, for a pole at the origin, is

$$\Phi_0 = \frac{1}{(j\omega)},$$

The Bode diagrams are thus as shown in Fig. 5.49.

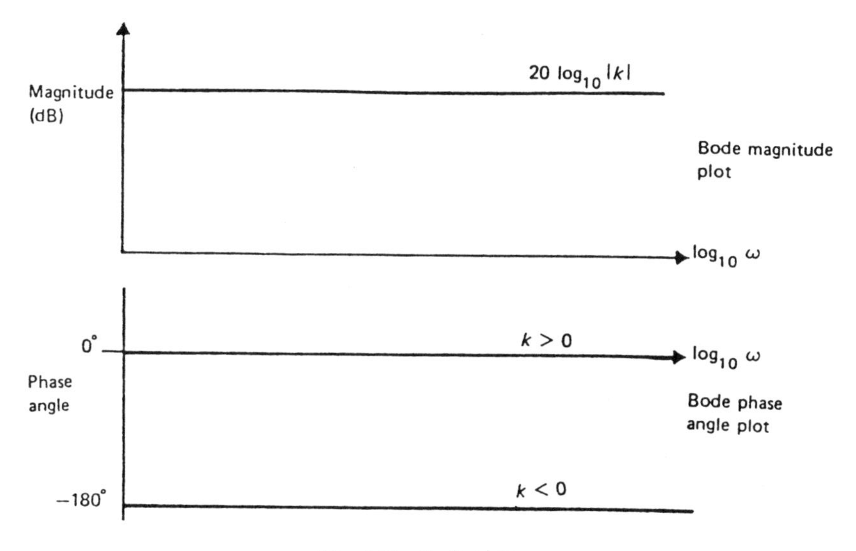

Fig. 5.48. Bode plots.

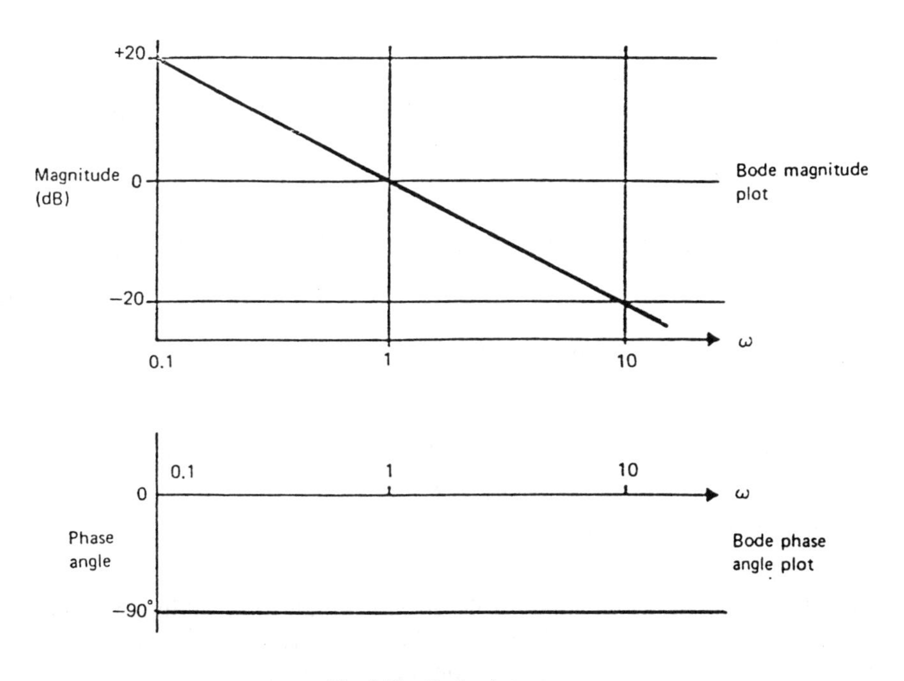

Fig. 5.49. Bode plots.

The Bode diagram for transfer functions of the form $(1 + j\omega T)$ and $1/(1 + j\omega T)$ can be represented most conveniently by straight line asymptotes. For example, consider the transfer function $1/(1 + j\omega T)$.

$$\left|\frac{1}{(1 + j\omega T)}\right| = \frac{1}{\sqrt{(1 + \omega^2 T^2)}}$$

so that as $\omega \to 0$, magnitude $\to 1$, and $\log_{10}$ gain $= 0$; and as $\omega \to$ a large value, magnitude $\to 1/(\omega T)$ and $\log_{10}$ gain $= -\log_{10}\omega T$.

These straight line asymptotes meet at $\omega T = 1$. $\omega = 1/T$ is the '*corner*' or '*break*' frequency. At this frequency.

$$\text{gain (dB)} = 20 \log_{10}\left|\frac{1}{1 + j}\right| = 20 \log\left(\frac{1}{\sqrt{2}}\right) = -20 \log \sqrt{2}$$

$$= -3 \text{ dB}.$$

For the phase angle diagram,

$$\text{Arg.} \quad \left(\frac{1}{1 + j\omega T}\right) = -\tan^{-1}\omega T = 0 \qquad \text{for } \omega T \ll 1, \text{ and}$$

$$= -90^\circ \qquad \text{for } \omega T \gg 1.$$

These results are shown in Fig. 5.50 opposite.

It can be seen that Bode diagrams provide a very flexible method for the analysis of control systems. If a system is modified, the additional transfer function terms can simply be added into the diagrams, and experimental results can be plotted so that from straight line approximations an equation for the transfer function can be found.

Example 74

Draw Bode diagrams for the open loop transfer function given below, and determine the gain and phase margins.

$$\Phi_0(j\omega) = \frac{5}{j\omega(1 + j\omega 0.6)(1 + j\omega 0.1)}$$

Bode magnitude diagram:

$$20 \log_{10}|\Phi_0(j\omega)| = 20 \log_{10}5 - 20 \log_{10}|j\omega| - 20 \log_{10}|1 + j\omega 0.6| - 20 \log_{10}|1 + j\omega 0.1|.$$

Initially, consider each term:

$$20 \log_{10}5 = 10 \times 0.698 = 14 \text{ dB}:$$

The Bode magnitude diagram for Φ_o given is obtained by adding graphically the diagrams shown on p. 276. Before doing this it is convenient to calculate the phase angle diagram so that it can be plotted on the same axes; the gain and phase margins can then easily be found.

Bode phase angle diagram.

$$\arg \Phi_0(j\omega) = -90^\circ - \arg \tan^{-1} 0.6\omega - \arg \tan^{-1}0.1\omega$$

Hence the table which follows on p. 276:

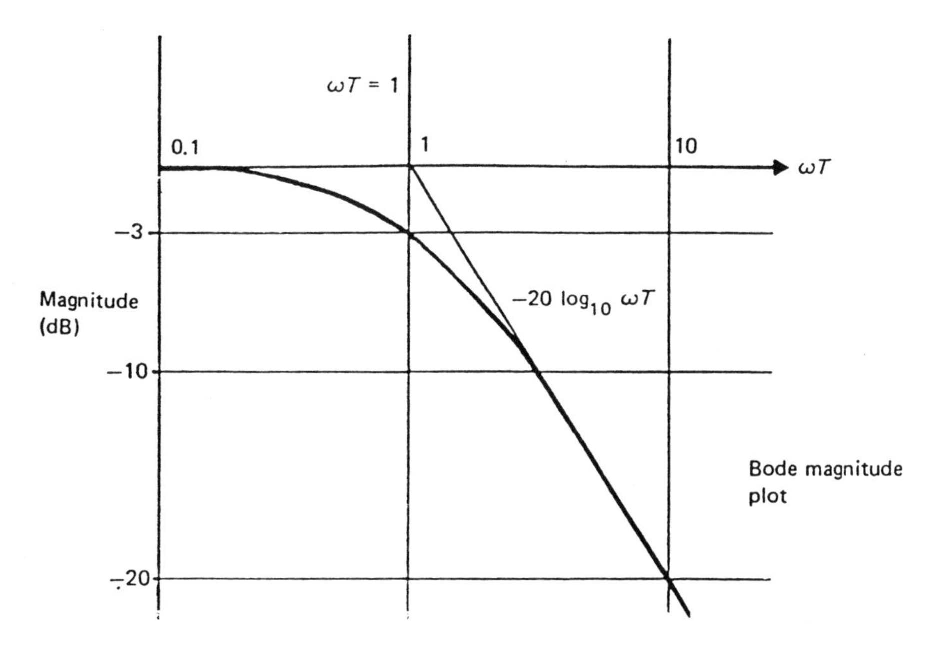

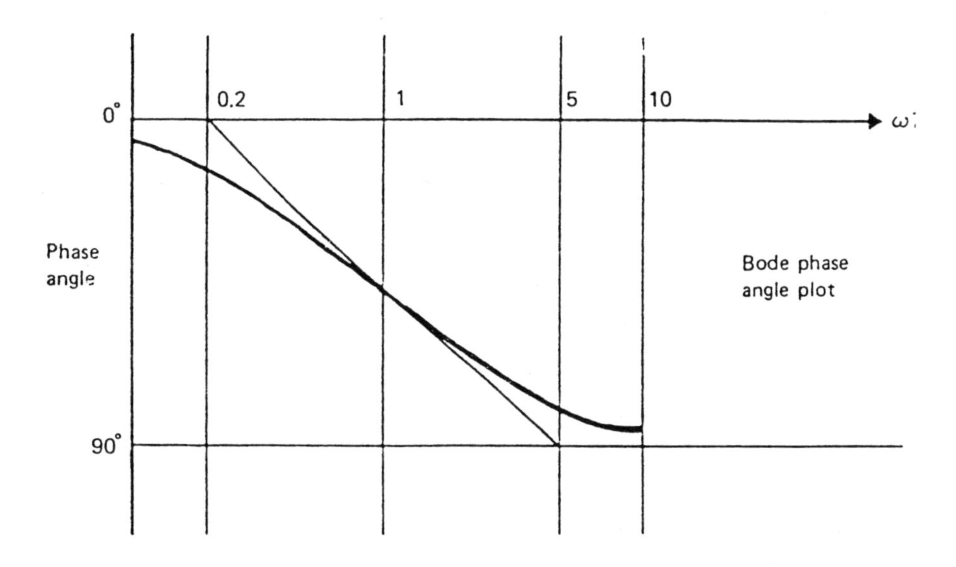

Fig. 5.50. Bode plots.

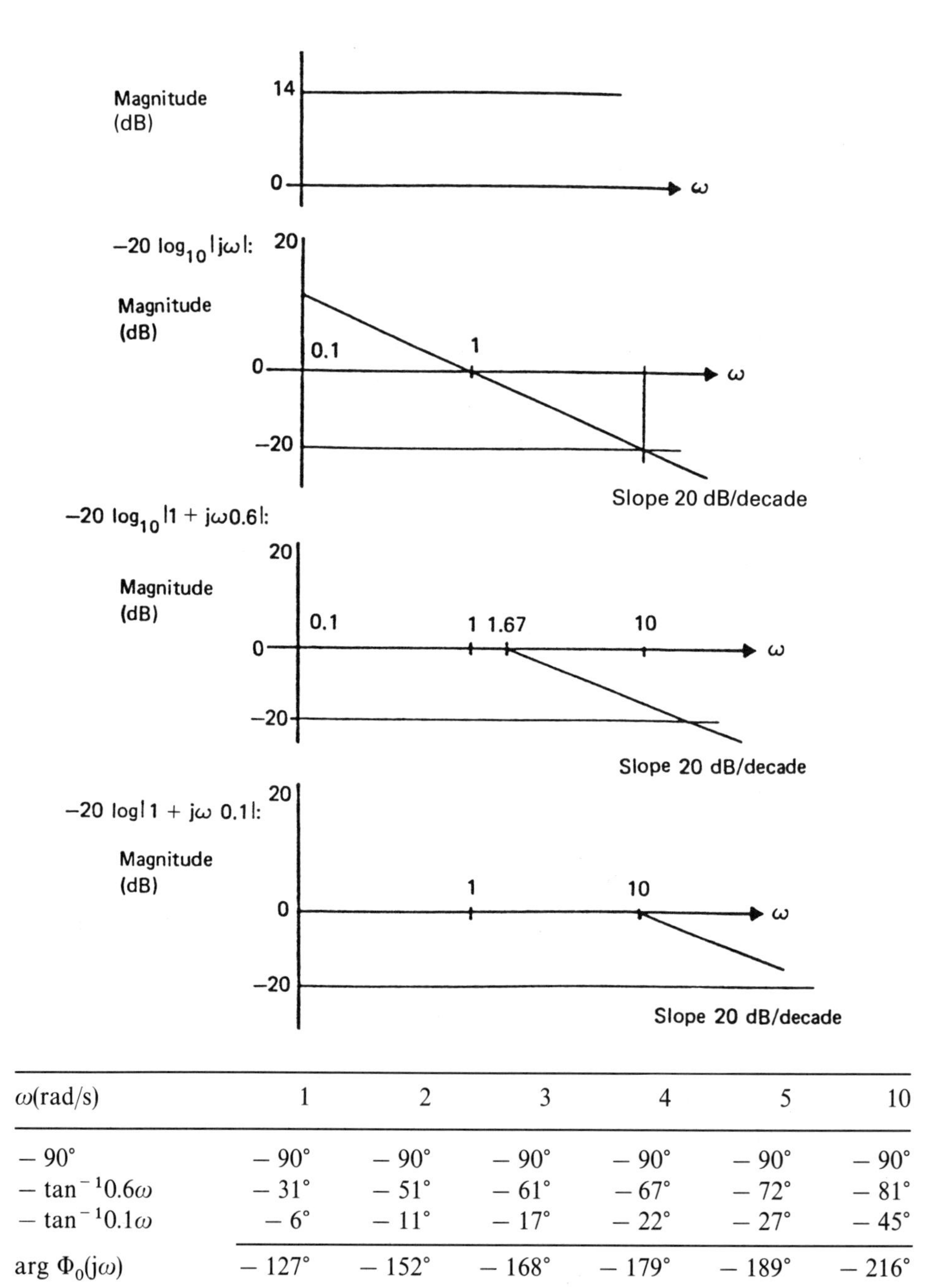

ω(rad/s)	1	2	3	4	5	10
$-90°$	$-90°$	$-90°$	$-90°$	$-90°$	$-90°$	$-90°$
$-\tan^{-1}0.6\omega$	$-31°$	$-51°$	$-61°$	$-67°$	$-72°$	$-81°$
$-\tan^{-1}0.1\omega$	$-6°$	$-11°$	$-17°$	$-22°$	$-27°$	$-45°$
$\arg \Phi_0(j\omega)$	$-127°$	$-152°$	$-168°$	$-179°$	$-189°$	$-216°$

The Bode phase angle diagrams can now be plotted. Linear–log graph paper should be used.

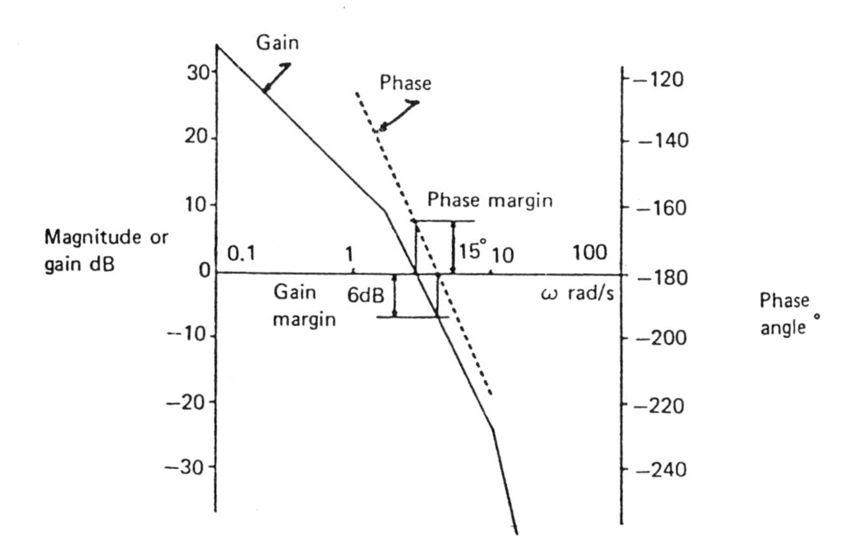

From the above diagram, when phase angle = $-180°$, Gain margin = 6 dB, and the phase margin is the angle when the gain is unity, that is, 0 dB. Hence, phase margin = 15°. That is, the system is stable in closed loop, but, because of the small phase margin, it will have a poor transient performance.

Example 75

The open loop transfer function of a control system is given by

$$\frac{30}{s(1 + 0.1s)(1 + 0.25s)}$$

(i) Plot the Bode amplitude and phase angle diagrams, determine the gain margin and the phase margin, and state whether the system is stable or not.

(ii) A phase lag network is introduced into the system with a transfer function

$$(1 + s)/(1 + 10s).$$

Draw the Bode diagrams for the modified system, find the new gain and phase margins, and comment on the system stability.

(i) Bode amplitude, or gain diagram:

$$20 \log_{10} \Phi_0(j\omega) = 20 \log_{10} 30 - 20 \log_{10} j\omega - 20 \log_{10}(1 + 0.1j\omega) - 20 \log_{10}(1 + 0.25j\omega).$$

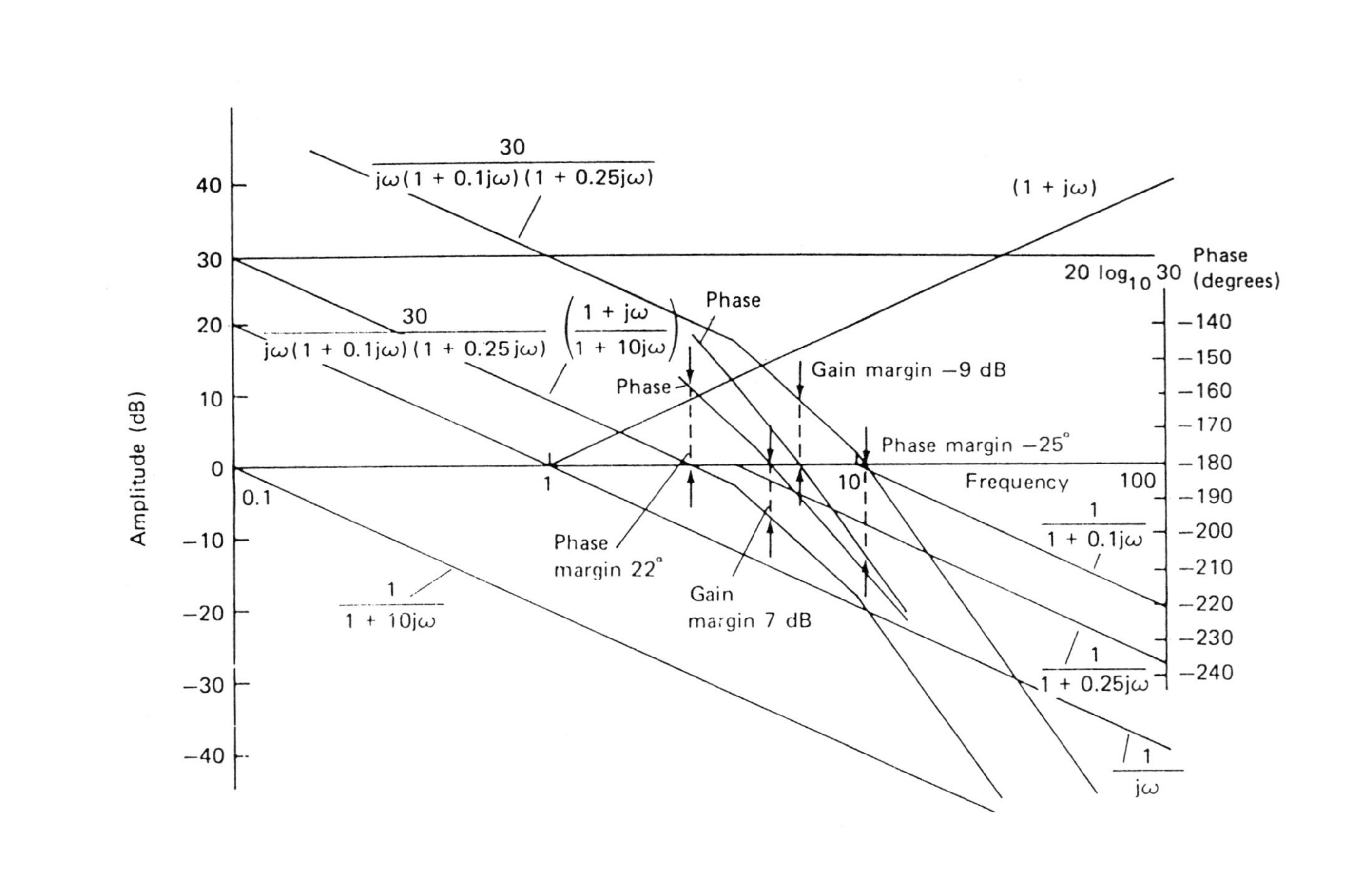

$\frac{30}{j\omega(1 + 0.1j\omega)(1 + 0.25j\omega)}$
$(1 + j\omega)$
Phase
20 $\log_{10}$ 30
Phase (degrees)
$\frac{30}{j\omega(1 + 0.1j\omega)(1 + 0.25j\omega)} \left(\frac{1 + j\omega}{1 + 10j\omega}\right)$
Phase
Gain margin −9 dB
Phase margin −25°
Frequency
Amplitude (dB)
40
30
20
10
0
−10
−20
−30
−40
0.1
1
10
100
−140
−150
−160
−170
−180
−190
−200
−210
−220
−230
−240
Phase margin 22°
Gain margin 7 dB
$\frac{1}{1 + 10j\omega}$
$\frac{1}{1 + 0.1j\omega}$
$\frac{1}{1 + 0.25j\omega}$
$\frac{1}{j\omega}$

$20 \log_{10} 30 = 29.5$, so this line can be drawn on the diagram directly. The other terms can be drawn on the diagram and added to give the amplitude diagram as shown.

The Bode phase diagram is drawn from the results of the following table:

$$\arg \Phi_0(j\omega) = -90° - \arg \tan 0.1\omega - \arg \tan 0.25\omega.$$

ωrad/s	3	5	10	15
$-90°$	$-90°$	$-90°$	$-90°$	$-90°$
$-\tan^{-1}0.1\omega$	$-17°$	$-27°$	$-45°$	$-56°$
$-\tan^{-1}0.25\omega$	$-37°$	$-51°$	$-68°$	$-75°$
$\arg \Phi_0(j\omega)$	$-144°$	$-168°$	$-203°$	$-221°$

From the diagrams, the system is unstable, gain margin $= -9$ dB and phase margin $= -25°$.

(ii) Phase lag network now added, so add to amplitude diagram for original system plot for $(1 + j\omega)$ and $1/(1 + 10j\omega)$, to give a new Bode amplitude diagram.

New phase adjustment as below:

ωrad/s	3	5	10	15
$-\tan^{-1}10\omega$	$-88°$	$-89°$	$-89°$	$-89°$
$+\tan^{-1}\omega$	$+71°$	$+79°$	$+84°$	$+86°$
	$-17°$	$-10°$	$-5°$	$-3°$
Original phase	$-144°$	$-168°$	$-203°$	$-221°$
$\arg \Phi_0(j\omega)$	$-161°$	$-178°$	$-208°$	$-224°$

The Bode phase diagram for the modified system can now be drawn. The system is now stable with gain margin = 7 dB and phase margin = 22°.

6

Problems

6.1 SYSTEMS HAVING ONE DEGREE OF FREEDOM

1. The figure shows a pendulum which consists of a light rigid rod of length L pivoted to a fixed point at one end and having a mass m fixed to its other end. A spring of stiffness k is attached as shown, at a distance a from the pivot. In the position shown the rod is vertical and the spring is horizontal and unloaded. Find the frequency of free oscillations of small amplitude in the plane of the diagram.

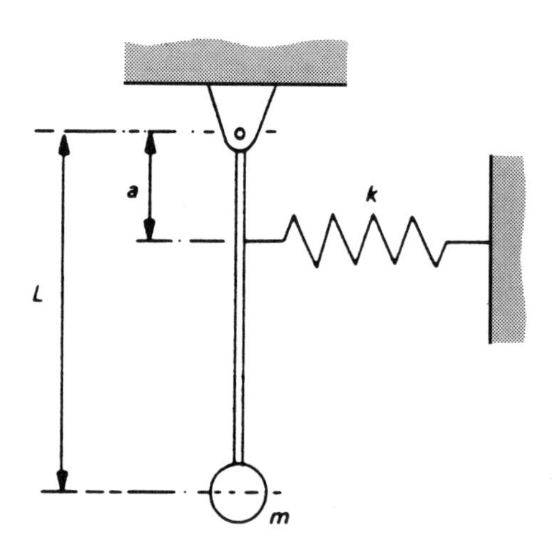

2. A structure is modelled by a rigid horizontal member of mass 3000 kg, supported at each end by a light elastic vertical member of flexural stiffness 2 MN/m.

   Find the frequency of small amplitude horizontal vibrations of the rigid member.
3. Part of a structure is modelled by a thin rigid rod of mass m pivoted at the lower end, and held in the vertical position by two springs, each of stiffness k, as shown.

   Find the frequency of small amplitude oscillation of the rod about the pivot.

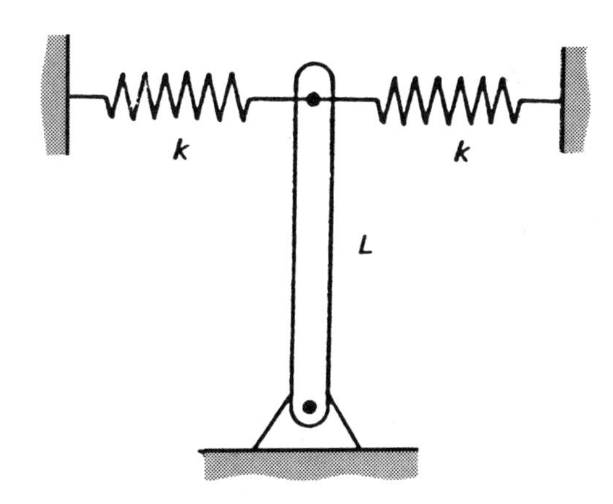

4. A wire is stretched between two fixed horizontal supports a distance L apart by a force T. A body of mass m is attached to the wire at a fixed distance x from one support. Find the natural frequency of small transverse vibrations of the body, assuming that T is constant.
5. The figure shows a simple model of a vehicle which is moving with a constant speed V over a surface with a sinusoidal profile. The mass of the vehicle is m and the stiffness of its suspension springs is k. The wavelength of the surface profile is L and the amplitude of its undulations is h.

   Derive an equation for the steady state vertical displacement of the vehicle as it moves over the sinusoidal surface in terms of its speed, natural frequency and the surface parameters.

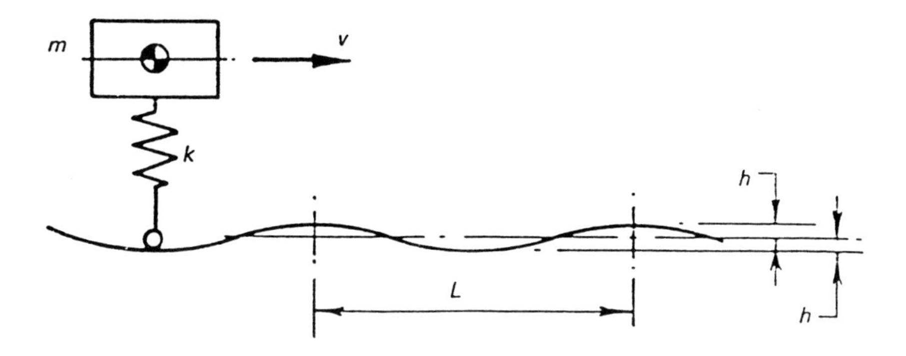

6. A spring of stiffness k is connected to a viscous damper in series, and released from rest with a displacement x_0 from the zero force position. Neglect the mass of the system and express the displacement x as a function of the time t after release.
7. A body supported by an elastic structure performs a damped oscillation of period 1 s, in a medium whose resistance is proportional to the velocity. At a given instant the amplitude was observed to be 100 mm and in 10 s this had reduced to 1 mm.

   What would be the period of the free vibration if the resistance of the medium were negligible?
8. A swing door is a uniform rectangular slab 0.9 m wide and mass 17 kg. It is pivoted about one vertical edge to a device which provides a spring moment proportional to the angle of the door from its closed position, acting in parallel with a viscous damper. When the door is held open at 90°, a moment of 16 N m is needed to keep it in that position, and when it is released from that position it swings through the closed position 2.3 s later.

   Find, to an accuracy of 5% or better, the damping ratio of the system.
9. A turbine disc of mass 20 kg has its mass centre on the axis of a central hole 40 mm in diameter. The disc is suspended from a horizontal knife edge passed through the central hole, and makes small free oscillations in the plane of the disc while immersed in a fluid which exerts a resistance proportional to the angular velocity of the disc.

   The period of the oscillations is found to be 1.6 s and the amplitude is observed to decrease sixfold in each complete oscillation. Write down the equation of motion and derive its solution, and hence calculate the mass moment of inertia of the disc about the axis of the central hole and the effective viscous damping coefficient.
10. The moving part of a galvanometer has a mass moment of inertia I of 1.5×10^{-6} kg m^2 and is connected to the frame by a spiral spring which has an effective stiffness k of 1.2×10^{-5} N m/rad and a viscous damper of coefficient c. Write down the equation of motion and derive from it the value of c which will make the movement just not oscillatory.

11. The figure shows a pendulum which consists of three light arms OA, OB and OC, rigidly attached to each other and pivoted at O, and of a mass m which is on the end of the vertical arm OA. The arm OB is horizontal and attached at its end to a vertical spring having a stiffness k while the arm OC is vertical and attached at its end to a horizontally positioned viscous damper whose damping coefficient is c.

    Determine the frequency of small amplitude undamped oscillations of the pendulum about O and find the critical damping coefficient in terms of the given parameters.

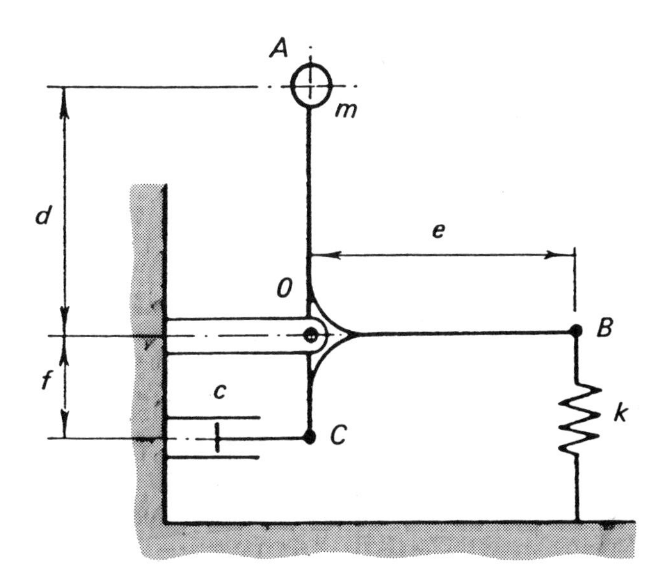

12. An indicator is operated by a solid cylindrical float of mass m, diameter d, and length D, which moves with its axis vertical in a coaxial vertical tube of diameter D, as shown. The float is connected by a rack and pinion to a rotating indicator disc, which can be treated as a uniform disc of mass m and radius r; r is also the pitch circle radius of the pinion. The link between the float and the rack is of negligible mass, as is the rack. In static equilibrium the float is half immersed in the liquid in the tube.

    Assuming that no liquid enters or leaves the tube and that viscous forces are negligible, find the frequency of small free oscillations of the system.

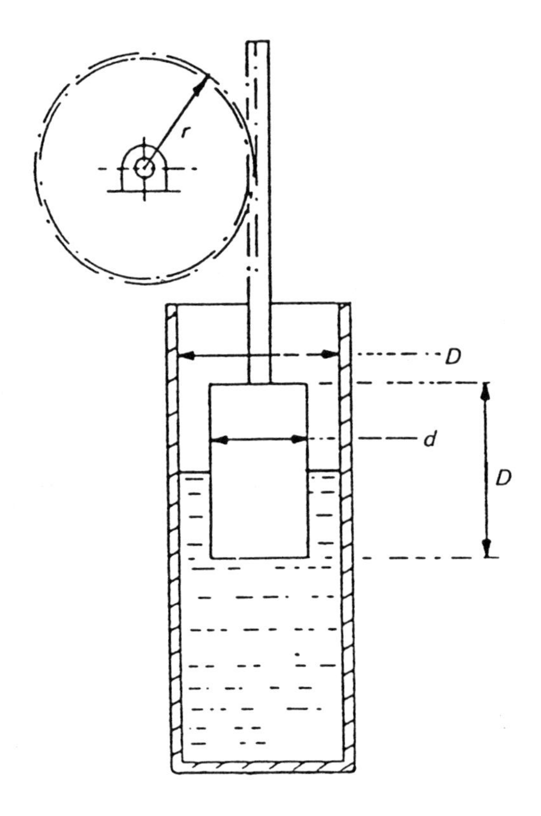

13. A uniform door of width 0.8 m and mass 20 kg is closed by a torsion spring of stiffness 100 N m/rad and a viscous damper of coefficient c.

    Write down the equation of motion of the door whilst closing, give the solution, and comment on the response of the door for amounts of damping less than, equal to, and greater than critical. Find the value of c for critical damping. With the damping made critical, the door is held open at 90° and released; find, to the nearest tenth of a second, the time taken for the door to turn through 89°.

14. A U-tube manometer with two vertical arms has an internal diameter of 6 mm, and contains mercury (density 13.6 g/cm^3), the total length of the mercury column being 0.8 m. Assuming the mercury to move as an incompressible body, the friction between the mercury and the inside of the tube can be expressed as a shear force c per unit wetted area per unit velocity.

    When the mercury column is displaced from its equilibrium position and then allowed to move freely, a damped oscillation is observed, the amplitude at the end of the eleventh cycle being one tenth of the amplitude at the end of the first cycle. Write the equation of motion, and hence find the value of c(in N s/m^3) and the natural frequency of the oscillation.

15. To determine the amount of damping in a bridge it was set into vibration in the fundamental mode by dropping a weight on it at centre span. The observed frequency was 1.5 Hz, and the amplitude was found to have decreased to 0.9 of the initial maximum after 2 s. The equivalent mass of the bridge (estimated by the Rayleigh energy method) was 10^5 kg.

    Assuming viscous damping and simple harmonic motion, calculate the damping coefficient, the logarithmic decrement, and the damping ratio.

16. When considering the vibrations of a structure, what is meant by the Q factor? Derive a simple relationship between the Q factor and the damping ratio for a single degree of freedom system with light viscous damping.

    Measurements of the vibration of a bridge section resulting from impact tests show that the period of each cycle is 0.6 s, and that the amplitude of the third cycle is twice the amplitude of the ninth cycle. Assuming the damping to be viscous, estimate the Q factor of the section.

    When a vehicle of mass 4000 kg is positioned at the centre of the section the period of each cycle increases to 0.62 s; no change is recorded in the rate of decay of the vibration. What is the effective mass of the section?

17. The figure shows a diagrammatic end view of one half of a swing-axle suspension of a motor vehicle which consists of a horizontal half-axle OA pivoted to the chassis at O, a wheel rotating about the centre line of the axle, and a spring of stiffness k and a viscous damper with a damping coefficient c both located vertically between the axle and the chassis. The mass of the half-axle is m_1 and its radius of gyration about O is h. The mass of the wheel is m_2 and it may be regarded as a thin uniform disc having an external radius r and located at a horizontal distance s from the pivot O. The spring and the damper are located at horizontal distances q and p from the pivot O, as shown.

Derive the equation for angular displacement of the axle-wheel assembly about the pivot O, and obtain from it an expression for the frequency of damped free oscillations of the assembly. Express this frequency in terms of the given parameters and the undamped natural frequency of the assembly.

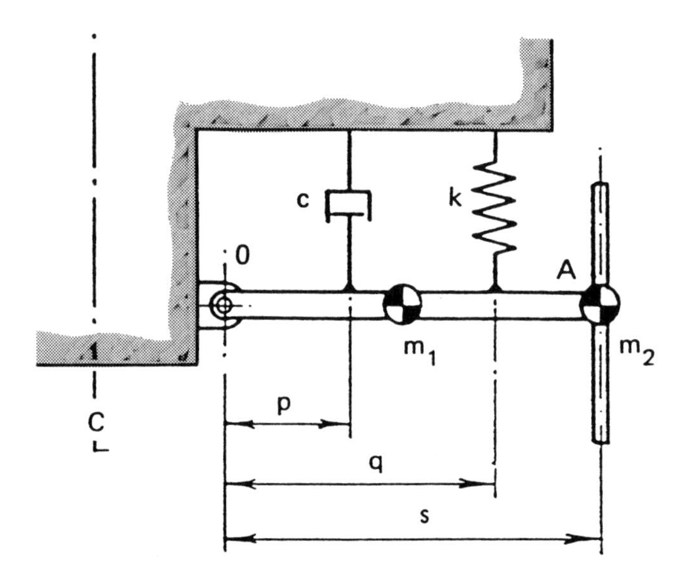

18. The T-shaped body shown pivots about a point O on a horizontal ground surface and is held upright, so that its mass centre G is a distance h vertically above O, by two springs pinned to it and to the ground. Each spring has a stiffness k and its vertical centre line is at a distance c from the pivot O. The T-shaped body has a mass m and a radius of gyration r about its mass centre.

    The body is acted on by a force F whose line of action is horizontal and at a height d above the ground, where $d > h$. Derive an expression for the rotation of the body if the force rises suddenly from zero to F, assuming that the angular displacement of the body is small.

    If the suddenly applied force F drops equally suddenly to zero after a time t_0 from its original application, derive the equation of the rotational motion of the body for times greater than t_0.

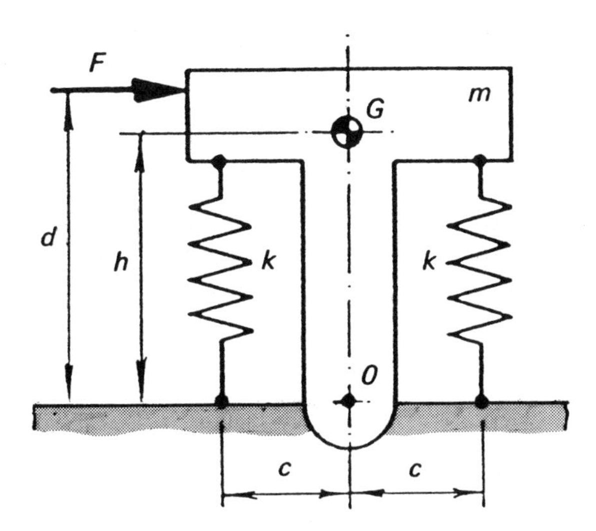

19. An aircraft ejector seat can be regarded as a horizontal platform which can be given a vertical acceleration, and the pilot can be treated (greatly simplified) as a rigid mass of 90 kg supported on the platform by springs of total effective stiffness 10.9 kN/m. When the pilot is to be ejected, the platform is given a constant vertically upward acceleration of 49 m/s^2 which lasts for 0.19 s, when the pilot is freed from the platform and continues to move through the air.

    Considering only the vertical component of motion and neglecting the effect of air resistance, find:

    (a) the maximum height to which the pilot rises above his original position of static equilibrium, and
    (b) the maximum acceleration to which he is subjected.

20. The natural frequency of the undamped spring-body system of the vibrometer shown is 5 Hz. When the vibrometer is fixed to a horizontal surface of a motor running at 750 rev/min, the absolute amplitude of vibration of the suspended body is observed to be 1 mm. What is the amplitude of vertical vibration of the motor, and at what motor speed will the amplitude of motor vibration be the same as the amplitude of vibration of the vibrometer body?

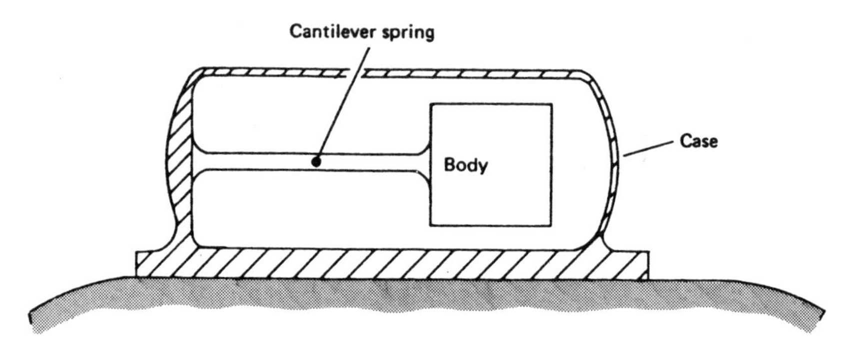

21. The figure shows a vibration exciter which consists of two contra-rotating wheels, each carrying an eccentric body of mass 0.1 kg at an effective radius of 10 mm from its axis of rotation. The vertical positions of the eccentric bodies occur at the same instant. The total mass of the exciter is 2 kg and damping is negligible.

    Find a value for the stiffness of the spring mounting so that a force of amplitude 100 N, due to rotor unbalance, is transmitted to the fixed support when the wheels rotate at 150 rad/s.

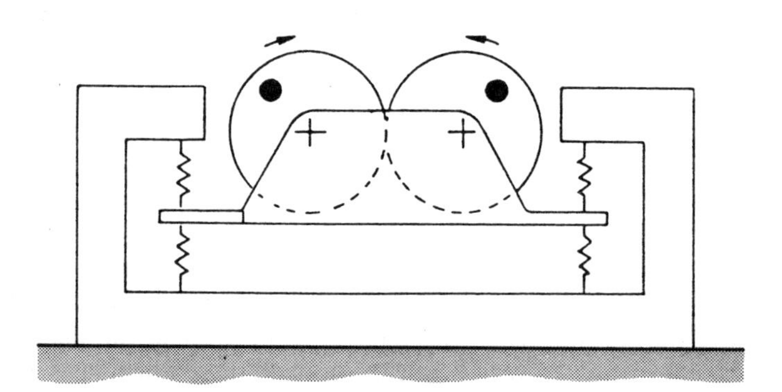

22. A motor generator set of total mass 365 kg is mounted on damped vibration isolators. It runs at 1450 rev/min, and the unit is observed to have a vibration amplitude of 0.76 mm. When the unit is disturbed at standstill, it is found that the resulting damped oscillations have a frequency of 23 Hz and a decay factor of 15.

Find (a) the coefficient of viscous damping, (b) the magnitude of the exciting force at the normal running speed, and (c) the amplitude of the forced vibration if damping were not present.

23. A piston is made to reciprocate in a vertical cylinder with absolute motion $X_p \cos \nu t$. The cylinder, of mass 2 kg, is attached to a platform of negligible mass and supported by springs of total stiffness 5 kN/m.

    If the viscous drag between the piston and cylinder is characterized by the coefficient $c = 200$ N s/m, find the amplitude X_c of the absolute motion of the cylinder, and its phase angle with respect to the piston motion. Take $X_p = 0.025$ m and $\nu = 25$ rad/s.

24. A machine of mass 520 kg produces a vertical disturbing force which oscillates sinusoidally at a frequency of 25 Hz. The force transmitted to the floor is to have an amplitude, at this frequency, not more than 0.4 times that of the disturbing force in the machine, and the static deflection of the machine on its mountings is to be as small as possible consistent with this.

    For this purpose, rubber mountings are to be used, which are available as units, each of which has a stiffness of 359 kN/m and a damping coefficient of 2410 N s/m. How many of these units are needed?

25. A fan is mounted on a spring and viscous damper in parallel so that only linear motion in the vertical direction occurs.

    Briefly derive an expression for the force transmitted to the ground through the spring and damper if the fan generates an harmonic disturbing force in the vertical direction.

    Given that the fan has a mass of 40 kg and a rotating unbalance of 0.01 kg m, determine the spring stiffness required for 10% force transmission. Take the damping ratio of the system to be 0.2 and the fan speed as 1480 rev/min.

    When running under these conditions what effect on the transmission would removal of the damping element have?

26. The basic element of many vibration-measuring devices is the seismic unit, which consists of a mass m supported from a frame by a spring of stiffness k in parallel with a damper of viscous damping coefficient c. The frame of the unit is attached to the structure whose vibration is to be determined, the quantity measured being z, the relative motion between the seismic mass and the frame. The motion of both the structure and the seismic mass is translation in the vertical direction only.

    Derive the equation of motion of the seismic mass, assuming that the structure has simple harmonic motion of circular frequency ν, and deduce the steady state amplitude of z.

    Given that the undamped natural circular frequency ω of the unit is much greater than ν, show why the unit may be used to measure the acceleration of the structure.

    Explain why in practice some damping is desirable.

    If the sensitivity of the unit (that is, the amplitude of z as a multiple of the amplitude of the acceleration of the frame) is to have the same value when $\nu = 0.2\,\omega$, as when $\nu \ll \omega$ find the necessary value of the damping ratio.

27. A machine of mass 940 kg produces a vertical disturbing force which oscillates sinusoidally, having an amplitude of 320 N at the machine's normal running speed of 550 rev/min. At this speed the force transmitted to the floor is to have an amplitude of not more than 100 N, and to achieve this the machine is to be mounted on viscously damped springs. The static deflection of the spring mountings is to be not more than 15 mm, and the amplitude of vibration at resonance is to be kept as low as possible.

    Find (a) the total spring stiffness and total viscous damping coefficient which will enable these conditions to be satisfied, and (b) the resonance speed of the machine when such mountings are fitted.

28. The vibration of the floor of a laboratory is found to be simple harmonic motion at a frequency in the range 15–60 Hz (depending on the speed of some nearby reciprocating plant). It is desired to install in the laboratory a sensitive instrument which requires insulating from the floor vibration. The instrument is to be mounted on a small platform which is supported by three similar springs resting on the floor, arranged to carry equal loads; the motion is restrained to occur in a vertical direction only. The combined mass of the instrument and the platform is 40 kg: the mass of the springs can be neglected, and the equivalent viscous damping ratio of the suspension is 0.2.

    Calculate the maximum value for the spring stiffness if the amplitude of the transmitted vibration is to be less than 10% of that of the floor vibration, over the given frequency range.

29. A new concert hall is to be protected from the ground vibrations from an adjacent highway by mounting the hall on rubber blocks. The predominant frequency of the sinusoidal ground vibrations is 40 Hz, and a motion transmissibility of 0.1 is to be attained at that frequency.

    Calculate the static deflection required in the rubber blocks, assuming that these act as linear, undamped springs.

30. A galvanometer consists of a coil and mirror assembly suspended on a torsional spring in a fixed enclosure filled with a liquid which gives viscous damping. The moment of inertia of the moving part about its axis of rotation is 2×10^{-9} kg m^2, the spring stiffness is 0.0116 N m/rad, and the damping coefficient is 5.29×10^{-6} N m s/rad.

    Find the natural frequency of the galvanometer, and the amplitude and phase errors in its indication, when supplied with a sinusoidal input signal at a frequency of 100 Hz and amplitude 100 mA.

31. The figure shows the essential elements of an instrument for recording vibrations, which consist of a frame A with a vertical pillar and light rod that pivots about a horizontal axis passing through the pillar at O. The rod has a concentrated mass m attached to its free end and is held in a horizontal position by a spring of stiffness K, while its vibrations are damped by a viscous damper whose damping coefficient is C.

    If the frame of the instrument is given a vertical displacement $a \sin vt$, consider the rotation of the rod about O and derive an equation for the maximum amplitude of the relative, steady-state, vertical displacements between the mass

m and the frame in terms of the given quantities and the natural frequency of the free vibrations of the instrument.

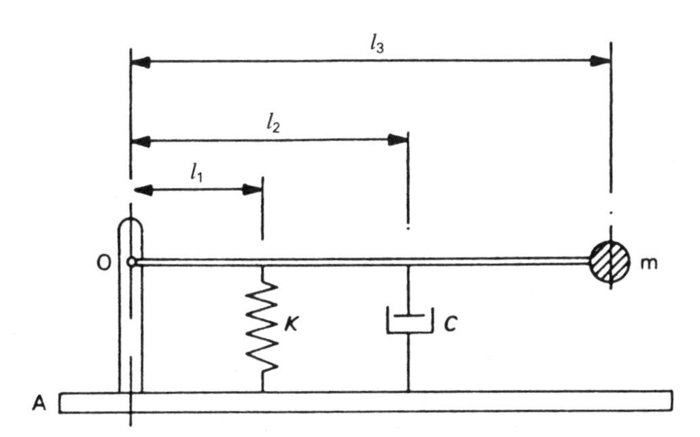

32. The figure shows, diagrammatically, a trailer which is towed at a constant velocity v along a test track with a sinusoidal surface. The trailer has a mass M and is connected by springs of total stiffness K to the wheels which have a total mass m. The wavelength of the track surface is L and the amplitude of its undulations is h.

    Derive an equation for the force exerted by the wheels on the track surface in terms of the given quantities, assuming that the only relative motion between the trailer and the wheels is in a vertical direction and that the diameter of the wheels is small compared with the undulations of the track.

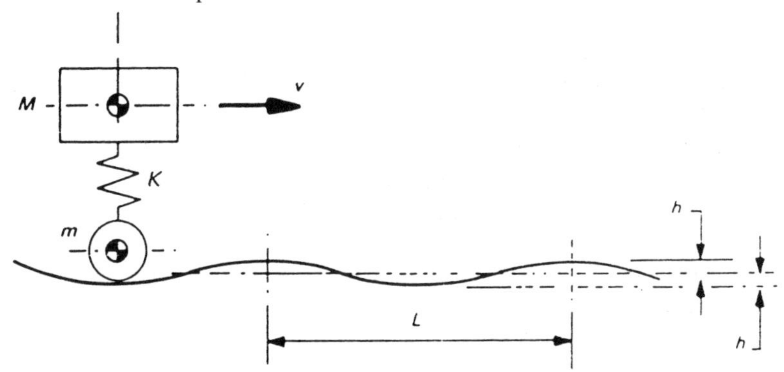

33. A two-wheeled trailer of sprung mass 700 kg is towed at 60 km/h along an undulating straight road whose surface may be considered sinusoidal. The distance from peak to peak of the road surface is 30 m and the height from hollow to crest 0.1 m. The effective stiffness of the trailer suspension is 60 kN/m, and the shock absorbers which provide linear viscous damping are set to give a damping ratio of 0.67.

    Assuming that only vertical motion of the trailer is excited, find the absolute amplitude of this motion and its phase angle relative to the undulations.

34. A wooden floor, 6 m by 3 m, is simply supported along the two shorter edges. The mass is 300 kg and the static deflection at the centre under the self-weight is 7 mm. It is proposed to determine the dynamic properties of the floor by dropping a sand bag of 50 kg mass on it at the centre, and to measure the response at that position with an accelerometer and a recorder.

    In order to select the instruments required, estimate:

    (a) the frequency of the fundamental mode of vibration which would be recorded;
    (b) the number of oscillations at the fundamental frequency for the signal amplitude on the recorder to be reduced to half, assuming a loss factor of 0.05; and
    (c) the height of drop of the sand bag so that the dynamic deflection shall not exceed 10 mm, and the corresponding maximum acceleration.

35. An instrument is supported by a spring and viscous damper in parallel so that only linear motion in the vertical direction occurs.

    Briefly derive an expression for the force transmitted to the support through the spring and damper, if the instrument generates an harmonic disturbing force $F \sin \nu t$ in the vertical direction.

    If the mass of the instrument is 7 kg, the damping ratio of the system is $\frac{1}{3}$ and the frequency of the disturbing force is 50 Hz determine the spring stiffness required for only 10 per cent force transmission to the support. Find also the value of the viscous damping coefficient.

6.2 SYSTEMS HAVING MORE THAN ONE DEGREE OF FREEDOM

36. A turbine rotor with mass moment of inertia 12 kg m^2 is connected to a generator by a straight shaft of length 1 m and torsional stiffness 600 kN m/rad. The generator armature has a mass moment of inertia of 5 kg m^2. Calculate the natural frequency of free torsional oscillation of the system and give the mode shape. Ignore damping.

37. A tractor and a trailer have masses of 1000 kg and 500 kg respectively. They are connected by a coupling which consists of two springs in series, one spring having a stiffness of 200 kN/m and the other of 50 kN/m; there is negligible damping.

    Find the frequency of free, small amplitude, longitudinal oscillations of the system when the tractor and trailer are in line and on a horizontal surface.

38. To analyse the vibrations of a two-storey building it is represented by the lumped mass system shown, where $m_1 = \frac{1}{2}m_2$ and $k_1 = \frac{1}{2}k_2$ (k_1 and k_2 represent the shear stiffnesses of the parts of the building shown).

    Calculate the natural frequencies of free vibrations and sketch the corresponding mode shapes of the building, showing the amplitude ratios.

    If a horizontal harmonic force $F_1 \sin \nu t$ is applied to the top floor, determine expressions for the amplitudes of the steady state vibration of each floor.

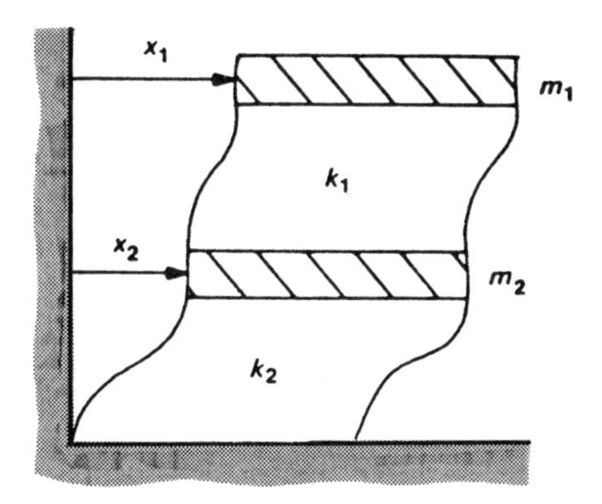

39. The rigid beam shown in its position of static equilibrium in the figure has a mass m and a mass moment of inertia $2ma^2$ about an axis perpendicular to the plane of the diagram and through its centre of gravity G.

    Assuming no horizontal motion of G, find the frequencies of small oscillations in the plane of the diagram and the corresponding positions of the nodes.

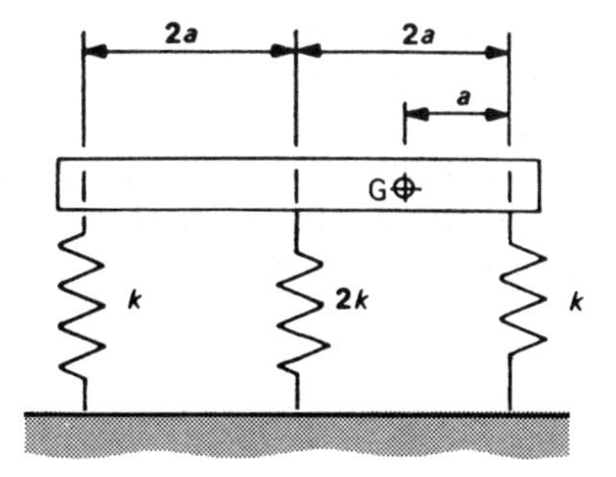

40.

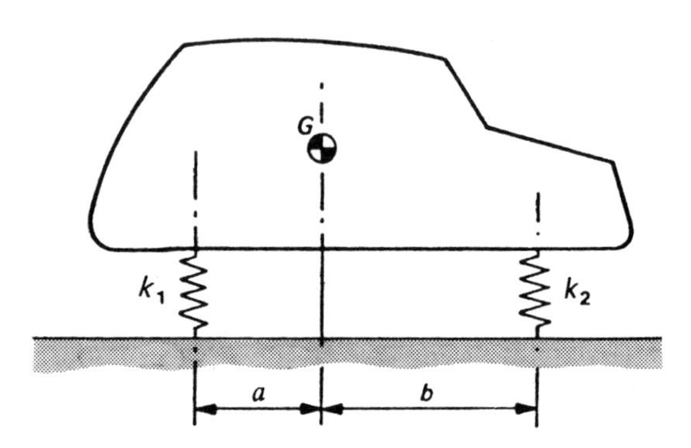

The figure shows a car mounted on its springs which are constrained so that the centre of gravity of the car G can move only vertically. The car has a mass m and a radius of gyration K_0 about an axis through G normal to the plane of the figure. The rear springs have a total stiffness k_1 and are at a distance a from G and the front springs have a total stiffness k_2 and are at a distance b from G.

Derive an equation for the natural frequencies of the vibrations of the car in terms of the natural frequencies of pure vertical, or bounce, vibrations and of pure rotational, or pitch, vibrations and the other parameters of the system.

Show that when there is no coupling between the bounce and pitch modes of vibration the natural frequencies in bounce and pitch are equal.

41. A vehicle has a mass of 2000 kg and a 3 m wheelbase. The mass moment of inertia about the centre of mass is 500 kg m^2, and the centre of mass is located 1 m from the front axle.

Considering the vehicle as a two degree of freedom system, find the natural frequencies and the corresponding modes of vibration, if the front and rear springs have stiffnesses of 50 kN/m and 80 kN/m respectively.

The expansion joints of a concrete road are 5 m apart. These joints cause a series of impulses at equal intervals to vehicles travelling at a constant speed. Determine the speeds at which pitching motion and up and down motion are most likely to arise for the above vehicle.

42. A film reel is driven by a tightly wound helical spring stretched round two grooved pulleys, as shown. The pulleys 1 and 2 have radii r_1 and r_2, angular displacements θ_1 and θ_2, and moments of inertia about their axes of rotation I_1 and I_2, respectively. The stiffness of each unsupported section of the spring is k.

    Derive an expression for the natural frequency of small amplitude free oscillations of the system. Assume that there is no damping and note that during oscillation $r_1\dot{\theta}_1 \neq r_2\dot{\theta}_2$.

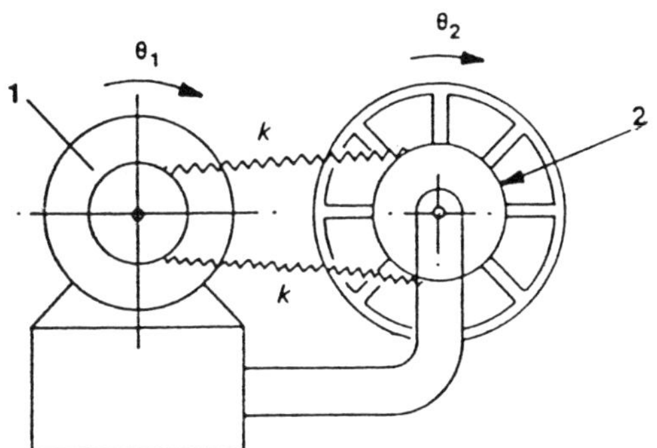

43. A small electronic package is supported by springs as shown. The mass of the package is m, each spring has a constant axial stiffness k, and damping is negligible.

    Considering motion in the plane of the figure only, and assuming that the amplitude of vibration of the package is small enough for the lateral spring forces to be negligible, write down the equations of motion and hence obtain the frequencies of free vibration of the package.

    Explain how the vibration mode shapes can be found.

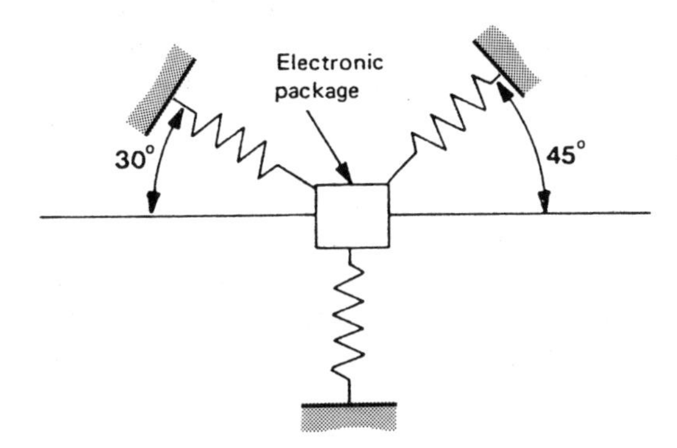

44. Explain, in one sentence each, what is meant by a *natural frequency* and *mode shape* of a dynamic system.
45. Part of a machine can be modelled by the system shown. Two uniform discs A and B, which are free to rotate about fixed parallel axes through their centres, are coupled by a spring. Similar springs connect the discs to the fixed frame as shown in the figure. Each of the springs has a stiffness of 2.5 kN/m which is the same in tension or compression. Disc A has a mass moment of inertia about

its axis of rotation of 0.05 kg m^2, and a radius of 0.1 m, whilst for the disc B the corresponding figures are 0.3 kg m^2 and 0.2 m. Damping is negligible.

Determine the natural frequencies of small amplitude oscillation of the system and the corresponding mode shapes.

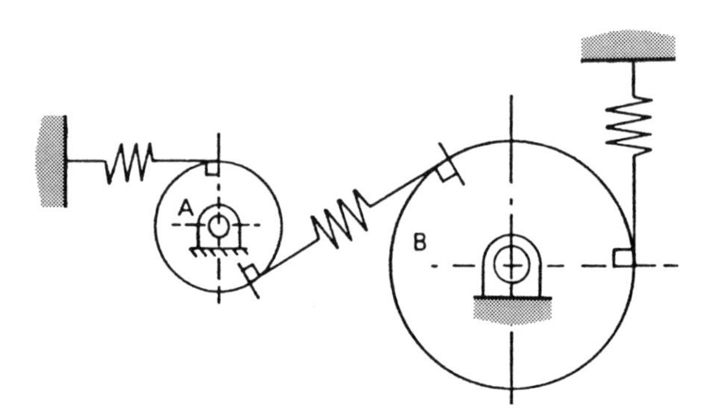

46. Briefly derive the equations which describe the operation of an undamped dynamic vibration absorber.

    A milling machine of mass 2700 kg demonstrates a large resonant vibration in the vertical direction at a cutter speed of 300 rev/min when fitted with a cutter having 20 teeth. To overcome this effect it is proposed to add an undamped vibration absorber.

    Calculate the minimum absorber mass and the relevant spring stiffness required if the resonance frequency is to lie outside the range corresponding to a cutter speed of 250 to 350 rev/min.

47. In a pumping station, a section of pipe resonated at a pump speed of 120 rev/min. To eliminate this vibration, it was proposed to clamp a spring–mass system to the pipe to act as an absorber. In the first test, an absorber mass of 2 kg tuned to 120 cycle/min resulted in the system having a natural frequency of 96 cycle/min.

    If the absorber system is to be designed so that the natural frequencies lie outside the range 85–160 cycle/min, what are the limiting values of the absorber mass and spring stiffness?

48. A certain machine of mass 300 kg produces an harmonic disturbing force $F \cos 15t$. Because the frequency of this force coincides with the natural frequency of the machine on its spring mounting an undamped vibration absorber is to be fitted.

    If no resonance is to be within 10% of the exciting frequency, find the minimum mass and corresponding stiffness of a suitable absorber. Derive your analysis from the equations of motion, treating the problem as one-dimensional.

49. A machine tool of mass 3000 kg has a large resonant vibration in the vertical direction at 120 Hz. To control this resonance, an undamped vibration absorber of mass 600 kg is fitted, tuned to 120 Hz.

    Find the frequency range in which the amplitude of the machine vibration is less with the absorber fitted than without.

50. The figure shows a body of mass m_1 which is supported by a spring of stiffness k_1 and which is excited by an harmonic force P sin νt. An undamped dynamic vibration absorber consisting of a mass m_2 and a spring of stiffness k_2 is attached to the body as shown.

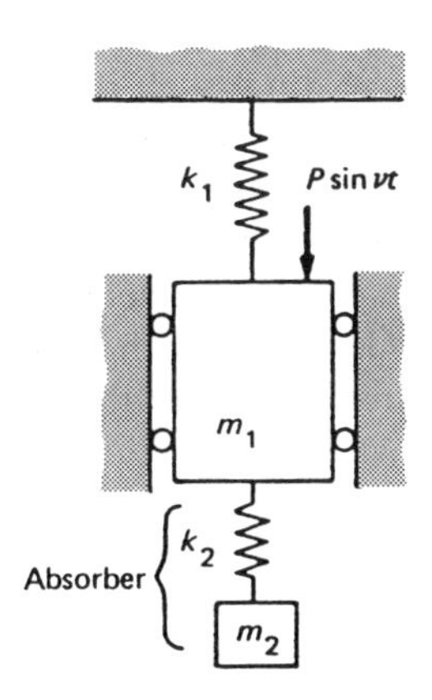

Derive an expression for the amplitude of the vibrations of the body.

The body shows a violent resonance at 152 Hz. As a trial remedy a vibration absorber is attached which results in a resonance frequency at 140 Hz. How many such absorbers are required if no resonance is to occur between 120 and 180 Hz?

51. (i) Show that the frequency at which an undamped vibration absorber is most effective (ω_a) is given by the expression

$$\omega_a = \frac{k_a}{m_a}$$

(where m_a and k_a are the mass and stiffness of the added absorber system) and is therefore independent of the properties of the system to which the absorber has been added. Also, derive an expression for the steady-state amplitude of the absorber mass when the system is being driven at its natural frequency ω_a.

(ii) In order to suppress vibration, a vibration absorber system is to be attached to a machine tool which operates over a range of speeds. The design of the absorber is chosen to be a light beam, which is rigidly fixed at one end to the machine tool, and a mass, which may be clamped at various positions along the length of the beam so as to tune the absorber to a required frequency.

Given that the beam is made of aluminium which has a Young's Modulus of 70 GN/m^2 and is of square section, 60 × 60 mm, and the absorber mass is 25 kg, calculate the minimum length of the beam required for the absorber to function over the frequency range 40–50 Hz. Ignore the mass of the beam itself.

Also, calculate how far from the fixed end of the beam the mass would have to be clamped in order to tune the absorber to the maximum frequency of its range of operation.

52. A machine of mass M produces an harmonic disturbing force at a frequency which coincides with the natural frequency of the machine on its spring mounting. A vibration absorber is to be fitted such that no resonance is to be within 20 per cent of the exciting frequency. Find the minimum mass of a suitable absorber.
53. The figure shows a steel shaft which can rotate in frictionless bearings, and which has three discs rigidly attached to it in the positions shown. Each of the three discs is solid, uniform, and of diameter 280 mm. The modulus of rigidity of the shaft material is 83 GN/m^2, and the density of the discs is 7000 kg/m^3.

   Find the frequencies of free torsional vibrations of the system.

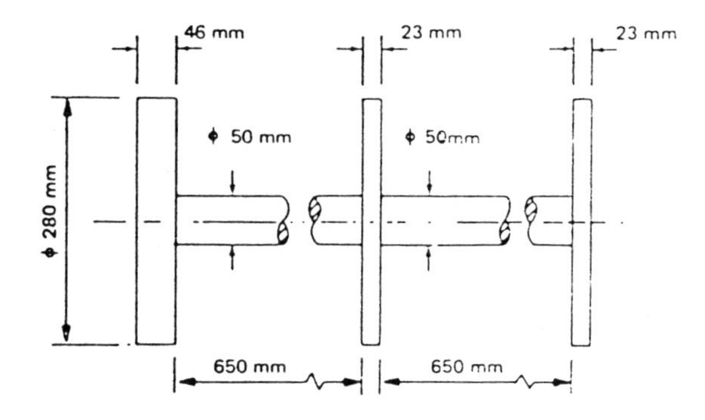

54. An aeroplane has a fuselage mass of 4000 kg. Each wing has an engine of mass 500 kg, and a fuel tank of mass 200 kg at its tip, as shown.

   Neglecting the mass of each wing, calculate the frequencies of flexural vibration in a vertical plane. Take the stiffness of the wing sections to be $3k$ and k as shown, where $k = 100$ kN/m.

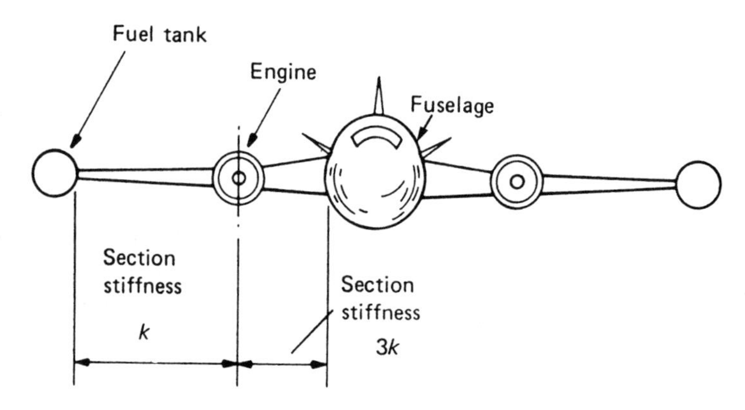

55. A manufacturer of delicate equipment ships his products, which have a mass m, by fitting them with a spring mounting of stiffness k. It is found that damage to the equipment is caused by large amplitude vibrations in the vertical direction. A special packing is therefore considered. The packing case, of mass m, is supported on a spring mounting of stiffness k; two pieces of equipment, each of mass m, are supported within the case by three springs each of stiffness k, as shown (idealized).

 Considering vibrations in the vertical direction only, find the natural frequencies of this system and hence comment on its suitability.

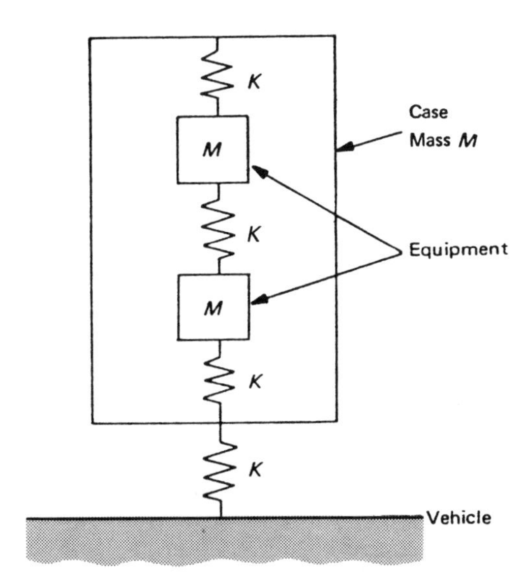

56. A marine propulsion installation is shown in the figure. For the analysis of torsional vibration, the installation can be modelled as the system shown, where the mass moments of inertia for the engine, gearbox, and propeller taken about the axis of rotation are I_E, I_G, and I_P respectively, and the stiffnesses of the gearbox and propeller shafts are k_G and k_P respectively. The numerical values are

$$I_E = 0.8 \text{ kg m}^2,$$
$$I_G = 0.3 \text{ kg m}^2,$$
$$I_P = 2.0 \text{ kg m}^2,$$
$$k_G = 400 \text{ kNm/rad},$$
$$k_P = 120 \text{ kNm/rad},$$

and damping can be neglected.

Calculate the natural frequencies of free torsional vibration and give the positions of the node for each frequency.

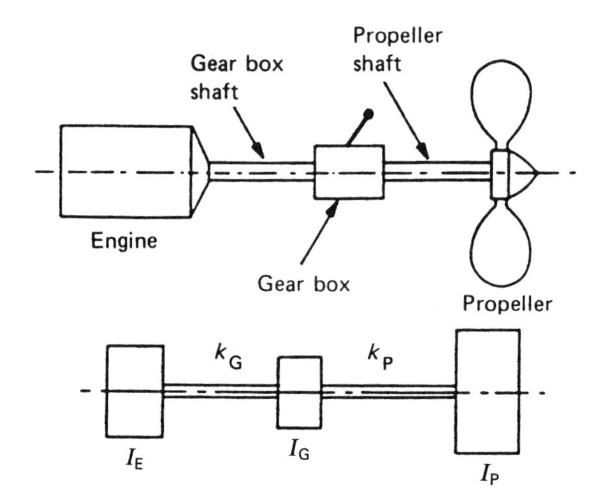

57. A machine is modelled by the system shown. The masses of the main elements are m_1 and m_2, and the spring stiffnesses are as shown. Each roller has a mass m, diameter d, and mass moment of inertia J about its axis, and rolls without slipping.

  Considering motion in the longitudinal direction only, use Lagrange's equation to obtain the equations of motion for small free oscillations of the system. Given that $m_1 = 4m$, $m_2 = 2m$, and $J = md^2/8$, deduce the natural frequencies of the system and the corresponding mode shapes.

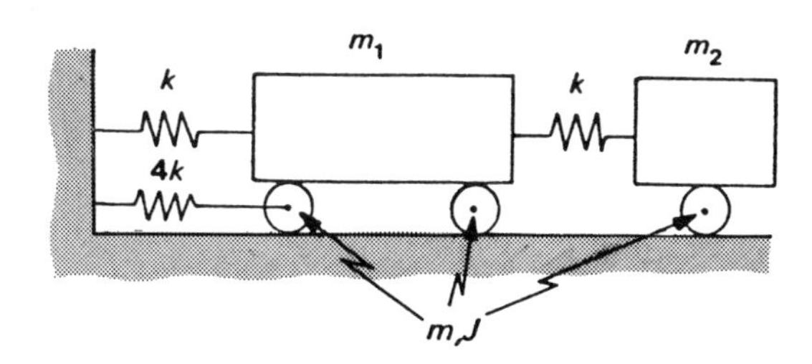

58. A simplified model for studying the dynamics of a motor vehicle is shown. The body has a mass M and a moment of inertia about an axis through its mass centre of I_G. It is considered to be free to move in two directions: vertical translation, and rotation in the vertical plane. Each of the unsprung wheel masses, m, is free to move in vertical translation only. Damping effects are ignored.
   (a) Derive equations of motion for this system. Define carefully the coordinates used.
   (b) Is it possible to determine the natural frequency of a 'wheel hop' mode without solving all the equations of motion? If not, suggest an approximation which might be made in order to obtain an estimate of the wheel hop frequency and calculate such an estimate given the following data:

   $k = 20$ kN/m; $K = 70$ kN/m; $m = 22.5$ kg.

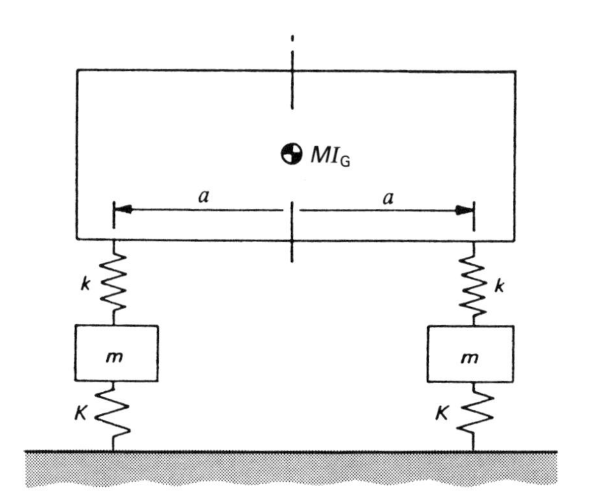

59. To analyse the vibration of a two-coach rail unit, it is modelled as the system shown. Each coach is represented by a rigid uniform beam of length l and mass m: the coupling is a simple ball-joint. The suspension is considered to be three similar springs, each of stiffness k, positioned as shown. Damping can be neglected.

Considering motion in the plane of the figure only, obtain the equations of motion for small amplitude free vibrations and hence obtain the natural frequencies of the system.

Explain how the mode shapes may be found.

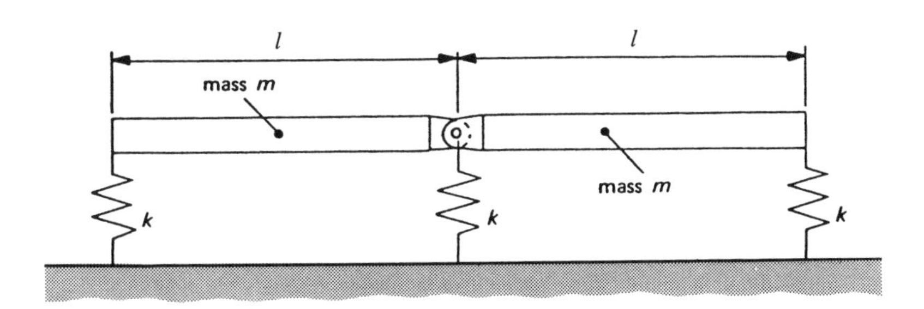

60. A structure is modelled by three identical long beams and rigid bodies, connected by two springs as shown. The rigid bodies are each of mass M and the mass of the beams is negligible. Each beam has a transverse stiffness K at its unsupported end, and the springs have stiffness k and $2k$ as shown.

Determine the frequencies and corresponding mode shapes of small amplitude oscillation of the bodies in the plane of the figure. Damping can be neglected.

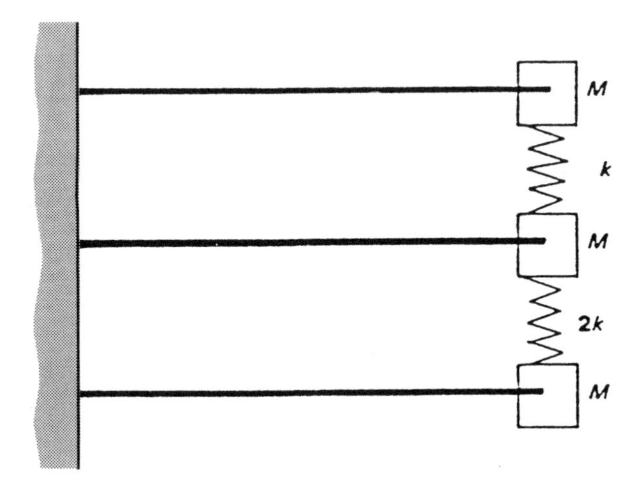

61. A tractor and trailer, each of mass M, are shown idealized in the figure. The trailer carries a heavy cylindrical drum of mass $\frac{2}{3}M$ and radius r; the drum can roll without slipping on the trailer. Both the tractor/trailer coupling and the trailer/drum attachment have a stiffness K.

Considering motion in the longitudinal (fore and aft) direction only, use Lagrange's equation to obtain the equations of motion for small free oscillations of the system. Hence deduce the natural frequencies of the system and the corresponding mode shapes.

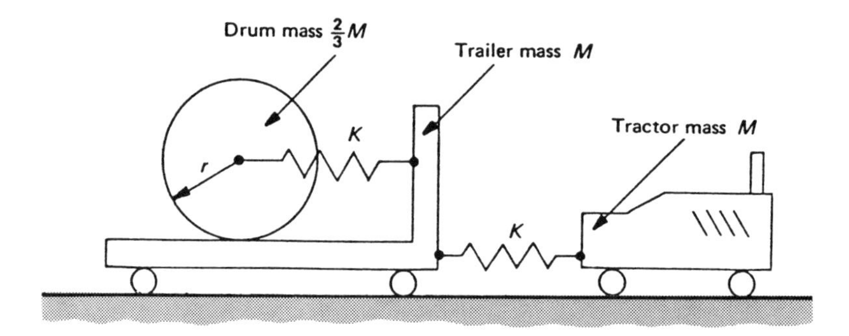

62. Part of a structure is modelled by the triple pendulum shown.

Obtain the equations of motion of small amplitude oscillation in the plane of the figure by using the Lagrange equation. Hence determine the natural frequencies of the structure and the corresponding mode shapes.

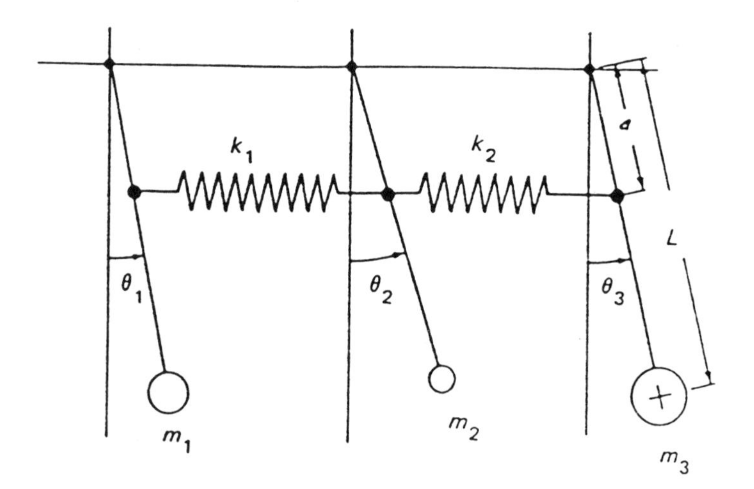

63. Vibrations of a particular structure can be analysed by considering the equivalent system shown. The bodies are mounted on small frictionless rollers whose mass is negligible, and motion occurs in a horizontal direction only.

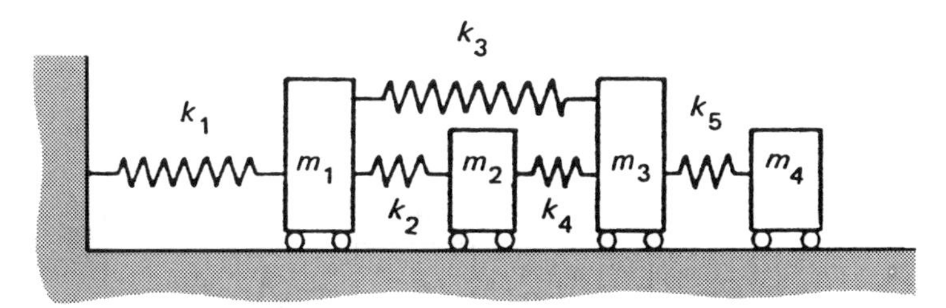

Write down the equations of motion of the system and determine the frequency equation in determinant form. Indicate how you would (a) solve the frequency equation, and (b) determine the mode shapes associated with each natural frequency.

Briefly describe how the Lagrange equation could be used to obtain the natural frequencies of free vibration of the given system.

64. A simply supported beam of negligible mass and of length l, has three bodies each of mass m attached as shown. The influence coefficients are, using standard notation.

$$\alpha_{11} = 3l^3/256\,EI \qquad \alpha_{31} = 2.33l^3/256\,EI$$
$$\alpha_{21} = 3.67l^3/256\,EI \qquad \alpha_{22} = 5.33l^3/256\,EI.$$

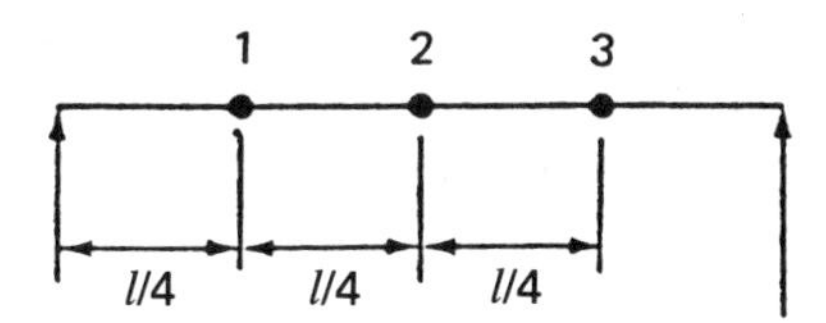

Write down the flexibility matrix, and determine by iteration the frequency of the first mode of vibration, correct to 2 significant figures, if $EI = 10\ \text{Nm}^2$, $m = 2$ kg, and $l = 1$ m.

Comment on the physical meaning of the eigenvector you have obtained, and use the orthogonality principle to obtain the frequencies of the higher modes.

65. Find the dynamic matrix of the system shown. Given that $k = 20$ kN/m and $m = 5$ kg, find the lowest natural frequency of the system and the associated mode shape.

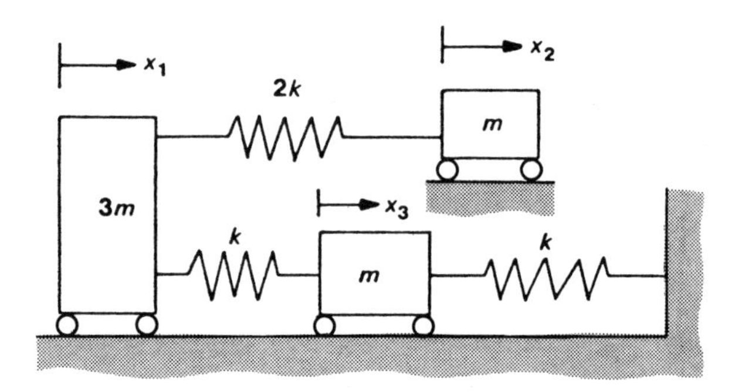

66. Define the term *degree of freedom* and explain the terms *normal mode of vibration* and *natural frequency* as applied to systems which have more than one degree of freedom.
67. Define the terms *direct receptance* and *cross receptance*, and explain how the receptance technique can be used in the analysis of complex dynamic systems.
68. Many dynamic systems are excited by time-dependent quantities which are not harmonic, yet harmonic frequency response methods are sometimes used in the analysis of the response of such systems. How, and under what conditions, can this be justified?
69. All real mechanical systems possess damping and yet, in the vibration analysis of such systems, damping is often neglected.

    Why is this done, and how is it justified?

70. A structure is modelled by the three degree system shown. Only translational motion in a vertical direction can occur.

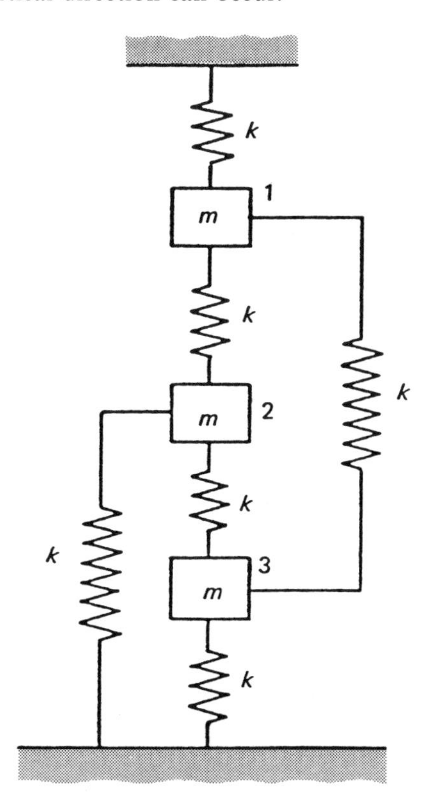

Show that the influence coefficients are

$$\alpha_{11} = \alpha_{22} = \alpha_{33} = \tfrac{1}{2}k$$

and

$$\alpha_{21} = \alpha_{31} = \alpha_{32} = \tfrac{1}{4}k,$$

and proceed to find the flexibility matrix. Hence obtain the lowest natural frequency of the system and the corresponding mode shape.

71. Find the driving point impedance of the system shown. The bodies move on frictionless rollers in a horizontal direction only.

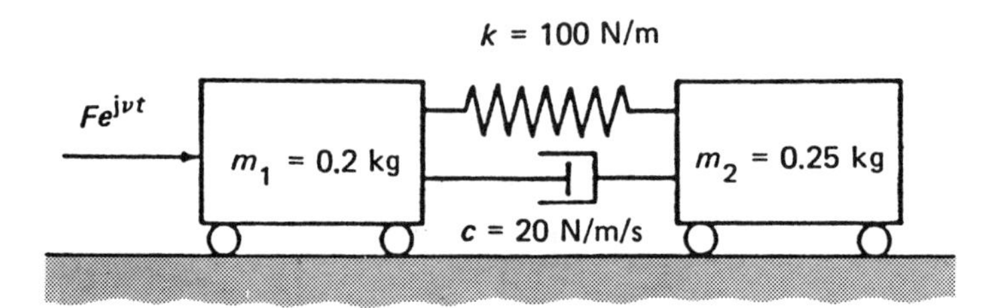

Hence show that the amplitude of body 1 is

$$\frac{\sqrt{((72\,000 + 2620\nu^2 + 0.2\nu^4)^2 + (20\nu^3)^2)}}{\nu^2(0.04\nu^4 + 1224\nu^2 + 32\,400)}\,F.$$

72.

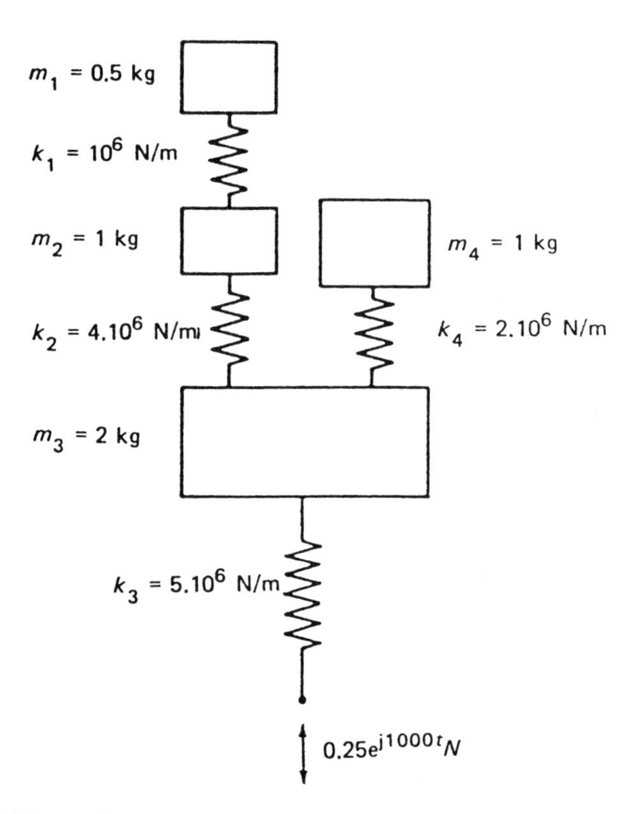

Find the driving point mobility of the system shown; only motion in the vertical direction occurs, and damping is negligible. Hence obtain the frequency equation: check your result by using a different method of analysis.

73. The receptance at a point in a structure is measured over a frequency range, and it is found that a resonance occurs in the excitation range. It is therefore decided to add an undamped vibration absorber to the structure.

    Sketch a typical receptance-frequency plot for the structure, and by adding the receptance plot of an undamped vibration absorber, predict the new natural frequencies. Show the effect of changes in the absorber mass and stiffness, on the natural frequencies, by drawing new receptance—frequency curves for the absorber.

6.3 SYSTEMS WITH DISTRIBUTED MASS AND ELASTICITY

74. The figure shows the model of a system consisting of an engine, a propeller and a shaft connecting them. The effective moments of inertia of the engine, propeller and shaft about the common axis of rotation are I_E, I_P and I_S, respectively. The shaft has a polar second moment of area J about the axis of rotation and the material from which it is made has a modulus of rigidity G. The other components of the system may be considered rigid and damping to be negligible.

    During free torsional oscillations of the system a node occurs along the shaft at a distance d from the disc representing the engine. Derive expressions for the natural frequency of the system and for the distance d in terms of the given quantities.

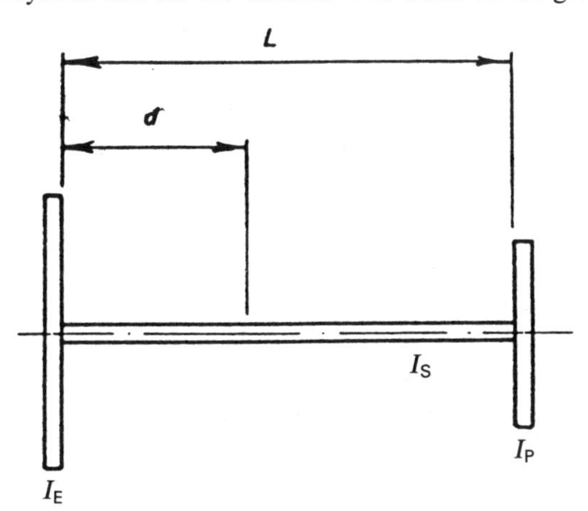

75. A uniform bar of length l is rigidly fixed at one end and is free at the other.

    Show that the frequencies of longitudinal vibration are

$$f = \frac{(n + \frac{1}{2})}{2l} c \text{ Hz}$$

where c is the velocity of longitudinal waves (E/ρ) in the bar, and $n = 0,1,2,\ldots.$

76. A solid steel shaft, 25 mm in diameter and 0.45 m long, is mounted in long bearings in a rigid frame at one end and has at its other end, which is unsupported, a steel flywheel. The flywheel can be treated as a rim 0.6 m in outer diameter and 20 mm square cross-section, with rigid spokes of negligible mass lying in the mid-plane of the rim.

    Find the frequency of free flexural vibrations.

77. A uniform horizontal steel beam is built in to a rigid structure at one end and pinned at the other end; the pinned end cannot move vertically but is otherwise unconstrained. The beam is 8 m long, the relevant flexural second moment of area of a cross-section is 4.3×10^6 mm^4, and the beam's own mass together with the mass attached to the beam is equivalent to a uniformly distributed mass of 600 kg/m.

    Using a combination of sinusoidal functions for the deflected shape of the beam, estimate the lowest natural frequency of flexural vibrations in the vertical plane.

78. In the figure, the bearings at each end of the shaft allow the shaft to tilt freely; the shaft is uniform, of material having a Young's modulus E, and with a flexural second moment of area I_s. The two discs, each having a mass m and a diametral mass moment of inertia I, are rigidly attached to the shaft in the positions shown.

    Neglecting the mass of the shaft, estimate the lowest frequency of free flexural vibrations when the shaft is not rotating.

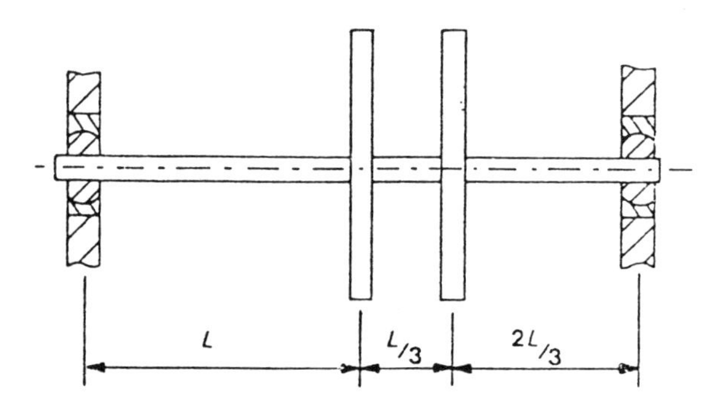

79. Part of the cooling system in a generating station consists of a steel pipe 80 mm in outer diameter, 5 mm thick, and 4 m long. The pipe may be assumed to be built in at each end so that the static deflection y at a distance x from one end of the pipe of length l is given by

$$y = \frac{mg}{24\ EI} x^2(l - x)^2,$$

    where m is the mass per unit length.

    Calculate the lowest natural frequency of transverse vibration of the pipe when full of liquid having a density of 930 kg/m^3. Take the density of steel as 7750 kg/m^3 and E as 200 GN/m^2.

80. A uniform horizontal beam, made of steel having Young's modulus equal to 207 GN/m^2, is supported as shown. The beam is rigidly built into a fixed support at A, and is free to rotate, but not to move vertically, at B. A mass of 560 kg is attached to the beam at C. The relevant flexural second moment of area of the beam is 2.2×10^{-5} m^4, and the beam's own mass and the load on it (excluding the mass at C) are equivalent to a uniformly distributed mass, moving with the beam, of 600 kg/m.

    Use Rayleigh's method to estimate the lowest natural frequency of flexural vibrations in the vertical plane.

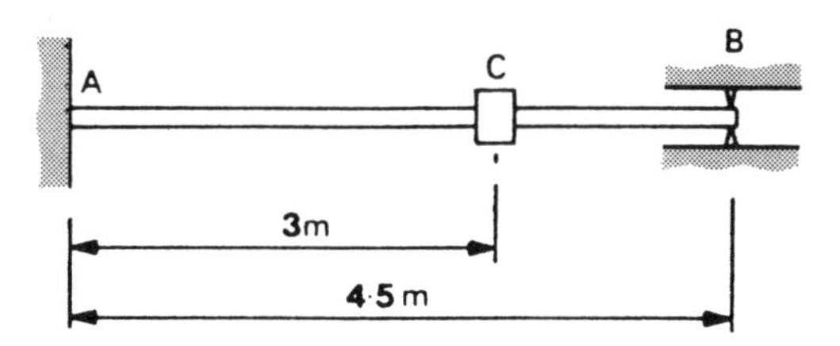

81. Derive the frequency equation for flexural vibration of a uniform beam which is pinned (simply supported) at one end and free at the other.

    Show that the fundamental mode of vibration has a natural frequency of zero and explain the physical significance of this mode.

    Obtain an approximate value for the natural frequency of the first bending mode of vibration and compare this with the corresponding value for a beam which is rigidly clamped at one end and free at the other.

82. A rotor of mass 90 kg is fixed at the mid-point of a shaft 1.8 m long. The ends of the shaft are freely supported in self-aligning bearings. Given that $EI = 10.8$ GN/m^2 for the shaft, and the motion is undamped, calculate the whirling speed.

    Damage to the rotor will occur if the amplitude of the axis of rotation of the rotor whilst running exceeds twice its eccentricity. Find the speed range which must be avoided in the absence of other restraints.

6.4 CONTROL SYSTEMS

83. Discuss the merits of open loop and closed loop control systems.
84. Draw block diagrams to represent
    (a) a system for automatically dipping the headlights of a motor car when approaching an oncoming vehicle, and
    (b) a servomechanism which could be used to assist the steering of a motor car.
85. For the hydraulic servo shown in Fig. 5.25, find the value of x_0 as a function of time following a sudden change in x_i of magnitude X.

86. Find the transfer function x_0/x_i for the hydraulic servo shown

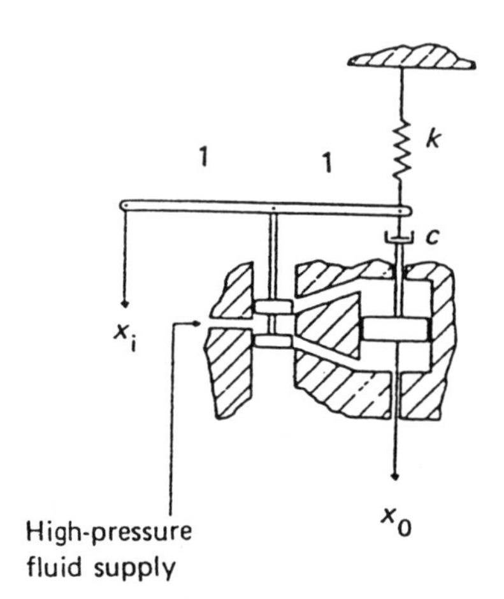

87. The temperature of a heat source is continuously monitored on a remote recorder where the rate of change of the indicated temperature at any instant is proportional to the difference between the real and recorded temperature.

    Show that the system can be described by a simple first order transfer function, and define the time constant required to achieve a steady state error of 1°C when the source temperature increases by 3°C per second.

88. A spool valve controls the flow of oil to a ram driving a load as shown below.

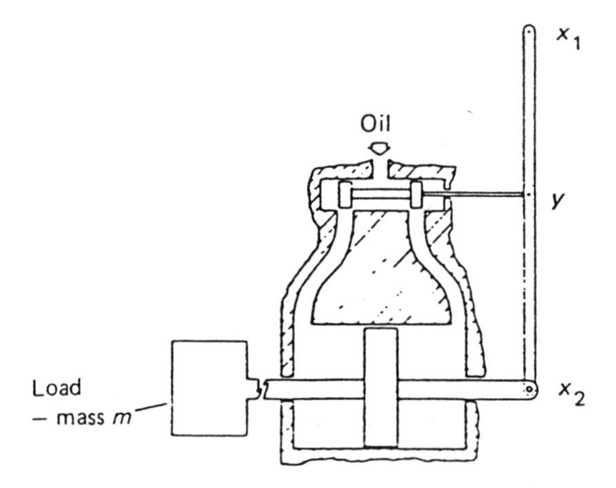

Given that

Q_0 is the flow through the valve per unit valve opening,
V_0 is the volume of oil in the system,

K is the bulk modulus of the oil,
M is the mass of the load,
A is the cross-sectional area of the ram, and
L is the leakage coefficient,

derive the TF for the system when the actuator (ram) position is fed back to the valve by means of a 1:1 linkage.

89. A simple position control system contains the following elements:
    a device to produce a signal proportional to the difference at any instant between the desired and actual positions;
    a linear amplifier; and
    a motor which exerts a torque proportional to the input signal to the motor
    The object whose position is to be controlled has a constant inertia I and is supported in a fixed frame by bearings which exert viscous damping of coefficient c.
    Draw a block diagram showing how these elements can be arranged to perform the intended function, and show the appropriate transfer function in each block. Find, in terms of I and c, the value of the factor relating the motor torque to the error signal which will ensure that when a step displacement is required the first overshoot will be one tenth the magnitude of the step.
90. The position of a gun turret, having a constant inertia J and rotating in bearings which give a viscous damping coefficient c, is to be controlled by an electric servo system which consists essentially of
    (a) a potentiometer to give a voltage proportional to the position demanded,
    (b) a second potentiometer to give a voltage proportional to the position achieved at any instant,
    (c) an amplifier which produces an output voltage proportional to the difference at any instant between these two voltages (that is, the error), and
    (d) a motor which exerts a torque proportional to the voltage applied to it.
    Draw a block diagram (not an electrical circuit diagram) showing how these elements would be arranged to give a working system, and show in each block the appropriate transfer function.
    During tests it is found that the steady-state response to a constant velocity input is acceptable, but that when a step displacement input is applied the first overshoot is larger than desired.
    Show, by adding to the block diagram and solving the relevant equations, that by adding to the amplifier input a signal proportional to the rate of change of the error, the overshoot to a step displacement input can be reduced without changing the steady-state response to a constant velocity input.
91. With reference to a position servo having linear characteristics, state the essential features of (a) output velocity feedback, (b) derivative of error control, and (c) integral of error control. State what kind of error can be eliminated by the use of each method.
    A linear position control system has an amplifier-plus-motor constant of k, an inertia I, and viscous damping (in fixed bearings) c. An integral of error

signal is to be added to the error signal before amplification.

Find the value of the constant by which the integral signal is multiplied which would make the system unstable.

92. A simple angular position control system contains a device to produce a signal proportional to the difference at any instant between the desired and actual positions, a linear amplifier, and a motor which exerts a torque proportional to the input signal to the motor. The load whose position is to be controlled has a constant mass moment of inertia I about its axis of rotation and is supported in a fixed frame by bearings which exert viscous damping of coefficient c.

Draw a block diagram showing how these elements can be arranged to perform the intended function, and show the appropriate transfer function in each block.

In a particular case for which $J = 100$ kg m^2 and $c = 50$ N m s/rad, when a step input is applied to the system whilst at rest the first overshoot is to be one fifth of the magnitude of the step.

Find the value of the factor which relates the motor torque to the error signal.

93. The output speed of an experimental turbine is dependent on the angular position of a regulator valve controlling the fluid input to the turbine. The angular position of the valve is controlled by an electromagnetic device which monitors the speed of the output shaft and produces a 2° change in valve position for each 1 rad/s change in turbine output speed. Take the fluid input system to be well represented by a first-order lag transfer function with a time constant of 0.2 s, and the inertia and viscous friction of the turbine system to be 5 kg m^2 and 20 N m per rad/s respectively.

(a) Draw a block diagram of the system and deduce its transfer function, if, with the output speed monitor disconnected, the turbine is found to change speed by 10 rad/s for a step change in regulator valve position of 5°.
(b) Calculate the frequency and decay rate of the transient response for a step change in the regulator valve position.
(c) Calculate the steady-state error in output speed for an increase in turbine load of 200 N m.

94. The angular position of a large turntable is remotely controlled in 30° steps. At present the transient response of the system to the step input is allowed to decay to within an acceptable error band before a braking system is applied to lock the table in position. The present error-actuated control system consists of a potentiometer giving 60 volt/rad error to an amplifier and motor unit working at 10 N m/volt. The motor drives the gearbox and turntable which have a total moment of inertia of 2400 kg m^2 and viscous bearing friction of 48 N m s/rad.

(a) If an acceptable position error is that the maxima of the error are less than $\pm 20\%$ of the input step, calculate the minimum time elapsed before the brake can be applied.
(b) Velocity feedback is to be employed to improve the response by reducing the time elapsed before the brake is applied. Calculate the velocity feedback requirements to achieve critical damping, and show that the minimum time

elapsed, before the brake can be applied, is less than $\frac{1}{3}$ of the time for the system without velocity feedback.

95. A certain servomechanism is required to control the angular position θ_o of a rotatable load of moment of inertia J. The rotation of the load is subject to a viscous friction torque c per unit angular velocity. The mechanism has velocity feedback such that the motor torque is

$$K_1\left(\varepsilon - K_2\frac{d\theta_o}{dt}\right)$$

where ε is the error between the desired angular position input θ_i and the output position θ_0, and K_1 and K_2 are constants.

(a) Draw a block diagram of the system and establish the equation of motion.
(b) Find the steady-state positional error when the input signal has constant velocity given by

$$\frac{d\theta_i}{dt} = \beta.$$

(c) When a step input is applied the response overshoot must not exceed 10%. Find the minimum value of the damping ratio required, and express this ratio in terms of the given parameters.

96. The angular position θ_0 of a turntable is controlled by a servomechanism which has positive acceleration feedback. The moment of inertia of the turntable about its axis of rotation is 5 kg m^2 and the effective viscous damping coefficient is 10 N m s/rad.

The motor torque is $20(\varepsilon + k\ddot{\theta}_0)$ N m where ε is the error, in radians, between the position of the input, θ_i, and the turntable position, θ_0, and $k\ddot{\theta}_o$ is the feedback signal proportional to the acceleration of the turntable.

(i) Draw a block diagram for the servomechanism and derive its equation of motion.
(ii) If a step input is applied, determine the steady-state error and the value of k required for the maximum output overshoot to be 5% of the steady-state value, and
(iii) Determine the steady-state position error if an external torque of 5 N m is suddenly applied to the turntable.

97. A linear remote position control system with negative output feedback consists of a potentiometer giving 8V/rad error to an amplifier and motor which applies a torque of 3 N m/V to the load. The load has an inertia of 6 kg m^2 and viscous friction of 12 N m s/rad.

(i) Draw a block diagram for the system and derive its equation of motion.
(ii) Calculate the maximum overshoot in the output response to a step input of 2 rad.
(iii) If a tachogenerator is employed to provide negative output velocity feedback, derive the new equation of motion and calculate the velocity feedback coefficient needed to give critical damping.

98. The angular position θ_0 of a turntable is controlled by a servomechanism which has positive acceleration feedback. The moment of inertia of the turntable about its axis of rotation is 4.2 kg m^2, and the effective viscous damping coefficient is 4 N m s/rad. The motor torque is given by

$$\text{Torque} = ((\theta_i - \theta_0) + 0.2\ddot{\theta}_0)16.$$

where θ_i is the input position, and $0.2\ddot{\theta}_0$ is the feedback signal proportional to the acceleration of the turntable.

 (i) Draw a block diagram for the servomechanism and derive its equation of motion.
 (ii) If a step input of 0.2 rad is applied, find the steady-state error, the frequency of the transient output oscillation, and the output position 0.5 s after the step has been applied.

99. An angular remote position feedback control system consists of a servomotor system, with a linear transfer function of $K_p = 32$ N m/rad error, driving a load of 0.5 kg m^2 which is subjected to a viscous resistance of 1 N m s/rad.

 (a) Sketch the block diagram for the system, derive the closed loop transfer function and calculate the steady-state error if the input is increased at a constant rate of 2 rad/s.
 (b) If a derivative of error controller, with a gain constant of K_d, is added in parallel with K_p, sketch the new block diagram, calculate the response to a step input, and define the value of K_d which will give critical damping.
 (c) Show that when an integral of error controller, with gain constant of K_i, is added in parallel with K_p and K_d, the velocity lag is reduced to zero, and define the range of K_i that will allow system stability.

100. An experimental fuel injection system can be approximated by a transfer function in the form:

$$\frac{K_i}{1 + sT},$$

and forms part of a system designed to control the speed of an engine. Feedback is obtained through a speed regulator system which alters the fuel input to the injection system by changing the angular position of a control valve. The regulator is designed to move the control valve K_r radians for each rad/s change in engine output speed.

 (a) Consider the engine to have inertia J, and viscous friction c, and draw a block diagram of the system when the input is an initial manual setting of the angular position of the control valve θ_i, and the output is the engine speed θ_0. Also deduce the system transfer function.
 (b) If the engine is subjected to a step change in θ_i, show that the time required for the output speed to settle down within $\pm 1\%$ of the step change is less than 1s under the following conditions:

$K_i = 200$ N m/rad,
$K_r = 0.2$ rad/(rad/s),
$T = 0.1$ s,
$J = 5$ kg m^2, and
$c = 10$ N m s/rad.

(c) Calculate the steady-state error in output speed when the engine load is increased by 100 N m.

101. A ship's rudder has a moment of inertia about its pivot of 370 kg m^2. When the ship is stationary, rotation of the rudder is resisted by a moment $M_1 = 800\ \mathrm{D}\theta_0$ N m, where θ_0 is the rudder angle in radians measured from the straight-ahead position. The position of the rudder is to be controlled by a simple servosystem which applies to the rudder a turning moment $M_2 = k(\theta_i - \theta_0)$ N m, where k is a constant and θ_i is the rudder angle demanded at any instant.

A step input of 0.1 radian is applied: find the value of k which will make the magnitude of the error $(\theta_i - \theta_0)$ equal to 0.01 radian at the first overshoot.

When the ship is moving, hydrodynamic forces on the rudder amount to an additional moment $M_3 = 3500\ \theta_0 + 700\ \mathrm{D}\theta_0$ N m: with the value of k found above, and the same step input, find the magnitude of the first overshoot and the steady-state error.

102. The angular position θ_0 of a turntable is controlled by a servomechanism which has acceleration feedback. The turntable has a moment of inertia of 30 kg m^2 and viscous damping of coefficient 20 N m s/rad. The motor torque is $10(\varepsilon + k(\mathrm{d}^2\theta_0/\mathrm{d}t^2))$ N m, where ε is the error, in radians, between the position of the input θ_i and the turntable position θ_0, and $k(\mathrm{d}^2\theta_0/\mathrm{d}t^2)$ is the feedback signal proportional to the acceleration of the turntable.

(a) Draw a block diagram for the system and derive its equation of motion.
(b) Find the value of k for the damping to be critical.
(c) Given that the input to the system is a constant angular velocity, derive an expression for the steady-state error. Determine the value of this error when the input is 2 revolutions per minute.

103. A simple linear servomechanism consists of a motor which supplies a torque equal to k times the error (θ) between input (θ_i) and output (θ_0). The motor drives a rotational load of mass moment of inertia I and equivalent viscous damping coefficient (less than critical) c.

Draw a block diagram for the mechanism and write down the equation of motion.

In a particular mechanism, $I = 80$ kg m^2, $k = 2$ kN m/rad error, and $c = 100$ N m s/rad. Given that a step rotation of 10° is applied to the input when the mechanism is at rest, find the time to reach zero error the first time, the time to reach the first overshoot value, and the magnitude of the first overshoot.

If, when the mechanism is at rest, the input is suddenly rotated at 10 rad/s, find the steady state error. How could this be altered to 10°?

104. A linear remote position angular control system with negative output feedback consists of a potentiometer giving 16 V/rad error to an amplifier, motor, and gearbox system which supplies 3 N m/V to the output shaft. The motor and gearbox system has an inertia of 12 kg m^2 and viscous friction of 24 N m s/rad.

   (a) Determine the maximum overshoot in the output response to a step input of 1 rad.
   (b) If a tachogenerator is employed to improve the response, derive the new transfer function and calculate the velocity feedback requirements to achieve critical damping.

105. What is meant by the *stability* of a dynamic system, and how does an unstable system respond to a stimulus?

106. A root locus diagram is often used to assess the dynamic behaviour of a system. What does each line or locus on such a diagram represent, and what aspects of the dynamic behaviour does it demonstrate?

107. For the control system shown in the figure in block diagram form, find the OLTF of the equivalent single loop system and sketch the root locus diagram for this system if $0 \leqslant K_1 \leqslant \infty$ and $K_2 = \frac{17}{32}$. Hence find the maximum value of the amplifier gain K_1 for stability.

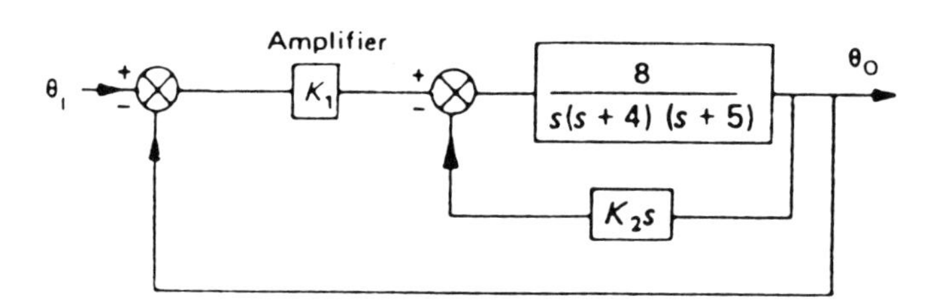

108. The control system for positioning the gun turret of a tank comprises the units shown.

    (a) Derive the open- and closed-loop transfer functions.
        (i) if G is constant, and
        (ii) if, to allow for build time of the field current,

$$G = G_0/(1 + sT_f)$$

    (b) Sketch the root locus for the system in each case, given that $c = 10$ V/rad/s; $K_0 = 0.2$ A/V; $K_1 = 1$ V/rad; $K_2 = 1$ kN m/A; $J = 400$ kg m^2; $T_f = 0.4$ s.

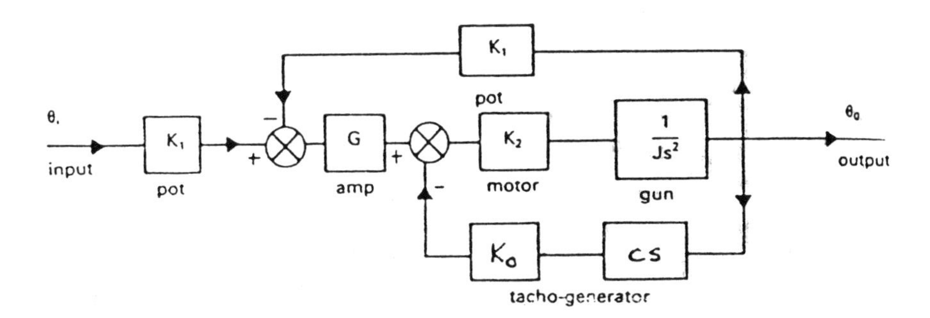

109. The simple unity feedback control system shown has three elements with individual functions as shown.

    (a) Sketch the root locus diagram of this system and determine the maximum value of the amplifier gain K for stability.
    (b) Find a value of K such that the closed loop system has a complex conjugate pole pair with damping factor 0.5.

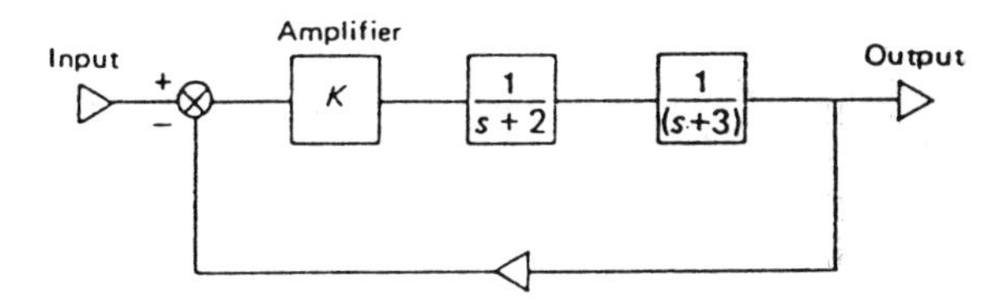

110. The block diagram of a certain position control system is shown. Both the input and output potentiometers have a gain K: the gain of the amplifier is G.

Simplify the system to one having a single unity feedback loop, and find the open loop transfer function.

By sketching the root locus diagram for this system, determine the maximum value of G for stability.

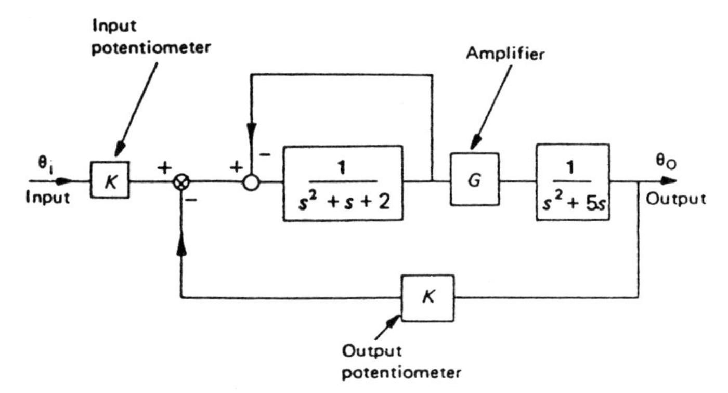

111. The main elements of a position control system are shown in the block diagram. The input and output potentiometers each have a gain K, and the gain of the amplifier is G.

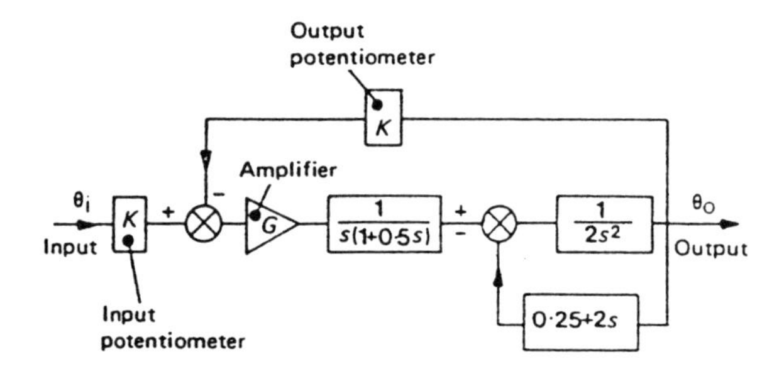

Find the open loop transfer function of the equivalent single loop system, and sketch the root locus diagram for this system. Hence determine the limiting value of the product GK for stability.

112. In a simple unity feedback control system the elements have individual functions as given in the figure below.

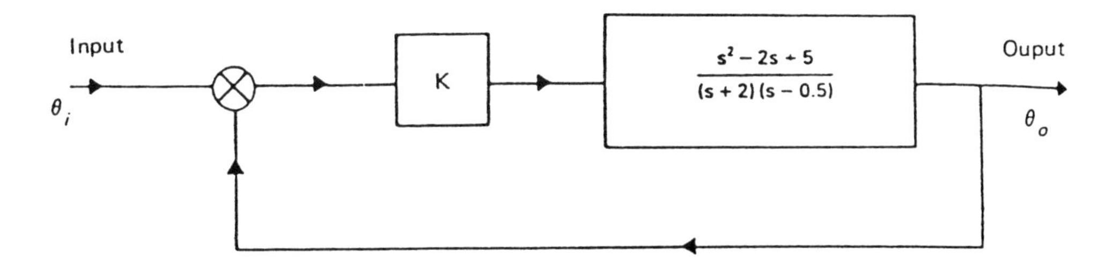

(a) Sketch the root locus diagram of this system for all positive values of the amplifier gain K, and determine the range of values for K for stability.
(b) Confirm this range of values by applying the Routh–Hurwitz criterion.
(c) Find the damping ratio of the system response when $K = 0.3$.

113. A dynamic system with overall gain K has an open loop transfer function

$$\frac{K}{s(1 + 0.1\ s)(1 + 0.05\ s)}.$$

Draw the root locus diagram as K increases to infinity. Find K_{max}, which is the maximum value K can have for the system to be stable, and the corresponding frequency of oscillation. Confirm the value of K_{max} by using the Routh–Hurwitz criterion.

The system can be modified by adding derivative action of time constant 0.5 s. Comment on the stability of the modified system, and find the system oscillation frequency and damping factor when $K = K_{max}$.

114. A certain controlled process can be represented by the system shown. The transfer function of the process is

$$\frac{10}{(1 + 0.2\ s)^2(1 + 0.05\ s)},$$

and the controller has a transfer function $G(s)$. The input to the controller is the error signal $(\theta_i - \theta_0)$.

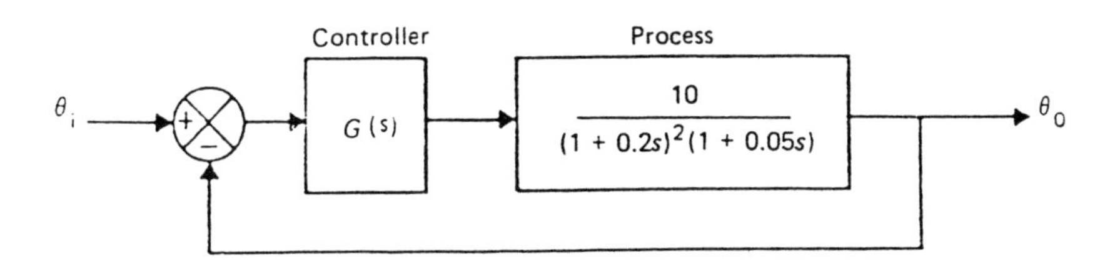

(a) Given that $G(s) = K$, sketch the root locus diagram for the system as K increases to infinity, and find the maximum value of K for the system to be stable, and the corresponding frequency of oscillation.
(b) The system is modified by using a new controller with $G(s) = K(1 + 0.2\ s)/(1 + 0.02\ s)$. Sketch the root locus diagram for the modified system as K increases to infinity, and briefly compare its performance with the original system when $K > 0$.
(c) Find the maximum value of K for the modified system to be stable, and check this result using the Routh–Hurwitz criterion.

115. The main elements of a position control system are shown in the figure. Both the input and output potentiometers have a gain K, and the gain of the amplifier is G.

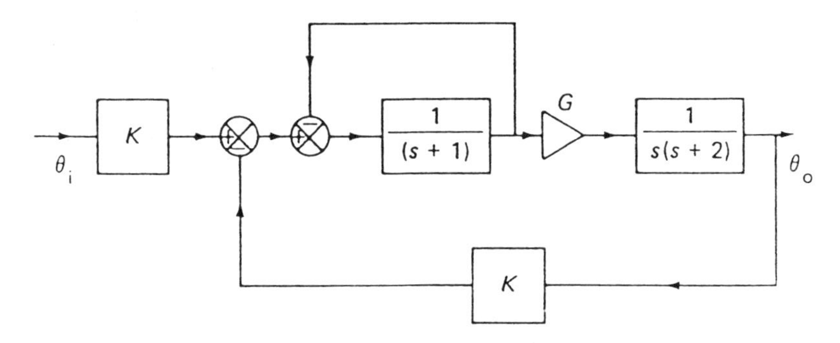

(a) Simplify the system to one having a single unity feedback loop and find the open loop transfer function.
(b) Sketch the root locus diagram for this system and hence find the maximum value of GK for the system to be stable.
(c) Confirm the result of part (b) by using the Routh–Hurwitz method and comment on the disadvantages of this method when compared to graphical methods such as root locus.

116. In a simple control system, feedback is used to maintain near-constant speed of an engine under varying load conditions. The system is shown in block diagram form.

The governor has proportional gain k_1, and the engine speed is sensed by an electrical tachometer with a first-order transfer function; the engine can be described by a second-order transfer function.

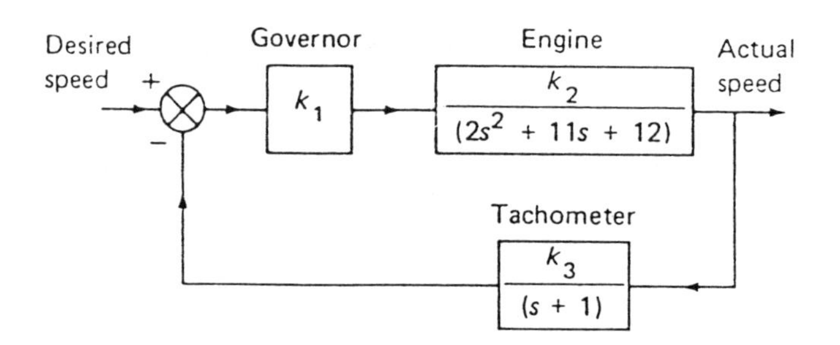

(a) Write down the open loop transfer function of the system, and put $K = k_1 k_2 k_3$.
(b) Sketch the root locus diagram for the system for values of $K > 0$, and determine the maximum value of K for stability. Confirm your answer by using the Routh–Hurwitz method.
(c) Find the value of K when complex conjugate roots exist whose real part is -0.7, and find the damping ratio ζ of the system with this gain.

117. A position control system is shown in block diagram form below.

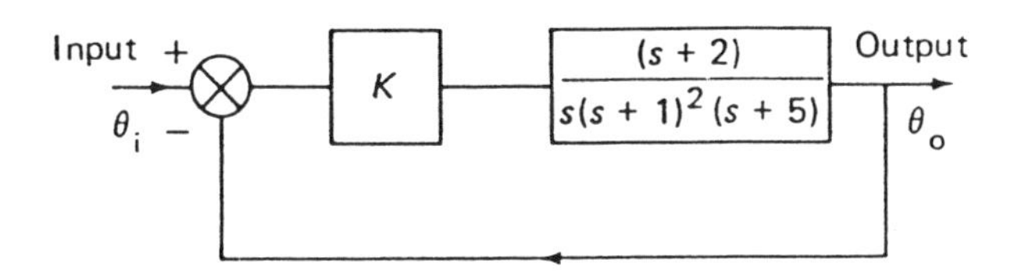

(a) Use the Routh–Hurwitz criterion to find the maximum value of gain K for stability.
(b) Because this value of K is too low for the application, a compensating subsystem with transfer function $(s + 1)/(s + 10)$ is to be inserted in the forward feed of the system.

Sketch the root locus diagram of the modified system for all values of the gain K, and find the range of value K can have for stability.

If the damped natural frequency of the system response is 1 rad/s, estimate the damping ratio.

118. Given the open loop transfer function of a dynamic system, explain how you would construct a Nyquist frequency response diagram. How would you determine from this diagram whether the system was stable or not?

119. A control system has an open loop transfer function given by

$$\Phi_0(s) = \frac{K}{(s + 1)(s^2 + 5s + 6)},$$

where K is an unspecified gain constant.

(a) Derive an expression for the overall or closed loop transfer function, $\Phi_c(s)$.
(b) For the particular case of $K = 90$, sketch the Nyquist frequency response diagram for the open-loop transfer function of this system and thus determine whether or not the system is stable.
(c) Hence, or otherwise, determine the range of values of K for which the system is stable.

120. A control system has an open loop transfer function

$$\Phi_0(s) = \frac{1}{s(1 + s)(1 + 2s)}.$$

Sketch the Nyquist frequency response diagram for $\Phi_0(s)$, comment on the stability of the system, and calculate the gain margin.

The system is modified by adding to the open loop derivative action with a time constant of 3 s. Write down the open loop transfer function of the modified system, and hence sketch the Nyquist frequency response diagram. Comment on the stability of the modified system.

121. A unity feedback control system has the following closed loop transfer function

$$\frac{K}{s(1 + sT)^2 + K}.$$

(a) Sketch the loop frequency response diagram and with $T = 0.25$ s use the Nyquist stability criterion to determine the upper limit of gain constant K for system stability.

(b) Using the Routh–Hurwitz method, confirm the upper limit of K and determine the lower limit.

(c) If $T = 0.1$ s and $K = 5$, determine the closed loop frequency and magnitude at a phase lag of 90°.

(d) Under the conditions of (c), determine the gain margin.

122. A process is controlled by adding negative feedback through an integral action controller as shown.

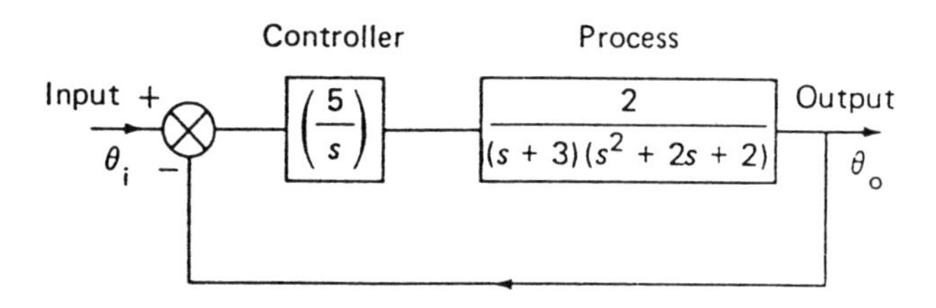

(a) Sketch the Nyquist frequency response diagram for the open loop transfer function (OLTF) of the system, and comment on the stability.

(b) The system is modified by adding proportional action so that the controller transfer function becomes $5(s + 1)/s$.

Sketch the Nyquist frequency response diagram for the OLTF of the modified system, comment on the stability and calculate the gain margin.

123. A control system has an open loop transfer function (OLTF) given by

$$\Phi_0(s) = \frac{k}{s^3 + 6s^2 + 8s}.$$

where k is a gain constant.

Sketch the Nyquist frequency response diagram for the OLTF of this system when $k = 20$ and thus determine whether or not the system is stable.

Also, find the maximum value which k can have for the system to be stable.

124. A control system has an open loop transfer function (OLTF) given by

$$\Phi_0(s) = \frac{2}{s(1 + s)(1 + 2s)}.$$

Sketch the Nyquist frequency response diagram for the OLTF of this system, and thus determine whether the system is stable or not.

The system is modified by adding to the open loop derivative action with a time constant of 0.5 s.

Write down the OLTF of the modified system, and hence sketch the Nyquist frequency response diagram. Comment on the stability of the modified system and calculate the gain margin.

125. How can a frequency response be represented by a Bode diagram? Indicate what *gain* and *phase margins* are in this context.

126. The OLTF of a control system is given by

$$\frac{7}{s(1 + 0.5\,s)(1 + 0.167\,s)}.$$

Plot the Bode diagrams and determine the gain and phase margins.

127. A servosystem has an OLTF

$$\frac{40}{s(1 + 0.0625\,s)(1 + 0.25\,s)}.$$

Plot the Bode diagram and determine the gain margin and phase margin and whether the system is stable.

A phase lag network is introduced into the servosystem. It has a transfer function.

$$\frac{(1 + 4\,s)}{(1 + 80\,s)}.$$

Draw the Bode diagram for the modified system, and find the new gain and phase margins.

128. The following data were collected from an open loop frequency response test on a unity feedback control system.

Frequency ω(rad/s)	0.1	0.4	1	2	4	10	20	40	100
Magnitude (dB)	40	28	20	13	4	−14	−30	−48	−72
Phase ϕ (degrees)	−90	−98	−112	−134	−168	−216	−242	−260	−270

Plot the Bode magnitude and phase diagrams and define the complete loop transfer function given that one component in the system is known to have a transfer function of the form $1/(1 + sT)^2$.

129. The forward path of a unity feedback control system has the following open loop transfer function.

$$GH(s) = \frac{K}{s(1 + T_1 s)(1 + T_2 s)}.$$

(a) Use the Bode method of asymptotes to plot magnitude and phase over the frequency range 1 to 1000 rad/s, and estimate the maximum value of K for system stability when $T_1 = 0.5$ s and $T_2 = 0.01$ s.

(b) The first-order lag term, $1/(1 + T_1 s)$, represents a control circuit in the system. If this is replaced by a simple time delay of T_1 s, show what effect this has on the Bode plots, and estimate the new maximum value of K for stability. Confirm your result by using the Nyquist method.

130. The data given in the following table represent the results of an open loop frequency response test on a unity feedback control system. The system is designed to attenuate the amplitude of input variations around a frequency of 4Hz.

Frequency (Hz)	0.1	0.2	0.4	1.0	2.0	4.0	10	20	40	100
Magnitude (dB)	−0.1	−0.4	−1.6	−5.6	−9.1	−10.6	−8.5	−5.0	−2.0	−0.7

(A magnitude of 0 dB was measured outside the range 0.01 Hz to 1000 Hz.)

(a) Use the data to plot the Bode magnitude diagram, and establish linear transfer functions which will accurately describe the open and closed loop system responses.

(b) Sketch the phase response of the open loop system and use your diagram to establish what approximate phase distortion exists between input frequencies of 1 and 10Hz.

131. A unity feedback closed loop control system has the following loop transfer function which, on test, exhibits unit gain at the phase cross-over frequency,

$$\frac{k}{s(1 + 0.1\,s)(1 + s)}.$$

(a) Sketch the Bode modulus and phase diagrams of this system and determine the value of gain constant k.

(b) A phase-lead network having the following form is to be used on the system;

$$\frac{(s + a)}{(s + b)}.$$

Show that the maximum phase-lead of this network is

$$\Phi_{max} = 90° - 2\tan^{-1}\sqrt{(a/b)}$$

and that it achieved at a frequency $\omega = \sqrt{(ab)}$.

(c) The phase-lead network is applied to the forward path of the system in such a way that the maximum phase change produced by the network occurs at the phase-crossover frequency. Calculate the parameters a and b of the network, and the modification needed to the gain constant k, to achieve a phase margin of 30°.

7

Answers and solutions to selected problems

1. FBDs are

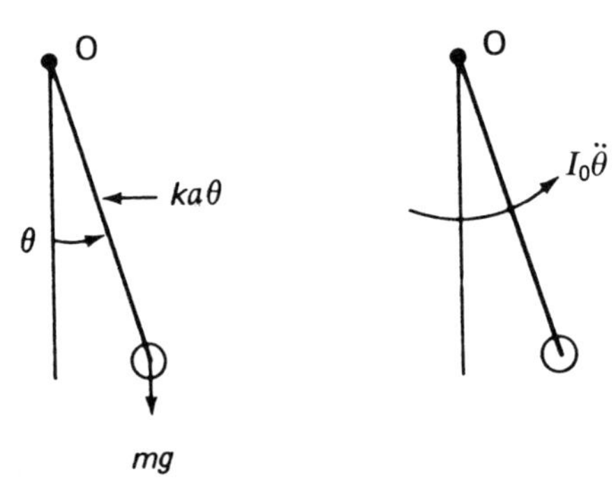

Moments about O gives

$$I_0\ddot{\theta} = -mgL\theta - ka^2\theta,$$

where

$$I_0 = mL^2.$$

Thus

$$\ddot{\theta} + \left(\frac{mgL + ka^2}{mL^2}\right)\theta = 0,$$

and

$$f = \frac{1}{2\pi}\sqrt{\left(\frac{mgL + ka^2}{mL^2}\right)}\ \text{Hz}.$$

2.

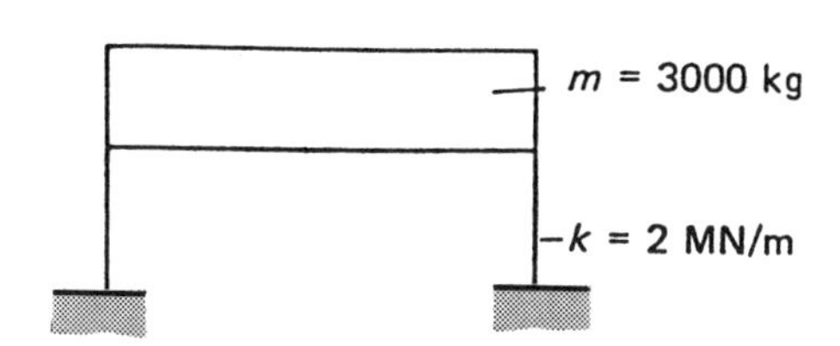

$$f = \frac{1}{2\pi}\sqrt{\frac{k}{m}} = \frac{1}{2\pi}\sqrt{\left(\frac{2 \times 2 \times 10^6}{3000}\right)} = \underline{5.8 \text{ Hz.}}$$

3. FBDs are

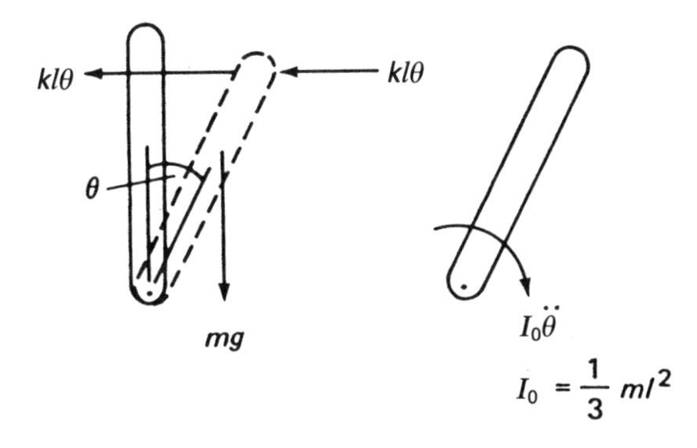

Moments about pivot gives

$$\tfrac{1}{3}ml^2\ddot{\theta} = mg\,\frac{l}{2}\,\theta - 2kl^2\theta,$$

so

$$\ddot{\theta} + \theta\left[\frac{2kl^2 - mgl/2}{\tfrac{1}{3}ml^2}\right] = 0.$$

and

$$\underline{f = \frac{1}{2\pi}\sqrt{\left(\frac{12kl - 3mg}{2ml}\right)} \text{ Hz.}}$$

4.

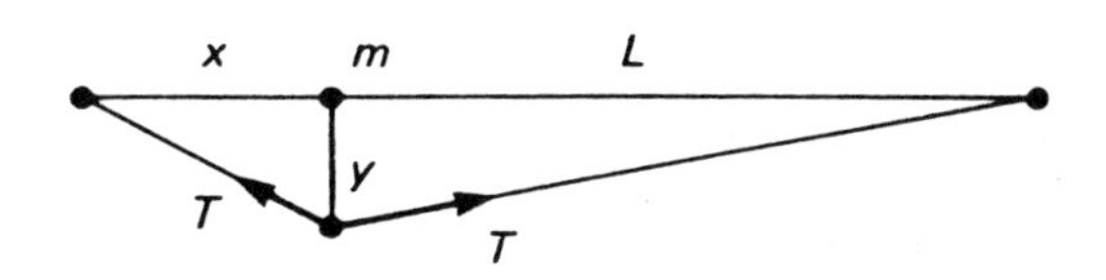

Equation of motion is

$$\frac{Ty}{x} + \frac{Ty}{L - x} = -m\ddot{y},$$

so

$$m\ddot{y} + \frac{TL}{x(L - x)}\, y = 0,$$

and

$$f = \frac{1}{2\pi}\sqrt{\left(\frac{TL}{x(L - x)}\right)}\ \text{Hz.}$$

5. For vehicle,

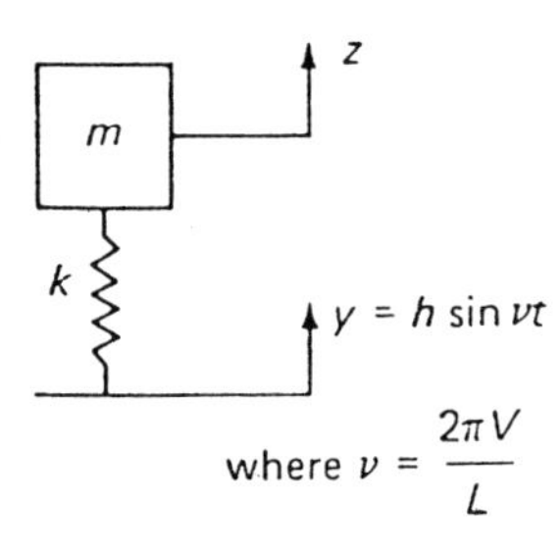

so

$$m\ddot{z} + k(z - h \sin \nu t) = 0.$$

Thus

$$m\ddot{z} + kz = kh \sin \nu t,$$

and

$$z = \left(\frac{kh}{k - m\nu^2}\right) \sin \nu t,$$

$$= \left(\frac{h}{1 - \left(\frac{\nu}{\omega}\right)^2}\right) \sin \nu t.$$

Hence

$$z = \left(\frac{h}{1 - \left(\frac{2\pi V}{L\omega}\right)^2}\right) \sin\left(\frac{2\pi V}{L}\right) t.$$

7. $$\Lambda = \frac{1}{10}\ln\frac{100}{1} = 0.46$$

$$\simeq 2\pi\zeta, \quad \text{so} \quad \zeta = 0.0732.$$

$$\omega_v = \omega\sqrt{(1-\zeta^2)}, \quad \text{so} \quad f_v = f\sqrt{(1-\zeta^2)},$$

and

$$\tau = \tau_v\sqrt{(1-\zeta^2)}$$
$$= 1\sqrt{(1-0.0732^2)}$$
$$\underline{= 0.997 \text{ s.}}$$

8. $\underline{\zeta = 0.72}$

9. FBDs:

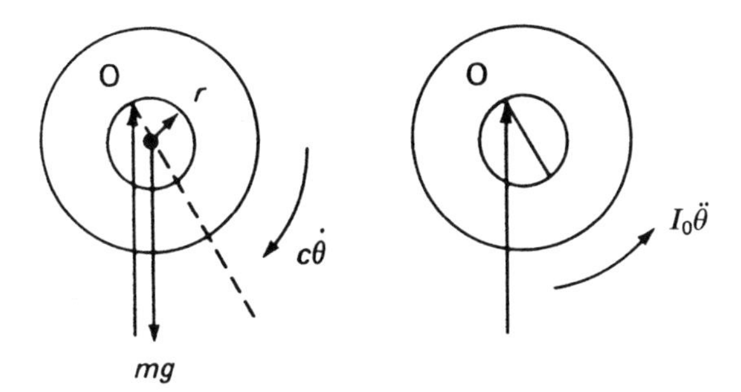

Moments about O gives

$$I_0\ddot{\theta} = -mgr\theta - c\dot{\theta},$$

so equation of motion is

$$I_0\ddot{\theta} + c\dot{\theta} + mgr\theta = 0.$$

Now if $\tau_v = 1.6$ s, and $\Lambda = \ln 6 = 1.792$,

$$1.792 = 2\pi\zeta/\sqrt{(1-\zeta^2)}$$

so $\zeta = 0.273$.

$$f_v = \frac{1}{\tau_v} = 0.625 \text{ Hz}, \qquad \text{so} \qquad \omega_v = 3.92 \text{ rad/s}.$$

$$\omega_v = \omega\sqrt{(1-\zeta^2)}, \qquad \text{so} \qquad \omega = 4.075 \text{ rad/s}.$$

Now,

$$\omega = \sqrt{\left(\frac{mgr}{I_0}\right)} = \sqrt{\left(\frac{20 \times 9.81 \times 0.02}{I_0}\right)} \text{ rad/s},$$

so

$$I_0 = 0.236 \text{ kg m}^2$$
$$= I_G + mr^2, \quad \text{and} \quad \underline{I_G = 0.228 \text{ kg m}^2.}$$
$$c_c = 2\sqrt{(mgr\, I_0)}$$
$$= 1.925 \text{ N m s/rad}, \qquad \text{Hence} \quad c = \zeta c_c = \underline{0.525 \text{ N m s/rad.}}$$

10. Equation of motion is $I_0\ddot{\theta} + c\dot{\theta} + k\theta = 0$. Require critical damping so

$$c_c = 2\sqrt{(kI_0)}$$
$$= 2\sqrt{(1.2 \times 10^{-5} \times 1.5 \times 10^{-6})}$$
$$= \underline{8.48 \times 10^{-6} \text{ N m s/rad.}}$$

11. FBDs without damping:

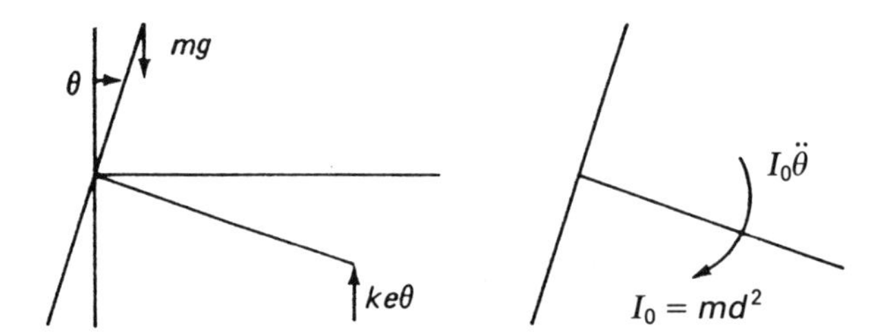

Equation of motion is $md^2\ddot{\theta} = mgd\theta - ke^2\theta$, so frequency of undamped oscillations is

$$\frac{1}{2\pi}\sqrt{\left(\frac{ke^2 - mgd}{md^2}\right)} \text{ Hz.}$$

With damping, equation of motion is

$$md^2\ddot{\theta} + cf^2\dot{\theta} + (ke^2 - mgd)\,\theta = 0,$$

so

$$\underline{c_c = \frac{2}{f^2}\sqrt{((ke^2 - mgd)\, md^2)}.}$$

12. FBDs:

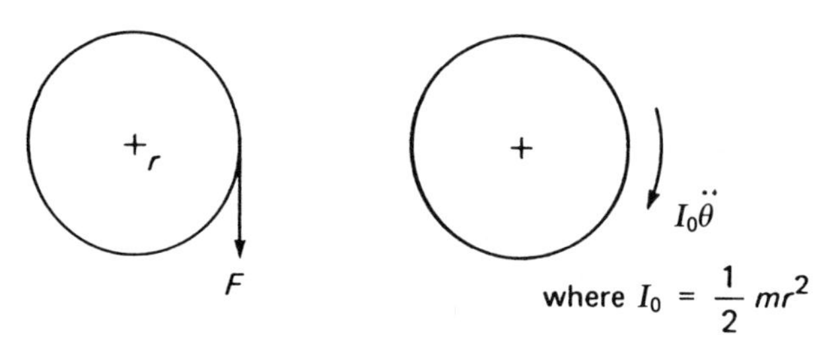

so

$$Fr = \tfrac{1}{2}mr^2\ddot{\theta}.$$

Thus $F = \tfrac{1}{2}m\ddot{x}$ because $r\ddot{\theta} = \ddot{x}$.
Body floats half submerged so

$$m = \frac{\pi}{4}\, d^2 D\rho_{\text{body}} = \frac{1}{2}\frac{\pi}{4}\, d^2 D\rho_{\text{liquid}},$$

and

$$\rho_{\text{b}} = \tfrac{1}{2}\rho_{\text{L}}.$$

When body displaced distance x downwards (say), liquid level rises by y where

$$\frac{\pi}{4}\, d^2 x = \frac{\pi}{4}\,(D^2 - d^2)\, y,$$

so

$$y = x\left(\frac{d^2}{D^2 - d^2}\right),$$

and

$$x + y = x\left(\frac{D^2}{D^2 - d^2}\right).$$

Equation of motion therefore

$$m\ddot{x} = -\tfrac{1}{2}m\ddot{x} - \rho_{\text{L}}\,\frac{\pi}{4}\, d^2\left(\frac{D^2}{D^2 - d^2}\right) x.$$

Hence

$$f = \frac{1}{2\pi}\sqrt{\left(\frac{4Dg}{3(D^2 - d^2)}\right)}\ \text{Hz}.$$

13.

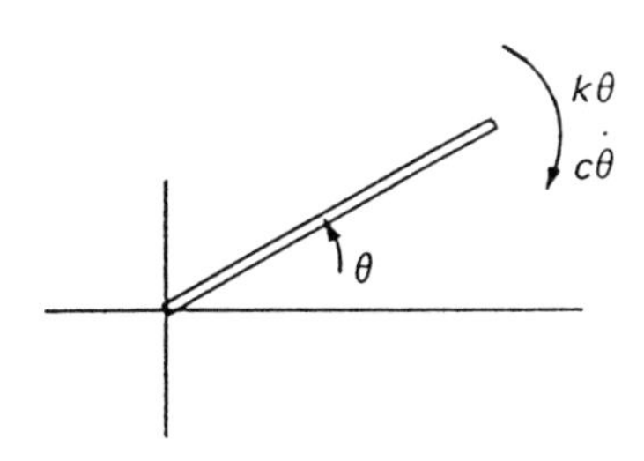

Equation of motion is $I\ddot{\theta} + c\dot{\theta} + k\theta = 0$, so that

$$\theta = \mathrm{e}^{-\zeta\omega t}\,(A \sin \omega_{\text{v}} t + B \cos \omega_{\text{v}} t),$$

where

$$\omega = \sqrt{\frac{k}{I}}, \quad c_c = 2\sqrt{(Ik)} \quad \text{and} \quad \zeta = \frac{c}{c_c}.$$

$$\omega_v = \omega\sqrt{(1 - \zeta^2)}.$$

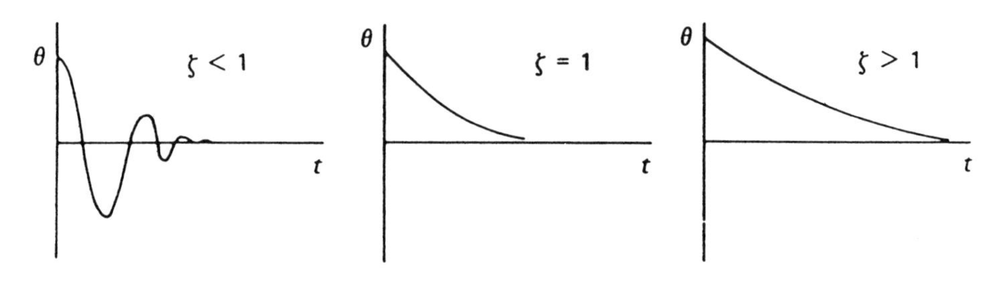

Now

$$I = 20 \times \frac{0.8^2}{3} = 4.27 \text{ kg m}^2,$$

and

$$k = 100 \text{ N m/rad},$$

so that

$$c_c = 2\sqrt{(100 \times 4.27)} = 41.3 \text{ N m s/rad}.$$

Also

$$\omega = \sqrt{\frac{100}{4.27}} = 4.84 \text{ rad/s}.$$

For critical damping,

$$\theta = (A + Bt)\,\mathrm{e}^{-\omega t}.$$

Initial conditions are $t = 0$, $\theta = \pi/2$, and $t = 0$, $\dot{\theta} = 0$. Hence

$$\theta = \frac{\pi}{2}(1 + 4.84t)\,\mathrm{e}^{-4.84t}.$$

Require

$$\theta = 0.01745.$$

Trial and error, or graphical solution gives $\underline{t = 1.3 \text{ s}}$.

14. $\underline{c = 7.9 \text{ N s/m}^3;\ f = 0.789 \text{ Hz}}$.

15. In 2 s execute 3 cycles, so

$$\Lambda = \tfrac{1}{3} \ln \frac{1}{0.9} = \underline{0.035.}$$

For small Λ, $\Lambda \simeq 2\pi\zeta$, hence $\underline{\zeta = 0.0056.}$

Also for small ζ, $\omega \simeq \omega_v$

so

$$c_c = 2\sqrt{(km)} = 2m\omega = 2 \times 10^5 \times 3\pi = \underline{10\ 600\ \text{N m s/rad.}}$$

16.

$$\frac{f_1}{f_2} = \frac{(1/2\pi)\sqrt{(k/m)}}{(1/2\pi)\sqrt{(k/(m+M))}}.$$

Thus

$$\frac{f_1}{f_2} = \sqrt{\frac{m+M}{m}} = \frac{\tau_2}{\tau_1},$$

so

$$\frac{0.62}{0.6} = \sqrt{\left(\frac{m+4000}{m}\right)},$$

and

$$\underline{m = 59\,000\ \text{kg.}}$$

17.

$$I_0 = m_1 h^2 + \frac{m_2 r^2}{4} + m_2 s^2.$$

Equation of motion is

$$I_0\ddot{\theta} + cp^2\dot{\theta} + kq^2\theta = 0.$$

When $c = 0$,

$$\omega = \sqrt{\left(\frac{kq^2}{I_0}\right)}\ \text{rad/s.}$$

With damping $\omega_v = \omega\sqrt{(1-\zeta^2)}$

where

$$\zeta = \frac{c}{c_c} = \frac{cp^2}{2\sqrt{(kq^2 I_0)}}.$$

Thus

$$\underline{\omega_v = \omega\sqrt{\left[1 - \left(\frac{c^2p^4}{4kq^2 I_0}\right)\right]}\ \text{rad/s.}}$$

18. FBDs are

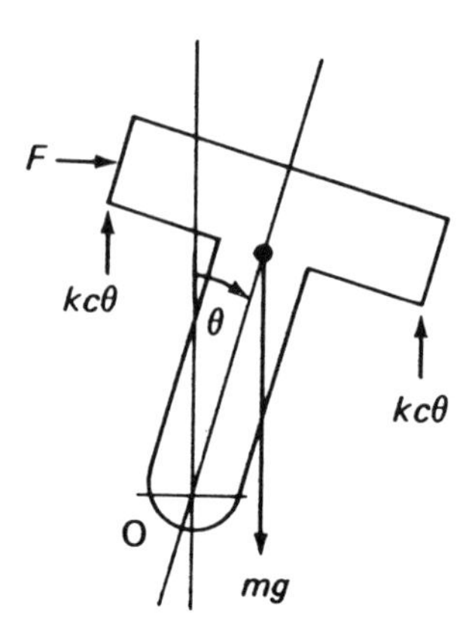

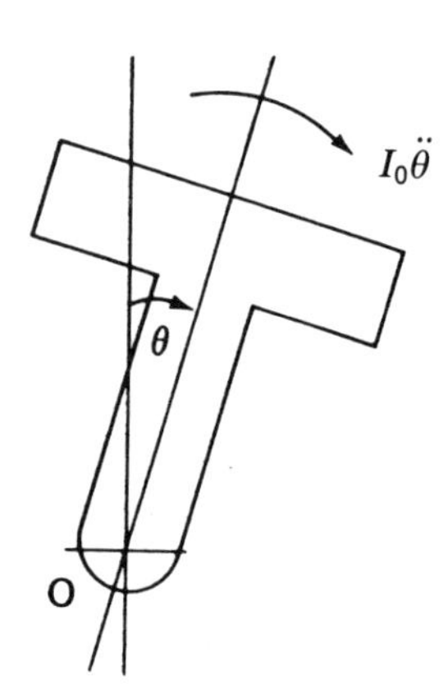

Equation of motion is $Fd + mgh\theta - 2kc^2\theta = I_0\ddot{\theta}$, or

$$\ddot{\theta} + \left(\frac{2kc^2 - mgh}{I_0}\right)\theta = \frac{Fd}{I_0}.$$

C.F. + P.I. give solution as

$$\theta = A \sin \omega t + B \cos \omega t + \frac{Fd}{\omega^2 I_0},$$

where

$$\omega = \sqrt{[(2kc^2 - mgh)/I_0]}.$$

Substitute initial conditions for

$$\underline{\theta = \frac{Fd}{\omega^2 I_0}(1 - \cos \omega t).}$$

19.

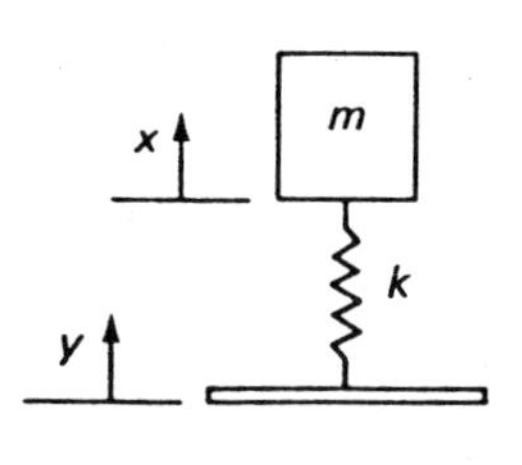

Equation of motion is

$$m\ddot{x} + kx = ky,$$

where $y = \frac{1}{2}at^2$. Hence

$$m\ddot{x} + kx = \frac{k}{2}at^2.$$

Solution is

$$x = A \cos \omega t + B \sin \omega t + \frac{at^2}{2} - \frac{am}{k}.$$

Initially $t = 0$ and $x = 0$ and $\dot{x} = 0$ so that

$$x = \frac{am}{k} (\cos \omega t - 1) + \frac{at^2}{2}.$$

When $t = 0.19$ s,

$$x = \frac{49 \times 90}{10.9 \times 10^3} \left(\cos \sqrt{\left(\frac{10.9 \times 10^3}{90} \right)} 0.19 - 1 \right) + \frac{49 \times 0.19^2}{2}$$

$$= 0.275 \text{ m},$$

and

$$\dot{x} = -\frac{am}{k} \omega \sin \omega t + at$$

$$= 5.45 \text{ m/s}.$$

Now for $t > 0.19$ s, initial velocity is 5.45 m/s, acc is -9.81 m/s^2 and if distance to max. height is s, then

$$0 = 5.54^2 - 2 \times 9.81 \; s$$

so

$$s = 1.51 \text{ m}.$$

Thus max. height of pilot $= 1.51 + 0.275 = \underline{1.785 \text{ m}}.$
Max. value of $\cos \omega t$ is 0.5 when $t = 0.19$ s, so

$$\ddot{x} = -\frac{am\omega^2}{k} \cos \omega t + a = \frac{3a}{2} = \underline{73.5 \text{ m/s}^2}.$$

20. $\underline{x = 5.25 \text{ mm}}$; $\underline{N = 424 \text{ rev/min.}}$

21. Equation of motion is

$$m\ddot{x} + kx = m_r r v^2 \sin vt,$$

so the amplitude of the motion is $m_r r v^2/(k - mv^2)$. Thus

$$k\left(\frac{m_r r v^2}{k - mv^2} \right) = \text{ transmitted force.}$$

Substituting values gives

$$k\left(\frac{0.2 \times 0.01 \times 150^2}{k - (2 \times 150^2)} \right) = 100.$$

Hence $\underline{k = 81.8 \text{ kN/m.}}$

22. Decay factor 15 so $\zeta\omega = 15$ and $\omega_v = \omega\sqrt{(1-\zeta^2)}$ where $\omega_v = 23.2\pi$ rad/s. Hence $\zeta = 0.103$ and $\omega = 145.6$ rad/s.
Since $\omega = \sqrt{(k/m)}$,

$$k = (145.6)^2 \times 365 = 7740 \text{ kN/m},$$

and

$$c_c = 2\sqrt{(km)} = 106\,310 \text{ N s/m}.$$

Thus

$$c = \zeta c_c = 10950 \text{ N s/m}.$$

At 1450 rev/min,

$$v = 1450 \cdot \frac{2\pi}{60} = 152 \text{ rad/s}.$$

Now

$$X = \frac{F}{\sqrt{[(k - mv^2)^2 + (cv)^2]}}$$

$$= \frac{F}{\sqrt{[(7740 \times 10^3 - 365 \times 152^2)^2 + (10950 \times 152)^2]}}.$$

Since $X = 0.76 \times 10^{-3}$m, $\underline{F = 1370 \text{ N.}}$

Without damping

$$X = \frac{F}{k - mv^2}$$

$$= \underline{2.85 \text{ mm.}}$$

23. $\underline{X_c = 0.02 \text{ m}}$; $\underline{X_c \text{ lags } X_p \text{ by } \tan^{-1}(\tfrac{3}{4})}$.

24. If n units are required, then

$$k = 359.10^3.n \text{ N/m},$$

$$c = 2410n \text{ N s/m}$$

$$\omega^2 = 359.10^3 n/520 = 691n \text{ s}^{-2}$$

$$c_c = 2\sqrt{(km)},$$

so

$$c_c^2 = 4 \times 359 \times 520 \times 10^3 n = 747 \times 10^6 n.$$

$$v = 25 \times 2\pi \text{ rad/s}, \quad \text{so} \quad v^2 = 2.46 \times 10^4/\text{s}^2.$$

Hence

$$\frac{F_T}{F_0} = 0.4, \qquad \text{and} \qquad \frac{v^2}{\omega^2} = 35.6/n.$$

Substitute values in

$$\left(\frac{F_T}{F_0}\right)^2 = \left[\frac{1 + \left(2\dfrac{c}{c_0}\dfrac{v}{\omega}\right)^2}{\left(1 - \dfrac{v^2}{\omega^2}\right)^2 + \left(2\dfrac{c}{c_0}\dfrac{v}{\omega}\right)^2}\right]$$

To give

$$n^2 + 6.42n - 114.6 = 0,$$

so

$$n = +8 \qquad \text{or} \qquad -14.5$$

Thus 8 units in parallel will give specified attenuation; more units would give less static deflection but more transmitted force.

25. $$(k - mv^2)^2 + c^2v^2 = 100\,(k^2 + c^2v^2),$$

so

$$99k^2 + 99c^2v^2 + 2kmv^2 - (mv^2)^2 = 0.$$

Now

$$c = 0.2c_c = 0.4\sqrt{(km)},$$

$$v = 154.8 \text{ rad/s},$$

and

$$m = 40 \text{ kg}.$$

Hence

$$k^2 + 17.3 \times 10^4 k - 93 \times 10^8 = 0,$$

and thus

$$\underline{k = 43 \text{ kN/m}.}$$

If

$$c = 0, \qquad \frac{F_T}{F_0} = \underline{0.0469.}$$

26.

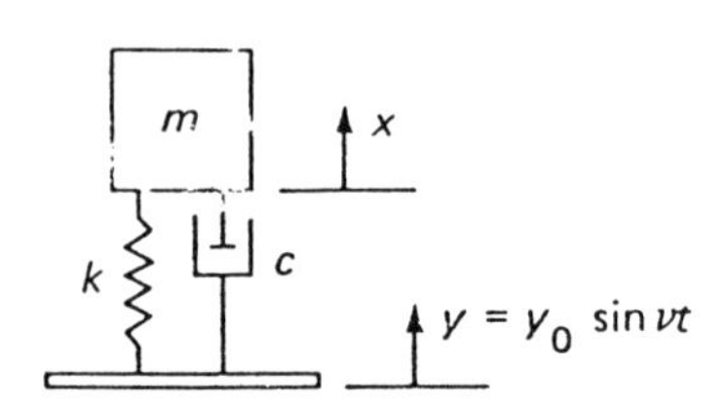

Equation of motion is $m\ddot{x} = k(y - x) + c(\dot{y} - \dot{x})$. If

$$z = y - x, \qquad \dot{z} = \dot{y} - \dot{x}, \qquad \text{and} \qquad \ddot{z} = \ddot{y} - \ddot{x},$$

then

$$m\ddot{z} + c\dot{z} + kz = m\ddot{y} = -m\nu^2 y_0 \sin \nu t.$$

Assume

$$z = z_0 \sin(\nu t + \phi),$$

then

$$-m\nu^2 z_0 \sin(\nu t + \phi) + cz_0\nu \cos(\nu t + \phi) + kz_0 \sin(\nu t + \phi) = -m\nu^2 y_0 \sin \nu t.$$

If

$$\omega = \sqrt{\frac{k}{m}} \qquad \text{and} \qquad \zeta = \frac{c}{c_c} = \frac{c\omega}{2k},$$

then

$$z_0 = \frac{(\nu/\omega)^2 \; y_0}{\sqrt{\left\{\left[1 - \left(\frac{\nu}{\omega}\right)^2\right]^2 + \left[2\zeta \; \frac{\nu}{\omega}\right]^2\right\}}}.$$

Now if $\omega \gg \nu$, $(\nu/\omega)^2 \ll 1$ and

$$z_0 \simeq \left(\frac{\nu}{\omega}\right)^2 y_0,$$

that is, the acceleration $\nu^2 y_0$ is measured.

As (ν/ω) increases $[1 - (\nu/\omega)^2]^2$ decreases, but the damping term in the denominator increases to compensate. If $\nu/\omega = 0.2$ and no error is required,

$$\frac{1}{\left[1-\left(\frac{v}{\omega}\right)^2\right]^2+\left[2\zeta\,\frac{v}{\omega}\right]^2}=1,\qquad\text{when}\qquad \frac{v}{\omega}=0.2,$$

so that

$$[1-0.04]^2+4\zeta^2[0.04]=1,$$

and hence $\underline{\zeta = 0.7.}$

27.
$$T_R=\frac{F_T}{F_0}=\frac{\sqrt{[1+(2\zeta\;v/\omega)^2]}}{\sqrt{\{[1-(v/\omega)^2]^2+[2\zeta\;v/\omega]^2]}}.$$

Here $F_T = 100$ N, $F_0 = 320$ N, $m = 940$ kg, $v = 57.6$ rad/s,

$$\omega=\sqrt{\frac{9.81}{0.015}}=25.6\text{ rad/s}\qquad\text{and}\qquad\left(\frac{v}{\omega}\right)^2=\left(\frac{57.6}{25.6}\right)^2=5.05.$$

Hence

$$\left(\frac{100}{320}\right)^2=\frac{1+20.2\zeta^2}{16.4+20.2\zeta^2},$$

so that $\zeta = 0.184$.
Now

$$k=\frac{940\times 9.81}{0.015}=\underline{615\text{ kN/m},}$$

Also

$$c_c=2\sqrt{(km)}=2\sqrt{615\times 10^3\times 940}$$
$$=48\text{ kN s/m},$$

so that

$$c=\zeta c_c=\underline{8850\text{ N s/m.}}$$

$$v=25.6\sqrt{(1-0.184^2)}=25.1\text{ rad/s.}$$

or $\underline{\text{241 rev/min.}}$

28.
$$\frac{X_T}{X_0}=\frac{\sqrt{[1+(2\zeta\;v/\omega)^2]}}{\sqrt{\{[1-(v/\omega)^2]^2+[2\zeta\;v/\omega]^2\}}}=\frac{1}{10}.$$

Substituting $\zeta = 0.2$ gives

$$\left(\frac{v}{\omega}\right)^4-17.84\left(\frac{v}{\omega}\right)^2-99=0,$$

so that

$$\left(\frac{\nu}{\omega}\right) = 4.72.$$

Limit at 15 Hz, so $\nu = 30\pi$ rad/s and

$$\omega = \frac{30\pi}{4.72} = 19.95 = \sqrt{\frac{k_T}{m}}.$$

Hence

$$k_T = 15.92 \text{ kN/m}$$

and

$$k = \frac{k_T}{3} = \underline{5.3 \text{ kN/m.}}$$

29. $\zeta = 0$ so $\frac{X_T}{X_0} = \frac{1}{1 - (\nu/\omega)^2} = 0.1.$

Hence

$$\frac{\nu}{\omega} = 3.32.$$

Since

$$\nu = 80\pi \text{ rad/s}, \qquad \omega = \frac{80\pi}{3.32} = \sqrt{\frac{g}{\delta}},$$

so that

$$\delta = \left(\frac{3.32}{80\pi}\right)^2 \times 9.81 = \underline{1.7 \text{ mm.}}$$

30. $\omega = \sqrt{\frac{k}{I}} = \sqrt{\frac{0.01158}{2.10^{-9}}} = 2400 \text{ rad/s},$

$$c_c = 2\sqrt{(kI)} = 9.84 \times 10^{-6} \text{ N m s/rad.}$$

Thus

$$\zeta^2 = \left(\frac{5.29}{9.84}\right)^2 = 0.29,$$

and

$$\omega_v = 2400\sqrt{(1 - 0.29)} = 2020 \text{ rad/s}$$

so

$$\text{frequency} = \underline{322 \text{ Hz.}}$$

For torsional system,

$$I\ddot{\theta} + c\dot{\theta} + k\theta = T \cos \Omega t,$$

so that

$$\theta = \frac{T/k \ \cos(\Omega t + \phi)}{\sqrt{\{[1 - \Omega^2/\omega^2]^2 + [2\zeta \ \Omega/\omega]^2\}}}.$$

Expect

$$\theta = \frac{T}{k},$$

so denominator gives amplitude error. Phase error is

$$\tan^{-1}\left[\frac{2\zeta(\Omega/\omega)}{1 - (\Omega/\omega)^2}\right].$$

Now

$$\frac{\Omega}{\omega} = \frac{100}{382}, \qquad \text{so} \quad \left(\frac{\Omega}{\omega}\right)^2 = 0.0686.$$

Hence

$$\begin{aligned}\text{denominator} &= \sqrt{[(1 - 0.0686)^2 + 4 \times 0.29 \times 0.0686]} \\ &= 0.975,\end{aligned}$$

so that amplitude error $= 2\frac{1}{2}\%$.
Phase error,

$$\begin{aligned}\phi &= \tan^{-1}\frac{2 \times 0.539 \times 100/382}{0.9314}, \\ &= \underline{17^\circ}.\end{aligned}$$

32. $$\underline{F = \left((M + m)g - \left[\frac{MK}{K - M(2\pi v/L)^2} + m\right] h(2\pi v/L)^2 \sin(2\pi v/L)t)\right).}$$

33. $x = \underline{0.056 \text{ m}}$; $\underline{\phi = 3.7^\circ}$.

35. Equation of motion is

$$m\ddot{x} + c\dot{x} + kx = F \sin vt,$$

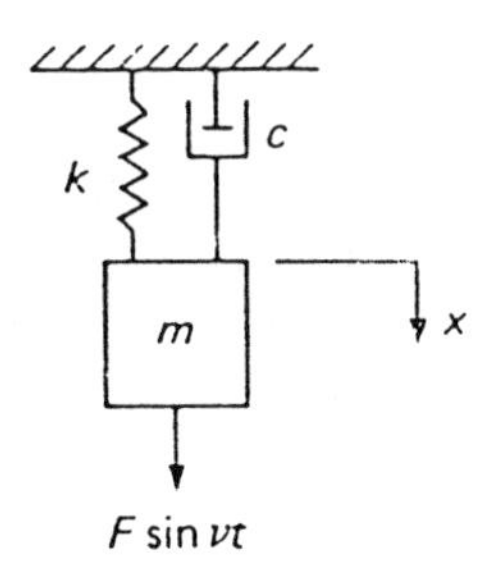

thus

$$x = \frac{F \ \sin(vt - \phi)}{\sqrt{[(k - mv^2)^2 + (cv)^2]}}.$$

Force transmitted to support,

$$\begin{aligned}F_T &= kx + c\dot{x} \\ &= \sqrt{[k^2 + (cv)^2)]} \ X.\end{aligned}$$

Hence

$$\frac{F_{\mathrm{T}}}{F} = \frac{\sqrt{\left[1 + \left(2\zeta \dfrac{\nu}{\omega}\right)^2\right]}}{\sqrt{\left[\left[1 - \left(\dfrac{\nu}{\omega}\right)^2\right]^2 + \left[2\zeta \dfrac{\nu}{\omega}\right]^2\right]}}.$$

Now,

$$\frac{F_{\mathrm{T}}}{F} = \frac{1}{10} \quad \text{and} \quad \zeta = \tfrac{1}{3},$$

thus

$$\left[1 - \left(\frac{\nu}{\omega}\right)^2\right]^2 + \left[2\zeta \frac{\nu}{\omega}\right]^2 = 100 + 100\left[2\zeta \frac{\nu}{\omega}\right]^2,$$

and

$$\left(\frac{\nu}{\omega}\right)^4 - 46\left(\frac{\nu}{\omega}\right)^2 - 99 = 0.$$

Hence

$$\nu = 6.93\sqrt{\frac{k}{m}} \text{ rad/s},$$

so

$$k = \frac{\nu^2 m}{6.93^2} = \frac{(2\pi \times 50)^2 \times 7}{6.93^2}$$
$$= \underline{14.38 \text{ kN/m.}}$$

Also

$$c_{\mathrm{c}} = 2\sqrt{(km)} = 2\sqrt{(14.38 \times 10^3 \times 7)} = 634.5 \text{ N/m/s.}$$

$$c = \frac{634.5}{3} = \underline{211.5 \text{ N s/m.}}$$

36.

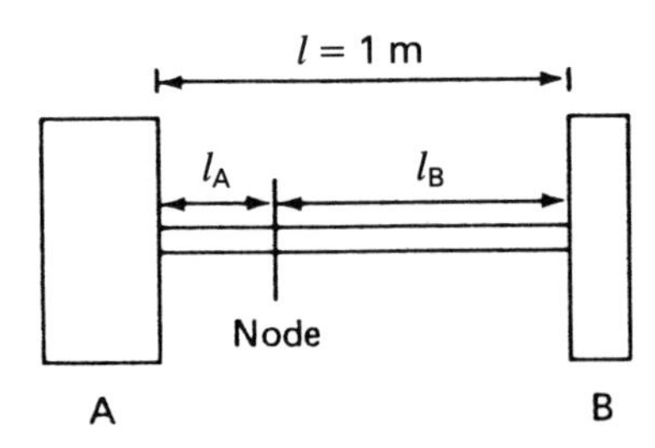

where

$$I_{\mathrm{A}} = 12 \text{ kg m}^2, \qquad I_{\mathrm{B}} = 5 \text{ kg m}^2$$

and

$$k = 600 \times 10^3 \text{ N m/rad.}$$

Now

$$f = \frac{1}{2\pi}\sqrt{\frac{k_A}{I_A}} = \frac{1}{2\pi}\sqrt{\frac{k_B}{I_B}} \text{ Hz,}$$

so $k_A I_B = k_B I_A$.

Also

$$\frac{1}{k_A} + \frac{1}{k_B} = \frac{1}{k}, \qquad \text{hence} \quad k = \frac{k_A k_B}{k_A + k_B},$$

and

$$k = \frac{k_A \times k_A \left(\dfrac{I_B}{I_A}\right)}{k_A + k_A \left(\dfrac{I_B}{I_A}\right)},$$

so that

$$k_A = \left(\frac{I_A + I_B}{I_B}\right) k,$$

and

$$f = \frac{1}{2\pi}\sqrt{\left[\left(\frac{I_A + I_B}{I_A I_B}\right) k\right]} \text{ Hz,}$$

$$= \frac{1}{2\pi}\sqrt{\left(\frac{17}{60} \times 600 \times 10^3\right)} = \underline{65.6 \text{ Hz.}}$$

Also,

$$l_A = \left(\frac{I_B}{I_A + I_B}\right) l = \frac{5}{17} = \underline{0.29 \text{ m,}}$$

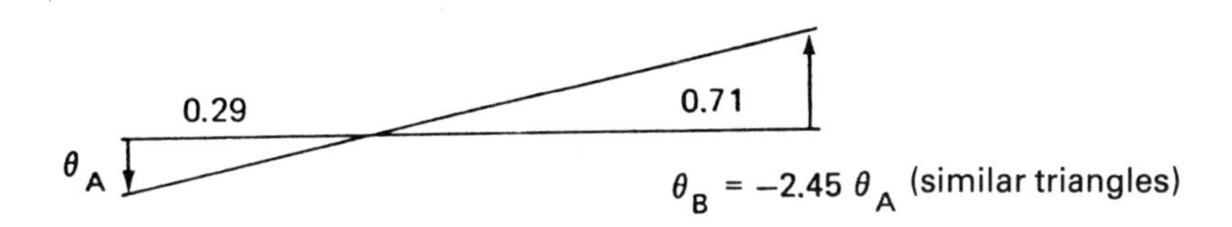

37.

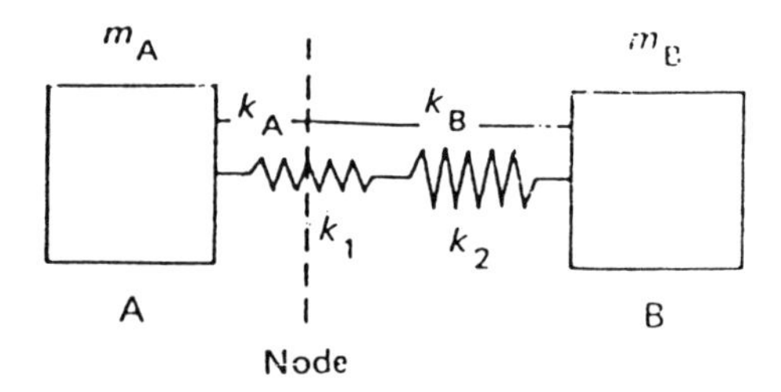

Frequency of oscillation $= \dfrac{1}{2\pi}\sqrt{\dfrac{k_A}{m_A}} = \dfrac{1}{2\pi}\sqrt{\dfrac{k_B}{m_B}}$, so

$$k_A m_B = k_B m_A,$$

or

$$k_A \times 500 = k_B \times 1000, \qquad \text{so that} \quad k_A = 2k_B.$$

Springs connected in series, so

$$\frac{1}{k} = \frac{1}{k_1} + \frac{1}{k_2} = \frac{1}{200} + \frac{1}{50},$$

so

$$k = 40 \text{ kN/m},$$

and

$$\frac{1}{k_A} + \frac{1}{k_B} = \frac{1}{40 \times 10^3}.$$

Since

$$k_A = 2k_B, \quad k_A = 120 \times 10^3 \text{ N/m}.$$

Thus

$$f = \frac{1}{2\pi}\sqrt{\frac{120 \times 10^3}{10^3}} = \underline{1.74 \text{ Hz.}}$$

38. Equations of motion are, for free vibration,

$$(k_1 - m_1\omega^2)\,X_1 + (-k_1)\,X_2 = 0,$$

and

$$(-k_1)\,X_1 + (k_1 + k_2 - m_2\omega^2)\,X_2 = 0.$$

Hence frequency equation is

$$(k_1 - m_1\omega^2)(k_1 + k_2 - m_2\omega^2) - (-k_1)^2 = 0,$$

or

$$\omega^4 (m_1 m_2) - \omega^2 (m_1 k_1 + m_1 k_2 + m_2 k_1) + k_1 k_2 = 0.$$

Now

$$m_1 = \tfrac{1}{2} m_2 \quad \text{and} \quad k_1 = \tfrac{1}{2} k_2,$$

Thus

$$2m_1^2 \omega^4 - 5 m_1 k_1 \omega^2 + 2k_1^2 = 0,$$

or

$$(2m_1 \omega^2 - k_1)(m_1 \omega^2 - 2k_1) = 0.$$

That is

$$\omega^2 = \frac{k_1}{2m_1} \quad \text{or} \quad \frac{2k_1}{m_1},$$

so that

$$\underline{f_1 = \frac{1}{2\pi}\sqrt{\frac{k_1}{2m_1}}\ \text{Hz}} \quad \text{and} \quad \underline{f_2 = \frac{1}{2\pi}\sqrt{\frac{2k_1}{m_1}}\ \text{Hz.}}$$

Now

$$\frac{X_1}{X_2} = \frac{k_1}{k_1 - m_1 \omega^2};$$

that is, at frequency f_1,

$$\frac{X_1}{X_2} = +0.5.$$

and at frequency f_2,

$$\frac{X_1}{X_2} = -1.0.$$

With harmonic force $F_1 \sin \nu t$ applied,

$$(k_1 - m_1 \omega^2) X_1 + (-k_1) X_2 = F_1,$$

and

$$(-k_1) X_1 + (k_1 + k_2 - m_2 \omega^2) X_2 = 0.$$

Hence

$$X_1 = \frac{(k_1 + k_2 - m_2 \omega^2)}{\Delta} F,$$

and

$$X_2 = \frac{k_1}{\Delta} F,$$

where

$$\Delta = m_1 m_2 \omega^4 - \omega^2(m_1 k_1 + m_1 k_2 + m_2 k_1) + k_1 k_2 .$$

39. FBD:

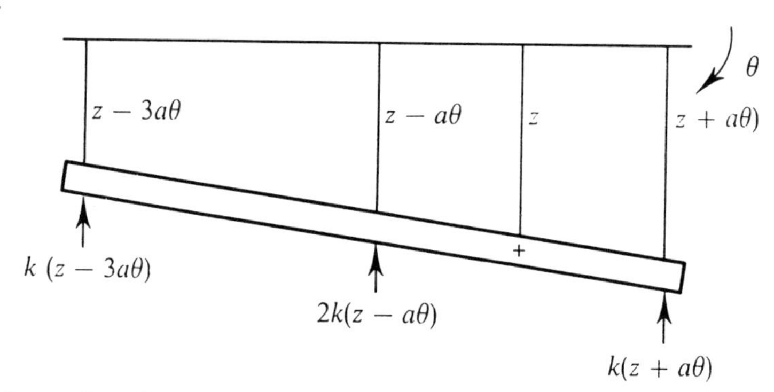

Equations of motion are

$$-k(z - 3a\theta) - 2k(z - a\theta) - k(z + a\theta) = m\ddot{z},$$

and

$$k\,(z - 3a\theta)3a + 2k(z - a\theta)a - k(z + a\theta)a = I_G\ddot{\theta}.$$

Substitute

$$I = 2ma^2, \quad z = A \sin \omega t \quad \text{and} \quad \theta = B \sin \omega t,$$

to give

$$(m\omega^2 - 4k)A + 4kaB = 0,$$

and

$$4kaA + (2ma^2\omega^2 - 12ka^2)B = 0,$$

so that the frequency equation is

$$(m\omega^2 - 4k)(2ma^2\omega^2 - 12ka^2) - 16k^2a^2 = 0.$$

Multiply out and factorize to give

$$\underline{f_1 = \frac{1}{2\pi}\sqrt{\frac{2k}{m}}\ \text{Hz}} \quad \text{and} \quad \underline{f_2 = \frac{1}{2\pi}\sqrt{\frac{8k}{m}}\ \text{Hz.}}$$

40. $$\underline{\omega^4 - \omega^2(\omega_B^2 + \omega_p^2) + \omega_B^2\omega_p^2 - (k_2 b - k_1 a)^2/m^2 K_0^2 = 0.}$$

41. System is

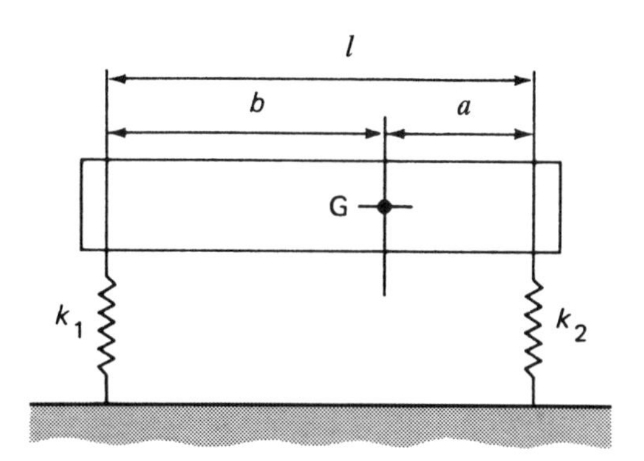

where

$$m = 2000 \text{ kg};\ l = 3 \text{ m};\ a = 1 \text{ m};\ b = 2 \text{ m};\ I_G = 500 \text{ kg m}^2;$$

$$k_2 = 50 \times 10^3 \text{ N/m};\ k_1 = 80 \times 10^3 \text{ N/m}.$$

FBDs:

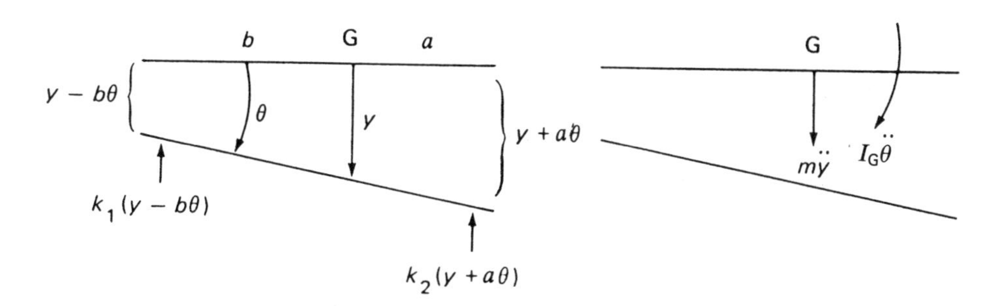

Equations of motion are

$$m\ddot{y} = -k_1(y - b\theta) - k_2(y + a\theta),$$

and

$$I_G\ddot{\theta} = k_1(y - b\theta)b - k_2(y + a\theta)a.$$

Substitute

$$y = Y \sin \omega t \qquad \text{and} \qquad \theta = \Theta \sin \omega t$$

and rearrange:

$$(k_1 + k_2 - m\omega^2)Y + (k_2 a - k_1 b)\Theta = 0,$$

and

$$(k_2a - k_1b)Y + (k_1b^2 + k_2a^2 - I_G\omega^2)\Theta = 0.$$

Hence the frequency equation is

$$(k_1 + k_2 - m\omega^2)(k_1b^2 + k_2a^2 - I_G\omega^2) - (k_2a - k_1b)^2 = 0.$$

Substituting numerical values and dividing by 10^3 gives

$$(130 - 2\omega^2)(370 - 0.5\omega^2) - (-110)^2 = 0,$$

or

$$\omega^4 - 805\omega^2 + 36\,000 = 0.$$

Hence

$$\omega^2 = \frac{805 \pm 710}{2} = 47.5 \quad \text{or} \quad 758\ (\text{rad/s})^2,$$

so that

$$\underline{f_1 = 1.09\ \text{Hz}} \qquad \text{and} \qquad \underline{f_2 = 4.38\ \text{Hz.}}$$

The mode shape is obtained from

$$\frac{Y}{\Theta} = \frac{k_1b - k_2a}{k_1 + k_2 - m\omega^2},$$

so that at f_1, $\omega^2 = 47.5$ and

$$\frac{Y}{\Theta} = \frac{160 \times 10^3 - 50 \times 10^3}{130 \times 10^3 - 2 \times 10^3 \times 47.5} = \underline{3.14,}$$

and at f_2, $\omega^2 = 758$ and

$$\frac{Y}{\Theta} = -\frac{110}{1386} = \underline{-0.079.}$$

Speeds are $\underline{V_P = 78\ \text{km/h}}$ and $\underline{V_T = 19.8\ \text{km/h.}}$

42. FBDs:

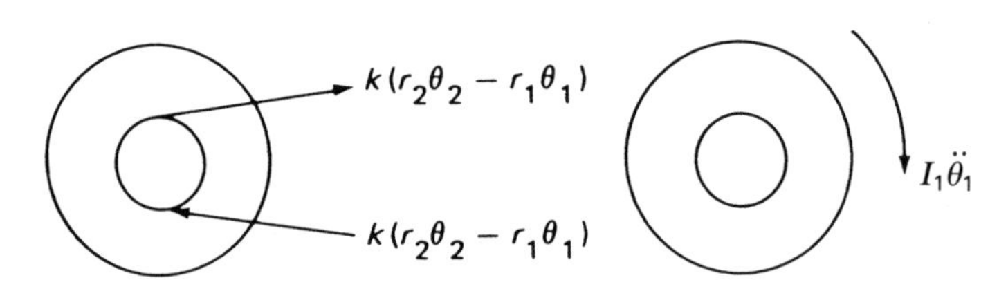

Equations of motion are

$$I_1\ddot{\theta}_1 = 2kr_1(r_2\theta_2 - r_1\theta_1),$$

and

$$I_2\ddot{\theta}_2 = -2kr_2(r_2\theta_2 - r_1\theta_1).$$

Assume

$$\theta_1 = A_1 \sin \omega t \quad \text{and} \quad \theta_2 = A_2 \sin \omega t.$$

Thus

$$A_1[2kr_1^2 - I_1\omega^2] + A_2[-2kr_1r_2] = 0,$$

and

$$A_1[-2kr_1r_2] + A_2[2kr_2^2 - I_2\omega^2] = 0.$$

Frequency equation is therefore

$$(2kr_1^2 - I_1\omega^2)(2kr_2^2 - I_2\omega^2) - (2kr_1r_2)^2 = 0,$$

or

$$I_1I_2\omega^4 - 2k(r_1^2I_2 + r_2^2I_1)\omega^2 = 0,$$

so

$$\omega = 0 \qquad \text{(rotation)},$$

or

$$\omega^2 = \frac{2k(r_1^2I_2 + r_2^2I_1)}{I_1I_2},$$

so that

$$f = \frac{1}{2\pi}\sqrt{\frac{2k(r_1^2I_2 + r_2^2I_1)}{I_1I_2}}\ \text{Hz.}$$

43. FBDs are as below:

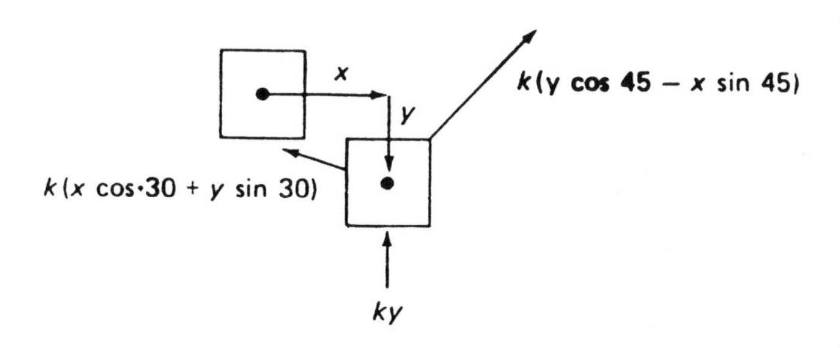

That is,

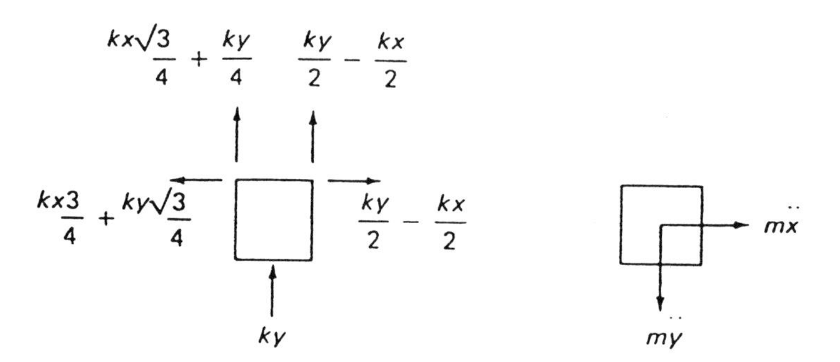

Equations of motion are

$$(\Sigma F_x) \qquad m\ddot{x} = \frac{ky}{2} - \frac{kx}{2} - kx\,\frac{3}{4} - ky\,\frac{\sqrt{3}}{4},$$

and

$$(\Sigma F_y) \qquad m\ddot{y} = -ky - \frac{ky}{2} + \frac{kx}{2} - kx\,\frac{\sqrt{3}}{4} - \frac{ky}{4}.$$

Assuming a solution of the form $x = X \sin \omega t$, $y = Y \sin \omega t$, these are

$$-m\omega^2 X + \frac{5k}{4}\,X + \left(\frac{\sqrt{3}-2}{4}\right) kY = 0,$$

and

$$\left(\frac{\sqrt{3}-2}{4}\right) kX - m\omega^2 Y + \frac{7}{4}\,kY = 0.$$

Hence

$$\left(\frac{5}{4}\,k - m\omega^2\right)\left(\frac{7}{4}\,k - m\omega^2\right) - \left(\left(\frac{\sqrt{3}-2}{4}\right)k\right)^2 = 0,$$

and

$$m^2\omega^4 - 3mk\omega^2 + \frac{k^2}{16}\,(28 + 4\sqrt{3}) = 0.$$

Hence

$$\frac{\omega^4}{\Omega^4} - 3\,\frac{\omega^2}{\Omega^2} + 2.183 = 0,$$

where

$$\Omega = \sqrt{\frac{k}{m}},$$

and

$$\frac{\omega}{\Omega} = 1.326 \quad \text{or} \quad 1.114.$$

Thus frequencies of vibration are therefore

$$\underline{\frac{1.11}{2\pi}\sqrt{\frac{k}{m}}\ \text{Hz}} \quad \text{and} \quad \underline{\frac{1.33}{2\pi}\sqrt{\frac{k}{m}}\ \text{Hz.}}$$

45. FBDs are

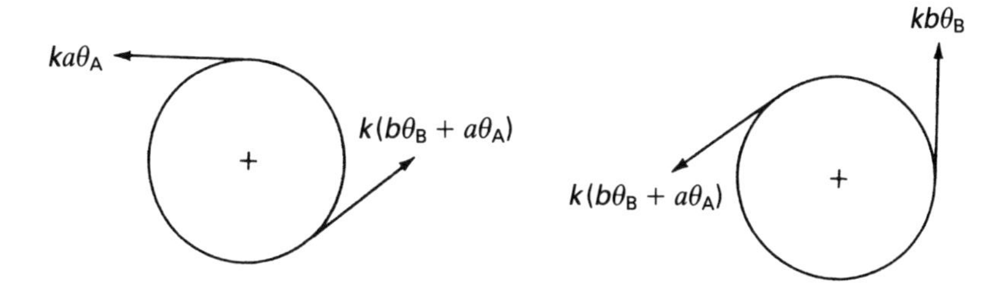

The equations of motion are

$$-ka^2\theta_A - kab\theta_B - ka^2\theta_A = I_A\ddot{\theta}_A,$$

and

$$-kab\theta_A - kb^2\theta_B - kb^2\theta_B = I_B\ddot{\theta}_B.$$

Substitute $\theta_A = A \sin \omega t$ and $\theta_B = B \sin \omega t$ to give

$$(-I_A\omega^2 + 2ka^2)A + kab\ B = 0,$$

and

$$kab\ A + (-I_A\omega^2 + 2kb^2)B = 0,$$

So the frequency equation is

$$I_A I_B\omega^4 - 2k(I_B a^2 + I_A b^2)\omega^2 + 3k^2a^2b^2 = 0.$$

Substitute numerical values to give

$$\omega = \underline{19.9 \text{ rad/s}} \quad \text{or} \quad \underline{35.7 \text{ rad/s.}}$$

At 19.9 rad/s, $A/B = -1.65$,
and at 35.7 rad/s $A/B = +3.68$.

46. From section 3.1.5,

$$\left(\frac{\Omega_1}{\omega}\right)^2 + \left(\frac{\Omega_2}{\omega}\right)^2 = 2 + \mu,$$

and $\Omega_1\Omega_2 = \omega^2$.
If $\Omega_1 = 250$, $\Omega_2 = 300^2/250 = 360$, and if $\Omega_2 = 350$, $\Omega_1 = 300^2/350 = 257$. Therefore require $\Omega_1 = 250$ and $\Omega_2 = 360$ to satisfy the frequency range criterion. (Ω_1 and Ω_2 are rev/min). Hence

$$\left(\frac{250}{300}\right)^2 + \left(\frac{360}{300}\right)^2 = 2 + \mu$$

and $\mu = 0.136$.
Hence

absorber mass = $\underline{367\ \text{kg}}$.

and

stiffness = $\underline{144.6 \times 10^6\ \text{N/m}}$.

47. Substitute numerical values into frequency equation to give $\underline{m = 9.8\ \text{kg}}$.
If $\Omega_1 = 85$, $\mu = 0.5$ so absorber mass = $\underline{4.9\ \text{kg}}$, and $k = \underline{773\ \text{N/m}}$.

48.
$$\left(\frac{\Omega}{\omega}\right)^2 = \frac{2+\mu}{2} \pm \sqrt{\left(\frac{\mu^2 + 4\mu}{4}\right)}.$$

If $\Omega_1 = 0.9\ \omega$, this gives $\mu = 0.0446$, and if $\Omega_2 = 1.1\ \omega$, $\mu = 0.0365$.

Limit therefore $\mu = 0.0446$ and absorber mass is $\underline{134\ \text{kg}}$ with stiffness $\underline{30.1\ \text{kN/m}}$.

49.

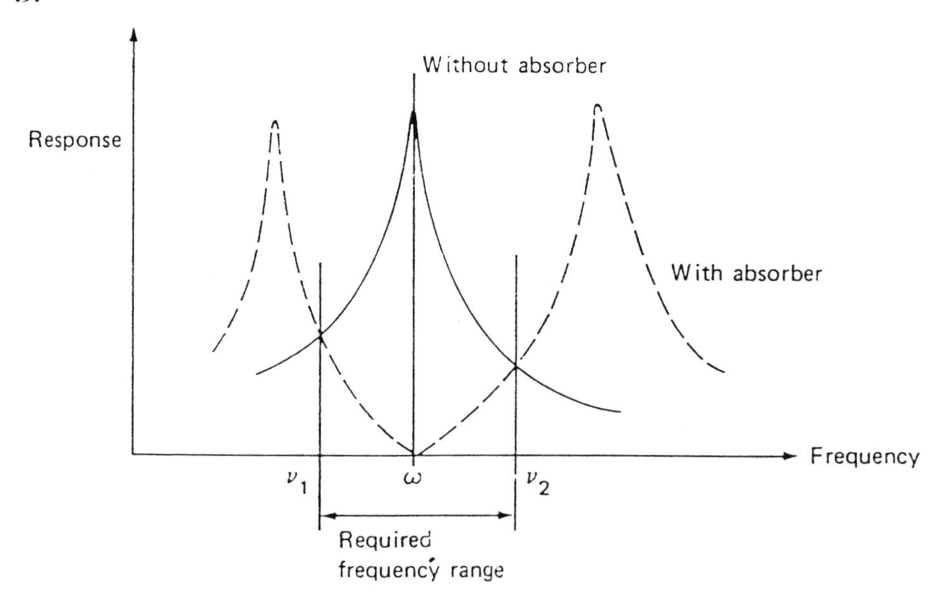

Require

$$\frac{F(K - mv^2)}{[(K + k) - Mv^2][k - m\omega^2] - k^2} = -\frac{F}{K - Mv^2}.$$ (Phase requires –ve sign)

Multiplying out and putting

$$\mu = \frac{m}{M} = 0.2$$

gives

$$2\left(\frac{v}{\omega}\right)^4 - \left(\frac{v}{\omega}\right)^2(4 + \mu) + 2 = 0,$$

so

$$\left(\frac{v}{\omega}\right)^2 = \frac{4 + \mu}{4} \pm \tfrac{1}{4}\sqrt{(\mu^2 + 8\mu)} = 1.05 \pm 0.32.$$

Thus

$$\left(\frac{v}{\omega}\right) = 1.17 \qquad \text{or} \qquad 0.855,$$

so

$$f_1 = 102 \text{ Hz} \qquad \text{and} \qquad \zeta_2 = 140 \text{ Hz}.$$

Frequency range is therefore 102 – 140 Hz.

50. $$\left(\frac{\Omega_1}{\omega}\right)^2 + \left(\frac{\Omega_2}{\omega}\right)^2 = 2 + \mu,$$

and $\Omega_1\Omega_2 = \omega_1^2$ (from section 3.1.5).

Now $\omega = 152$ Hz, $\Omega_1 = 140$ Hz so $\Omega_2 = 152^2/140 = 165$ Hz; hence

$$\left(\frac{140}{152}\right)^2 + \left(\frac{165}{152}\right)^2 = 2 + \mu,$$

and

$$\mu = 0.0266.$$

Require $\omega = 152$ Hz, $\Omega_1 = 120$ Hz so $\Omega_2 = 192$ Hz (which meets frequency range criterion). Hence

$$\left(\frac{120}{152}\right)^2 + \left(\frac{192}{152}\right)^2 = 2 + \mu^1$$

so

$$\mu^1 = 0.219.$$

Therefore require $0.219/0.0266 = 8.2$, that is 9 absorbers.

51. Cantilever absorber:

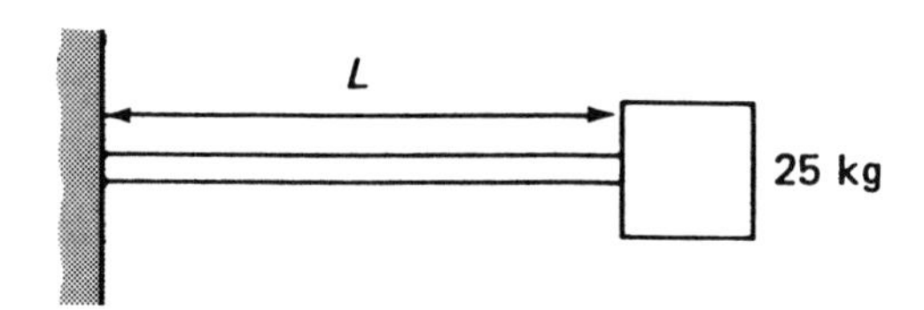

$$\text{Beam stiffness at free end} = \frac{3EI}{L^3} = k.$$

Thus

$$k = \frac{3 \times 70 \times 10^9 \times (0.06)^4}{L^3 \times 12}.$$

Design based on 40 Hz frequency so

$$k = (2\pi \times 40)^2 \times 25.$$

Hence

$$\underline{L = 0.524 \text{ m.}}$$

When $f = 50$ Hz, calculation gives $\underline{L = 0.452 \text{ m.}}$

52. From section 3.1.5,

$$\Omega_1\Omega_2 = \omega^2$$

and

$$\left(\frac{\Omega_1}{\omega}\right)^2 + \left(\frac{\Omega_2}{\omega}\right)^2 = 2 + \mu.$$

Thus if $\Omega_2 = 1.2\ \omega$, $\Omega_1 = 0.833\ \omega$, which is not within 20% of ω $(=\nu)$.
So require $\Omega_1 = 0.8\ \omega$ and then $\Omega_2 = 1.25\ \omega$. Hence

$$0.8^2 + 1.25^2 = 2 + \mu,$$

so that $\mu = 0.2$, and

$$\text{mass of absorber} = \underline{0.2 \text{ M.}}$$

53.

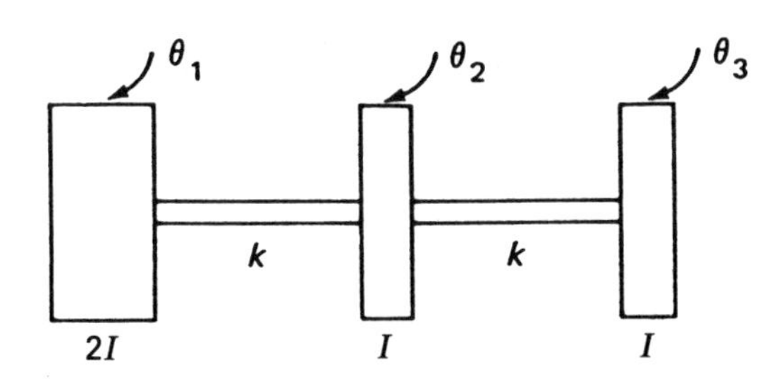

Equations of motion are

$$k(\theta_2 - \theta_1) = 2I\ddot{\theta}_1,$$

$$k(\theta_1 - \theta_2) + k(\theta_3 - \theta_2) = I\ddot{\theta}_2,$$

and

$$k(\theta_2 - \theta_3) = I\ddot{\theta}_3.$$

Substitute $\theta_1 = A \sin \omega t$, $\theta_2 = B \sin \omega t$ and $\theta_3 = C \sin \omega t$ and eliminate A, B and C to give frequency equation as

$$2\omega^4 - 7\frac{k}{I}\omega^2 + 4\frac{k^2}{I^2} = 0.$$

Hence

$$\omega^2 = 0.72\frac{k}{I} \quad \text{or} \quad 2.78\frac{k}{I}.$$

Now

$$k = \frac{\text{GJ}}{l} = \frac{83 \times 10^9 \times \pi \times 0.05^4}{0.65 \times 32} = 7.83 \times 10^4 \text{ N m/rad},$$

and

$$I = \frac{\pi \times 0.28^4}{32} \times 0.023 \times 7000 = 97.1 \times 10^{-3} \text{ kg m}^2.$$

Hence

$$\underline{f_1 = 121.6 \text{ Hz}} \quad \text{and} \quad \underline{f_2 = 239 \text{ Hz.}}$$

where $f = \omega/2\pi$.

54. Consider half of aircraft:

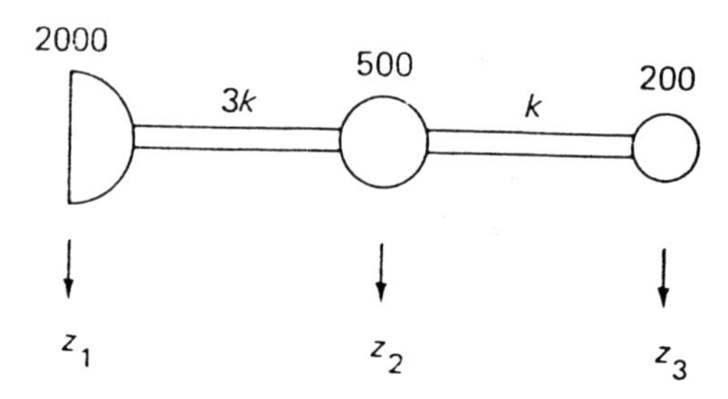

Equations of motion are

$$3k(z_2 - z_1) = 2000\,\ddot{z}_1,$$

$$3k(z_1 - z_2) + k(z_3 - z_2) = 500\,\ddot{z}_2,$$

and

$$k(z_2 - z_3) = 200\,\ddot{z}_3.$$

Substitute

$$z_1 = A_1 \sin \omega t, \qquad z_2 = A_2 \sin \omega t \qquad \text{and} \qquad z_3 = A_3 \sin \omega t,$$

and eliminate A_1, A_2, A_3 to give frequency equation as

$$2 \times 10^4 \omega^4 - 290\,\omega^2 k + 0.81\,k^2 = 0.$$

Hence

$$\omega^2 = 379 \quad \text{or} \quad 1074,$$

and

$$\underline{f_1 = 3.09 \text{ Hz}} \quad \text{and} \quad \underline{f_2 = 5.22 \text{ Hz.}}$$

55. FBDs

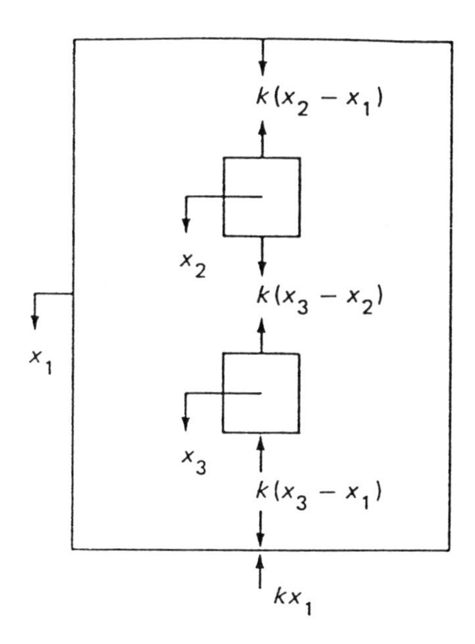

Equations of motion are

$$m\ddot{x}_1 = k(x_2 - x_1) + k(x_3 - x_1) - k_1 x_1,$$

$$m\ddot{x}_2 = k(x_3 - x_2) - k(x_2 - x_1),$$

and

$$m\ddot{x}_3 = -k(x_3 - x_2) - k(x_3 - x_1).$$

Substitute $x_i = X_i \sin \omega t$, so that

$$(3k - m\omega^2)\, X_1 - kX_2 - kX_3 = 0,$$

$$-kX_1 + (2k - m\omega^2)\, X_2 - kX_3 = 0,$$

and

$$-kX_1 - kX_2 + (2k - m\omega^2)\, X_3 = 0.$$

Hence frequency equation is

$$(3k - m\omega^2)^2(k - m\omega^2) - 2k^2(3k - m\omega^2) = 0.$$

and either $3k - m\omega^2 = 0$ so that $\omega = \sqrt{3}\sqrt{(k/m)}$, or

$$k^2 - 4mk\omega^2 + m^2\omega^4 = 0.$$

Thus

$$\omega = \underline{0.52\sqrt{(k/m)}}, \qquad \underline{1.73\sqrt{(k/m)}} \qquad \text{or} \qquad \underline{1.93\sqrt{(k/m)}}\ \text{rad/s}$$

which are all well clear of $\sqrt{(k/m)}$, so system suitable.

56. Model system as follows:

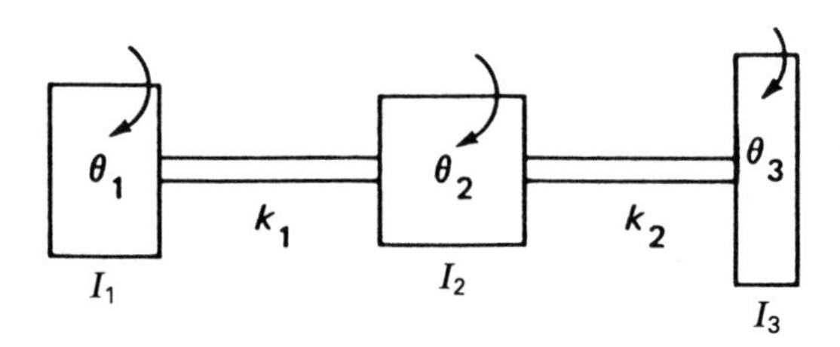

Equations of motion are

$$I_1\ddot{\theta}_1 = -k_1(\theta_1 - \theta_2),$$

$$I_2\ddot{\theta}_2 = k_1(\theta_1 - \theta_2) - k_2(\theta_2 - \theta_3),$$

and

$$I_3\ddot{\theta}_3 = k_2(\theta_2 - \theta_3).$$

Substituting $\theta_i = \Theta_i \sin \omega t$ gives

$$\Theta_1[k_1 - I_2\omega^2] + \Theta_2[-k_1] = 0,$$

$$\Theta_1[-k_1] + \Theta_2[k_1 + k_2 - I_2\omega^2] + \Theta_3[-k_2] = 0,$$

and

$$\Theta_2[-k_2] + \Theta_3[k_2 - I_3\omega^2] = 0.$$

The frequency equation is therefore

$$(k_1 - I_1\omega^2)[(k_1 + k_2 - I_2\omega^2)(k_2 - I_3\omega^2) - k_2^2] + k_1[(k_2 - I_3\omega^2)(-k_1)] = 0.$$

That is

$$\omega^2[I_1I_2I_3\omega^4 - \omega^2(k_1I_2I_3 + k_1I_1I_3 + k_2I_1I_3 + k_2I_1I_2) + k_1k_2(I_1 + I_2 + I_3)] = 0$$

so that either $\omega = 0$ (rigid body rotation) or $[\ldots] = 0$.
Substituting numerical values gives

$$0.48\ \omega^4 - 1100 \times 10^3\ \omega^2 + 149 \times 10^9 = 0,$$

so that

$$\omega_1 = 380\ \text{rad/s} \qquad \text{and} \qquad \omega_2 = 1460\ \text{rad/s}.$$

That is

$$\underline{f_1 = 60.5 \text{ Hz}} \quad \text{and} \quad \underline{f_2 = 232 \text{ Hz}.}$$

At $f_1 = 60.5$ Hz,

$$\frac{\theta_1}{\theta_2} = \frac{k_1}{k_1 - I_1\omega^2} = +1.4,$$

and

$$\frac{\theta_3}{\theta_2} = \frac{k_2}{k_2 - I_3\omega^2} = -0.697.$$

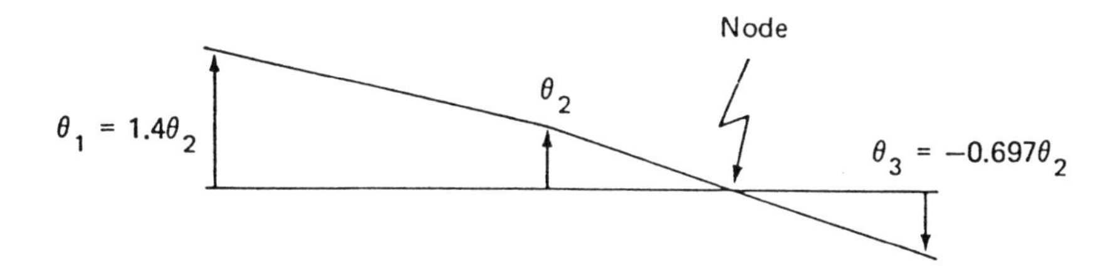

and at $f_2 = 232$ Hz,

$$\frac{\theta_1}{\theta_2} = -0.304,$$

and

$$\frac{\theta_3}{\theta_2} = -0.028.$$

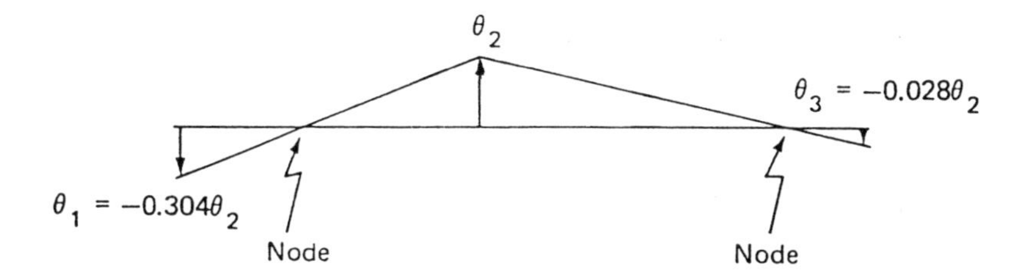

57.
$$T = \tfrac{1}{2}m_1\dot{x}_1^2 + \tfrac{1}{2}m_2\dot{x}^2 + \tfrac{1}{2}m\left(\frac{\dot{x}_1}{2}\right)^2 2 + \tfrac{1}{2}m\left(\frac{\dot{x}_2}{2}\right)^2 + \tfrac{1}{2}J\left(\frac{\dot{x}_1}{d}\right)^2 2 + \tfrac{1}{2}J\left(\frac{\dot{x}_2}{d}\right)^2$$

and

$$V = \tfrac{1}{2}kx_1^2 + \tfrac{1}{2}\cdot 4k\left(\frac{x_1}{2}\right)^2 + \tfrac{1}{2}k(x_1 - x_2)^2.$$

Apply the Lagrange equation

$$\frac{\mathrm{d}}{\mathrm{d}t}\left(\frac{\partial T}{\partial \dot{x}_1}\right) + \frac{\partial V}{\partial x_1} = 0:$$

$$\frac{\mathrm{d}}{\mathrm{d}t}\left(\frac{\partial T}{\partial \dot{x}_1}\right) = m_1\ddot{x}_1 + \frac{m}{2}\ddot{x}_1 + \frac{2J}{d^2}\ddot{x}_1,$$

and

$$\frac{\partial V}{\partial x_1} = kx_1 + kx_1 + \tfrac{1}{2}k(2x_1 - x_2) = 3kx_1 - kx_2.$$

Hence equation of motion is

$$\ddot{x}_1\left(m_1 + \frac{m}{2} + \frac{2J}{d^2}\right) + 3kx_1 - kx_2 = 0.$$

Similarly other equation of motion is

$$\ddot{x}_2\left(m_2 + \frac{m}{4} + \frac{2J}{d^2}\right) + k(x_2 - x_1) = 0.$$

If $m_1 = 4m$, $m_2 = 2m$ and $J = md^2/8$, equations become

$$\ddot{x}_1(4m + \frac{m}{2} + \frac{m}{4}\Big) + 3kx_1 - kx_2 = 0,$$

and

$$\ddot{x}_2\left(2m + \frac{m}{4} + \frac{m}{8}\right) + kx_2 - kx_1 = 0.$$

Assume $x_i = X_i \sin \omega t$, so that

$$X_1\left[3k - \frac{19}{4}m\omega^2\right] + X_2[-k] = 0,$$

and

$$X_1[-k] + X_2\left[k - \frac{19}{8}m\omega^2\right] = 0.$$

The frequency equation is therefore

$$\left(3k - \frac{19}{4}m\omega^2\right)\left(k - \frac{19}{8}m\omega^2\right) - (-k)^2 = 0,$$

which is

$$361\left(\frac{\omega}{\Omega}\right)^4 - 380\left(\frac{\omega}{\Omega}\right)^2 + 64 = 0,$$

where

$$\Omega = \sqrt{\frac{k}{m}}.$$

Hence

$$\left(\frac{\omega}{\Omega}\right)^2 = \frac{380 \pm \sqrt{(380^2 - 4 \times 361 \times 64)}}{361 \times 2},$$

and

$$\underline{\omega = 0.459 \sqrt{\frac{k}{m}}} \quad \text{or} \quad \underline{0.918 \sqrt{\frac{k}{m}} \text{ rad/s.}}$$

For the mode shape,

$$\frac{X_1}{X_2} = \frac{k}{3k - \frac{19}{4}\, m\omega^2}.$$

When

$$\omega = 0.459 \sqrt{\frac{k}{m}} \text{ rad/s}, \quad \frac{X_1}{X_2} = \frac{k}{3k - \frac{19}{4} \times 0.21k} = \underline{+0.5,}$$

and when

$$\omega = 0.918 \sqrt{\frac{k}{m}} \text{ rad/s}, \quad \frac{X_1}{X_2} = \frac{k}{3k - \frac{19}{4} \times 0.843k} = \underline{-1.0.}$$

58. FBDs:

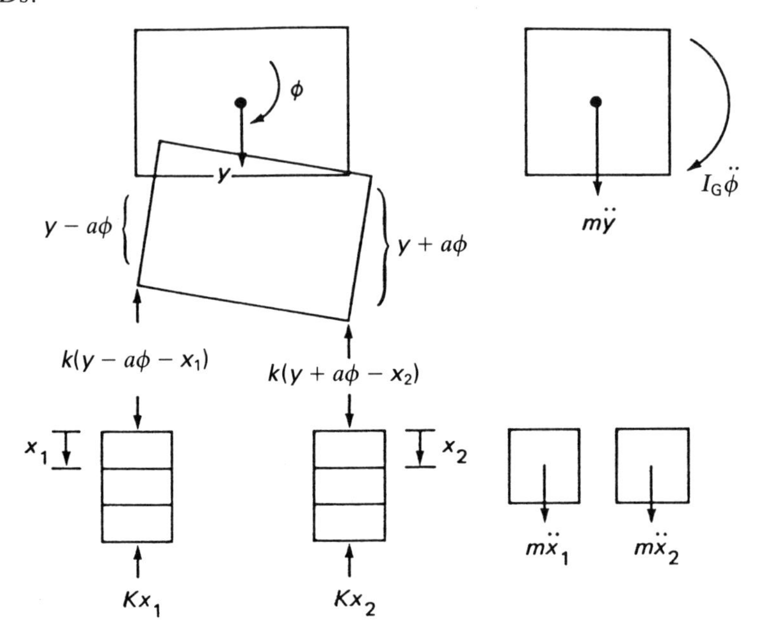

Equations of motion are:

$$m\ddot{y} = -k(y - a\phi - x_1) - k(y + a\phi - x_2),$$

$$I_G\ddot{\phi} = k(y - a\phi - x_1)a - k(y + a\phi - x_2)a,$$

$$m\ddot{x}_1 = k(y - a\phi - x_1) - Kx_1,$$

and

$$m\ddot{x}_2 = k(y + a\phi - x_2) - Kx_2.$$

Assume for wheel hop that body does not move; then

$$y = \phi = 0 \qquad \text{and} \qquad f = \frac{1}{2\pi}\sqrt{\frac{K + k}{m}},$$

$$= \frac{1}{2\pi}\sqrt{\frac{9.0 \times 10^3}{22.5}} = \underline{10.1 \text{ Hz.}}$$

59.

$$f_1 = \frac{1}{2\pi}\sqrt{\left(\frac{3k}{m} - \frac{\sqrt{3}k}{m}\right)}\text{Hz};$$

$$f_2 = \frac{1}{2\pi}\sqrt{\left(\frac{3k}{m}\right)}\ \text{Hz};$$

$$f_3 = \frac{1}{2\pi}\sqrt{\left(\frac{3k}{m} + \frac{\sqrt{3}k}{m}\right)}\ \text{Hz}.$$

60. Assume $x_1 > x_2 > x_3$. FBDs are then as follows:

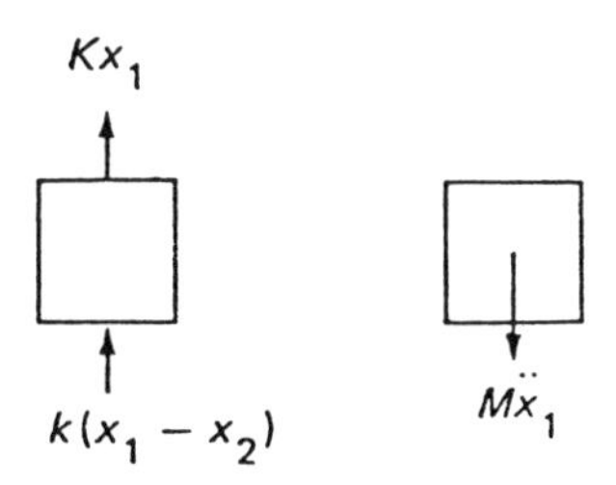

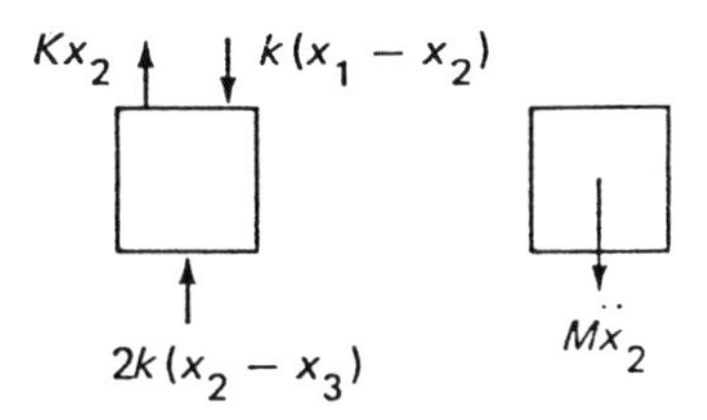

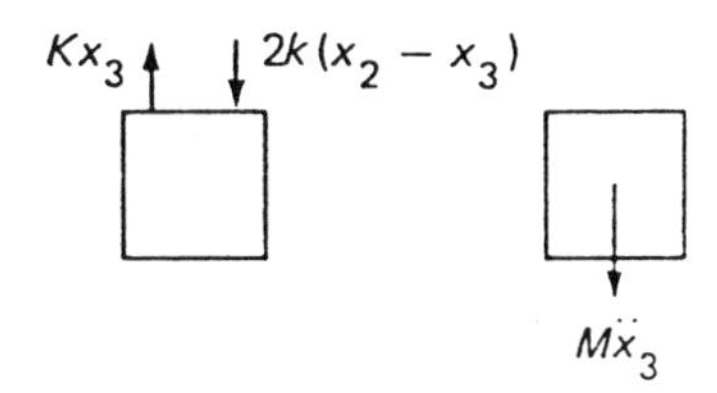

The equations of motion are therefore

$$-Kx_1 - k(x_1 - x_2) = M\ddot{x}_1,$$

$$k(x_1 - x_2) - Kx_2 - 2k(x_2 - x_3) = M\ddot{x}_2,$$

and

$$2k(x_2 - x_3) - Kx_3 = M\ddot{x}_3.$$

Substituting $x_i = X_i \sin \omega t$ and rearranging gives:

$$X_1[K + k - M\omega^2] + X_2[-k] + X_3[0] = 0,$$

$$X_1[-k] + X_2[K + 3k - M\omega^2] + X_3[-2k] = 0,$$

and

$$X_1[0] + X_2[-2k] + X_3[K + 2k - M\omega^2] = 0.$$

The frequency equation is therefore

$$\begin{vmatrix} K+k-M\omega^2 & -k & 0 \\ -k & K+3k-M\omega^2 & -2k \\ 0 & -2k & K+2k-M\omega^2 \end{vmatrix} = 0.$$

That is,

$$(K+k-M\omega^2)[(K+3k-M\omega^2)(K+2k-M\omega^2)-4k^2]$$

$$+k[-k(K+2k-M\omega^2)=0,$$

or

$$M^3\omega^6 - \omega^4(3M^2K+6M^2k)+\omega^2(3MK^2+12MKk+6Mk^2)$$

$$-(K^3+6K^2k+6Kk^2)=0.$$

The solutions to this equation give the natural frequencies.

61. $\underline{f_1 = 0 \text{ Hz};}$ $\underline{1{:}1{:}1;}$

$$\underline{f_2 = \frac{1}{2\pi}\sqrt{\left(\frac{K}{M}\right)} \text{ Hz};} \quad \underline{1{:}0{:}-1.5;}$$

$$\underline{f_3 = \frac{1}{2\pi}\sqrt{\left(\frac{24K}{11M}\right)} \text{ Hz};} \quad \underline{1{:}-\frac{13}{11}{:}\frac{3}{11}}$$

62. $T = \frac{1}{2}m_1L^2\dot{\theta}_1^2 + \frac{1}{2}m_2L^2\dot{\theta}_2^2 + \frac{1}{2}m_3L^2\dot{\theta}_3^2$

$$V = m_1gL(1-\cos\theta_1)+m_2gL(1-\cos\theta_2)+m_3gL(1-\cos\theta_3)$$

$$+\tfrac{1}{2}k_1(a\sin\theta_2 - a\sin\theta_1)^2 + \tfrac{1}{2}k_2(a\sin\theta_3 - a\sin\theta_2)^2.$$

For small oscillations,

$$1-\cos\theta \simeq \frac{\theta^2}{2} \quad \text{and} \quad \sin\theta \simeq \theta.$$

Apply Lagrange equation with $q_i = \theta_1, \theta_2, \theta_3$ in turn to obtain the equations of motion.

63. Assume $x_1 > x_2 > x_3 > x_4$. FBDs are then as follows:

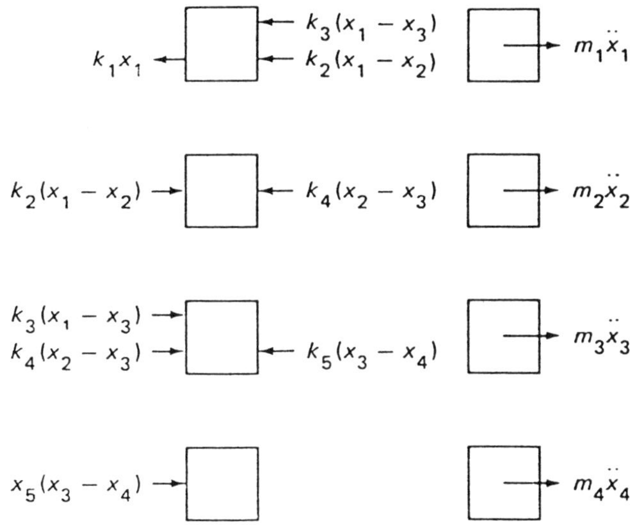

The equations of motion are therefore

$$k_1x_1 + k_2(x_1 - x_2) + k_3(x_1 - x_3) = -\mathrm{m}_1\ddot{x}_1,$$

$$-k_2(x_1 - x_2) + k_4(x_2 - x_3) = -m_2\ddot{x}_2,$$

$$-k_3(x_1 - x_3) - k_4(x_2 - x_3) + k_5(x_3 - x_4) = -m_3\ddot{x}_3,$$

and

$$-k_5(x_3 - x_4) = -m_4\ddot{x}_4.$$

Substitute $x_i = X_i \sin \omega t$:

$$k_1X_1 + k_2(X_1 - X_2) + k_3(X_1 - X_3) = m_1\omega^2X_1,$$

$$-k_2(X_1 - X_2) + k_4(X_2 - X_3) = m_2\omega^2X_2,$$

$$-k_3(X_1 - X_3) - k_4(X_2 - X_3) + k_5(X_3 - X_4) = m_3\omega^2X_3,$$

and

$$-k_5(X_3 - X_4) = m_4\omega^2X_4.$$

Thus

$$X_1[k_1 + k_2 + k_3 - m_1\omega^2] + X_2[-k_2] + X_3[-k_3] + X_4[0] = 0,$$

$$X_1[-k_2] + X_2[k_2 + k_4 - m_2\omega^2] + X_3[-k_4] + X_4[0] = 0.$$

$$X_1[-k_3] + X_2[-k_4] + X_3[k_3 + k_4 + k_5 - m_3\omega^2] + X_4[-k_5] = 0,$$

and

$$X_1[0] + X_2[0] + X_3[-k_5] + X_4[k_5 - m_4\omega^2] = 0.$$

Frequency equation is, therefore:

$$\begin{vmatrix} k_1 + k_2 + k_3 - m_1\omega^2 & -k_2 & -k_3 & 0 \\ -k_2 & k_2 + k_4 - m_2\omega^2 & -k_4 & 0 \\ -k_3 & -k_4 & k_3 + k_4 + k_5 - m_3\omega^2 & -k_5 \\ 0 & 0 & -k_5 & k_5 - m_4\omega^2 \end{vmatrix} = 0.$$

65. Assume $x_1 > x_2 > x_3$. FBDs are

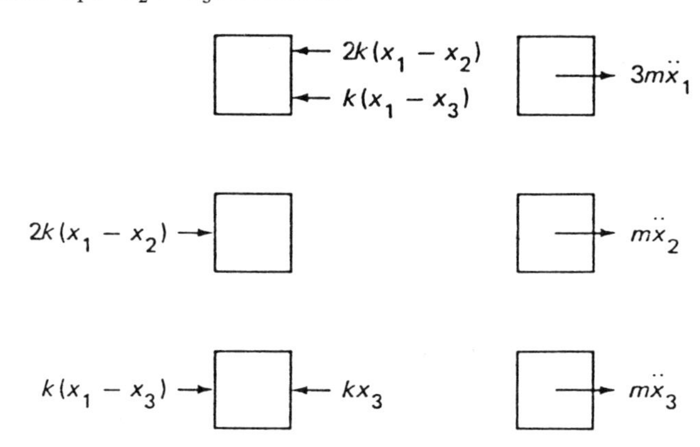

Equations of motion are therefore

$$-2k(x_1 - x_2) - k(x_1 - x_3) = 3m\ddot{x}_1,$$

$$2k(x_1 - x_2) = m\ddot{x}_2,$$

and

$$k(x_1 - x_3) - kx_3 = m\ddot{x}_3.$$

Putting $x_i = X_i \sin \omega t$ and rearranging gives

$$-3kX_1 + 2kX_2 + kX_3 = -3m\omega^2 X_1,$$

$$2kX_1 - 2kX_2 = -m\omega^2 X_2,$$

and

$$kX_1 - 2kX_3 = -m\omega^2 X_3.$$

That is

$$\begin{bmatrix} \frac{k}{m} & -\frac{2k}{3m} & -\frac{k}{3m} \\ -\frac{2k}{m} & \frac{2k}{m} & 0 \\ -\frac{k}{m} & 0 & \frac{2k}{m} \end{bmatrix} \begin{Bmatrix} X_1 \\ X_2 \\ X_3 \end{Bmatrix} = \omega^2 \begin{Bmatrix} X_1 \\ X_2 \\ X_3 \end{Bmatrix}$$

70.

$$\begin{bmatrix} \alpha_{11} & \alpha_{12} & \alpha_{13} \\ \alpha_{21} & \alpha_{22} & \alpha_{23} \\ \alpha_{31} & \alpha_{32} & \alpha_{33} \end{bmatrix} \begin{Bmatrix} X_1 \\ X_2 \\ X_3 \end{Bmatrix} = \frac{1}{m\omega^2} \begin{Bmatrix} X_1 \\ X_2 \\ X_3 \end{Bmatrix}$$

Hence

$$\begin{bmatrix} 0.5 & 0.25 & 0.25 \\ 0.25 & 0.5 & 0.25 \\ 0.25 & 0.25 & 0.5 \end{bmatrix} \begin{Bmatrix} X_1 \\ X_2 \\ X_3 \end{Bmatrix} = \frac{k}{m\omega^2} \begin{Bmatrix} X_1 \\ X_2 \\ X_3 \end{Bmatrix}$$

or

$$\begin{bmatrix} 2 & 1 & 1 \\ 1 & 2 & 1 \\ 1 & 1 & 2 \end{bmatrix} \begin{Bmatrix} X_1 \\ X_2 \\ X_3 \end{Bmatrix} = \frac{4k}{m\omega^2} \begin{Bmatrix} X_1 \\ X_2 \\ X_3 \end{Bmatrix}$$

For lowest natural frequency assume mode shape 1, 1, 1:

$$\begin{bmatrix} 2 & 1 & 1 \\ 1 & 2 & 1 \\ 1 & 1 & 2 \end{bmatrix} \begin{Bmatrix} 1 \\ 1 \\ 1 \end{Bmatrix} = \begin{Bmatrix} 4 \\ 4 \\ 4 \end{Bmatrix} = 4 \begin{Bmatrix} 1 \\ 1 \\ 1 \end{Bmatrix}.$$

Hence correct assumption and

$$\frac{k}{m\omega^2} = 1,$$

so

$$\underline{f = \frac{1}{2\pi}\sqrt{\frac{k}{m}}\ \text{Hz.}}$$

74.

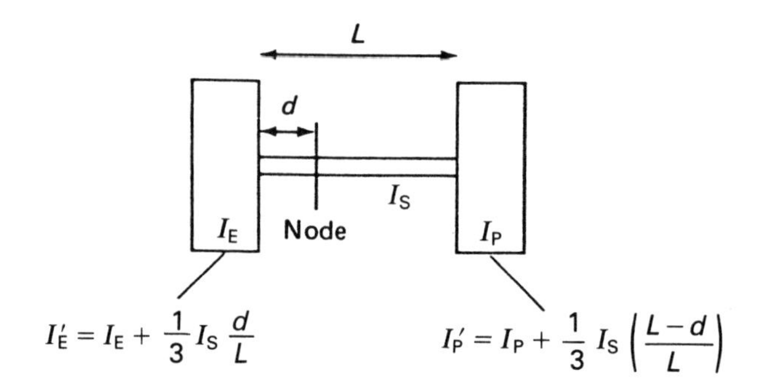

Now

$$f = \frac{1}{2\pi}\sqrt{\frac{k_E}{I'_E}} = \frac{1}{2\pi}\sqrt{\frac{GJ}{I'_E d}} = \frac{1}{2\pi}\sqrt{\frac{GJ}{I'_P(L-d)}}\ \text{Hz},$$

where G is modulus of rigidity for the shaft, and
J is polar second moment of area of shaft about axis of rotation.

Thus

$$I'_E d = I'_P (L-d),$$

or

$$\left(I_E + \tfrac{1}{3} I_S \frac{d}{L}\right) d = \left(I_P + \tfrac{1}{3} I_S \left(\frac{L-d}{L}\right)\right)(L-d).$$

so

$$d\,(I_E + I_P + \tfrac{2}{3} I_S) = (I_P + \tfrac{1}{3} I_S)\, L$$

Hence

$$d = \left(\frac{I_P + \frac{1}{3} I_S}{I_E + I_P + \frac{2}{3} I_S}\right) L$$

76.

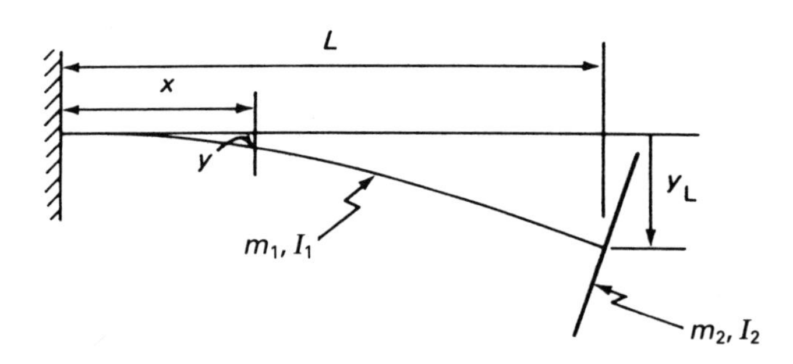

Assume

$$y = y_L\left(1 - \cos\left(\frac{\pi}{2}\frac{x}{L}\right)\right),$$

then

$$\frac{d^2y}{dx^2} = -\left(\frac{\pi}{2L}\right)^2 \cos\left(\frac{\pi}{2}\frac{x}{L}\right),$$

Now

$$\omega^2 = \frac{\displaystyle\int EI\ (d^2y/dx^2)^2\ dx}{\displaystyle\int y^2\ dm},$$

where

$$\int_0^L EI\left(\frac{d^2y}{dx^2}\right)^2 dx = EI_1\int_0^L \left(\frac{\pi}{2L}\right)^4 \cos^2\left(\frac{\pi}{2}\frac{x}{L}\right) dx$$

$$= EI_1\left(\frac{\pi}{2L}\right)^4 y_L^2\,\frac{L}{2}.$$

and

$$\int y^2\ dm = \int_0^L \frac{m_1}{L} y^2\ dx + m_2 y_L^2 + I_2\left(\frac{dy_L}{dx}\right)^2,$$

$$= \frac{m_1}{L} y_L^2 \int_0^L \left[1 - 2\cos\left(\frac{\pi}{2}\frac{x}{L}\right) + \cos^2\left(\frac{\pi}{2}\frac{x}{L}\right)\right] dx$$

$$+ m_2 \dot{y}_L^2 + I_2\left(\frac{\pi}{2L}\right)^2 \dot{y}_L^2,$$

$$= \frac{m_1}{L} y_L^2\left[L - \frac{4L}{\pi} + \frac{L}{2}\right].$$

Substituting numerical values gives

$$E = 207 \times 10^9\ \text{N/m}^2,$$

$$L = 0.45\ \text{m},$$

$$I_1 = \frac{\pi}{64} \times 2.5^4 \times 10^{-8} = 1.916 \times 10^{-8}\ \text{m}^4,$$

$$m_1 = 7850 \times \frac{\pi}{4} \times (0.025)^2 \times 0.45 = 1.732\ \text{kg},$$

$$m_2 = 7850 \times \pi \times 0.58(0.02)^2 = 5.71\ \text{kg},$$

and

$$I_2 = \tfrac{1}{2} \times 5.71 \times \left(\frac{0.58}{2}\right)^2 = 0.24 \text{ kg m}^2,$$

so that

$$\omega^2 = \frac{1.325 \times 10^5}{9.026},$$

and $\omega = 121$ rad/s, so that $\underline{f = 19.3 \text{ Hz.}}$

77.

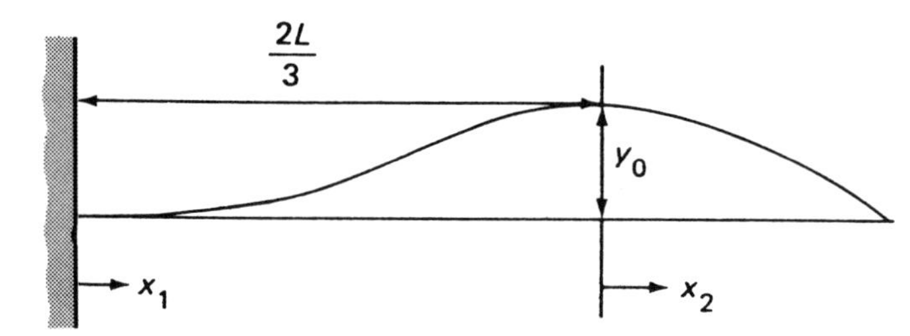

Assume

$$\text{for } 0 < x_1 < \frac{2L}{3}, \qquad y = \frac{y_0}{2}\left(1 - \cos\frac{3\pi}{2L}x_1\right),$$

and

$$\text{for } 0 < x_2 < \frac{L}{3}, \qquad y = y_0 \cos\frac{3\pi}{2L}x_2 .$$

Now

$$\omega^2 = \frac{\displaystyle\int EI\ (\mathrm{d}^2y/\mathrm{d}x^2)^2\ \mathrm{d}x}{\displaystyle\int y^2\ \mathrm{d}m},$$

where

$$\int_0^L \left(\frac{\mathrm{d}^2y}{\mathrm{d}x^2}\right)^2 \mathrm{d}x$$

$$= y_0^2\left[\int_0^{2L/3} \frac{1}{4}\left(\frac{3\pi}{2L}\right)^4 \cos^2\frac{3\pi}{2L}x_1\ \mathrm{d}x_1 + \int_0^{L/3}\left(\frac{3\pi}{2L}\right)^4 \cos^2\frac{3\pi}{2L}x_2\ \mathrm{d}x_2\right]$$

$$= y_0^2\left(\frac{3\pi}{2L}\right)^4 \frac{L}{4},$$

and

$$\int_0^L y^2 \, \mathrm{d}m = m y_0^2 \left[\int_0^{2L/3} \frac{1}{4} \left(1 - 2 \cos \frac{3\pi}{2L} x_1 + \cos^2 \frac{3\pi}{2L} x_1 \right) \mathrm{d}x_1 + \int_0^{L/3} \cos^2 \frac{3\pi}{2L} x_2 \, \mathrm{d}x_2 \right]$$

$$= \frac{5mL}{12} y_0^2 .$$

Substituting numerical quantities gives $f = 1.65$ Hz.

78.

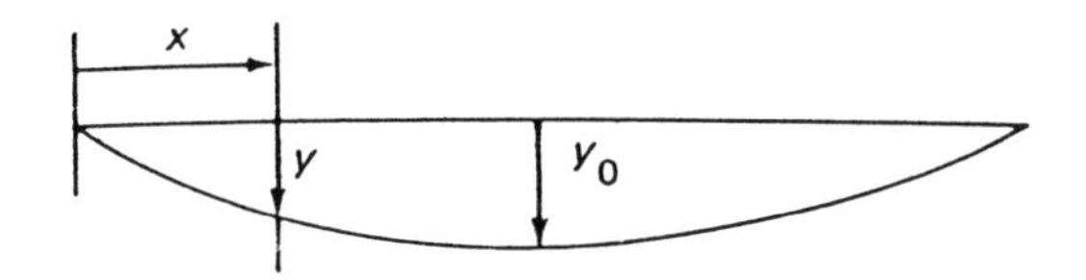

Assume

$$y = y_0 \sin \frac{\pi x}{2L}.$$

Then

$$\int EI \left(\frac{\mathrm{d}^2 y}{\mathrm{d}x^2} \right)^2 \mathrm{d}x = \int_0^{2L} EI_{\mathrm{S}} \left(\frac{\mathrm{d}^2 y}{\mathrm{d}x^2} \right)^2 \mathrm{d}x$$

$$= EI_{\mathrm{S}} \int_0^{2L} y_0^2 \left(\frac{\pi}{2L} \right)^4 \sin^2 \frac{\pi x}{2L} \, \mathrm{d}x,$$

$$= EI_{\mathrm{S}} \left(\frac{\pi}{2L} \right)^4 y_0^2 L$$

and

$$\int y^2 \, \mathrm{d}m = m y_L^2 + m y_{2L/3}^2 + I y_{2L/3}'^2 ,$$

where

$$y_{2L/3} = y_0 \sin \frac{\pi}{3} = y_0 \frac{\sqrt{3}}{2}$$

and

$$y' = y_0 \frac{\pi}{2L} \cos \frac{\pi x}{2L},$$

so

$$y'_{2L/3} = y_0 \frac{\pi}{2L} \times \frac{1}{2}.$$

Hence

$$\int y^2 \, \mathrm{d}m = y_0^2 \left[\frac{7m}{4} + \frac{I\pi^2}{16L^2} \right],$$

and

$$\omega^2 = \frac{EI_S(\pi/2L)^4 L}{[7m/4 + \pi^2 I/(16L^2)]}.$$

79.

$$I = \frac{\pi}{64}(80^4 - 70^4) = 832 \times 10^3 \text{ mm}^4,$$

and

$$\text{mass/length} = \frac{\pi}{4}(0.08^2 - 0.07^2)\, 7750 + \frac{\pi}{4}(0.07)^2\, 930$$

$$= 12.72 \text{ kg/m}.$$

Now

$$\omega^2 = g \frac{\int_0^L x^2(L-x)^2 \, \mathrm{d}x}{(mg/24EI)\int_0^L x^4(L-x)^4 \, \mathrm{d}x},$$

$$= \frac{24EI}{M} \frac{L^5/30}{L^9/630} = \frac{24EI}{m} \frac{21}{L^4},$$

$$= \frac{24 \times 200 \times 10^9 \times 832 \times 10^3 \times 21}{12.72 \times 10^{12} \times 4^4} \text{ s}^{-2}$$

Hence $\omega = 161$ rad/s and $f = 25.6$ Hz.

80.

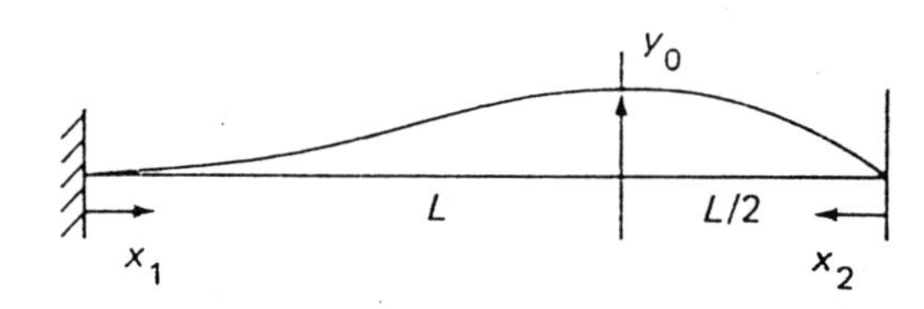

For

$$0 < x_1 < L: y = \frac{y_0}{2}\left(1 - \cos\frac{\pi x_1}{L}\right),$$

and for

$$0 < x_2 < \frac{L}{2}: y = y_0 \sin\frac{\pi x_2}{L}.$$

Thus

$$\int EI\left(\frac{d^2y}{dx^2}\right)^2 dx = EI\int_0^L \frac{y_0^2}{4}\left(\frac{\pi}{L}\right)^4 \cos^2\frac{\pi x_1}{L}\, dx_1$$

$$+ EI\int_0^{L/2} y_0^2\left(\frac{\pi}{L}\right)^4 \sin^2\frac{\pi x_2}{L}\, dx_2$$

$$= EIy_0^2\left(\frac{\pi}{L}\right)^4 \frac{3L}{8}.$$

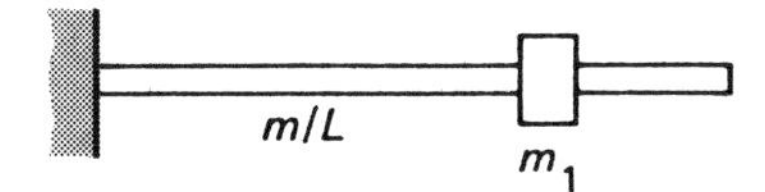

and

$$\int y^2\, dm = \frac{m}{L}\int_0^L \frac{y_0^2}{4}\left(1 - \cos\frac{\pi x_1}{L}\right)^2 dx_1$$

$$+ \int_0^{L/2} \frac{m}{2L} y_0^2 \sin^2\frac{\pi x_2}{L}\, dx_2 + m_1 y_0^2$$

$$= y_0^2\,(0.625\, m + m_1).$$

Substitute numerical values to get ω^2
Hence <u>$f = 9.5$ Hz.</u>

86. If spool valve displacement $= y$,
and spring displacement $= p$.

For control rod, $y = \dfrac{x_i + p}{2}$,

Spring/damper force balance gives $kp = CD(p - x_0)$, and flow equation gives $by = ADx_0$. Eliminating p and y gives

$$\underline{\frac{x_0}{x_i} = \frac{b(cD - k)}{2AcD^2 - D(2Ak + bc)}}.$$

87. $$D\theta_o = k(\theta_i - \theta_o),$$

so

$$\frac{\theta_o}{\theta_i} = \frac{1}{1 + \tau D}, \qquad \text{where} \qquad \tau = \frac{1}{k}.$$

If

$$\theta_i = \beta t, \quad \theta_o = (1 - \tau D + \ldots)\,\beta t,$$
$$= \beta t - \tau\beta.$$

Thus

$$\theta_{SS} = \tau\beta.$$

If $\theta_{SS} = 1°$ when $\beta = 3°/\text{s}$, $\underline{\tau = \tfrac{1}{3}\text{s}.}$

88. The flow Q into the main cylinder is used to move the piston, Q_V, leak Q_L, and compress oil Q_C. Thus

$$Q = Q_V + Q_L + Q_C,$$

and

$$Q = Q_o y.$$

Flow equation gives

$$Q_V = AD x_2,$$

and

$$Q_L = L\,\delta p,$$

where δp is the pressure difference across the piston and accelerates the load. That is,

$$MD^2 x_2 = A\,\delta p,$$

so

$$Q_L = \frac{LM}{A} D^2 x_2.$$

Now bulk modulus is the pressure change/volumetric strain. That is $K\,dV = V\,dp$ where dp is the pressure charge causing a volumetric strain dV/V. But $Q_C = dV/dt$, and since $V_o/2$ is the volume of half the cylinder, $dp/dt = (2k/V_o)Q_c$. The same rate of pressure charge occurs in the other half of the cylinder, so the rate of change of pressure differential is given by

$$\frac{d}{dt}\,\delta p = \frac{4K}{V_o}\,Q_c.$$

Since

$$\delta p = \frac{M}{A}\,\mathrm{D}^2 x_2, \qquad Q_c = \frac{MV_o}{4KA}\,\mathrm{D}^3 x_2.$$

Now

$$Q = Q_0\, y,$$

so

$$y = \frac{A}{Q_0}\,\mathrm{D}x_2 + \frac{LM}{Q_0 A}\,\mathrm{D}^2 x_2 + \frac{MV_0}{4KAQ_0}\,\mathrm{D}^3 x_2.$$

For the control rod, $y = \frac{1}{2}(x_1 - x_2)$, so

$$\frac{x_2}{x_1} = \frac{1}{\dfrac{MV_0}{2KAQ_0}\,\mathrm{D}^3 + \dfrac{2LM}{AQ_0}\,\mathrm{D}^2 + \dfrac{2A}{Q_0}\,\mathrm{D} + 1}.$$

89.

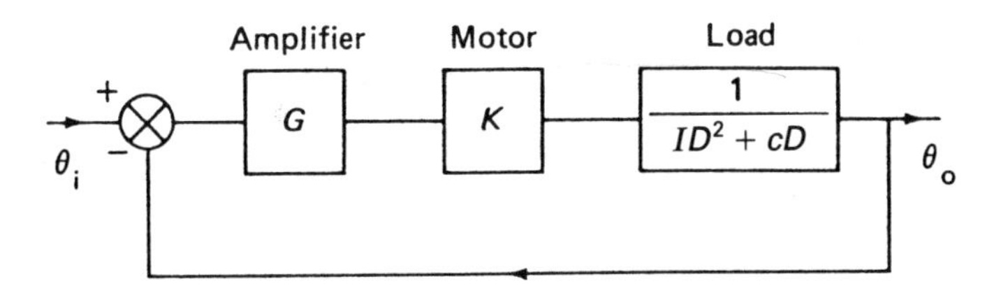

Equation of motion is

$$(I\mathrm{D}^2 + c\mathrm{D} + GK)\theta_o = KG\theta_i.$$

Now

$$\text{overshoot} = 0.1 = \mathrm{e}^{-\zeta\pi/\sqrt{(1-\zeta^2)}}$$

Thus

$$\frac{\zeta\pi}{\sqrt{(1-\zeta^2)}} = 2.3,$$

and

$$\zeta = 0.59.$$

Also

$$\zeta = \frac{c}{c_c} = \frac{c}{2\sqrt{(GKI)}},$$

so

$$GK = \frac{c^2}{4\zeta^2 I} = \underline{0.718\,\frac{c^2}{I}}.$$

90.

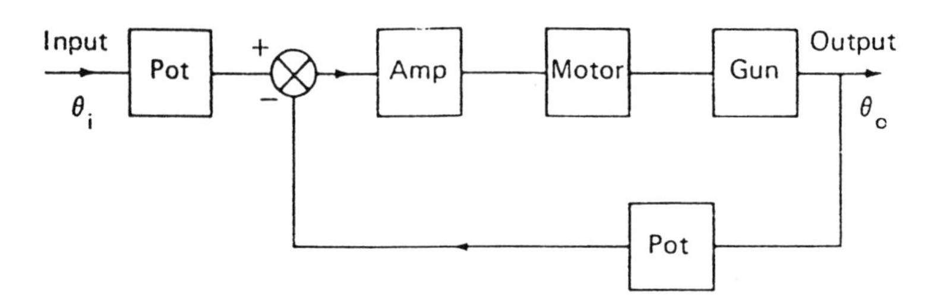

This system is equivalent to

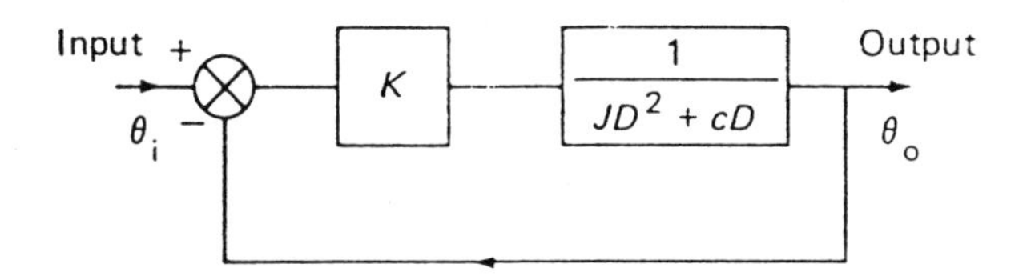

Equation of motion is $(JD^2 + cD + K)\theta_o = K\theta_i$. Modified system is

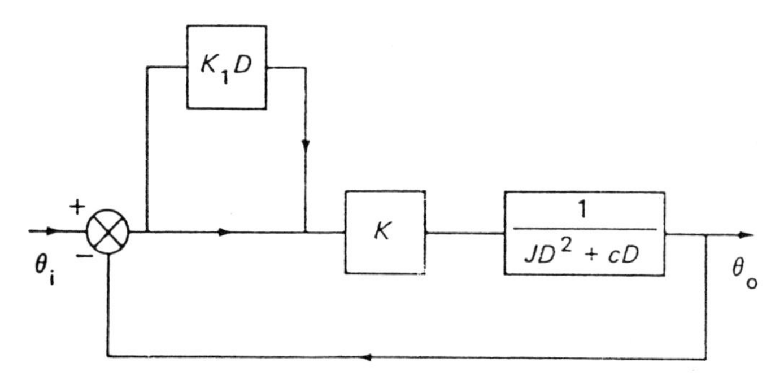

Equation of motion is

$$[(\theta_i - \theta_o) + K_1 D\theta_i - K_1 D\theta_o]K = (JD^2 + cD)\theta_o,$$

That is,

$$[JD^2 + cD + KK_1 D + K]\theta_o = (KK_1 D + K)\theta_i.$$

With a step input $\theta_i = X$,

P.I. gives

$$\theta_o = \left(1 - \left(K_1 + \frac{c}{K}\right) D + \ldots\right)(1 + K_1 D)X,$$

$$= X.$$

C.F. has damping term increased to $(c + KK_1)$ so overshoot is reduced.
With a ramp input $\theta_i = \beta t$, P.I. gives

$$\theta_o = \left(1 - \left(K_1 + \frac{c}{K}\right) D + \ldots\right)(1 + K_1 D)\ \beta t$$

$$= \left(1 - \left(K_1 + \frac{c}{K}\right) D + \ldots\right)(\beta t + K_1 \beta)$$

$$= \beta t + K_1 \beta - \left(K_1 + \frac{c}{K}\right) \beta$$

$$= \beta t - \frac{c}{K} \beta.$$

so steady-state error is

$$\frac{c}{K} \beta, \qquad \text{as before.}$$

92.

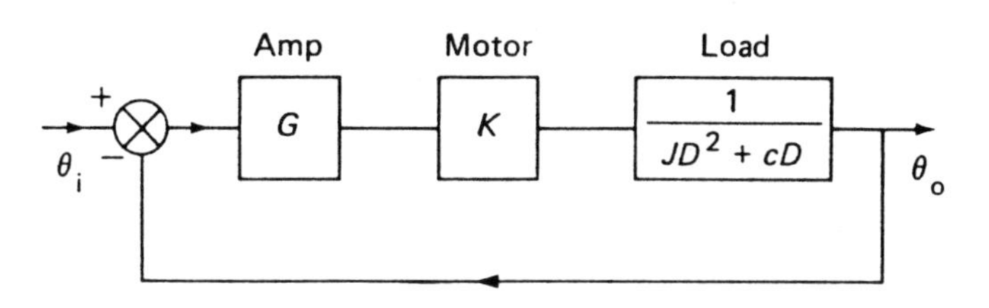

For first overshoot from step input,

$$\tfrac{1}{5} = e^{-\zeta\pi/\sqrt{(1-\zeta^2)}},$$

so

$$\frac{\zeta\pi}{\sqrt{(1 - \zeta^2)}} = \log_e 5 = 1.6094.$$

Hence

$$\zeta^2 = 0.20788.$$

Now

$$\zeta = \frac{c}{c_c} = \frac{c}{2\sqrt{(JKG)}}$$

so

$$GK = \frac{c^2}{4I\zeta^2} = \frac{50^2}{4 \times 100 \times 0.20788} = \underline{30 \text{ Nm/rad.}}$$

93. (a) $\underline{K = 2292 \text{ N m/rad;}}$

(b) $\underline{\omega = 8.93 \text{ rad/s}, \ \delta = 4.5 \text{ rad/s;}}$

(c) $\underline{\dot{\theta}_{SS} = 2 \text{ rad/s.}}$

94. (a) $\underline{t = 19.6\text{s;}}$

(b) $\underline{K = 192 \text{ V/rad/s.}}$

95.

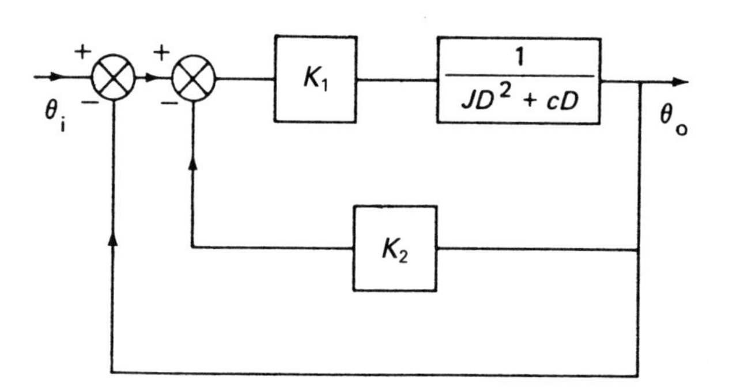

Equation of motion is

$$(JD^2 + (c + K_1K_2)D + K_1)\theta_o = K_i\theta_i.$$

Velocity input $\theta_i = \beta t$ gives P.I as

$$\theta_o = \left(1 - \left(\frac{c + K_1K_2}{K_1}\right) D + \ldots\right) \beta t,$$

$$= \beta t - \left(\frac{c + K_1K_2}{K_1}\right) \beta,$$

so steady-state error is

$$\left(\frac{c + K_1K_2}{K_1}\right) \beta.$$

With a step input, from the equation of motion

$$\zeta = \frac{c + K_1 K_2}{2\sqrt{(K_1 J)}}$$

For 10% limit,

$$e^{-\zeta\pi/\sqrt{(1-\zeta^2)}} = 0.1,$$

so

$$-\frac{\zeta\pi}{\sqrt{(1-\zeta^2)}} = \log_e 0.1 = -2.3026.$$

Hence

$$\zeta = 0.591,$$

so that

$$\frac{c + K_1 K_2}{2\sqrt{(K_1 J)}} = 0.591.$$

98.

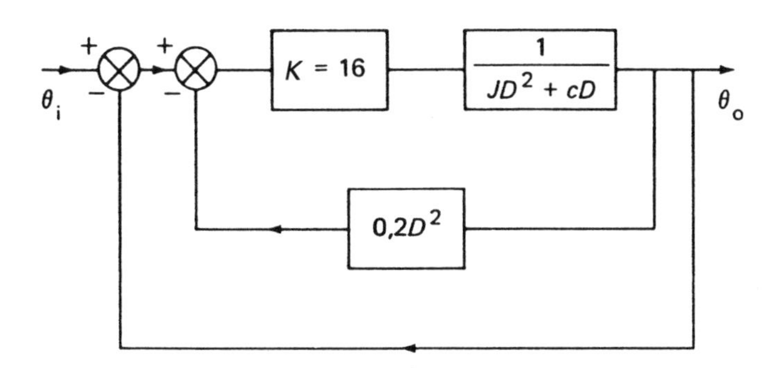

(i)

Equation of motion is

$$K(\theta_i - \theta_o + 0.2D^2\theta_o) = (JD^2 + cD)\,\theta_o,$$

or

$$[(J - 0.2K)D^2 + cD + K]\,\theta_o = K\theta_i.$$

(ii) For step input θ_i, P.I. gives

$$\theta_o = \left[1 - \frac{c}{K}D + \ldots\right]\theta_i = \theta_i.$$

That is, there is no steady-state error. C.F. gives

$$\theta_o = Ae^{-\zeta\omega t}\sin(\omega\surd(1-\zeta^2)\,t+\phi),$$

so transient output frequency is $\omega\surd(1-\zeta^2)$, where

$$\omega = \sqrt{\left(\frac{K}{J-0.2K}\right)} = \sqrt{\frac{16}{4.2-0.2\times 16}} = 4 \text{ rad/s},$$

and

$$\zeta = \frac{c}{2\surd[K(J-0.2K)]} = \frac{4}{2\surd[16(4.2-0.2\times 16)]} = \tfrac{1}{2},$$

so

$$\text{transient frequency} = 4\surd[1-(\tfrac{1}{2})^2] = 3.46 \text{ rad/s} = \underline{0.55 \text{ Hz.}}$$

Complete solution for θ_o is P.I. plus CF;

$$\theta_o = \theta_i + Ae^{-\zeta\omega t}\sin(\omega\surd(1-\zeta^2)\,t+\phi).$$

Now at $t = 0$, $\theta_o = 0$, so $0 = \theta_i + A\sin\phi$, and at $t = 0$, $D\theta_o = 0$ so $0 = A - \zeta\omega\sin\phi + A\omega\surd(1-\zeta^2)\cos\phi$.
Hence

$$\tan\phi = \frac{\surd(1-\zeta^2)}{\zeta} \qquad \text{and} \qquad \sin\phi = \surd(1-\zeta^2).$$

Also

$$A = -\frac{\theta_i}{\surd(1-\zeta^2)}.$$

Thus

$$\theta_o = \theta_i - \frac{\theta_i}{\surd(1-\zeta^2)}e^{-\zeta\omega t}\sin(\omega\surd(1-\zeta^2)\,t+\phi),$$

where

$$\sin\phi = \surd(1-\zeta^2) = \frac{\surd 3}{2}; \qquad \text{hence } \phi = \frac{\pi}{3}.$$

When $t = 0.5$ s and $\theta_i = 0.2$ rad

$$\theta_o = 0.2 - \frac{0.2}{\surd[1-(\frac{1}{2})^2]}e^{-\frac{1}{2}\times 4\times 0.5}\sin\left(3.46\times 0.5+\frac{\pi}{3}\right),$$

$$= \underline{0.17 \text{ rad.}}$$

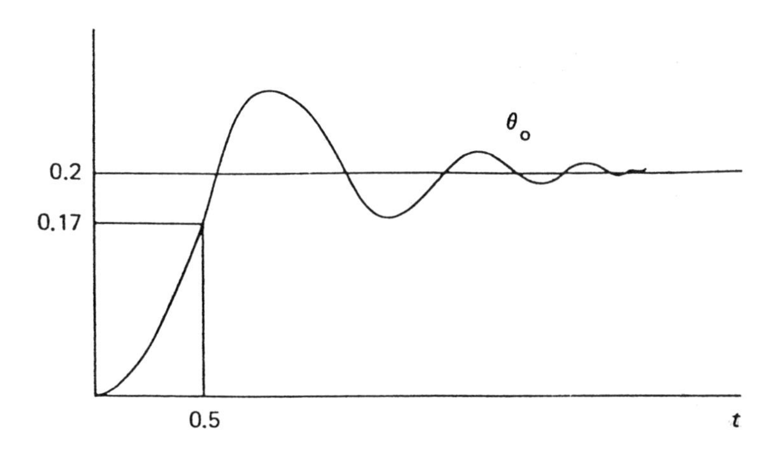

99.

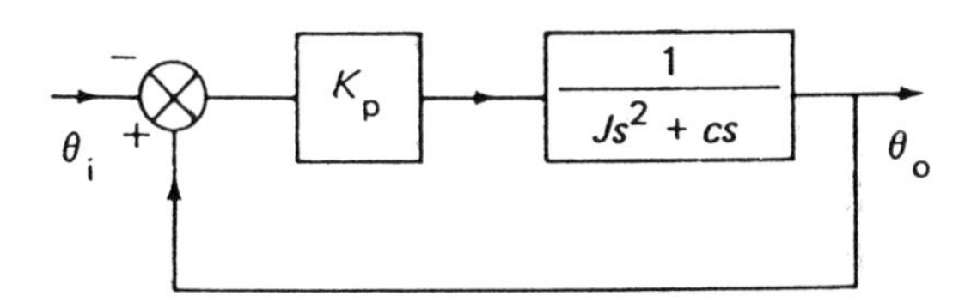

where s is the Laplace operator.

$$\text{CLTF,} \qquad \frac{\theta_o}{\theta_i} = \frac{K_p}{Js^2 + cs + K_p} = \frac{32}{0.5s^2 + s + 32}.$$

Ramp input gives

$$\theta_i = \frac{2}{s^2},$$

so

$$\text{error} = \theta_i - \theta_o = \left(1 - \frac{64}{s^2 + 2s + 64}\right)\frac{2}{s^2} = \frac{1}{s}\left(\frac{2s + 4}{s^2 + 2s + 64}\right).$$

F.V.T. gives steady state error as

$$\lim_{s \to 0}\left[sE(s)\right] = \lim_{s \to 0}\left[\frac{2s + 4}{s^2 + 2s + 64}\right] = \underline{\tfrac{1}{16}\text{ rad.}}$$

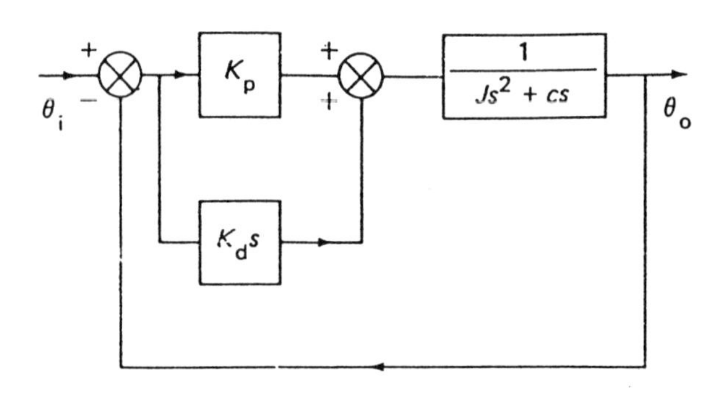

$$\frac{\theta_o}{\theta_i} = \frac{K_d s + K_p}{Js^2 + (K_d + c)\ s + K_p}.$$

For critical damping

$$(K_d + c) = 2\sqrt{(K_p J)},$$

so

$$K_d + 1 = 2\sqrt{(32 \times 0.5)},$$

so $\underline{K_d = 7 \text{ N m s/rad.}}$

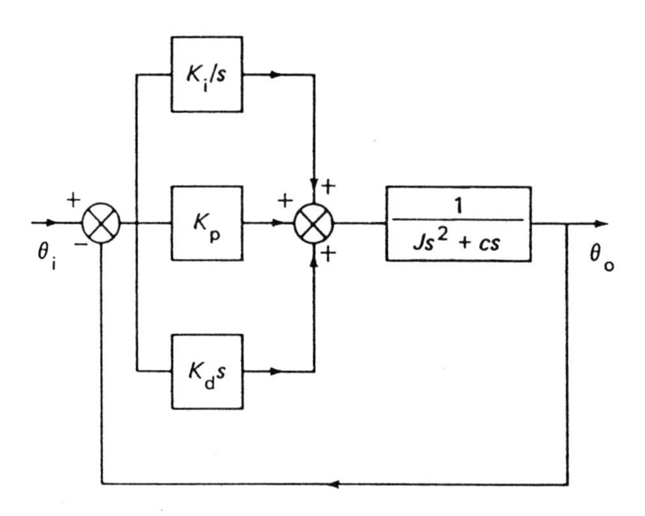

Now

$$\frac{\theta_o}{\theta_i} = \frac{K_d s^2 + K_p s + K_i}{Js^3 + (K_d + c)s^2 + K_p s + K_i},$$

so

$$\theta_i - \theta_o = \frac{(Js^3 + cs^2)\theta_i}{Js^3 + (K_d + c)s^2 + K_p s + K_i}.$$

F.V.T. gives velocity lag

$$= \lim_{s \to 0} \left[\frac{s(Js^3 + cs^2)(\theta_i/s^2)}{Js^3 + (K_d + c)s^2 + K_p s + K_i} \right] = 0.$$

100.

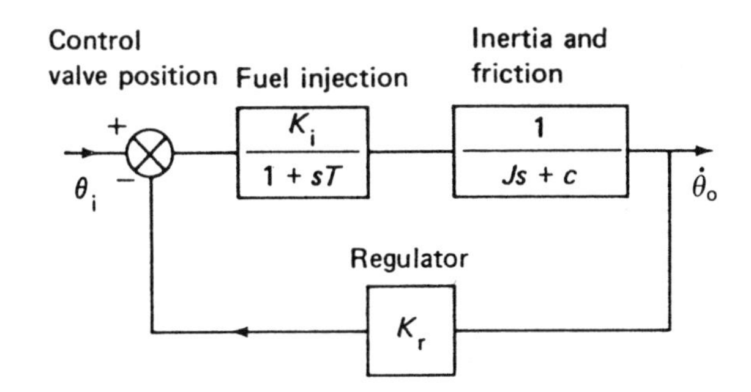

$$\frac{\dot{\theta}_o}{\theta_i} = \frac{K_i}{(1 + sT)(Js + c) + K_i K_r},$$

$$= \frac{K_i/TJ}{s^2 + \left(\dfrac{J + Tc}{TJ}\right)s + \left(\dfrac{K_i K_r + c}{TJ}\right)},$$

$$= \frac{K\omega^2}{s^2 + 2\zeta\omega s + \omega^2},$$

where

$$\omega = \sqrt{\left(\frac{K_i K_r + c}{TJ}\right)} = \sqrt{\frac{200 \times 0.2 + 10}{0.1 \times 5}} = 10 \text{ rad/s},$$

and

$$2\zeta\omega = \frac{J + Tc}{TJ} = \frac{5 + 0.1 \times 10}{0.1 \times 5} = 12,$$

so that

$$\zeta = 0.6; \qquad \text{and} \qquad K\omega^2 = \frac{K_i}{TJ}$$

and

$$K = 4.$$

Hence

$$\frac{\dot{\theta}_o}{\theta_i} = \frac{400}{s^2 + 12s + 100}.$$

For a step input

$$\theta_i(s) = \frac{X}{s},$$

so using Laplace tables

$$\dot{\theta}_o(t) = 400X\left[\frac{1}{\omega^2} - \frac{1}{\omega^2\sqrt{(1-\zeta^2)}}\,e^{-\zeta\omega t}\sin(\omega\sqrt{(1-\zeta^2)}t + \phi)\right].$$

Hence

$$\text{error} = \frac{400X}{\omega^2}\left[\frac{1}{\sqrt{(1-\zeta^2)}}\,e^{-\zeta\omega t}\sin(\omega\sqrt{(1-\zeta^2)}t + \phi)\right].$$

Maximum error when $\omega\sqrt{(1-\zeta^2)}t = n\pi$, that is when $t = n\pi/8$. With

$$n = 1,\ \text{error} = \frac{400X}{\omega^2}[-0.0947],$$

$$n = 2,\ \text{error} = \frac{400X}{\omega^2}[0.00898],$$

which is less than 1% with $t = 2\pi/8$ ($t < 1$ s).
With load applied,

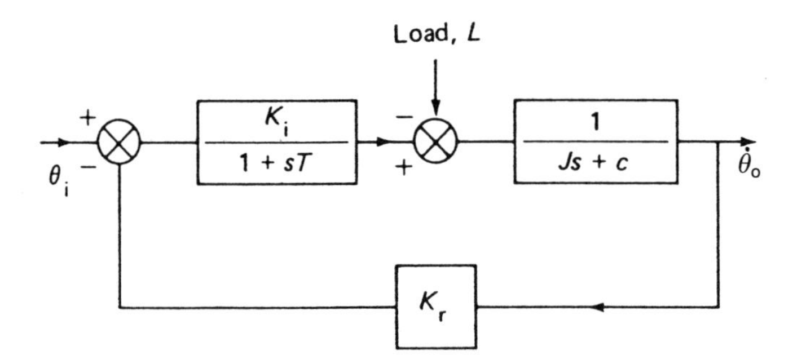

so

$$\dot{\theta}_o = \frac{\theta_i K_i - L(1 + sT)}{(1 + sT)(Js + c) + K_i K_r}.$$

Hence error in

$$\dot{\theta}_o = \frac{\theta_i K_i}{(1 + sT)(Js + c) + K_i K_r} - \dot{\theta}_o = \frac{L(1 + sT)}{(1 + sT)(Js + c) + K_i K}$$

F.V.T for $L = 100/s$ gives

$$\text{error} = \frac{100}{c + K_1 K_r} = \underline{2 \text{ rad/s.}}$$

101.

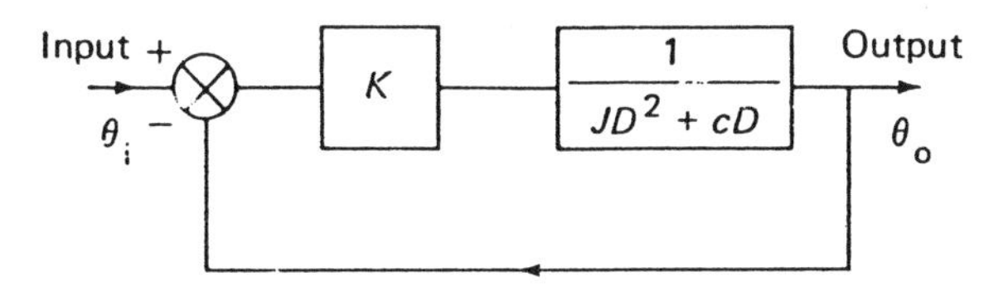

Equation of motion is $(JD^2 + cD + K)\theta_o = K\theta_i$. With step input X, overshoot is $X\ e^{-\zeta\pi/\sqrt{(1-\zeta^2)}}$. Thus

$$0.01 = 0.1e^{-\zeta\pi/\sqrt{(1-\zeta^2)}}.$$

Hence

$$\zeta = 0.592.$$

Also

$$\zeta = \frac{c}{2\sqrt{(JK)}}, \qquad \text{so} \quad K = \frac{c^2}{4J\zeta^2},$$

and

$$K = \frac{800^2}{4 \times 370 \times 0.592^2} = \underline{1236 \text{ N m/rad.}}$$

With hydrodynamic forces,

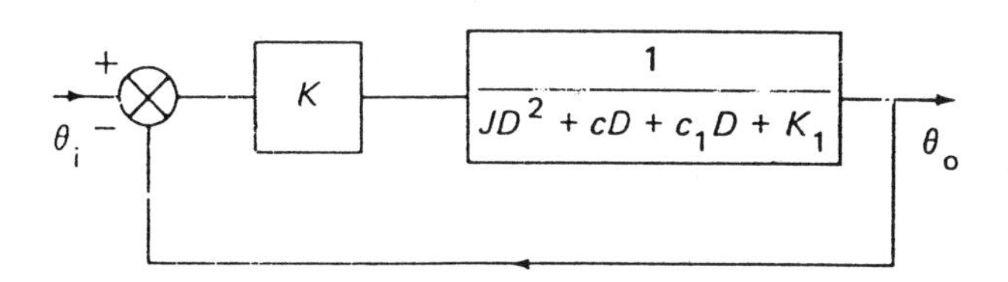

so that

$$[JD^2 + (c + c_1)D + (K + K_1)]\theta_o = K\theta_i.$$

Now, steady-state output with step input

$$X = \left(\frac{K}{K + K_1}\right)X,$$

so steady-state error $= 0.1 - \left(\frac{1236}{4736}\right) 0.1$

$$= 0.074 \text{ rad.}$$

First overshoot is

$$\left(\frac{K}{K + K_1}\right) X \, e^{-\zeta\pi/\surd(1-\zeta^2)},$$

where

$$K + K_1 = 4736 \text{ N m/rad,}$$

$$c + c_1 = 1500 \text{ N m s/rad,}$$

$$c_c = 2\surd[(K + K_1)J] = 2\surd(4736 \times 370)$$

$$= 2647 \text{ N m s/rad}$$

and

$$\zeta = \frac{1500}{2647} = 0.567.$$

Thus $e^{-\zeta\pi/\surd(1-\zeta^2)} = e^{-2.17} = 0.114$, and first overshoot is

$$0.114 \times 0.1 \times \frac{1236}{4736} = \underline{0.0029 \text{ rad.}}$$

102.

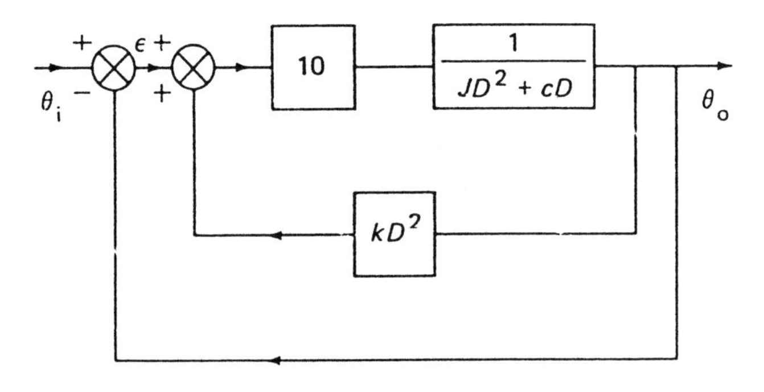

Equation of motion given by

$$10(\varepsilon + kD^2\theta_o)\left(\frac{1}{JD^2 + cD}\right) = \theta_o,$$

that is

$$(J - 10k)D^2\theta_o + cD\theta_o + 10\theta_o = 10\theta_i.$$

For critical damping

$$c = 2\sqrt{[10(J - 10k)]},$$

so

$$20 = 2\sqrt{[10(30 - 10k)]},$$

hence

$$k = 2s^2.$$

Now

$$\theta_i = \left[\left(\frac{J}{10} - k\right)D^2 + \frac{c}{10}D + 1\right]\theta_o,$$

so that if $\theta_i = \beta t$,

$$\theta_o = \left[1 + \frac{c}{10}D + \left(\frac{J}{10} - k\right)D^2\right]^{-1}\beta t$$

$$= \beta t - \frac{c}{10}\beta.$$

That is, the steady-state error is

$$\frac{c}{10}\beta.$$

Thus

$$\theta_{ss} = \frac{20}{10} \times \frac{2\pi^2}{60} \text{ rad} = 0.419 \text{ rad} = \underline{24°.}$$

103.

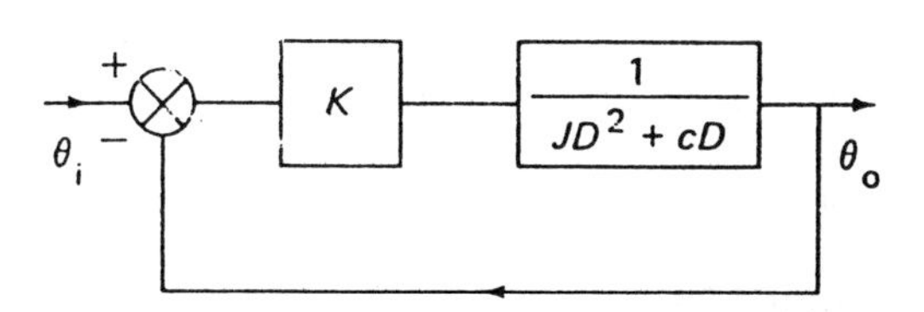

Equation of motion is $(JD^2 + cD + K)\theta_o = K\theta_i$. Step input applied so

$$\theta_o = \theta_i - \frac{\zeta}{\sqrt{(1-\zeta^2)}} e^{-\zeta\omega t} [\sin(\omega\sqrt{(1-\zeta^2)}t + \phi],$$

where

$$\tan\phi = \frac{\sqrt{(1-\zeta^2)}}{\zeta}.$$

For no error, $\theta_o = \theta_i$, so that

$$\sin\left[\omega\sqrt{(1-\zeta^2)}t + \tan^{-1}\frac{\sqrt{(1-\zeta^2)}}{\zeta}\right] = 0.$$

Now

$$\zeta = \frac{c}{c_c} = \frac{c}{2\sqrt{(KJ)}} = \frac{100}{2\sqrt{(2000\cdot 80)}} = \tfrac{1}{8},$$

and

$$\omega = \sqrt{\frac{K}{J}} = \sqrt{\frac{2000}{80}} = 5 \text{ rad/s}.$$

Thus

$$\left[(5\sqrt{[1-(\tfrac{1}{8})^2]})t + \tan^{-1}\frac{\sqrt{[1-(\tfrac{1}{8})^2]}}{\tfrac{1}{8}}\right] = 0, \pi, \ldots.$$

so that

$$\left(\frac{5\sqrt{63}}{8}\right)t + 1.445 = 0, \pi, \ldots.$$

Hence $\underline{t = 0.342 \text{ s.}}$

The time to the first overshoot is

$$\frac{\pi}{\omega\sqrt{(1-\zeta^2)}} = \frac{\pi\cdot 8}{5\sqrt{63}} = \underline{0.633 \text{ s.}}$$

The magnitude of the first overshoot is

$$\theta_i e^{-\zeta\pi/\sqrt{(1-\zeta^2)}}.$$

Now

$$e^{-\zeta\pi/\sqrt{(1-\zeta^2)}} = e^{-(\pi\times 8)/(8\times\sqrt{63})} = 0.673.$$

Since $\theta_i = 10°$, first overshoot is $\underline{6.73°.}$

With a ramp input, steady-state error is $(c/K)\,\beta$. That is,

$$\frac{100}{2000} \times 10 = \underline{0.5 \text{ rad.}}$$

104.

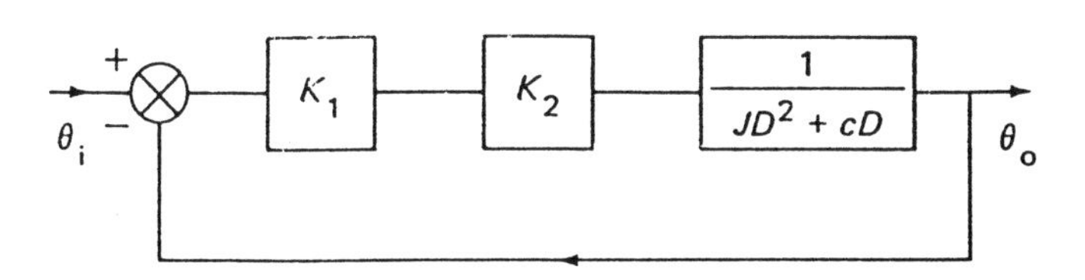

Let $K = K_1K_2 = 16 \times 3 = 48$ Nm/rad.
With a step input,

$$\theta_o = \theta_i[1 + e^{-\zeta\pi/\sqrt{(1-\zeta^2)}}],$$

where

$$\zeta = \frac{c}{c_c} = \frac{c}{2\sqrt{(JK)}} = \frac{24}{2\sqrt{(48 \times 12)}} = \tfrac{1}{2},$$

so that

$$e^{-\zeta\pi/\sqrt{(1-\zeta^2)}} = e^{-1.814} = 0.1629.$$

Therefore the maximum overshoot is $\underline{0.163 \text{ rad.}}$

With tachogenerator

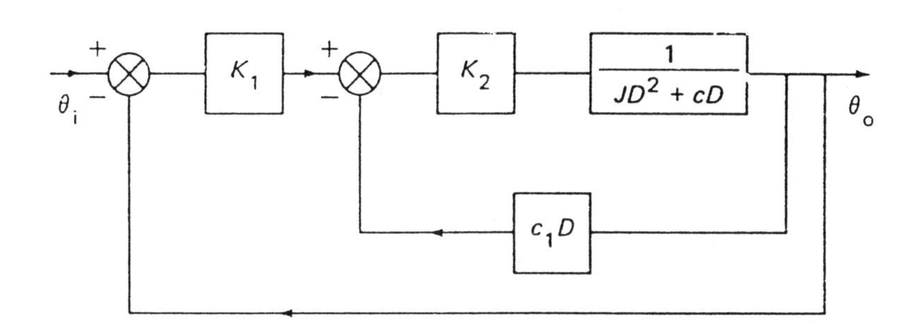

The equation of motion is

$$[(\theta_i - \theta_o)K_1 - c_1 D\theta_o]K_2 = [JD^2 + cD]\theta_o.$$

Hence

$$[JD^2 + (c + c_1K_2)D + K_1K_2]\theta_o = [K_1K_2]\theta_i.$$

Transfer function

$$\frac{\theta_o}{\theta_i} = \frac{K_1 K_2}{JD^2 + (c + c_1 K_2)D + K_1 K_2}.$$

For critical damping

$$c + c_1 K_2 = 2\sqrt{(K_1 K_2 J)},$$

or

$$24 + c_1 3 = 2\sqrt{(3 \times 16 \times 12)} = 48.$$

Hence $\underline{c_1 = 8 \text{ V/rad/s.}}$

107. Equation of motion is

$$[K_1(\theta_i - \theta_o) - K_2 s\theta_o]\left[\frac{8}{2s(s+4)(s+5)}\right] = \theta_o,$$

so that

$$\frac{\theta_i}{\theta_o} = \frac{s(s+4)(s+5) + 8(K_1 + K_2 s)}{8K_1},$$

and

$$\text{OLTF} = \frac{8K_1}{s(s+4)(s+5) + 8K_1 s}.$$

If

$$K_2 = \tfrac{17}{32}, \quad \Phi_o = \frac{8K_1}{s(s^2 + 9s + \frac{97}{4})}.$$

Apply rules to draw root locus diagram:
Poles at $s = 0$ and $-\frac{9}{2} \pm j2$.
Asymptotes intersect at -3.
Breakaway points at $3s^2 + 18s + \frac{97}{4} = 0$; that is $s = -4$ or -2.
Intersection with imaginary axis given by putting $s = j\omega$ in

$$s^3 + 9s^2 + \tfrac{97}{4}s + 8K_1 = 0.$$

Equating real and imaginary parts gives

$$\omega = 0,\ K_1 = 0 \qquad \text{and} \qquad \omega^2 = \tfrac{97}{4},\ K_1 = 27.3.$$

Hence maximum value of K_1 for stability is $\underline{27.3.}$

The root locus diagram can now be drawn:

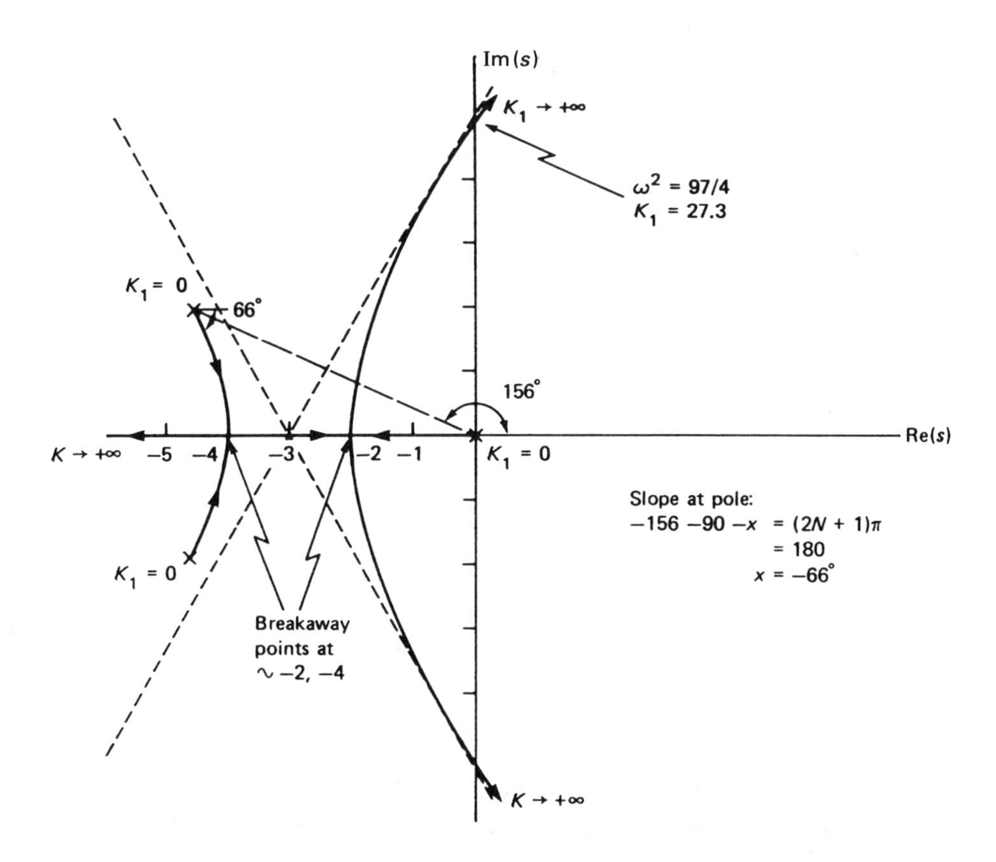

Now

$$s^3 + 9s^2 + (20 + 8K_2)s + 8K_1 = 0.$$

If pole at $-2 + j2$ (or can take $-$ve sign)

$$s^2 = -8j \qquad \text{and} \qquad s^3 = 16j + 16.$$

Hence

$$16j + 16 - 72j - 40 - 16K_2 + 40j + 16K_2 j + 8K_1 = 0.$$

Equate real and imaginary parts to give

$$16 - 72 + 40 + 16K_2 = 0 \qquad \text{so} \quad \underline{K_2 = 1,}$$

and

$$16 - 40 - 16K_2 + 8K_1 = 0 \qquad \text{so} \quad \underline{K_1 = 5}$$

(can check by substitution: other pole at -5).

108. If G is constant,

$$\{G[K_1\theta_i - K_1\theta_o] - K_0cs\theta_o\}\frac{K_2}{Js^2} = \theta_o.$$

Hence

$$\text{CLTF} = \Phi_c(s) = \frac{\theta_o}{\theta_i} = \frac{GK_1K_2}{Js^2 + K_0K_2cs + GK_1K_2},$$

and

$$\text{OLTF} = \Phi_o(s) = \frac{\theta_o}{\theta_i - \theta_o} = \frac{GK_1K_2}{Js^2 + K_0K_2cs}.$$

And if

$$G = \frac{G_0}{1 + sT_f},$$

$$\Phi_c(s) = \frac{G_0K_1K_2}{(Js^2 + K_0K_2cs)(1 + sT_f) + G_0K_1K_2},$$

$$= \frac{G_0K_1K_2}{JT_fs^3 + (J + K_0K_2cT_f)s^2 + K_0K_2cs + G_0K_1K_2},$$

and

$$\Phi_o(s) = \frac{G_0K_1K_2}{JT_fs^3 + (J + K_0K_2cT_f)s^2 + K_0K_2cs}.$$

The root locus diagrams can now be drawn by applying the rules.

With G constant, there are poles at 0 and -5 and a breakaway point at -2.5:

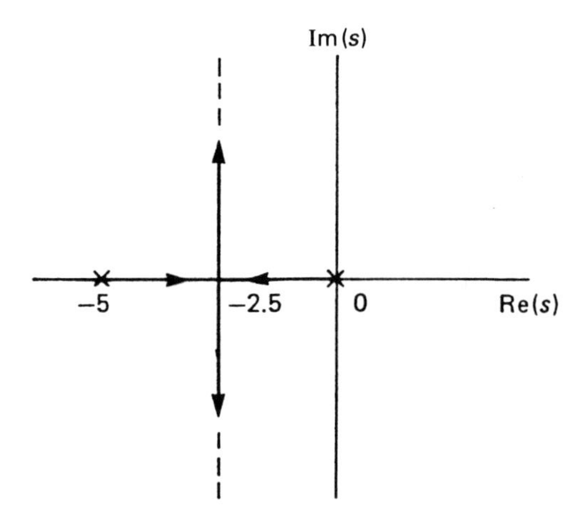

The system is always stable provided $G > 0$.
With

$$G = \frac{G_0}{1 + sT_f},$$

$$\Phi_o(s) = \frac{25G}{4s^3 + 30s^2 + 50s}.$$

There are three loci, poles at 0, $-\frac{5}{2}$ and -5, and breakaway points at approx -1 and -4:

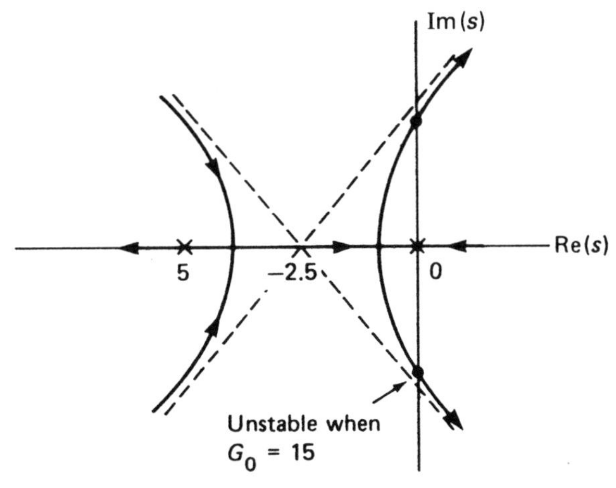

109. (a) $\underline{K = 30}$; (b) $\underline{K = 72.}$

110. If signal in feedback loop is x,

$$(K\theta_i - K\theta_o - x)\left(\frac{1}{s^2 + s + 2}\right) = x,$$

and

$$\theta_0 = \frac{Gx}{s^2 + 5s}.$$

Hence

$$\theta_o = \frac{G}{(s^2 + 5s)} \frac{K(\theta_i - \theta_o)}{(s^2 + s + 3)}.$$

and OLTF,

$$\left(\frac{\theta_o}{\theta_i - \theta_o}\right) = \frac{GK}{s(s + 5)(s^2 + s + 3)}.$$

Plot root locus by using the rules:
4 loci:

Poles at

$$s = 0 \quad -5, \quad -\tfrac{1}{2} \pm j\sqrt{\tfrac{11}{4}}.$$

Asymptotes intersect at $-\frac{3}{2}$.
Breakaway point at $\sim 3\frac{1}{2}$.
Substituting $s = j\omega$ gives $GK = 13\frac{3}{4}$.
Hence root locus diagram can be drawn:

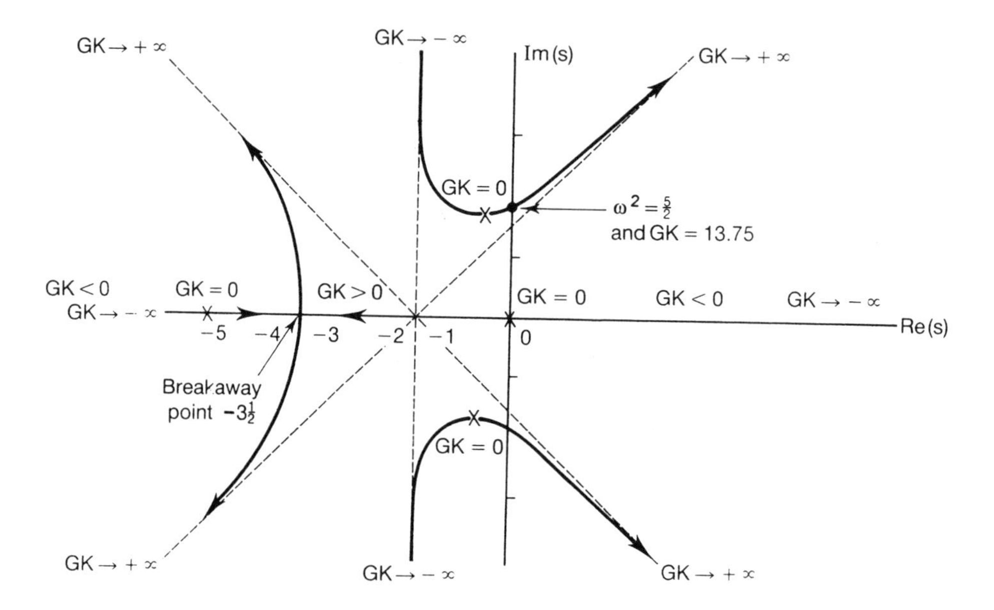

For stability, loci must lie on LHS of s-plane so that $GK > 0$ and $\underline{GK < 13\frac{3}{4}}$,

111. For system shown, by considering signal in each loop

$$\left[\frac{GK(\theta_i - \theta_o)}{s(1 + 0.5s)} - (0.25 + 2s)\theta_o\right]\frac{1}{2s^2} = \theta_o,$$

so that the equivalent single loop system OLTF is

$$\frac{\theta_o}{\theta_i - \theta_o} = \frac{GK}{s(1 + 0.5s)(0.25 + 2s + 2s^2)}.$$

The root locus diagram can be drawn by applying the rules:
4 loci
Poles at $s = 0, -2, -0.15, -0.85$.
Asymptotes intersect at $-\frac{3}{4}$.

Breakaway points at ~1.6, −0.04 and −0.6.
Substituting $s = j\omega$ gives maximum GK for stability as 0.17 when $\omega^2 = 0.25/3$.
Hence the root locus diagram is as follows:

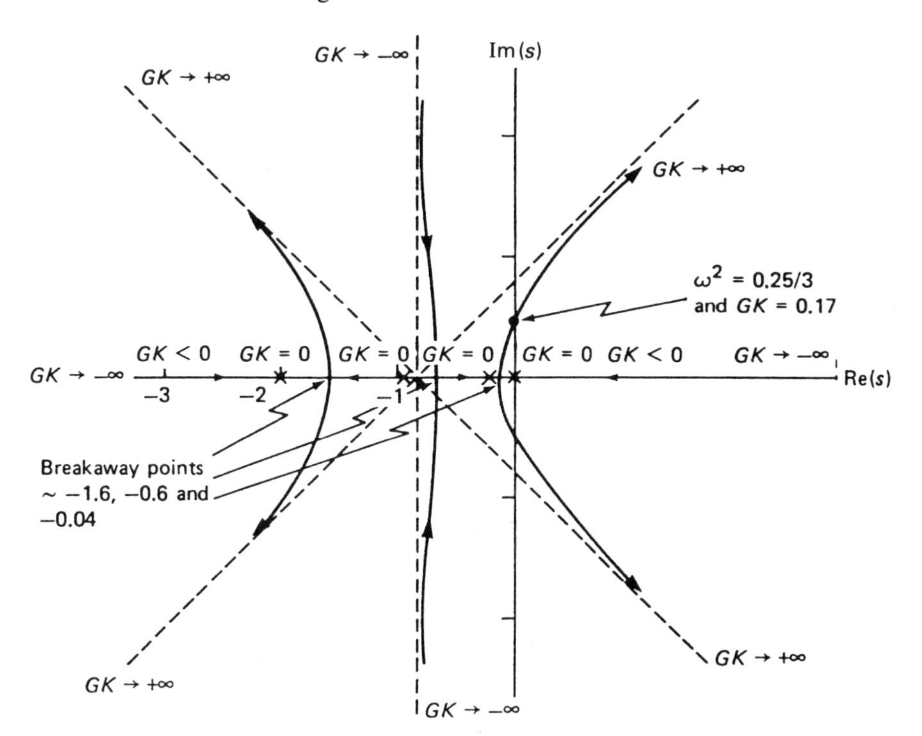

Maximum value of GK for stability is <u>0.17.</u>

112. (a) $\underline{0.75 > K > 0.2}$ (c) $\underline{\zeta = 0.558.}$

113. To draw the root locus diagram, apply the rules:
3 loci:
Poles at $s = 0, -10, -20$.
Asymptotes intersect at -10.
Breakaway points at -4.2 and -15.8.
K_{max} given by substituting $s = j\omega$ in

$$0.005s^3 + 0.15s^2 + s + K = 0,$$

Equating real and imaginary parts gives

$$\omega = 0 \quad \text{and} \quad K = 0, \quad \text{and} \quad \omega^2 = 200 \quad \text{and} \quad K = 30.$$

Thus for stability

$$K_{max} = 30 \quad \text{and} \quad f = \frac{\sqrt{200}}{2\pi} = 2.26 \text{ Hz}.$$

The root locus diagram can now be drawn:

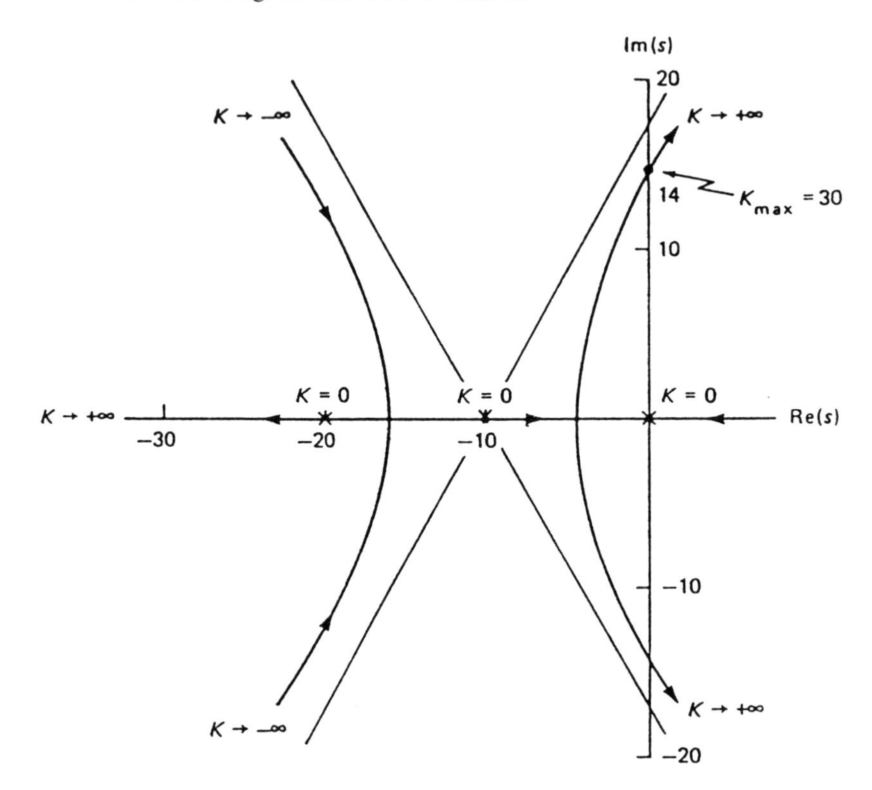

For Routh–Hurwitz criterion, equation is

$$0.005s^3 + 0.15s^2 + s + K = 0$$

so condition one requires $K > 0$, and for condition two, $a_0 = K$, $a_1 = 1$, $a_2 = 0.15$, $a_3 = 0.005$. So that

$$R_1 = +1 \qquad \text{and} \qquad R_2 = \begin{vmatrix} 1 & K \\ 0.005 & 0.15 \end{vmatrix} = 0.15 - 0.005K$$

For R_2 to be +ve, $K_{max} = 30$, as before.

Modified system has OLTF

$$\frac{K(1 + 0.5s)}{s(1 + 0.1s)(1 + 0.05s)}.$$

That is a zero now exists at $s = -2$, and the asymptotes intersect at -14.

The root locus diagram is therefore as follows:

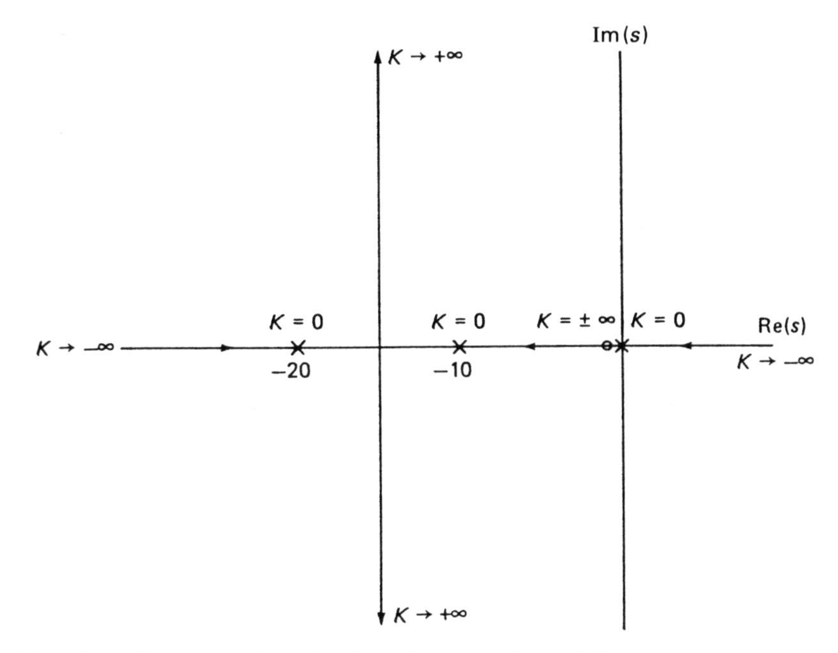

Therefore the system is stable for all +ve values of K.
When $K = K_{max} = 30$,

$$30 + 15s = -0.005s^3 - 0.15s^2 - s,$$

or

$$0.005s^3 + 0.15s^2 + 16s + 30 = 0.$$

Trial and error gives $s = -1.9$. Hence

$$5s^2 + 140s + 16000 = 0,$$

and

$$s = -14 \pm j55.$$

That is, a damped oscillation frequency of 55 rad/s or 8.75 Hz, with damping factor 14.

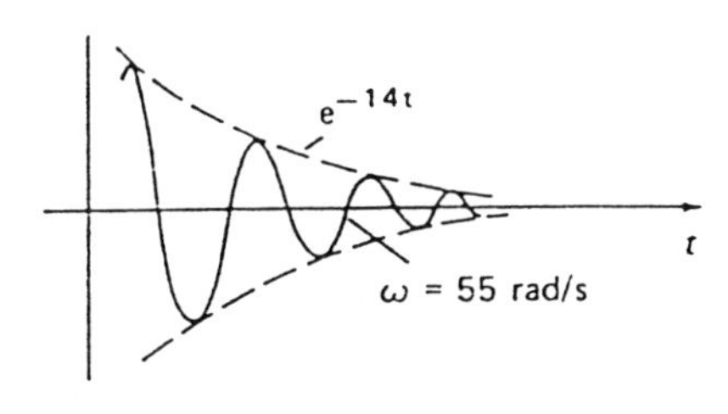

114. $$\Phi_0(s) = \frac{10K}{(1 + 0.2s)^2(1 + 0.05s)}.$$

To draw the root locus diagram, apply the rules:
3 loci:
3 poles at $s = -5, -5, -20$.
Asymptotes intersect at -10.
Breakaway points. Put $dK/ds = 0$ to get

$$s^2 + 20s + 75 = 0$$

so

$$s = -5 \quad \text{or} \quad -15.$$

Intersection with imaginary axis when $s = j\omega$:

$$10K + 1 + 0.45j\omega - 0.06\omega^2 - 0.002j\omega^3 = 0.$$

Equate imaginary parts:

$$0.45\omega - 0.002\omega^3 = 0$$

that is $\omega = 0$ or 15 rad/s.
Equate real parts:

$$10K + 1 - 0.06\omega^2 = 0$$

so $K = -0.1$ or 1.25.
Hence maximum value of K for stability is <u>1.25.</u>
The root locus diagram can now be drawn:

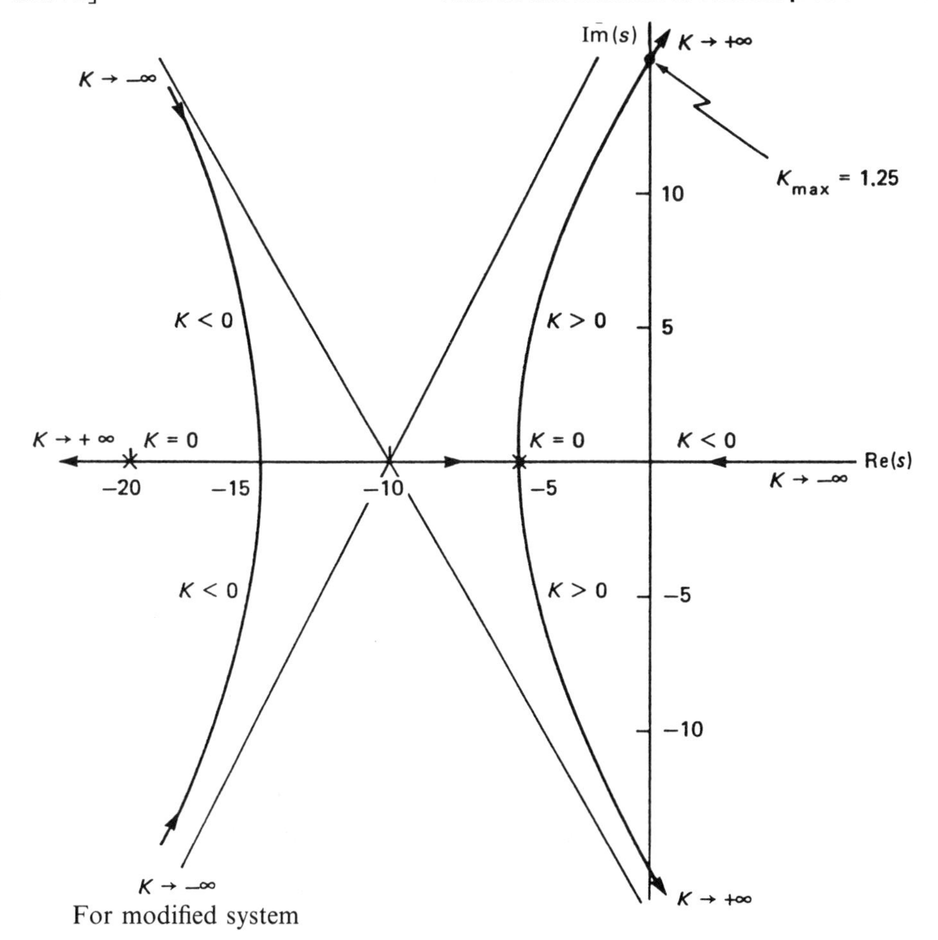

For modified system

$$\Phi_0(s) = \frac{10K}{(1 + 0.02s)(1 + 0.2s)(1 + 0.05s)}.$$

Root locus diagram can be drawn by applying the rules:
3 loci:
3 poles at $s = -5, -20, -50$.
Asymptotes intersect at -25.
Breakaway points. Put $\mathrm{d}K/\mathrm{d}s = 0$ to get

$$s^2 + 50s + 450 = 0,$$

so

$$s = -12 \quad \text{or} \quad -37.$$

Intersection with imaginary axis when $s = \mathrm{j}\omega$.
This gives $\omega = 0$ and $K = -0.1$
and $\omega^2 = 1350$ and $K = 1.925$.
Hence maximum value of K for stability is 1.925.

To apply the Routh–Hurwitz criterion,

$$s^3 + 75s^2 + 1350s + 5000 + 5000 \times 10K = 0$$

Condition (1) requires $K = -0.1$, and for condition (2),

$$a_3 = 1, \quad a_2 = 75, \quad a_1 = 1350, \quad \text{and} \quad a_0 = 5000 + 50000K.$$

That is

$$5000 + 50000K \geqslant 75 \times 1350$$

or

$$K_{max} = 1.925, \qquad \text{as before.}$$

The root locus diagram for the modified system can now be drawn:

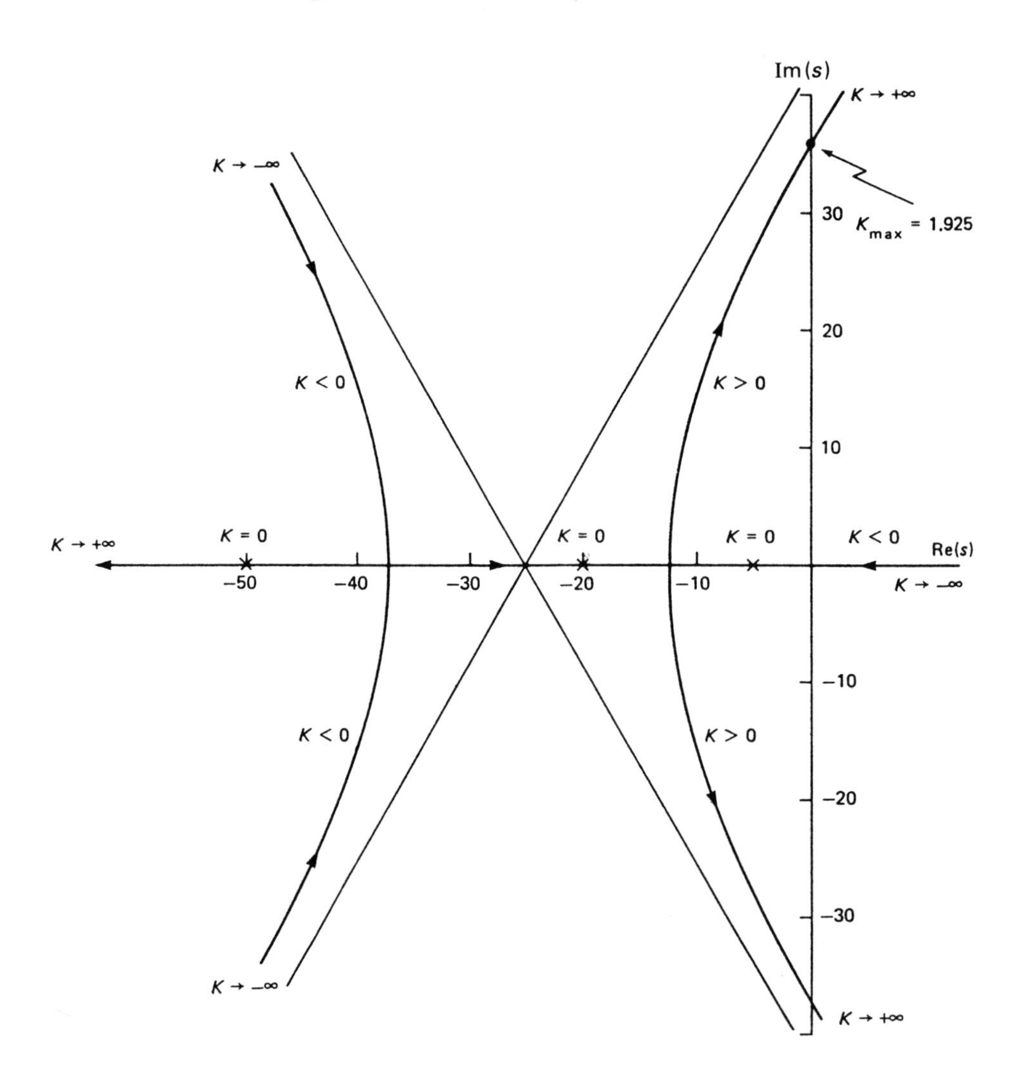

Compared with the original system, the *RL*diagram is similar but moved to the left on the Re(s) axis. Hence the response is more highly damped for a given value of K. System has a damped, stable response at previous K_{max} (1.25). Frequency of oscillatory response at new K_{max} (1.925) is 37 rad/s, and much higher than previous value of 15 rad/s.

116. By considering signal flow in block diagram and writing down the transfer functions, the characteristic equation is found to be

$$(s + 1)(2s^2 + 11s + 12) + K = 0.$$

Root locus is drawn by applying the rules:
3 loci —:
Poles at -1, $-\frac{3}{2}$, -4.
Asymptotes intersect at $-\frac{13}{6}$.
Breakaway points at $-\frac{5}{4}$ and -3.
Substituting $s = j\omega$ gives $K_{max} = 137.5$.
Root locus diagram is as follows: only positive frequency and $K > 0$ is shown:

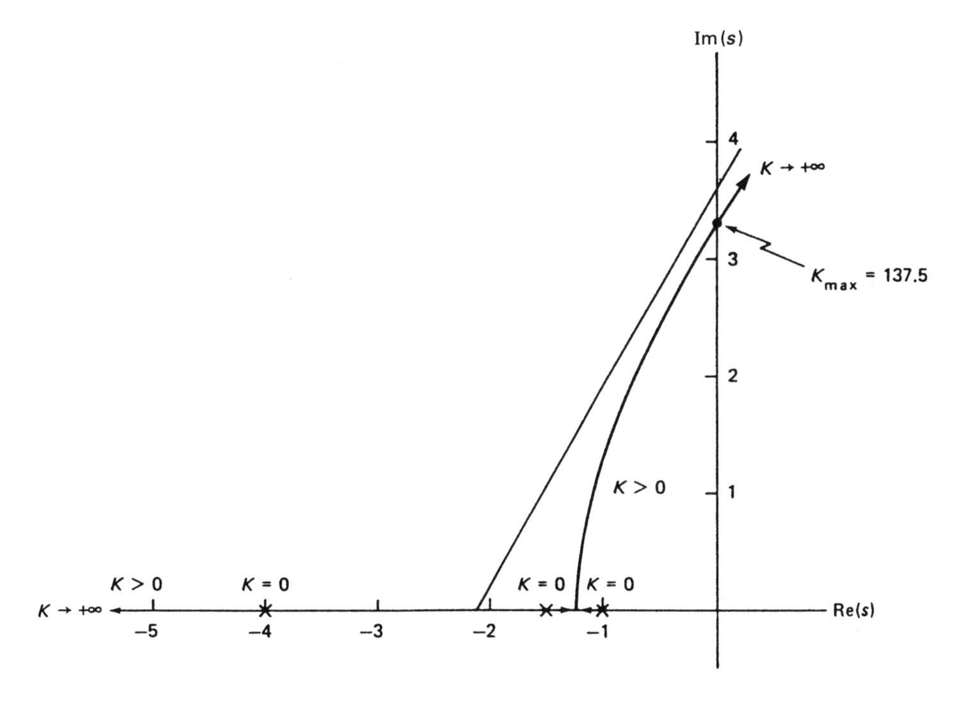

117. The characteristic equation is

$$1 + \frac{K(s + 2)}{s(s + 1)^2(s + 5)} = 0.$$

That is,

$$s^4 + 7s^3 + 11s^2 + (5 + K)s + 2K = 0.$$

For Routh–Hurwitz criteria,
Condition 1 requires $K > 0$.
For condition 2,

$$R_1 = 5 + K \qquad \text{which is +ve for } K > 0,$$

$$R_2 = \begin{vmatrix} 5+K & 2K \\ 7 & 11 \end{vmatrix} = 55 - 3K,$$

which is +ve for $K < \frac{55}{3}$

$$R_3 = \begin{vmatrix} 5+K & 2K & 0 \\ 7 & 11 & 5+K \\ 0 & 1 & 7 \end{vmatrix}$$

$$= 360 - 5K + 72K - K^2 - 98K.$$

Limit when $K^2 + 31K - 360 = 0$. That is $K = 9$ or -40.
Thus for stability, $\underline{9 \geqslant K > 0.}$

For the modified system,

$$\text{OLTF} = \frac{K(s+2)}{s(s+1)(s+5)(s+10)}$$

The root locus diagram can be drawn by applying the rules:
4 loci–:
Poles at $s = 0, -1, -5, -10$.
Zero at -2,
Asymptotes intersect at $-4\frac{2}{3}$.
Breakaway points found by trial and error at $-\frac{1}{2}, -7\frac{1}{2}$.
Intersection with imaginary axis found by substituting $s = j\omega$ to give $K_{max} = 526$ at 6 rad/s.
The root locus diagram can now be drawn:

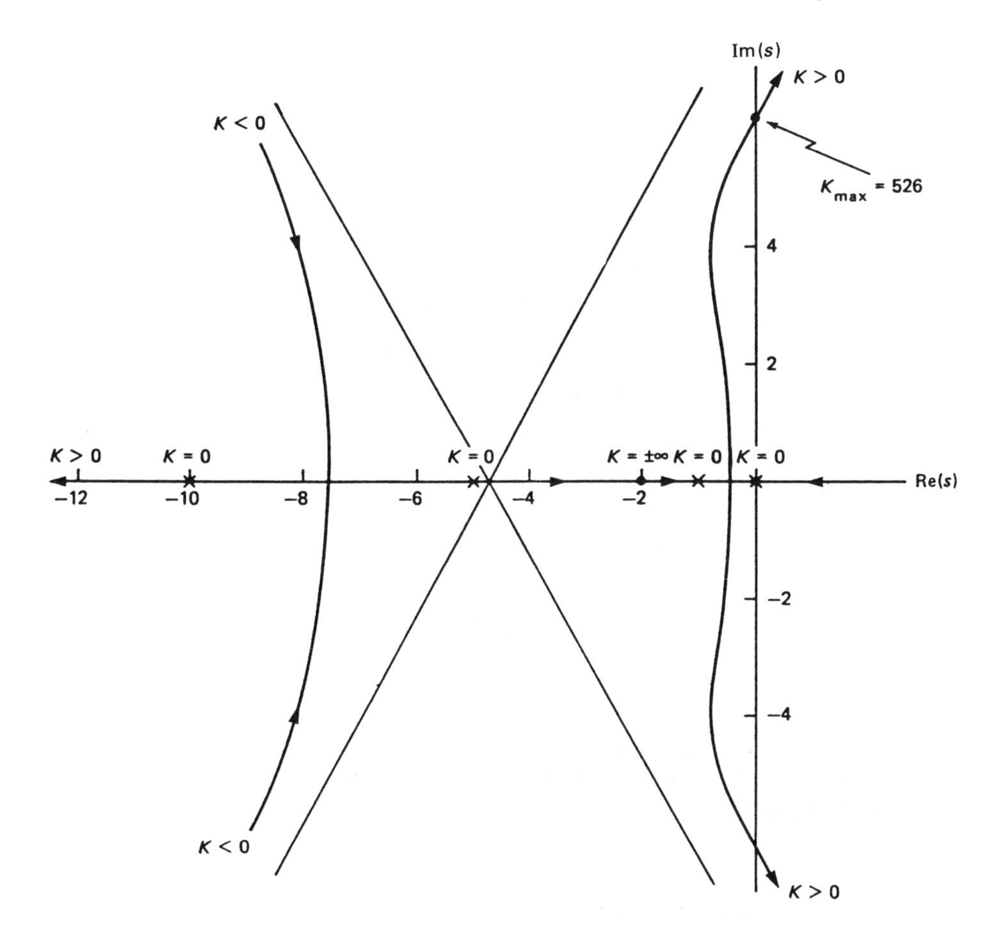

For the damping ratio at 1 rad/s,

$$\zeta = \cos\theta$$

$$\simeq \frac{0.55}{\sqrt{(1^2 + 0.55^2)}}.$$

Hence $\underline{\zeta = 0.48.}$

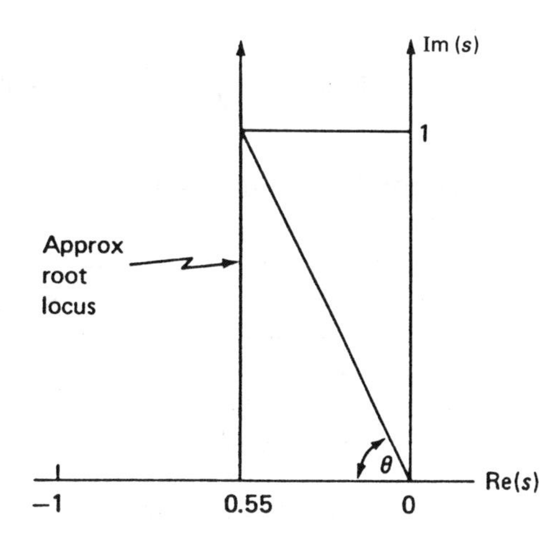

119. $$\Phi_c = \frac{\Phi_0}{1+\Phi_0} = \frac{K}{s^3+6s^2+11s+6+K}.$$

When $K = 90$,

$$\Phi_c(j\omega) = \frac{90}{-j\omega^3 - 6\omega^2 + 11j\omega + 6}$$

$$= \frac{90[6(1-\omega^2) - j\omega(11-\omega^2)]}{[36(1-\omega^2) + \omega^2(11-\omega^2)^2]}.$$

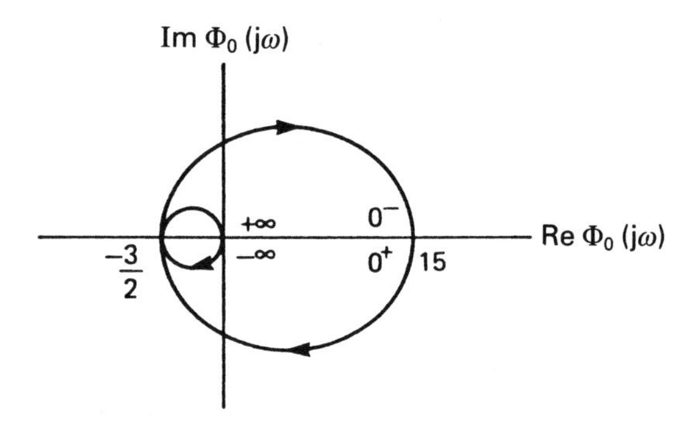

$-1,0$ enclosed so system unstable.
To pass through or enclose $(-1,0)$, $K \leqslant 60$. Since when $K = 90$ point $(-\frac{3}{2},0)$ passed through.

120. $$\Phi_0 = \frac{1}{s(1+s)(1+2s)}.$$

Put $s = j\omega$, rationalize and split into real and imaginary parts:

$$\Phi_0(j\omega) = \frac{-3\omega^2}{4\omega^6 + 5\omega^4 + \omega^2} + j\left[\frac{-\omega(1-2\omega^2)}{4\omega^6+5\omega^4+\omega^2}\right].$$

Hence the following table:

ω	$\mathrm{Re}\Phi_0(j\omega)$	$\mathrm{Im}\Phi_0(j\omega)$
$-\infty$	0	0
$-$ve large	$-$ve small	$-$ve small
$-$ve small	$-$ve	$+$ve
0^-	-3	∞

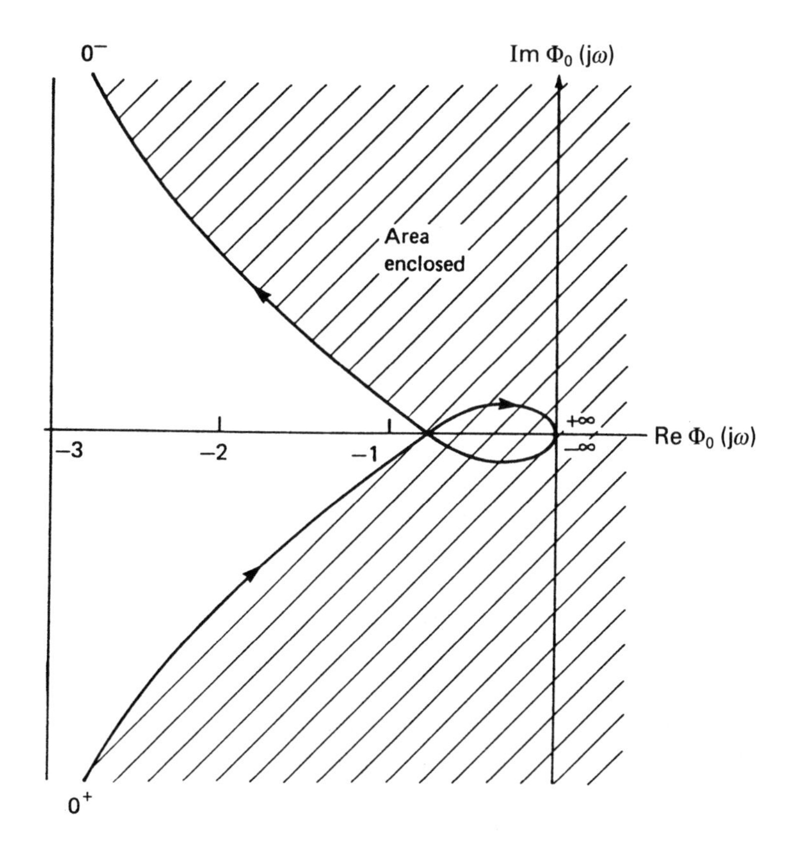

When

$$\mathrm{Im}\Phi_0(j\omega) = 0, \quad \omega = 0 \quad \text{or} \quad \omega^2 = \tfrac{1}{2}.$$

When

$$\omega^2 = \tfrac{1}{2},\ \mathrm{Re}\Phi_0(j\omega) = \frac{-3(\frac{1}{2})}{4(\frac{1}{2})^3 + 5(\frac{1}{2})^2 + (\frac{1}{2})} = -\tfrac{2}{3}.$$

Since the loop does not enclose $(-1,0)$ the system is stable

$$\text{Gain margin} = \frac{1}{2/3} = \frac{3}{2}.$$

For the modified system,

$$\Phi_0 = \frac{(1 + 3s)}{s(1 + s)(1 + 2s)}.$$

Putting $s = \mathrm{j}\omega$, rationalizing and finding real and imaginary parts gives

$$\Phi_0(\mathrm{j}\omega) = \frac{-6\omega^4}{4\omega^6 + 5\omega^4 + \omega^2} + \mathrm{j}\left[\frac{-\omega(1 + 7\omega^2)}{4\omega^6 + 5\omega^4 + \omega^2}\right].$$

Hence the following table:

ω	$\mathrm{Re}\Phi_0(\mathrm{j}\omega)$	$\mathrm{Im}\Phi_0(\mathrm{j}\omega)$
$-\infty$	0	0
$-$ve large	$-$ve	$+$ve
$-$ve small	$-$ve	$+$ve
0^-	0	∞

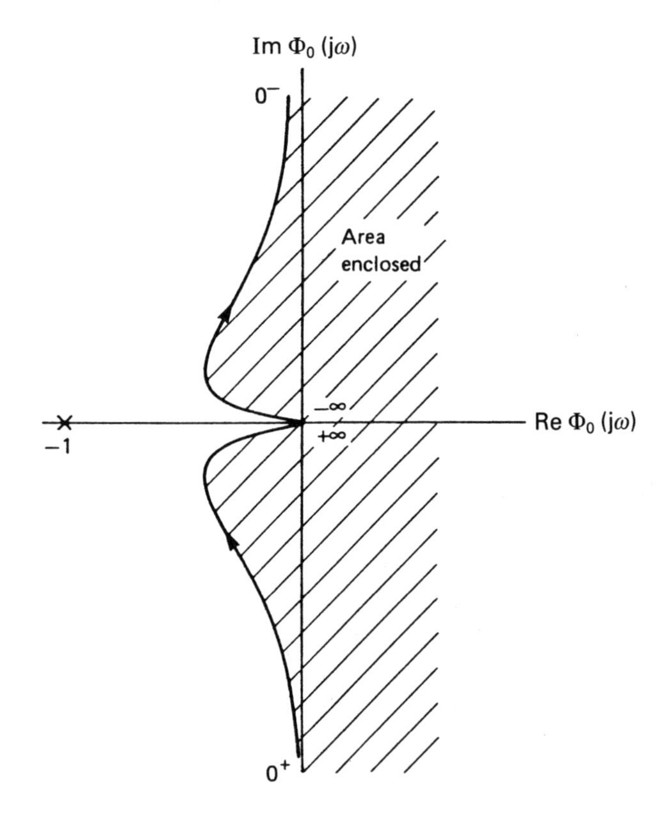

System is stable for all values of gain because $(-1,0)$ can never be enclosed.

121. (a) $K_{max} = 8$

(b) $K_{min} = 0$

(c) $\omega = 5$ rad/s, magnitude $= 1.33$

(d) Gain margin $= 0.75$.

122.
$$\text{OLTF} = \frac{10}{s(s+3)(s^2+2s+2)} = \frac{10}{s^4+5s^3+8s^2+6s}\cdot$$

Put $s = j\omega$ and rationalize to get

$$\text{Re}\Phi_0(j\omega) = \frac{10\omega^2(\omega^2-8)}{\omega^8+9\omega^6+4\omega^4+36\omega^2}$$

and

$$\text{Im}\Phi_0(j\omega) = \frac{-10\omega(6-5\omega^2)}{\omega^8+9\omega^6+4\omega^4+36\omega^2}\cdot$$

Hence the following table:

ω	$\text{Re}\Phi_0(j\omega)$	$\text{Im}\Phi_0(j\omega)$
$-\infty$	0	0
−ve large	+ve	−ve
$-\sqrt{8}$	0	-0.104
$-\sqrt{\frac{6}{5}}$	-1.23	0
−ve small	−ve	+ve
0^-	-2.2	∞

The Nyquist diagram can now be plotted.

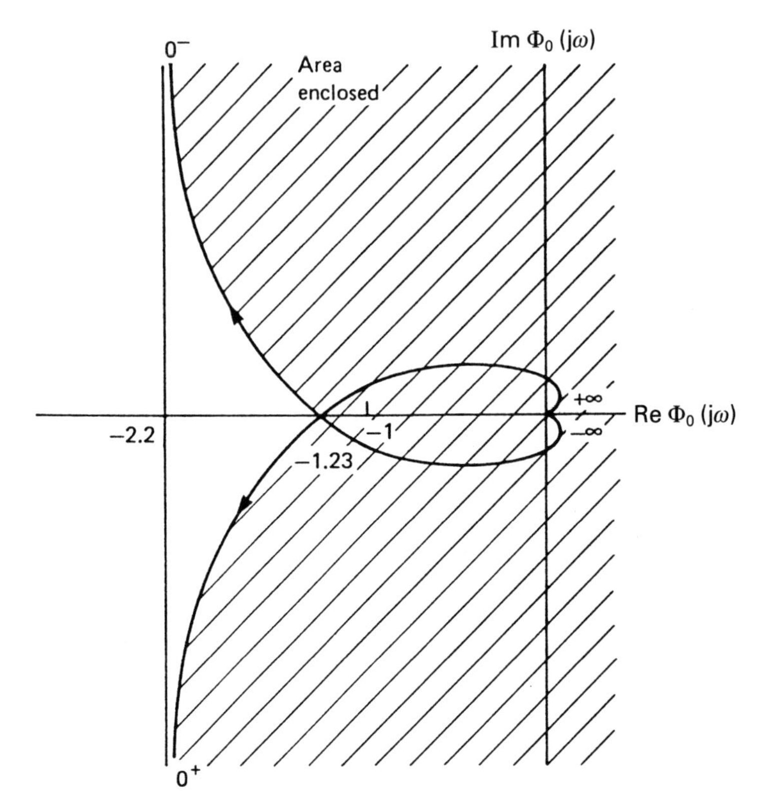

The system is unstable because $(-1{,}0)$ point is enclosed.
With proportional control added,

$$\text{OLTF} = \frac{10(s+1)}{s(s+3)(s^2+2s+2)}.$$

Substituting $s = j\omega$, rationalizing and dividing into real and imaginary parts gives

$$\text{Re}\Phi_0(j\omega) = \frac{-40\omega^4 - 20\omega^2}{\omega^8 + 9\omega^6 + 4\omega^4 + 36\omega^2},$$

and

$$\text{Im}\Phi_0(j\omega) = \frac{10\omega^5 - 30\omega^3 - 60\omega}{\omega^8 + 9\omega^6 + 4\omega^4 + 36\omega^2}.$$

By drawing up a frequency table, the following Nyquist diagram is obtained:

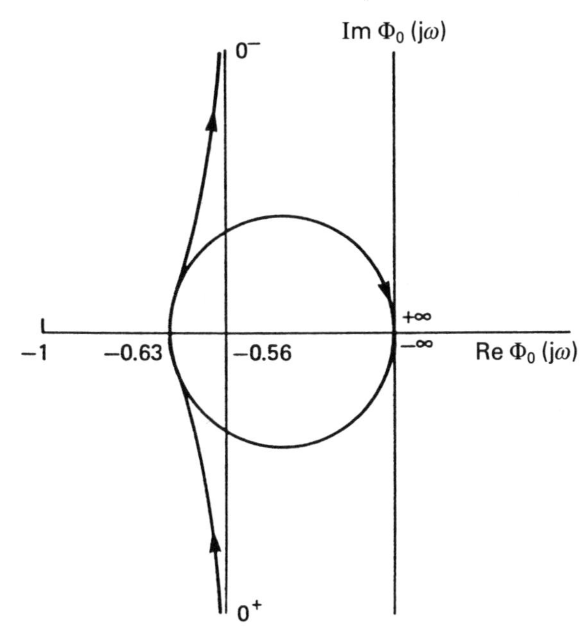

The system is now stable because $(-1{,}0)$ is not enclosed.
Gain margin $= \frac{1}{0.63} = \underline{1.59}$ with satisfactory phase margin.

123. Substitute $s = \mathrm{j}\omega$ into $\Phi_0(s)$, rationalize and split into real and imaginary parts to give

$$\mathrm{Re}\Phi_0(\mathrm{j}\omega) = \frac{-6K\omega^2}{(-6\omega^2)^2 + (8\omega - \omega^3)^2},$$

and

$$\mathrm{Im}\Phi_0(\mathrm{j}\omega) = \frac{-K(8\omega - \omega^3)}{(-6\omega^2)^2 + (8\omega - \omega^3)^2}.$$

Hence the following diagram:

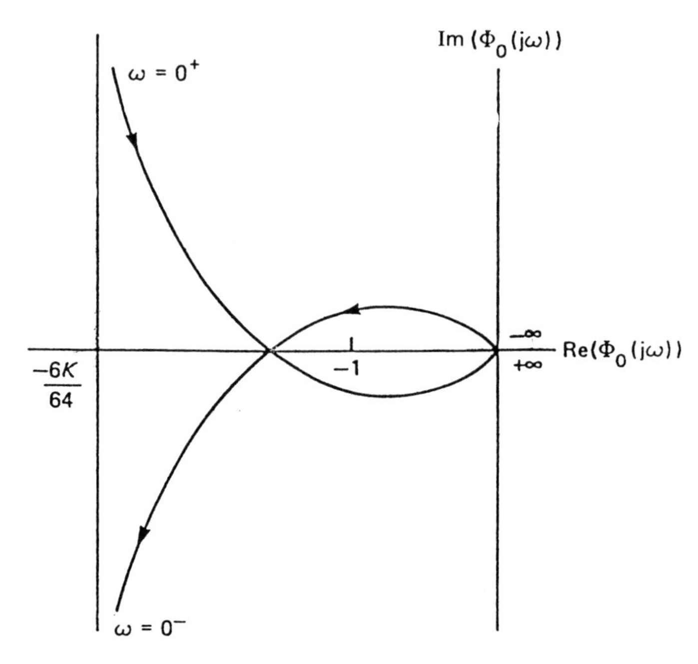

If $\text{Im}\Phi_0(j\omega) = 0$, $\omega^2 = 8$, and then

$$\text{Re}\Phi_0(j\omega) = -\frac{K}{48}.$$

That is, K_{max} for stability is 48, so system stable when $K = 20$.

124. Substituting $s = j\omega$ into the OLTF, rationalizing and splitting into real and imaginary parts gives

$$\text{Re}\Phi_0(j\omega) = \frac{-6\omega^2}{4\omega^6 + 5\omega^4 + \omega^2},$$

and

$$\text{Im}\Phi_0(j\omega) = \frac{-2(\omega - 2\omega^3)}{4\omega^6 + 5\omega^4 + \omega^2}.$$

Hence the following table:

ω	$\text{Re}\Phi_0(j\omega)$	$\text{Im}\Phi_0(j\omega)$
$-\infty$	0	0
$-$ve large	$-$ve	$-$ve
$-$ve small	$-$ve	+ve
0^-	-6	∞
$\sqrt{\frac{1}{2}}$	$-\frac{4}{3}$	0

The Nyquist loop can now be drawn:

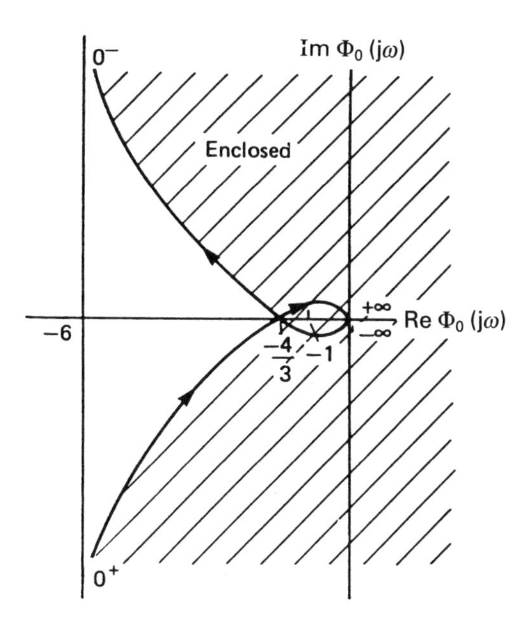

Since the $(-1, 0)$ point is enclosed the system is unstable.

For the modified system,

$$\Phi_0(s) = \frac{2(1 + 0.5s)}{s(s + 1)(2s + 1)}.$$

Substituting $s = j\omega$, rationalizing and splitting into real and imaginary parts gives

$$\mathrm{Re}\Phi_0(j\omega) = \frac{-5\omega^2 - 2\omega^4}{4\omega^6 + 5\omega^4 + \omega^2},$$

and

$$\mathrm{Im}\Phi_0(j\omega) = \frac{-2\omega + \omega^3}{4\omega^6 + 5\omega^4 + \omega^2}.$$

Hence the following table:

ω	$\text{Re}\Phi_0(j\omega)$	$\text{Im}\Phi_0(j\omega)$
$-\infty$	0	0
$-$ve large	$-$ve	$-$ve
$-$ve small	$-$ve	$+$ve
0^-	-5	∞
$-\sqrt{\frac{5}{2}}$	0	$+$ve
$-\sqrt{2}$	$-\frac{1}{3}$	0

The Nyquist loop can now be drawn:

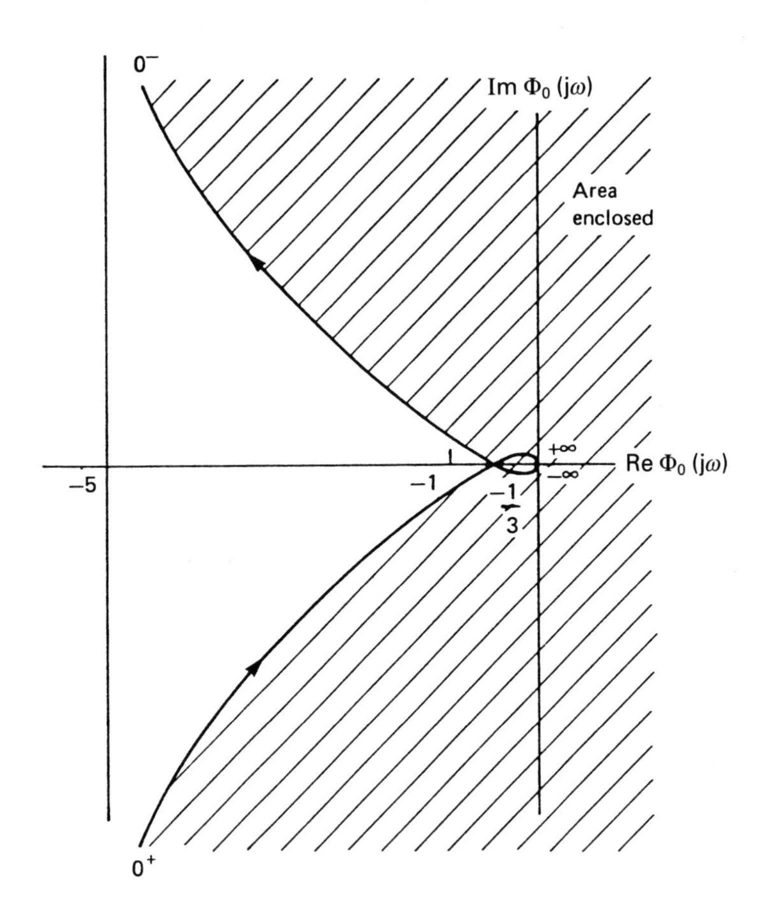

$(-1,0)$ is not enclosed, therefore the system is stable.

$$\text{Gain margin} = \frac{1}{1/3} = \underline{3.}$$

126. $$\Phi_0(j\omega) = \frac{7}{j\omega(1 + 0.5j\omega)(1 + 0.167j\omega)}.$$

Magnitude:

$$20 \log|\Phi_0(j\omega)| = 20 \log 7 - 20 \log j\omega - 20 \log|1 + 0.5j\omega| - 20 \log|1 + 0.167j\omega|.$$

Now $20 \log 7 = 17$ dB $\quad -20 \log j\omega$ is plotted

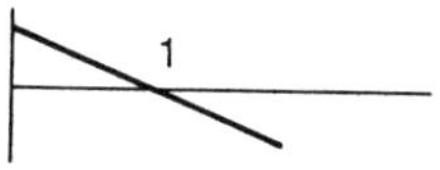

$-20 \log|1 + 0.5j\omega|$ is plotted

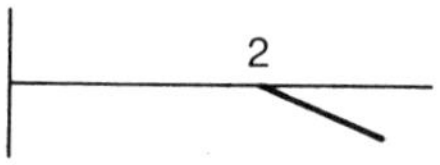

$-20 \log|1 + 0.167j\omega|$ is plotted

These plots are all made on Log–Linear graph paper and added to give the Bode gain (or amplitude) plot.
Phase:

$$\langle\Phi_0(j\omega) = -90° - \tan^{-1}(0.5\omega) - \tan^{-1}(0.167\omega)$$

ω	1	2	4	6	10
$-90°$	$-90°$	$-90°$	$-90°$	$-90°$	$-90°$
$-\tan^{-1}(0.5\omega)$	$-27°$	$-45°$	$-63°$	$-72°$	$-79°$
$-\tan^{-1}(0.167\omega)$	$-9°$	$-18°$	$-34°$	$-45°$	$-59°$
	$-126°$	$-153°$	$-187°$	$-207°$	$-228°$

Hence the Bode phase plot can be drawn, see over leaf.

If the magnitude and phase plots are drawn on the same frequency axis, it can be seen that the system is unstable with

gain margin $= -1$ dB,

and

phase margin $= -4°$.

127. $$\Phi_0(j\omega) = \frac{40}{j\omega(1 + 0.0625j\omega)(1 + 0.25\omega)}.$$

Magnitude:

$$20 \log|\Phi_0(j\omega)| = 20 \log 40 - 20 \log j\omega - 20 \log|1 + 0.0625j\omega| - 20 \log|1 + 0.25j\omega|.$$

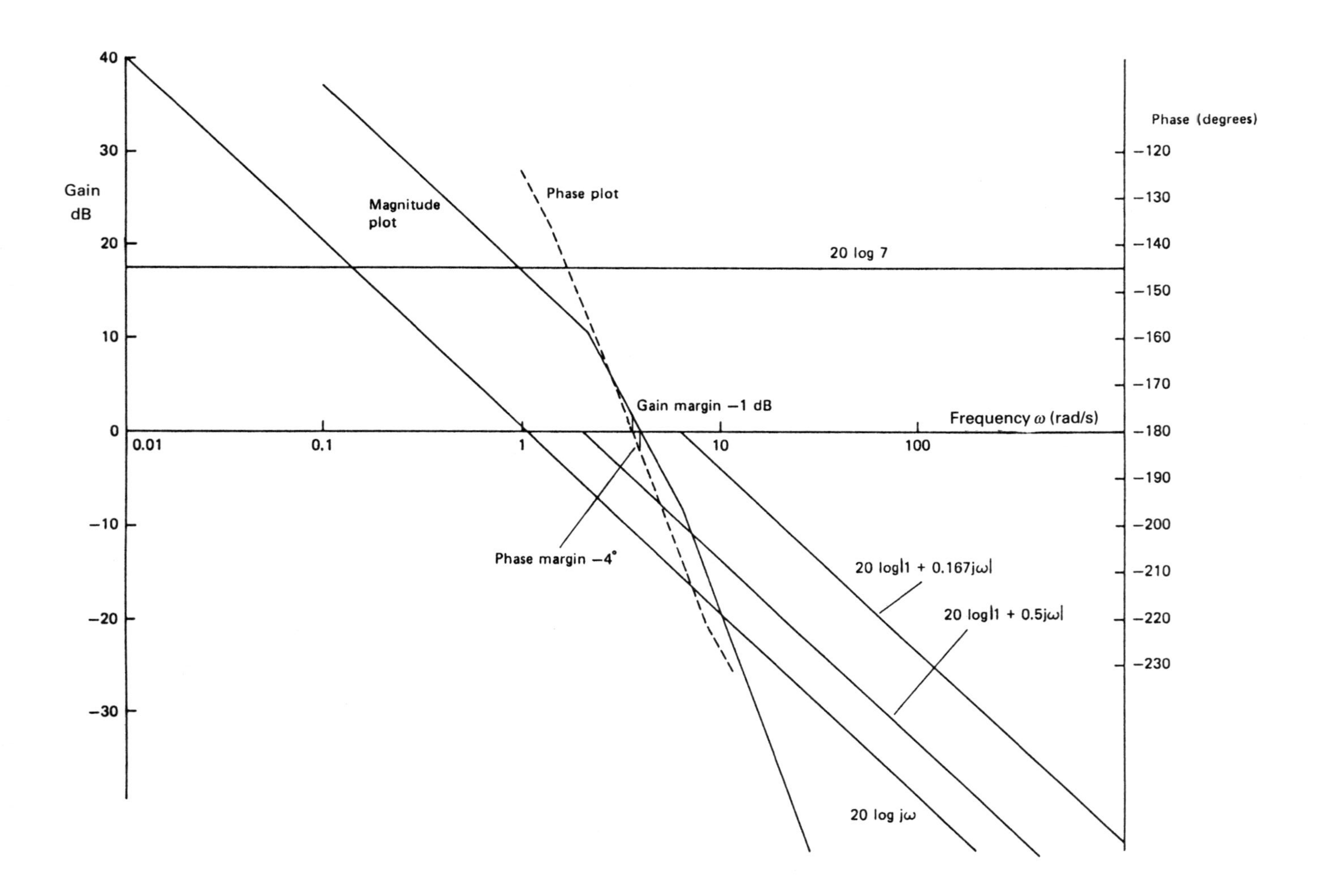

Gain
dB
40
30
20
10
0
−10
−20
−30
0.01
0.1
1
10
100
Frequency ω (rad/s)
Phase (degrees)
−120
−130
−140
−150
−160
−170
−180
−190
−200
−210
−220
−230
Magnitude
plot
Phase plot
20 log 7
Gain margin −1 dB
Phase margin −4°
20 log|1 + 0.167jω|
20 log|1 + 0.5jω|
20 log jω

Draw individual plots on log – linear graph paper and add to give Bode magnitude plot.
Phase:

$\langle\Phi(j\omega) = -90° - \tan^{-1} 0.0625\omega - \tan^{-1} 0.25j\omega$

ω	1	5	10	20
$-90°$	$-90°$	$-90°$	$-90°$	$-90°$
$-\tan^{-1} 0.0625\omega$	$-4°$	$-17°$	$-32°$	$-51°$
$-\tan^{-1} 0.25\omega$	$-14°$	$-51°$	$-68°$	$-79°$
	$-120°$	$-158°$	$-200°$	$-220°$

Hence Bode phase angle plot can be drawn.
From plots, system is unstable with

$$\text{gain margin} = -7 \text{ dB},$$

and

$$\text{phase margin} = -21°,$$

Phase lag network introduces new terms to be added into existing plots. Found that system now stable with

$$\text{gain margin} = 18 \text{ dB},$$

and

$$\text{phase margin} = 50°.$$

129. Draw magnitude plots for

$$\frac{1}{j\omega}, \frac{1}{1 + T_1 j\omega} \quad \text{and} \quad \frac{1}{1 + T_2 j\omega}.$$

Sketch in modulus for $K = 1$.
Calculate a few phase values.
Cross over occurs at 14 rad/s where K_{max} can be found from $20 \log|K| = -40$. Hence

$$K_{max} = 100.$$

When

$$\frac{1}{1 + T_1 s}$$

term is replaced by a time delay term,

Sketch magnitude and phase plots.
At phase cross over $\omega = 3.1$ rad/s and $K = 3.2$.

130. 50°.

131. (a) $K = 11$.

(c) $a = 1.83$ rad/s; $b = 5.48$ rad/s; $k' = 19.05$.

Bibliography

Anand, D. K., *Introduction to Control System*, 2nd edn, Pergamon Press, 1984.

Atkinson, P., *Feedback Control Theory for Engineers*, 2nd edn, Heinemann, 1977.

Beards, C. F., *Structural Vibration Analysis*, Ellis Horwood, 1983.

Beards, C. F., *Vibrations and Control System*, Ellis Horwood, 1988.

Bickley, W. G. and Talbot, A., *Vibrating Systems*, Oxford University Press, 1961.

Bishop, R. E. D. Gladwell, G. M. L. and Michaelson, S., *The Matrix Analysis of Vibration*, Cambridge University Press, 1965.

Bishop, R. E. D. and Johnson, D. C., *The Mechanics of Vibration*, Cambridge University Press, 1960/1979.

Blevins, R. D., *Formulas for Natural Frequency and Mode Shape*, Van Nostrand, (1979).

Brogan, W. L., *Modern Control Theory*, Prentice-Hall, 1982.

Buckley, R. V., *Control Engineering*, Macmillan, 1976.

Burghes, D. and Graham, A., *Introduction to Control Theory Including Optimal Control*, Ellis Horwood, 1980.

Chesmond, C. J., *Basic Control System Technology*, Edward Arnold, 1990.

Close, C. M. and Frederick, D. K., *Modeling and Analysis of Dynamic Systems*, Houghton Mifflin, 1978.

Collar, A. R. and Simpson, A., *Matrices and Engineering Dynamics*, Ellis Horwood, 1987.

Crandall, S. H., *Random Vibration*, Technology Press and John Wiley, 1958.

Crandall, S. H. and Mark, W. D., *Random Vibration in Mechanical Systems*, Academic Press, 1963.

Davenport, W. B., *Probability and Random Processes*, McGraw-Hill, 1970.

Den Hartog, J. P., *Mechanical Vibrations*, McGraw-Hill, 1956.

Dorf, R. C., *Modern Control Systems*, 5th edn, Addison-Wesley, 1989, *Solution Manual*, 1989.

Dransfield, P. and Habner, D. F., *Introducing Root Locus*, Cambridge University Press, 1973.
Eveleigh, V. W., *Control Systems Design*, McGraw-Hill, 1972.
Franklin, G. F., Powell, J. D. and Emami-naeini, A., *Feedback Control of Dynamic Systems*, 2nd edn, Addison-Wesley, 1991.
Franklin, G. F., Powell, J. D. and Workman, M. L., *Digital Control of Dynamic Systems*, 2nd edn, Addison-Wesley, 1990.
Guy, J. J., *Solution of Problems in Automatic Control*, Pitman, 1966.
Healey, M., *Principles of Automatic Control*, Hodder and Stoughton, 1975.
Helstrom, C. W., *Probability and Stochastic Processes for Engineers*, Macmillan, 1984.
Huebner, K. H., *The Finite Element Method for Engineers*, Wiley, 1975.
Irons, B. and Ahmad, S., *Techniques of Finite Elements*, Ellis Horwood, 1980.
Jacobs, O. L. R., *Introduction to Control Theory*, Oxford University Press, 1974.
James, M. L., Smith, G. M., Wolford, J. C. and Whaley, P. W., *Vibration of Mechanical and Structural Systems*, Harper Row, 1989.
Lalanne, M. Berthier, P. and Der Hagopian, J., *Mechanical Vibrations for Engineers*, Wiley, 1983.
Lazan, B. J., *Damping of Materials and Members in Structural Mechanics*, Pergamon (1968).
Leff, P. E. E., *Introduction to Feedback Control Systems*, McGraw-Hill, 1979.
Marshall, S. A., *Introduction to Control Theory*, Macmillan, 1978.
Meirovitch, L., *Elements of Vibration Analysis*, 2nd edn, McGraw-Hill, 1986.
Nashif, A. D., Jones, D. I. G. and Henderson, J. P., *Vibration Damping*, Wiley, 1985.
Newland, D. E., *An Introduction to Random Vibrations and Spectral Analysis*, 2nd edn, Longman, 1984.
Newland, D. E., *Mechanical Vibration Analysis and Computation*, Longman, 1989.
Nigam, N. C., *Introduction to Random Vibrations*, Massachusetts Institute of Technology Press, 1983.
Piszek, K. and Niziol, J., *Random Vibrations of Mechanical Systems*, Ellis Horwood, 1986.
Power, H. M. and Simpson, R. J., *Introduction to Dynamics and Control*, McGraw-Hill, 1978.
Prentis, J. M. and Leckie, F. A., *Mechanical Vibrations; An Introduction to Matrix Methods*, Longman, 1963.
Rao, S. S., *Mechanical Vibrations*, Addison-Wesley, 1986, 2nd edn, 1990; *Solutions Manual*, 1990.
Raven, F. H., *Automatic Control Engineering*, 4th edn, McGraw-Hill, 1987.
Richards, R. J., *An Introduction to Dynamics and Control*, Longman, 1979.
Robson, J. D., *An Introduction to Random Vibration*, Edinburgh University Press, 1963.
Schwarzenbach, J. and Gill, K. F., *System Modelling and Control*, 2nd edn, Arnold, 1984.
Sinha, N. K., *Control Systems*, Holt, Rinehart and Winston, 1986.
Smith, J. D., *Vibration Measurement and Analysis*, Butterworths, 1989.
Snowdon, J. C., *Vibration and Shock in Damped Mechanical Systems*, Wiley, 1968.
Steidel, R. F., *An Introduction to Mechanical Vibrations*, 3rd edn, Wiley, 1989.
Thompson, S., *Control Systems, Engineering and Design*, Longman, 1989.

Thomson, W. T., *Theory of Vibration with Applications*, 3rd edn, Unwin Hyman, 1989.
Timoshenko, S. P., Young, D. H. and Weaver, W., *Vibration Problems in Engineering*, 4th edn, Wiley, 1974.
Tse, F. S., Morse, I. E. and Hinkle, R. T., *Mechanical Vibrations, Theory and Applications*, 2nd edn, Allyn and Bacon, 1983; *Solutions Manual*, 1978.
Tuplin, W. A., *Torsional Vibration*, Pitman, 1966.
Walshaw, A. C., *Mechanical Vibrations with Applications*, Ellis Horwood, 1984.
Welbourn, D. B., *Essentials of Control Theory*, Edward Arnold, 1963.
Willems, J. L., *Stability Theory of Dynamical Systems*, Nelson, 1970.

Index

Notes